Properties of Concrete

To Old Hallam

Properties of Concrete

A M Neville

MC, TD, DSc(Eng), PhD, MSc, CEng, PEng, FICE, FIStructE,
MSocCE(France), FAmSocCE, FACI, FIArb

*Professor and Head of Department of Civil Engineering, University of Leeds,
formerly Dean of Engineering, University of Calgary*

Pitman Publishing

PITMAN PUBLISHING LIMITED
39 Parker Street, London WC2B 5PB

Associated Companies
Copp Clark Ltd, Toronto · Pitman Publishing Corporation/Fearon
Publishers Inc, Belmont, California · Pitman Publishing Co SA
(Pty) Ltd, Johannesburg · Pitman Publishing New Zealand Ltd,
Wellington · Pitman Publishing Pty Ltd, Melbourne

© A. M. Neville 1973, 1975, 1977

Second Edition 1973, Reprinted (with corrections) 1975, 1977

ISBN 0 273 36150 3

Reproduced and printed by photolithography and bound in
Great Britain at The Pitman Press, Bath

Preface to the Second (Metric) Edition

IN the ten years since its first appearance this book was reprinted five times, both in a hard cover and in a paperback edition. This can be taken to mean that concrete continues to be a most important structural material and that interest in it is unabated. But knowledge progresses, and this is why a new up-dated edition is being brought out.

This edition not only takes notice of research and development up to the present but also deals with some topics absent from the first edition. Specifically, I have added sections on fatigue, impact strength, air and vapour permeability, erosion resistance, cavitation resistance, electrical properties, acoustic properties, properties of concrete as a radiation shield, shotcrete, very high strength concrete and on some specialized concretes. Some of the existing topics have also been expanded.

In these times of change in units of measurement I thought it best to write the new edition both in the S.I. and in the British units, now paradoxically known as customary American. All the data, diagrams and tables are therefore conveniently presented for readers, progressive or traditionalist, in all countries.

A.M.N.

Preface

CONCRETE and steel are the two most commonly used structural materials. They sometimes complement one another, and sometimes compete with one another so that structures of a similar type and function can be built in either of these materials. And yet, the engineer often knows less about the concrete of which the structure is made than about the steel.

Steel is manufactured under carefully controlled conditions, its properties are determined in a laboratory and described in a manufacturer's certificate. Thus the designer need only specify the steel as complying with a relevant standard, and the site engineer's supervision is limited to the workmanship of the connexions between the individual steel members.

On a concrete building site the situation is totally different. It is true that the quality of cement is guaranteed by the manufacturer in a manner similar to steel and provided a suitable cement is chosen it is hardly ever a cause of faults in a concrete structure. But it is concrete and not cement that is the building material. The structural members are generally made *in situ*, and their quality is almost exclusively dependent on the workmanship of concrete making and placing.

The disparity in the methods of steel and concrete making is, therefore, clear, and the importance of the control of the quality of concrete work on the site is apparent. Furthermore, as the trade of a concretor has not yet the education and the tradition of some of the other building trades, an engineer's supervision on the site is essential. These facts must be borne in mind by the designer, as careful and intricate design can be easily vitiated if the properties of the actual concrete differ from those assumed in the design calculations.

From the above it must not be concluded that making good concrete is difficult. "Bad" concrete—often a substance of the consistence of soup, hardening into a honeycombed, non-homogeneous mass—is made simply by mixing cement, aggregate and water. Surprisingly, the ingredients of a good concrete are exactly the same, and it is only the "know-how", often without additional cost of labour, that is responsible for the difference.

What, then, is good concrete? There are two overall criteria: the concrete has to be satisfactory in its hardened state, and also in its fresh state while being transported from the mixer and placed in the formwork. The requirements in the fresh state are that the consistence of the mix be such that it can be compacted by the means desired without excessive effort, and also that the mix be cohesive enough for the method of placing used not to produce segregation with a consequent lack of homogeneity of the finished product.

The usual primary requirement of a good concrete in its hardened state is a satisfactory compressive strength. This is aimed at not only so as to ensure that the concrete can withstand a prescribed compressive stress but also because many other desired properties of concrete are concomitant with high strength. The various properties of concrete—density, durability, tensile strength, impermeability, resistance to abrasion, resistance to sulphate attack, and many others—are discussed in the appropriate chapters.

Interest in these properties of concrete has recently been heightened since modern specifications tend to state requirements for particular properties of concrete rather than simply to stipulate the quality and quantity of the constituent materials. A knowledge of the properties of concrete thus makes possible the selection of a more suitable and more economical mix. Interest in concrete making has also been aided by the development of equipment which leads to improved uniformity of concrete, with the associated economic and technical advantages.

In a book of this size it is not possible to cover the whole field of concrete: the author selects what he considers most important or most interesting or simply what he knows most about but the emphasis is on an integrated view of the properties of concrete and on the underlying scientific reasons, for, as Henri Poincaré said, an accumulation of facts is no more a science than a heap of stones is a house.

A.M.N.

Acknowledgements

PROFESSOR J. W. H. KING developed through his teaching and inspiration my interest in concrete and I would like to express my sincere gratitude to him.

I would like to thank Dr. A. R. Collins, Dr. P. E. Halstead, and Mr. B. W. Shacklock of the Cement and Concrete Association in London for reading parts of the manuscript.

Dr. J. F. Kirkaldy kindly provided the data of Table 3.7.

The copyright of the following illustrations and tables rests with the Crown and my thanks are due to the Controller of H.M. Stationery Office and the Department of Scientific and Industrial Research for their permission to reproduce them: Figures 1.12, 3.1, 3.13, 3.14, 4.1, 5.14, 5.15, 7.24, 8.7, 8.10, 8.25, 8.26, 8.31, and 10.3, and Tables 3.8, 3.9, 3.18, 3.21, 3.24, 3.25, 3.26, 4.2, 7.11, 9.3, 10.2, 10.7, and 10.8.

The following have made material from their publications available to me, for which I thank them: the National Bureau of Standards, Washington D.C.; the U.S. Bureau of Reclamation; the American Society for Testing and Materials (A.S.T.M.); the Cement and Concrete Association; the Portland Cement Association; the American Concrete Institute; The Institution of Civil Engineers (London); the Council of The Institution of Structural Engineers (London); Department of Mines, Ottawa; Edward Arnold (Publishers) Ltd., London; the Editorial Director of *Civil Engineering and Public Works Review*; Reinhold Publishing Corporation, Book Division (New York); Butterworths Scientific Publications (London); the Editorial Director of *Deutscher Ausschuss für Stahlbeton*.

The full details of the sources can be found at the end of each chapter; the reference numbers appear with the captions to the illustrations and the headings to the tables.

Tables from B.S. 812: 1967, *Methods for sampling and testing of mineral aggregates, sand and fillers*, and B.S. 882, 1201: 1965, *Aggregates from natural sources for concrete (including granolithic)*, are reproduced by kind permission of the British Standards Institution, 2 Park Street, London, W.1, from whom copies of the complete standards may be purchased.

Contents

Splitting Test—Influence of Rate of Application of Load on Strength—Influence of Moisture Condition during Test—Influence of Size of Specimen on Strength: Specimen size and aggregate size—Test Cores—Accelerated Curing Test—Rebound Hammer Test—Penetration Resistance Test—Ultrasonic Pulse Test—Electrodynamic Determination of the Modulus of Elasticity—Tests on the Composition of Hardened Concrete: Cement content; Determination of the original water/cement ratio; Physical methods—Variation in Test Results: Distribution of strength; Standard deviation.

Values of the strength of concrete, unless otherwise stated, are those of standard test cubes

1. Portland Cement

CEMENT, in the general sense of the word, can be described as a material with adhesive and cohesive properties which make it capable of bonding mineral fragments into a compact whole. This definition embraces a large variety of cementing materials.

For constructional purposes the meaning of the term cement is restricted to the bonding materials used with stones, sand, bricks, building blocks, etc. The principal constituents of this type of cement are compounds of lime, so that in building and civil engineering we are concerned with calcareous cement. The cements of interest in the making of concrete have the property of setting and hardening under water by virtue of a chemical reaction with it and are, therefore, called hydraulic cements.

Hydraulic cements consist mainly of silicates and aluminates of lime, and can be classified broadly as natural cements, Portland cements and aluminous cements. The present chapter deals with the manufacture of Portland cement and its structure and properties both when unhydrated and in a hardened state. The different types of Portland and other cements are described in Chapter 2.

Historical Note

The use of cementing materials is very old. The ancient Egyptians used calcined impure gypsum. The Greeks and the Romans used calcined limestone and later learned to add to lime and water, sand and crushed stone or brick and broken tiles. This was the first concrete in history. Lime mortar does not harden under water, and for construction under water the Romans ground together lime and a volcanic ash or finely ground burnt clay tiles. The active silica and alumina in the ash and the tiles combined with the lime to produce what became known as pozzolanic cement from the name of the village of Pozzuoli, near Vesuvius, where the volcanic ash was first found. The name pozzolanic cement is used to this day to describe cements obtained simply by the grinding of natural materials at normal temperature. Some of the Roman structures in which masonry was bonded by mortar, such as the Coliseum in Rome and the Pont du Gard near Nîmes, have survived to this day, with the cementitious material still hard and firm. In the ruins at Pompeii, the mortar is often less weathered than the rather soft stone.

The Middle Ages brought a general decline in the quality and use of cement, and it is only in the 18th century that an advance in the knowledge of cements can be reported. John Smeaton, commissioned in 1756 to rebuild the Eddystone Lighthouse, off the Cornish coast, found that the best mortar was produced when pozzolana was mixed with limestone containing a high proportion of clayey matter. By recognizing the rôle of the clay, hitherto considered undesirable, Smeaton was the first to understand the chemical properties of hydraulic lime.

There followed a development of other hydraulic cements, such as the "Roman cement" obtained by Joseph Parker by calcining nodules of argillaceous limestone, culminating in the patent for "Portland cement" taken out by Joseph Aspdin, a Leeds builder, in 1824. This cement was prepared by heating a mixture of finely divided clay and hard limestone in a furnace until CO_2 had been driven off; this temperature was much lower than that necessary for clinkering. The prototype of modern cement was made in 1845 by Isaac Johnson who burnt a mixture of clay and chalk until clinkering, so that the reactions necessary for the formation of strongly cementitious compounds took place.

The name Portland cement, given originally due to the resemblance of the colour and quality of set cement to Portland stone—a limestone quarried in Dorset—has remained to this day to describe a cement obtained by intimately mixing together calcareous and argillaceous, or other silica-, alumina-, and iron oxide-bearing materials, burning them at a clinkering temperature, and grinding the resulting clinker. This is the definition of the current British Standard (B.S. 12: 1958) which stipulates also that no materials other than gypsum and water may be added after burning.

Manufacture of Portland Cement
From the definition of Portland cement given above it can be seen that it is made primarily from a calcareous material, such as limestone or chalk, and from alumina and silica found as clay or shale. Marl, a mixture of calcareous and argillaceous materials, is also used. In Great Britain, chalk is found in the South-East, and for this reason the largest concentration of cement works is near the mouth of the Thames and on the banks of the Medway. Limestone occurs in many parts of the South-West, the Midlands, Northern England and Wales, and argillaceous deposits are found throughout the country.

The process of manufacture of cement consists essentially of grinding the raw materials, mixing them intimately in certain proportions and burning in a large rotary kiln at a temperature of approximately 1,300 to 1,400°C when the material sinters and partially fuses into balls known as clinker. The clinker is cooled and ground to a fine powder, with some gypsum added, and the resulting product is the commercial Portland cement so widely used throughout the world.

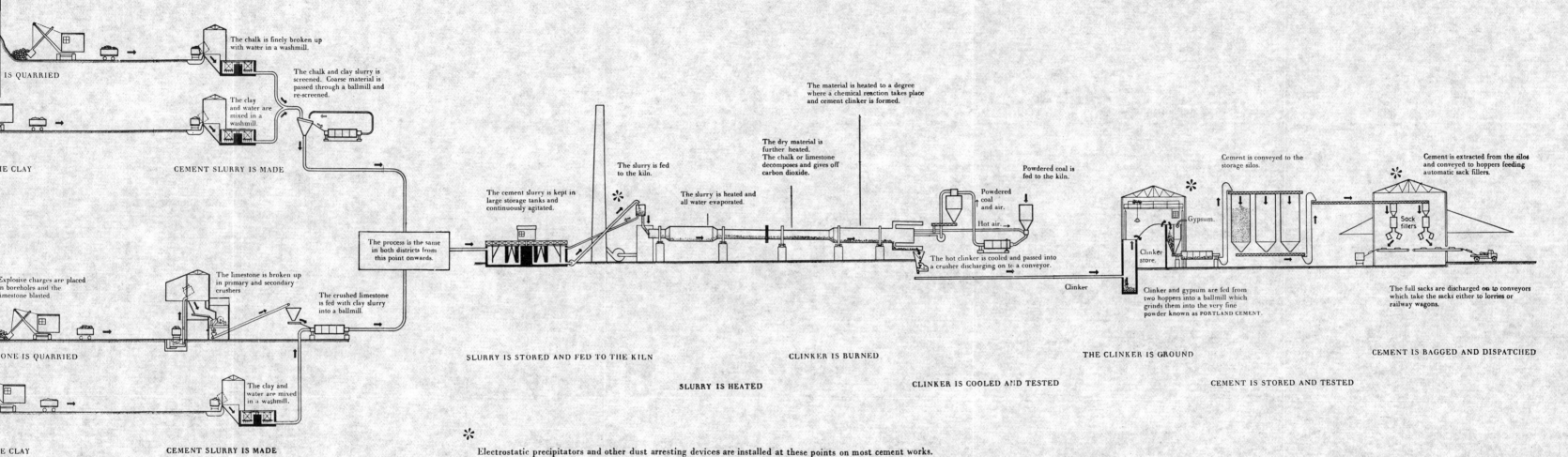

IS QUARRIED

The chalk is finely broken up with water in a washmill.

The chalk and clay slurry is screened. Coarse material is passed through a ballmill and re-screened.

HE CLAY

The clay and water are mixed in a washmill.

CEMENT SLURRY IS MADE

The material is heated to a degree where a chemical reaction takes place and cement clinker is formed.

The dry material is further heated. The chalk or limestone decomposes and gives off carbon dioxide.

The slurry is fed to the kiln.

The slurry is heated and all water evaporated.

Powdered coal is fed to the kiln.

Cement is conveyed to the storage silos.

Cement is extracted from the silos and conveyed to hoppers feeding automatic sack fillers.

The cement slurry is kept in large storage tanks and continuously agitated.

Powdered coal and air.

Hot air.

Gypsum.

Sack fillers

The process is the same in both districts from this point onwards.

Explosive charges are placed in boreholes and the limestone blasted.

The limestone is broken up in primary and secondary crushers

The crushed limestone is fed with clay slurry into a ballmill.

The hot clinker is cooled and passed into a crusher discharging on to a conveyor.

Clinker store.

Clinker

Clinker and gypsum are fed from two hoppers into a ballmill which grinds them into the very fine powder known as PORTLAND CEMENT.

The full sacks are discharged on to conveyors which take the sacks either to lorries or railway wagons.

ONE IS QUARRIED

SLURRY IS STORED AND FED TO THE KILN

CLINKER IS BURNED

THE CLINKER IS GROUND

CEMENT IS BAGGED AND DISPATCHED

The clay and water are mixed in a washmill.

E CLAY

CEMENT SLURRY IS MADE

SLURRY IS HEATED

CLINKER IS COOLED AND TESTED

CEMENT IS STORED AND TESTED

Electrostatic precipitators and other dust arresting devices are installed at these points on most cement works.

Fig. 1.1. *Diagrammatic representation of the wet process of manufacture of cement*

Some details of the manufacture of cement will now be given, and these can best be followed with reference to the diagrammatic representation of the process given in Fig. 1.1.

The mixing and grinding of the raw materials can be done either in water or in a dry condition, hence the names "wet" and "dry" processes. The actual methods of manufacture depend also on the nature of the raw materials used.

Let us consider first the wet process. When chalk is used it is finely broken up and dispersed in water in a washmill; this is a circular pit with revolving radial arms carrying rakes which break up the lumps of solid matter. The clay is also broken up and mixed with water, usually in a similar washmill. The two mixtures are now pumped so as to mix in predetermined proportions and pass through a series of screens. The resulting cement slurry flows into storage tanks.

When limestone is used it has to be blasted, then crushed, usually in two progressively smaller crushers, and then fed into a ball mill with the clay dispersed in water. There, the comminution of the limestone (to the fineness of flour) is completed, and the resultant cement slurry is pumped into storage tanks. From here onwards the process is the same regardless of the original nature of the raw materials.

The slurry is a liquid of creamy consistency, with a water content of between 35 and 50 per cent, and only a small fraction of material—about 2 per cent—larger than a 90 μm (No. 170) B.S. sieve size. There is usually a number of storage tanks in which the slurry is kept, the sedimentation of the suspended solids being prevented by mechanical stirrers or bubbling by compressed air. The lime content of the slurry is governed by the proportioning of the original calcareous and argillaceous materials as mentioned earlier. Final adjustment in order to achieve the required chemical composition can be made by blending slurries from different storage tanks, sometimes using an elaborate system of blending tanks.

Finally, the slurry with the desired lime content passes into the rotary kiln. This is a large, refractory-lined steel cylinder, up to 5 m (or 16 ft) in diameter, sometimes as long as 150 m (or 500 ft), slowly rotating about its axis, which is slightly inclined to the horizontal. The slurry is fed in at the upper end while pulverized coal is blown in by an air blast at the lower end of the kiln, where the temperature reaches 1,400 to 1,500°C. The coal, which must not have too high an ash content, deserves a special mention because up to as much as 350 kg (7 cwt) of coal is used to make one tonne of cement. This is worth bearing in mind when considering the price of cement. Oil or natural gas can be used instead of coal.

The slurry, in its movement down the kiln, encounters a progressively higher temperature. At first, the water is driven off and CO_2 is liberated; further on, the dry material undergoes a series of chemical reactions until finally, in the hottest part of the kiln, some 20 to 30 per cent of the material becomes liquid, and lime, silica and alumina recombine. The mass then fuses into balls, 3 to 25 mm ($\frac{1}{8}$ to 1 in.) in diameter, known as clinker. The clinker drops into coolers, which are of various types and often provide means for an exchange of heat with the air subsequently used for the combustion of the pulverized coal. A large kiln can produce over 700 tonnes of cement a day.

The cool clinker, which is characteristically black, glistening, and hard, is interground with gypsum in order to prevent flash-setting of the cement. The grinding is done in a ball mill consisting of several compartments with progressively smaller steel balls. In some plants a closed-circuit grinding system is used: the cement discharged by the mill is passed through a separator, fine particles being removed to the storage silo by an air current, while the coarser particles are passed through the mill once again. Closed-circuit grinding avoids the production of a large amount of excessively fine material or of a small amount of too coarse material, faults often encountered with open-circuit grinding.

Once the cement has been satisfactorily ground, when it will have as many as $1 \cdot 1 \times 10^{12}$ particles per kg (5×10^{11} per lb), it is ready for packing in the familiar paper bags* or in drums, or for transport in bulk.

In the dry and semi-dry processes, the raw materials are crushed and fed in the correct proportions into a grinding mill, where they are dried and reduced in size to a fine powder. The dry powder, called raw meal, is then pumped to a blending silo, and final adjustment is now made in the proportions of the materials required for the manufacture of cement. To obtain a uniform and intimate mixture the raw meal is blended, usually by means of compressed air inducing an upward movement of the powder and decreasing its apparent density. The air is pumped over one quadrant of the silo at a time and this permits the apparently heavier material from the non-aerated quadrants to move laterally into the aerated quadrant. Thus the aerated material tends to behave almost like a liquid, and by aerating all quadrants in turn for a total period of about one hour a uniform mixture is obtained. In some cement works continuous blending is used.

The blended meal is now sieved and fed into a rotating dish called a granulator, water weighing about 12 per cent of the meal being added at the same time. In this manner hard pellets about 15 mm ($\frac{1}{2}$ in.) in diameter are formed. This is necessary, as powder fed direct into a kiln would not permit the air flow and exchange of heat necessary for the chemical reactions of formation of cement clinker.

The pellets are baked hard in a pre-heating grate by means of hot gases

---

* A standard bag in the United Kingdom contains 50 kg (previously 112 lb) of cement; a U.S. sack weighs 94 lb; others are: Canada 80 lb, New Zealand $93\frac{1}{3}$ lb, Australia 40 kg, South Africa 50 kg, China 45 kg.

from the kiln. The pellets then enter the kiln, and subsequent operations are the same as in the wet process of manufacture. Since, however, the moisture content of the pellets is only 12 per cent as compared with the 40 per cent moisture content of the slurry used in the wet process, the dry-process kiln is considerably smaller. The amount of heat required is also very much lower as only some 12 per cent of moisture has to be driven off, but additional heat has already been used in removing the original moisture content of the raw materials (usually 6 to 10 per cent). The process is thus quite economical, but only when the raw materials are comparatively dry. In such a case the total coal consumption can be as little as 100 kg (2 cwt) per tonne of cement.

The difficulties in the control of dry mixing and blending until recently prevented a wider use of the dry process, but it has been used in the United States and Germany, and in 1957 the first British semi-dry process cement works was opened. Small vertical kilns are used in several countries: they produce up to 150 tonnes of cement a day.

It should be explained that an intimate mixture of the raw materials is necessary because a part of the reactions in the kiln must take place by diffusion in solid materials, and a uniform distribution of materials is essential to ensure a uniform product.

There are also other processes of manufacture of cement, of which one, using gypsum instead of lime, perhaps deserves mention. Gypsum, clay and coke with sand and iron oxide are burnt in a rotary kiln, the end products being Portland cement and sulphur dioxide which is further converted into sulphuric acid.

Chemical Composition of Portland Cement

We have seen that the raw materials used in the manufacture of Portland cement consist mainly of lime, silica, alumina and iron oxide. These compounds interact with one another in the kiln to form a series of more complex products, and, apart from a small residue of uncombined lime which has not had sufficient time to react, a state of chemical equilibrium is reached. However, equilibrium is not maintained during cooling, and the rate of cooling will affect the degree of crystallization and the amount of amorphous material present in the cooled clinker. The properties of this amorphous material, known as glass, differ considerably from those of crystalline compounds of a nominally similar chemical composition. Another complication arises from the interaction of the liquid part of the clinker with the crystalline compounds already present.

Nevertheless, cement can be considered as being in frozen equilibrium, i.e. the cooled products are assumed to reproduce the equilibrium existing at the clinkering temperature. This assumption is, in fact, made in the calculation of the compound composition of commercial cements: the "potential" composition is calculated from the measured quantities of oxides present in the clinker as if full crystallization of equilibrium products had taken place.

Four compounds are usually regarded as the major constituents of cement: they are listed in Table 1.1 together with their abbreviated symbols. This shortened notation, used by cement chemists, describes each oxide by one letter, viz.: $CaO = C$; $SiO_2 = S$; $Al_2O_3 = A$; and $Fe_2O_3 = F$. Likewise, H_2O in hydrated cement is denoted by H.

Table 1.1: *Main Compounds of Portland Cement*

Name of compound	Oxide composition	Abbreviation
Tricalcium silicate	$3CaO.SiO_2$	C_3S
Dicalcium silicate	$2CaO.SiO_2$	C_2S
Tricalcium aluminate	$3CaO.Al_2O_3$	C_3A
Tetracalcium aluminoferrite	$4CaO.Al_2O_3.Fe_2O_3$	C_4AF

In reality, the silicates in cement are not pure compounds, but contain minor oxides in solid solution. These oxides have significant effects on the atomic arrangements, crystal form and hydraulic properties of the silicates.

The calculation of the potential composition of Portland cement is based on work of R. H. Bogue and others, and is often referred to as "Bogue composition." There are also other methods of calculating the composition[1.1] but the subject is not considered to be within the scope of the present work.

Bogue's[1.2] equations for the percentages of main compounds in cement are given below. The symbols in brackets represent the percentage of the given oxide in the total weight of cement.

$$C_3S = 4 \cdot 07(CaO) - 7 \cdot 60(SiO_2) - 6 \cdot 72(Al_2O_3) - 1 \cdot 43(Fe_2O_3) - 2 \cdot 85(SO_3)$$
$$C_2S = 2 \cdot 87(SiO_2) - 0 \cdot 754(3CaO.SiO_2)$$
$$C_3A = 2 \cdot 65(Al_2O_3) - 1 \cdot 69(Fe_2O_3)$$
$$C_4AF = 3 \cdot 04(Fe_2O_3)$$

In addition to the main compounds listed in Table 1.1, there exist *minor compounds*, such as MgO, TiO_2, Mn_2O_3, K_2O and Na_2O; they usually amount to not more than a few per cent of the weight of cement. Two of the minor compounds are of interest: the oxides of sodium and potassium, Na_2O and K_2O, known as *the alkalis* (although other alkalis also exist in cement). They have been found to react with some aggregates, the products of the reaction causing distintegration of the concrete, and have also been observed to affect the rate of the gain of strength of cement.[1.3] It should, therefore, be pointed out that the term "minor compounds" refers primarily to their quantity and not necessarily to their importance. The quantity of the alkalis and of Mn_2O_3 can be rapidly determined using a spectrophotometer.

The compound composition of cement has been established largely on the basis of studies of phase equilibria of the ternary systems C–A–S and C–A–F, and the quaternary system $C-C_2S-C_5A_3-C_4AF$, and others. The course of melting or crystallization was traced and the compositions of liquid and solid phases at any temperature were computed. In addition to the methods of chemical analysis, the actual composition of clinker can be determined by a microscope examination of powder preparations and their identification by the measurement of the refractive index. The amounts of the two silicates can be measured by means of a Shands micrometer using a thin section (similar to those used in petrographic studies) in transmitted light. Polished and etched sections can also be used both in reflected and transmitted light. Other methods include the use of X-ray powder diffraction to identify the crystalline phases and also to study the crystal structure of some of the phases, and of differential thermal analysis. Another development is the use of the electron microscope which produces a high magnification by employing an electron beam instead of light waves. Estimating the composition of cement is aided by new and more rapid methods of determining the elemental composition, such as X-ray fluorescence and electron probe microanalysis.

C_3S, which is normally present in the largest amounts, occurs as small, equidimensional colourless grains. On cooling below 1,250°C it decomposes slowly, but if cooling is not too slow, C_3S remains unchanged and is relatively stable at ordinary temperatures.

C_2S is known to have three or possibly even four forms. α-C_2S, which exists at high temperatures, inverts to the β-form at 1,456°C.* β-C_2S undergoes further inversion to γ-C_2S at 675°C* but at the rates of cooling of commercial cements, β-C_2S is preserved in the clinker. β-C_2S forms rounded grains, usually showing twinning.

C_3A forms rectangular crystals, but C_3A in frozen glass forms an amorphous interstitial phase.

C_4AF is really a solid solution ranging from C_2F to C_6A_2F, but the description C_4AF is a convenient simplification.[1.4]

The actual quantities of the various compounds vary considerably from cement to cement, and indeed different types of cement are obtained by suitable proportioning of the materials. In the United States an attempt was at one time made to control the properties of cements required for different purposes by specifying the limits of the four major compounds, as calculated from the oxide analysis. This procedure would cut out numerous physical tests normally performed, but unfortunately the calculated compound composition is not sufficiently accurate nor does it

---

* It now appears that the inversion temperatures are in fact lower.

take into account all the relevant properties of cement, and cannot therefore serve as a substitute for direct testing of the required properties.

A general idea of the composition of cement can be obtained from Table 1.2, which gives the oxide composition limits of Portland cements. Table 1.3 gives the oxide composition of a typical cement and the calculated compound composition,[1.5] obtained by means of Bogue's equations on page 6.

Table 1.2: *Approximate Composition Limits of Portland Cement*

Oxide	Content, per cent
CaO	60–67
SiO_2	17–25
Al_2O_3	3–8
Fe_2O_3	0·5–6·0
MgO	0·1–4·0
Alkalis	0·2–1·3
SO_3	1–3

Table 1.3: *Oxide and Compound Compositions of a Typical Portland Cement*[1.5]

Typical oxide composition per cent		Hence, calculated compound composition (using formulae of p. 6) per cent	
CaO	63	C_3A	10·8
SiO_2	20	C_3S	54·1
Al_2O_3	6	C_2S	16·6
Fe_2O_3	3	C_4AF	9·1
MgO	1½	Minor compounds	—
SO_3	2		
K_2O	1		
Na_2O			
Others	1		
Loss on ignition	2		
Insoluble residue	½		

Two terms used in Table 1.3 require explanation. The insoluble residue, determined by treating with hydrochloric acid, is a measure of adulteration of cement, largely arising from impurities in gypsum. B.S. 12: 1958 limits the insoluble residue to 1·5 per cent of the weight of cement. The loss on ignition shows the extent of carbonation and hydration of free lime and free magnesia due to the exposure of cement to the atmosphere. The maximum loss on ignition (at 1,000°C) permitted by B.S. 12:1958 is 3 per cent for cements in a temperate climate and 4 per cent for cements in the tropics. Since hydrated free lime is innocuous (*see* p. 50), for a given free lime content of cement, a greater loss on ignition is really advantageous.

It is interesting to observe the large influence of a change in the oxide composition on the compound composition of cement. Some data of Czernin's[1.5] are given in Table 1.4; column (1) shows the composition of a fairly typical rapid hardening cement. If the lime content is decreased by

Table 1.4: *Influence of Change in Oxide Composition on the Compound Composition*[1.5]

Oxide	Percentage in Cement No.		
	(1)	(2)	(3)
CaO	66·0	63·0	66·0
SiO_2	20·0	22·0	20·0
Al_2O_3	7·0	7·7	5·5
Fe_2O_3	3·0	3·3	4·5
Others	4·0	4·0	4·0
Compound			
C_3S	65	33	73
C_2S	8	38	2
C_3A	14	15	7
C_4AF	9	10	14

3 per cent, with corresponding increases in the other oxides [column (2)], a considerable change in the $C_3S:C_2S$ ratio results. Column (3) shows a change of $1\frac{1}{2}$ per cent in the alumina and iron contents compared with the cement of column (1). The lime and silica contents are unaltered and yet the ratio of the silicates, as well as the contents of C_3A and C_4AF, is greatly affected. It is apparent that the significance of the control of the oxide composition of cement cannot be over-emphasized. Within the usual range of ordinary and rapid hardening Portland cements the sum of the contents of the two silicates varies only within narrow limits, so that the variation in composition depends largely on the ratio of CaO to SiO_2 in the raw materials.

It may be convenient at this stage to summarize the pattern of formation and hydration of cement; this is shown schematically in Fig. 1.2.

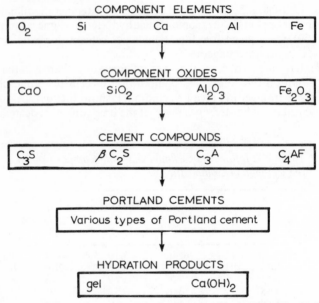

Fig. 1.2. *Schematic representation of the formation and hydration of Portland cement*

Hydration of Cement

The reactions by virtue of which Portland cement becomes a bonding agent take place in a water-cement paste. In other words, in the presence of water, the silicates and aluminates listed in Table 1.1 form products of hydration, which in time produce a firm and hard mass—the hardened cement paste.

There are two ways in which compounds of the type present in cement may react with water. In the first, a direct addition of some molecules of water takes place, this being a true reaction of hydration. The second type of reaction with water is hydrolysis. It is convenient and usual, however, to apply the term hydration to all reactions of cement with water, i.e. to both true hydration and hydrolysis.

Le Chatelier was the first to observe, some ninety years ago, that the products of hydration of cement are chemically the same as the products of hydration of the individual compounds under similar conditions. This was later confirmed by Steinour[1.6] and by Bogue and Lerch,[1.7] with the proviso that the products of reaction may influence one another or may themselves interact with the other compounds in the system. The two calcium silicates are the main cementitious compounds in cement, and the physical behaviour of cement during hydration is similar to that of these two compounds alone.[1.8] The hydration of the individual compounds will be described in more detail in the succeeding sections.

The products of hydration of cement have a low solubility in water as shown by the stability of the hardened cement paste in contact with water. The hydrated cement bonds firmly to the unreacted cement, but the exact way in which this is achieved is not certain. It is possible that the newly produced hydrate forms an envelope which grows from within by the action of water that has penetrated the surrounding film of hydrate. Alternatively, the dissolved silicates may pass through the envelope and precipitate as an outer layer. A third possibility is for the colloidal solution to be precipitated throughout the mass after the condition of saturation has been reached, the further hydration continuing within this structure.

Whatever the mode of precipitation of the products of hydration, the rate of hydration decreases continuously, so that even after a long time there remains an appreciable amount of unhydrated cement. For instance, after 28 days in contact with water, grains of cement have been found to have hydrated to a depth of only 4 μm,[1.9] and 8 μm after a year. Powers[1.10] calculated that complete hydration under normal conditions is possible only for cement particles smaller than 50 μm, but full hydration has been obtained by grinding cement in water continuously for five days.

Microscopic examination of hydrated cement shows no evidence of channelling of water into the grains of cement to hydrate selectively the more reactive compounds (e.g. C_3S) which may lie in the centre of the particle. It would seem, then, that hydration proceeds by a gradual reduction in the size of the cement particle. In fact, unhydrated grains of coarse cement were found to contain C_3S as well as C_2S at the age of several months,[1.11] and it is probable that small grains of C_2S hydrate before the hydration of large grains of C_3S has been completed. The various compounds in cement are generally intermixed in all grains, and some investigations have suggested that the residue of a grain after a given period of hydration has the same percentage composition as the whole of the original grain.[1.12] However, the composition of the residue does change throughout the period of cement hydration,[1.49] and especially during the first 24 hours selective hydration may take place.

The main hydrates can be broadly classified as calcium silicate hydrates and tricalcium aluminate hydrate. C_4AF is believed to hydrate into tricalcium aluminate hydrate and an amorphous phase, probably $CaO.Fe_2O_3$.aq. It is possible also that some Fe_2O_3 is present in solid solution in the tricalcium aluminate hydrate.

The progress of hydration of cement can be determined by different means, such as the measurement of : (*a*) the amount of $Ca(OH)_2$ in the paste; (*b*) the heat evolved by hydration; (*c*) the specific gravity of the paste; (d) the amount of chemically combined water; (e) the amount of unhydrated cement present (using X-ray quantitative analysis); (f) and also indirectly from the strength of the hydrated paste. Recently, thermogravimetric techniques and continuous X-ray diffraction scanning of wet pastes undergoing hydration[1.50] have been successfully used in studying early reactions.

Calcium Silicate Hydrates

When hydration takes place in a limited amount of water, as is the case in cement paste and in concrete, C_3S is believed to undergo hydrolysis producing a calcium silicate of lower basicity, ultimately $C_3S_2H_3$, with the released lime separating out as $Ca(OH)_2$. There exists, however, a considerable uncertainty as to whether C_3S and C_2S result ultimately in the same hydrate. It would appear to be so from considerations of the heat of hydration[1.6] and of the surface area of the products of hydration,[1.13] but physical observations indicate that there may be more than one—possibly several—distinct calcium silicate hydrates. Lea[1.1] considers the possibility of these hydrates being salts of orthosilicic acid (H_4SiO_4); thus four lime/silica ratios would be possible: 1:2, 1:1, 3:2, and 2:1. Furthermore, other ratios would be encountered if some of the lime were absorbed or were held in solid solution, and there is strong evidence that the ultimate product of hydration of C_2S has a lime/silica ratio of 1·65. This may be due to the fact that the hydration of C_3S is controlled by the rate of diffusion of ions through the overlying hydrate films while the hydration of C_2S is controlled by its slow intrinsic rate of reaction.[1.14] Furthermore, temperature may affect the products of hydration of the two silicates as the permeability of the gel is affected by temperature.

The rates of hydration of C_3S and C_2S in a pure state differ considerably, as shown in Fig. 1.3.

The overall composition of the silicate hydrates is approximately $C_3S_2H_3$ and they are sometimes referred to as tobermorite because of structural similarity to a naturally occurring mineral of this name. Since the crystals formed by hydration are imperfect and extremely small the mole ratio of water to silica need not be a whole number. Making the approximate *assumption* that $C_3S_2H_3$ is the final product of hydration of both C_3S and C_2S the reactions of hydration can be written (as a guide, although not as exact stoichiometric equations) as follows—

For C_3S:

$$2C_3S + 6H \rightarrow C_3S_2H_3 + 3Ca(OH)_2$$

The corresponding weights involved are:

$$100 + 24 \rightarrow 75 + 49$$

For C_2S:

$$2C_2S + 4H \rightarrow C_3S_2H_3 + Ca(OH)_2$$

The corresponding weights are:

$$100 + 21 \rightarrow 99 + 22$$

Thus on a weight basis both silicates require approximately the same amount of water for their hydration, but C_3S produces more than twice as much $Ca(OH)_2$ as is formed by the hydration of C_2S.

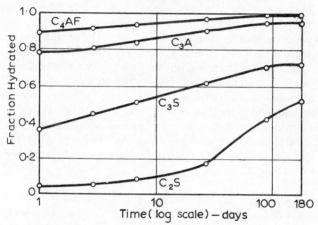

Fig. 1.3. *Rate of hydration of pure compounds*[1.47]

It is likely that in commercial cements the calcium silicates in fact contain small impurities of some of the oxides present in the clinker. The "impure" C_3S is known as alite and the "impure" C_2S as belite. These impurities have a strong effect on the properties of the calcium silicate hydrates (*see* p. 41).

The physical properties of the calcium silicate hydrates are of interest in connexion with the setting and hardening properties of cement. These hydrates are usually described as completely amorphous but an electron microscope shows their crystalline character. It is interesting to note that one of the hydrates, denoted by Taylor[1.15] as CSH(I), has a layer structure similar to that of some clay minerals, e.g. montmorillonite and halloysite. The individual layers in the plane of the *a* and *b* axes are well crystallized while the distances between them are less rigidly defined. Such a lattice would be able to accommodate varying amounts of lime without fundamental change—a point relevant to the varying lime/silica ratios mentioned above. In fact, powder diagrams have shown that lime in excess of one molecule per molecule of silica is held in a random manner.[1.15] Steinour[1.16] described this as a merger of solid solution and adsorption.

The use of Calcium 45 tracers* has shown that the calcium silicates do not hydrate in the solid state but the anhydrous silicates probably first pass into solution and then react to form less soluble hydrated silicates which separate out of the supersaturated solution.[1.17] This is the type of mechanism of hydration first envisaged by Le Chatelier in 1881.

Studies by Bernal[1.18] indicate that the calcium silicate hydrates are in the form of very thin fibrous crystals with a short fibre repeat unit of 3·65 Å,† this being interpreted to mean that there exist silicate tetrahedra joined by hydrogen bonds. Other studies have indicated the existence of fibrous particles with sheaf-like ends, similar to the swelling clay mineral halloysite. A variety of transitional forms are believed to exist, including some small spherical particles, but they all become finally converted to the fibrous form and to sheets or foils aggregated as fluffy masses.[1.19]

It is interesting to observe that calcium silicate hydrates show a strength development similar to that of Portland cement.[1.20] A considerable strength is possessed long before the reaction of hydration is complete and it would thus seem that a small amount of the hydrate binds together the unhydrated remainder; further hydration results in little or no increase in strength.

Tricalcium Aluminate Hydrate and the Action of Gypsum

The amount of C_3A present in most cements is comparatively small but its behaviour and structural relationship with the other phases in cement make it of interest. The tricalcium aluminate hydrate forms a prismatic dark interstitial material, possibly with other substances in solid solution, and is often in the form of flat plates individually surrounded by the calcium silicate hydrates.

The reaction of pure C_3A with water is very violent and leads to immediate stiffening of the paste, known as *flash set*. To prevent this from happening, gypsum $(CaSO_4.2H_2O)$ is added to cement clinker. Gypsum and C_3A react to form insoluble calcium sulphoaluminate $(3CaO.Al_2O_3.3CaSO_4.31H_2O)$ but eventually a tricalcium aluminate hydrate is formed, although it is probable that this is preceded by a metastable $3CaO.Al_2O_3.CaSO_4.12H_2O$, produced at the expense of the original high-sulphate calcium sulphoaluminate.[1.6] As more C_3A comes into solution the composition changes, the sulphate content decreasing continuously. The rate of reaction of the aluminate is high and, if this readjustment in composition is not rapid enough, direct

---

* A radioactive ion behaves chemically like its non-reactive isotope, but its movement can be traced.

† $1 \text{ Å} = 10^{-10} \text{ m}$.

hydration of C_3A is likely. In particular, a peak in the rate of heat development normally observed within five minutes of adding water to cement
means that some calcium aluminate hydrate is formed directly during that
period, the conditions for the retardation by gypsum not yet having been
established.

The stable form of the calcium aluminate hydrate ultimately existing
in the hydrated cement paste is probably the cubic crystal C_3AH_6, but it
is possible that hexagonal C_4AH_{12} crystallizes out first and later changes
to the cubic form. Thus the final form of the reaction can be written—

$$C_3A + 6H \rightarrow C_3AH_6$$

(This again is an approximation and not a stoichiometric equation.)

The molecular weights show that 100 parts of C_3A react with 40 parts
of water by weight, which is a much higher proportion of water than that
required by the silicates.

The presence of C_3A in cement is undesirable: it contributes little or
nothing to the strength of cement except at early ages, and when hardened
cement paste is attacked by sulphates, expansion due to the formation of
calcium sulphoaluminate from C_3A may result in a disruption of the
hardened paste. However, C_3A acts as a flux and thus reduces the temperature of burning of clinker and facilitates the combination of lime
and silica; for these reasons C_3A is useful in the manufacture of cement.
C_4AF also acts as a flux. It may be noted that if some liquid were not
formed during burning, the reactions in the kiln would progress much
more slowly and would probably be incomplete.

Gypsum reacts not only with C_3A: with C_4AF it forms calcium
sulphoferrite as well as calcium sulphoaluminate, and its presence may
accelerate the hydration of the silicates.

The amount of gypsum added to the cement clinker has to be very
carefully watched; in particular, an excess of gypsum leads to an expansion and consequent disruption of the set cement paste. The optimum
gypsum content is determined by observation of the generation of the
heat of hydration. Generally, the immediate peak in the rate of heat
evolution is followed by a second peak some 4 to 8 hours after the water
has been added to cement, and with the correct amount of gypsum there
should be little C_3A available for reaction after all the gypsum has combined, and no further peak in the heat liberation should occur. Thus, an
optimum gypsum content leads to a desirable rate of early reaction and
prevents local high concentration of products of hydration (*see* p. 44).

The amount of gypsum required increases with the C_3A content and
also with the alkali content of the cement. Increasing the fineness of
cement has the effect of increasing the quantity of C_3A available at early
ages, and this raises the gypsum requirement.

The amount of gypsum added to cement clinker is usually expressed as the weight of SO_3 present; this is limited by B.S. 12 : 1958 to a maximum of 2·5 per cent when the C_3A content is not more than 7 per cent, and to 3·0 per cent when the amount of C_3A exceeds 7 per cent.

Setting

This is the term used to describe the stiffening of the cement paste, although the definition of the stiffness of the paste which is considered set is somewhat arbitrary. Broadly speaking, setting refers to a change from a fluid to a rigid state. Although during setting the paste acquires some strength, for practical purposes it is convenient to distinguish setting from hardening, which refers to the gain of strength of a set cement paste.

In practice, the terms initial set and final set are used to describe arbitrarily chosen stages of setting. The method of measurement of these setting times is described on page 48.

It seems that setting is caused by a selective hydration of cement compounds: the two first to react are C_3A and C_3S. The flash-setting properties of the former were mentioned in the preceding section but the addition of gypsum delays the formation of calcium aluminate hydrate and it is thus C_3S that sets first. Neat C_3S mixed with water also exhibits an initial set but C_2S stiffens in a more gradual manner.

In a properly retarded cement the framework of the hydrated cement paste is established by the calcium silicate hydrate, while if C_3A were allowed to set first a rather porous calcium aluminate hydrate would form. The remaining cement compounds would then hydrate within this porous framework and the strength characteristics of the cement paste would be adversely affected.

Apart from the rapidity of formation of crystalline products, the development of films around cement grains and a mutual coagulation of components of the paste have also been suggested as factors in the development of set.

The setting process is accompanied by temperature changes in the cement paste: initial set corresponds to a rapid rise in temperature and final set to the peak temperature. At this time a sharp drop in electrical conductivity also takes place, and attempts have been made to measure setting by electrical means.

The setting time of cement decreases with a rise in temperature, but above about 30°C (85°F) a reverse effect may be observed.[1.1] At low temperatures setting is retarded. The influence of temperature on setting time is indicated in Fig. 7.7.

False Set

False set is the name given to the abnormal premature stiffening of cement within a few minutes of mixing with water. It differs from *flash set* in that no appreciable heat is evolved, and remixing the cement paste without addition of water restores plasticity of the paste until it sets in the normal manner and without a loss of strength.

Some of the causes of false set are to be found in the dehydration of gypsum when interground with too hot a clinker: hemihydrate ($CaSO_4.\frac{1}{2}H_2O$) or anhydrite ($CaSO_4$) are formed, and when the cement is mixed with water these hydrate to form gypsum. Thus plaster set takes place with a resulting stiffening of the paste.

Another cause of false set may be associated with the alkalis in the cement. During storage they may carbonate, and alkali carbonates react with $Ca(OH)_2$ liberated by the hydrolysis of C_3S to form $CaCO_3$. This precipitates and induces a rigidity of the paste.

It has also been suggested that false set can be due to the activation of C_3S by aeration at moderately high humidities. Water is adsorbed on the grains of cement, and these freshly activated surfaces can combine very rapidly with more water during mixing: this rapid hydration would produce false set.[1.21]

Laboratory tests at cement works generally ensure that cement is free from false set. If, however, false set is encountered, it can be dealt with by remixing the concrete without adding any water. Although this is not easy, workability will be improved and the concrete can be placed in the normal manner.

Fineness of Cement

It may be recalled that one of the last steps in the manufacture of cement is the grinding of clinker mixed with gypsum. Since the hydration starts at the surface of the cement particles it is the total surface area of cement that represents the material available for hydration. Thus the rate of hydration depends on the fineness of the cement particles, and for a rapid development of strength high fineness is necessary (*see* Fig. 1.4).

On the other hand, the cost of grinding to a higher fineness is considerable, and also the finer the cement the more rapidly it deteriorates on exposure to the atmosphere. Finer cement leads to a stronger reaction with alkali-reactive aggregate,[1.44] and makes a paste, though not necessarily concrete, exhibiting a higher shrinkage and a greater proneness to cracking. However, fine cement bleeds less than a coarser one.

An increase in fineness increases the amount of gypsum required for proper retardation as in a finer cement more C_3A is available for early hydration. The water content of a paste of standard consistence is greater the finer the cement, but conversely an increase in fineness of cement slightly improves the workability of a concrete mix. This anomaly may be due partly to the fact that the tests for consistence of cement paste and workability measure different properties of fresh paste; also, accidental air affects the workability of cement paste, and cements of different fineness may contain different amounts of air.

We can see then that fineness is a vital property of cement and has to be carefully controlled. In the past, the fraction of cement retained on a 90 μm (No. 170) B.S. test sieve* was determined, and the maximum residue was limited to 10 per cent by weight for ordinary and 5 per cent for rapid hardening Portland cement. A cement satisfying these conditions would not contain an excess of large grains which, because of their comparatively small surface area per unit weight, would play only a small rôle in the process of hydration and development of strength.

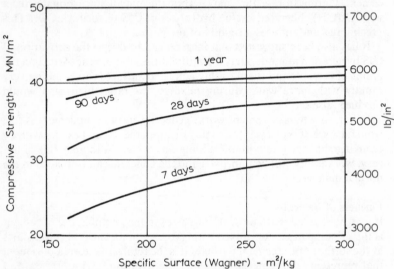

Fig. 1.4. *Relation between strength of concrete at different ages and fineness of cement*[1.43]

However, the sieve test yields no information on the size of grains smaller than 90 μm (No. 170) B.S. sieve, and it is the finer particles that play the greatest part in the early hydration. Attempts to use smaller sieves, down to 53 μm (No. 300), have generally been unsuccessful because of clogging of such extremely fine mesh.

For these reasons B.S. 12:1958 prescribes a test for fineness by determination of the specific surface of cement expressed as the total surface area in square centimetres per gram.† A direct approach is to measure the particle size distribution by sedimentation or elutriation: these methods are based on the dependence of the rate of free fall of particles on their diameter. Stokes' law gives the terminal velocity of fall under gravity of a spherical particle in a fluid medium. This medium must of course be

---

* For size of openings of different sieves see Table 3.14, page 147.

† It is expected that on metrication this will be expressed in kilogrammes per square metre.

chemically inert with respect to cement. It is also important to achieve a satisfactory dispersion of cement particles as partial flocculation would produce a decrease in the apparent specific surface.

A development of these methods is the Wagner turbidimeter used in the United States (A.S.T.M. Standard C 115–70). In this test the concentration of particles in suspension at a given level in kerosene is determined using a beam of light, the percentage of light transmitted being measured by a photo-cell. The turbidimeter gives generally consistent results but an error is introduced by assuming a uniform size distribution of particles smaller than 7·5 μm. It is precisely these finest particles that contribute most to the specific surface of cement and the error is especially significant with the finer cements used nowadays. However, an improvement on the standard method is possible if the concentration of particles 5 μm in size is determined and a modification of calculations is made.[1.51] A typical curve of particle size distribution is shown in Fig. 1.5, which

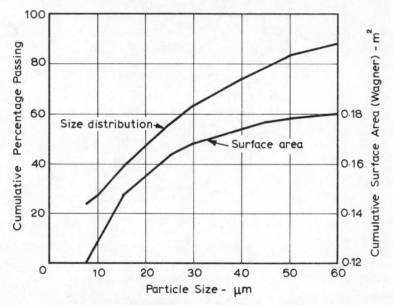

Fig. 1.5. *Example of particle size distribution and cumulative surface area contributed by particles up to any size for 1 gram of cement*

gives also the corresponding contribution of these particles to the total surface area of the sample. As mentioned on page 4, the particle size distribution depends on the method of grinding and varies, therefore, from plant to plant.

It must be admitted, however, that it is not quite clear what is a "good" grading of cement: should all the particles be of the same size or should their distribution be such that they are able to pack densely?

A more recent method of determination of the specific surface of cement is the air permeability method, using an apparatus developed by Lea and Nurse. This is the method of measurement prescribed by B.S. 12:1958. It is based on the relation between the flow of a fluid through a granular bed and the surface area of the particles comprising the bed. From this the surface area per unit weight of the bed material can be related to the permeability of a bed of a given porosity, i.e. containing a fixed volume of pores in the total volume of the bed.

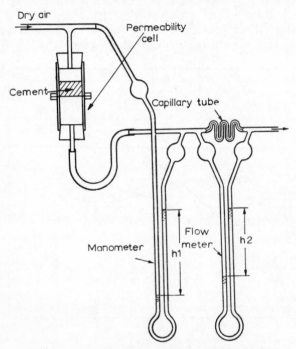

Fig. 1.6. *Lea and Nurse permeability apparatus*

The permeability apparatus is shown diagrammatically in Fig 1.6. Knowing the density of cement the weight required to make a bed of porosity of 0·475 and 1 cm thick can be calculated. This amount of cement is placed in a cylindrical container, a stream of dry air is passed through the cement bed at a constant velocity and the resulting pressure drop is measured by a manometer connected to the top and bottom of the bed.

The rate of airflow is measured by a flowmeter consisting of a capillary placed in the circuit and a manometer across its ends.

An equation developed by Carman gives the specific surface in square centimetres per gram as

$$S_w = \frac{14}{\rho(1-\varepsilon)} \sqrt{\frac{\varepsilon^3 A h_1}{KLh_2}}$$

where ρ = density of cement (g/cm³)

ε = porosity of the cement bed (0·475 in the B.S. test)

A = cross-sectional area of the bed (5·066 cm²)

L = height of the bed (1 cm)

h_1 = pressure drop across the bed

h_2 = pressure drop across the flowmeter capillary (between 25 and 55 cm of kerosene)

and K = the flowmeter constant.

For a given apparatus and porosity the expression simplifies to

$$S_w = \frac{K_1}{\rho} \sqrt{\frac{h_1}{h_2}}$$

where K_1 is a constant.

In the United States a modification of the Lea and Nurse method, developed by Blaine, is used. Here, the air does not pass through the bed at a constant rate but a known volume of air passes at a prescribed average pressure, the rate of flow diminishing steadily. The time t for the flow to take place is measured, and for a given apparatus and a standard porosity of 0·500, the specific surface is given by—

$$S = K_2\sqrt{t}$$

where K_2 is a constant.

The Lea and Nurse, and Blaine methods give values of specific surface in close agreement with one another but very much higher than the value obtained by the Wagner method. This is due to Wagner's assumptions about the size distribution of particles below 7·5 μm, mentioned earlier. The actual distribution in this range is such that the average value of 3·75 μm, assumed by Wagner, underestimates the surface area of these particles. In the air permeability method the surface area of *all* particles is measured directly, and the resulting value of the specific surface is about 1·8 times higher than the value calculated by the Wagner method. The actual range of the conversion factor varies between 1·6 and 2·2, depending on the fineness of the cement and its gypsum content.

Either method gives a good picture of the *relative* variation in the fineness of cement, and for practical purposes this is sufficient. The

Wagner method is somewhat more informative in that it gives an indication of the particle size distribution. An absolute measurement of the specific surface can be obtained by the nitrogen adsorption method, based on the work of Brunauer, Emmett and Teller.[1.45] While in the air permeability methods only the continuous paths through the bed of cement contribute to the measured area, in the nitrogen adsorption method the "internal" area is also accessible to the nitrogen molecules. For this reason the measured value of the specific surface is considerably higher than that determined by the air permeability methods. Some typical values are given in Table 1.5.

Table 1.5: *Specific Surface of Cement Measured by Different Methods*[1.1]

Cement	Specific surface, m²/kg measured by—		
	Wagner method	Lea and Nurse method	Nitrogen adsorption
A	180	260	790
B	230	415	1000

B.S. 12:1958 lays down the minimum specific surface (determined by the Lea and Nurse method) as 225 m²/kg for ordinary Portland cement, and 325 m²/kg for rapid hardening Portland cement.

Other minimum requirements are 225 m²/kg for Portland blast-furnace cement (B.S. 146:1958) and 320 m²/kg for low heat Portland cement (B.S. 1370:1958).

During the last twenty years there has been a tendency to grind cement more finely, and the commercial ordinary Portland cement produced in England is generally finer than the minimum laid down by B.S. 12:1958: about 300 m²/kg is a typical value.

Aluminous cement is normally coarser than Portland cements. A minimum of 225 m²/kg is specified by B.S. 915:1947, but slightly higher values are generally encountered in practice.

Structure of Hydrated Cement

Many of the mechanical properties of hardened cement and concrete appear to depend not so much on the chemical composition of the hydrated cement as on the physical structure of the products of hydration, viewed at the level of colloidal dimensions. For this reason it is important to have a good picture of the physical properties of the cement gel.

Fresh cement paste is a plastic network of particles of cement in water, but once the paste has set its apparent or gross volume remains approximately constant. At any stage of hydration the hardened paste consists of hydrates of the various compounds, referred to collectively as gel, of crystals of $Ca(OH)_2$, some minor components, unhydrated cement, and the residue of the water-filled spaces in the fresh paste. These voids are called capillary pores, but within the gel itself there exist interstitial voids, called gel pores. There are thus in hydrated paste two distinct classes of pores represented diagrammatically in Fig. 1.7.

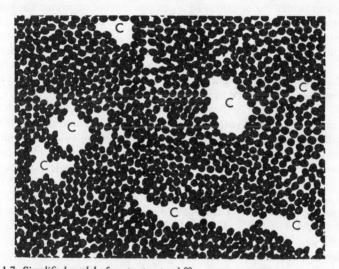

Fig. 1.7. *Simplified model of paste structure*[1.22]
Solid dots represent gel particles; interstitial spaces are gel pores; spaces such as those marked C are capillary cavities. Size of gel pores is exaggerated.

Since most of the products of hydration are colloidal, during hydration the surface area of the solid phase increases enormously, and a large amount of free water becomes adsorbed on this surface. If no water movement to or from the cement paste is permitted the reactions of hydration use up the water until too little is left to saturate the solid surfaces, and the relative humidity within the paste decreases. This is known as self-desiccation. Since gel can form only in water-filled space, self-desiccation leads to a lower hydration compared with a moist-cured paste. However, in self-desiccated pastes with water/cement ratios in excess of 0·5 the amount of mixing water is sufficient for hydration to proceed at the same rate as when moist-cured.

Volume of Products of Hydration

The gross space available for the products of hydration consists of the absolute volume of the dry cement together with the volume of water added to the mix. The small loss of water due to bleeding and the contraction of the paste while still plastic will be ignored at this stage. The water bound chemically by C_3S and C_2S was shown to be very approximately 24 and 21 per cent of the weight of the two silicates respectively. The corresponding figures for C_3A and C_4AF are 40 and 37 per cent* respectively.

As mentioned earlier, these figures are not accurate as our knowledge of stoichiometry of the products of hydration of cement is inadequate to state the amounts of water combined chemically. It is preferable, therefore, to consider non-evaporable water as determined by a given method (*see* p. 35). This water, determined under specified conditions,[1.48] is taken as 23 per cent of the weight of anhydrous cement (although in Type II cement the value may be as low as 18 per cent).

The specific gravity of the products of hydration of cement is such that they occupy a greater volume than the absolute volume of unhydrated cement but smaller than the sum of volumes of the dry cement and the non-evaporable water by approximately 0·254 of the volume of the latter. An average value of specific gravity of the products of hydration (including pores in the densest structure possible) in a saturated state is 2·16.

As an example let us consider the hydration of 100 g of cement. Taking the specific gravity of dry cement as 3·15, the absolute volume of unhydrated cement is $100/3\cdot15 = 31\cdot8$ ml. The non-evaporable water is, as we have said, about 23 per cent of the weight of cement, i.e. 23 ml. The solid products of hydration occupy a volume equal to the sum of volumes of anhydrous cement and water less 0·254 of the volume of non-evaporable water, i.e.—

$$31\cdot8 + 0\cdot23 \times 100(1 - 0\cdot254) = 48\cdot9 \text{ ml}$$

Since the paste in this condition has a characteristic porosity of about 28 per cent, the volume of gel water, w_g, is given by

$$\frac{w_g}{48\cdot9 + w_g} = 0\cdot28$$

whence $w_g = 19\cdot0$ ml

and the volume of hydrated cement is $48\cdot9 + 19\cdot0 = 67\cdot9$ ml.

---

* This is on the assumption that the final reaction of hydration of C_4AF is, in approximate terms—

$$C_4AF + 2Ca(OH)_2 + 10H \rightarrow C_3AH_6 + C_3FH_6$$

Summarizing, we have—

Weight of dry cement	$= 100{\cdot}0$ g
Absolute volume of dry cement	$= 31{\cdot}8$ ml
Weight of combined water	$= 23{\cdot}0$ g
Volume of gel water	$= 19{\cdot}0$ ml
Total water in the mix	$= 42{\cdot}0$ ml
Water/cement ratio by weight	$= 0{\cdot}42$
Water/cement ratio by volume	$= 1{\cdot}32$
Volume of hydrated cement	$= 67{\cdot}9$ ml
Original volume of cement and water	$= 73{\cdot}8$ ml
Decrease in volume due to hydration	$= 5{\cdot}9$ ml
Volume of products of hydration of 1 ml of dry cement	$= 2{\cdot}1$ ml

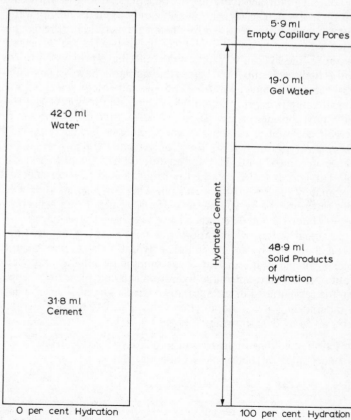

Fig. 1.8. *Diagrammatic representation of volume changes due to hydration of paste with a water/cement ratio of 0·42*

It should be noted that the hydration was assumed to take place in a sealed test tube with no water movement to or from the system. The volumetric changes are shown in Fig. 1.8. The "decrease in volume" of 5·9 ml represents the empty capillary space distributed throughout the hydrated paste.

The figures given above are only approximate but had the total amount of water been lower than about 42 ml it would have been inadequate for full hydration as gel can form only when sufficient water is available both for the chemical reactions and for the filling of the gel pores being formed. The gel water, because it is held firmly, cannot move into the capillaries so that it is not available for hydration of the still unhydrated cement.

Thus when hydration in a sealed specimen has progressed to a stage when the combined water has become about one-half of the original water content no further hydration will take place. It follows also that full hydration in a sealed specimen is possible only when the mixing water is at least twice the water required for chemical reaction, i.e. the mix has a water/cement ratio of about 0·5 by weight. In practice, in the example given above, the hydration would not in fact have progressed to completion as hydration stops even before the capillaries have become empty. It has been found that hydration becomes very slow when the water vapour pressure falls below about 0·8 of the saturation pressure.[1.23]

Let us now consider the hydration of a paste cured under water so that water can be imbibed as some of the capillaries become emptied by hydration. As shown before, 100 g of cement (31·8 ml) will on full hydration occupy 67·9 ml. Thus for no unhydrated cement to be left and no capillary pores to be present the original mixing water should be approximately $(67·9 - 31·8) = 36·1$ ml. This corresponds to a water/cement ratio of 1·14 by volume or 0·36 by weight. From other work, values of about 1·2 and 0·38 respectively have been suggested.[1.22]

If the actual water/cement ratio of the mix, allowing for bleeding, is less than about 0·38 by weight, complete hydration is not possible as the volume available is insufficient to accommodate all the products of hydration. It will be recalled that hydration can take place only in water within the capillaries. For instance, if we have a mix of 100 g of cement (31·8 ml) and 30 g of water, the water would suffice to hydrate x gram of cement, given by the following:

Contraction in volume on hydration $= 0·23x \times 0·254 = 0·0585x$

Volume occupied by solid products of hydration $=$

$$\frac{x}{3·15} + 0·23x - 0·0585x = 0·489x$$

Porosity $= \dfrac{w_g}{0·489x + w_g} = 0·28$

and total water $= 0·23x + w_g = 30$

Hence $x = 71\cdot5$ g $= 22\cdot7$ ml
and $w_g = 13\cdot5$ g
Thus, the volume of hydrated cement $= 0\cdot489 \times 71\cdot5 + 13\cdot5$
$$= 48\cdot5 \text{ ml}$$

The volume of unhydrated cement $= 31\cdot8 - 22\cdot7 = 9\cdot1$ ml
Therefore, the volume of empty capillaries $=$

$$(31\cdot8 + 30) - (48\cdot5 + 9\cdot1) = 4\cdot2 \text{ ml}$$

If water is available from outside some further cement can hydrate, its quantity being such that the products of hydration occupy $4\cdot2$ ml more than the volume of dry cement. We found that $22\cdot7$ ml of cement hydrates to occupy $48\cdot5$ ml, i.e. the products of hydration of 1 ml of cement occupy $48\cdot5/22\cdot7 = 2\cdot13$ ml. Thus $4\cdot2$ ml would be filled by the hydration of y ml of cement such that $(4\cdot2 + y)/y = 2\cdot13$; hence $y = 3\cdot7$ ml. Thus the volume of still unhydrated cement is $31\cdot8 - (22\cdot7 + 3\cdot7) = 6$ ml, which weighs 19 g. In other words, 19 per cent of the original weight of cement has remained unhydrated and can never hydrate since the gel already occupies all the space available, i.e. the gel/space ratio (*see* p. 240) of the hydrated paste is $1\cdot0$.

It may be added that unhydrated cement is not detrimental to strength and, in fact, among pastes all with a gel/space ratio of $1\cdot0$ those with a higher proportion of unhydrated cement (i.e. a lower water/cement ratio) have a higher strength, possibly because in such pastes the layers of hydrated paste surrounding the unhydrated grains are thinner.[1.24] Abrams obtained strengths of the order of 280 MN/m² (40,000 lb/in²) using mixes with a water/cement ratio of $0\cdot08$ by weight, but clearly considerable pressure is necessary to obtain a properly consolidated mix of such proportions. More recently, Lawrence[1.52] made compacts of cement powder in a die assembly under a very high pressure (up to 672 MN/m² (or 97,500 lb/in²)), using the techniques of powder metallurgy. Upon subsequent hydration for 28 days, compressive strengths up to 375 MN/m² (or 54,500 lb/in²) and tensile strengths up to 25 MN/m² (or 3600 lb/in²) were measured. The porosity of such mixes and therefore the "equivalent" water/cement ratio are very low.

On the other hand, if the water/cement ratio is higher than about $0\cdot38$ all the cement can hydrate but capillary pores will also be present. Some of the capillaries will contain excess water from the mix, others will fill by imbibing water from outside. Fig. 1·9 shows the relative volumes of unhydrated cement, products of hydration and capillaries for mixes of different water/cement ratios.

As a more specific example let us consider the hydration of a paste with a water/cement ratio of $0\cdot475$, sealed in a tube. Let the weight of dry cement be 126 g, which corresponds to 40 ml. The volume of water is

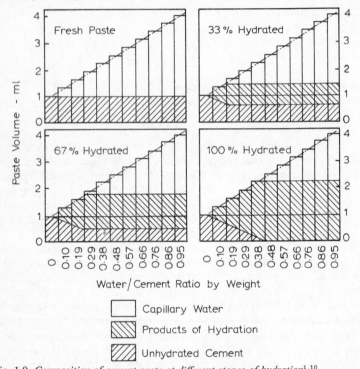

Fig. 1.9. *Composition of cement paste at different stages of hydration*[1.10]

The percentage indicated applies only to pastes with enough water-filled space to accommodate the products at the degree of hydration indicated

then $0.475 \times 126 = 60$ ml. These mix proportions are shown in the left-hand diagram of Fig 1.10, but in reality the cement and water are of course intermixed, the water forming a capillary system between the unhydrated cement particles.

Let us now consider the situation when the cement has hydrated fully. The non-evaporable water is $0.23 \times 126 = 29.0$ ml and the gel water is w_g such that—

$$\frac{w_g}{40 + 29.0\,(1 - 0.254) + w_g} = 0.28$$

whence, the volume of gel water is 24.0 ml, and the volume of hydrated cement is 85.6 ml. There are thus $60 - (29.0 + 24.0) = 7.0$ ml of water left as capillary water in the paste. In addition, $100 - (85.6 + 7.0) = 7.4$ ml form empty capillaries. If the paste had access to water during curing these capillaries would have filled with imbibed water.

This then is the situation at 100 per cent hydration when the gel/space ratio is 0·856, as shown in the right-hand diagram of Fig. 1.10. As a further illustration, the centre diagram shows the volumes of different components when only half the cement has hydrated. The gel/space ratio is then

$$\frac{\frac{1}{2}[40 + 29(1 - 0\cdot254) + 24]}{100 - 20} = 0\cdot535$$

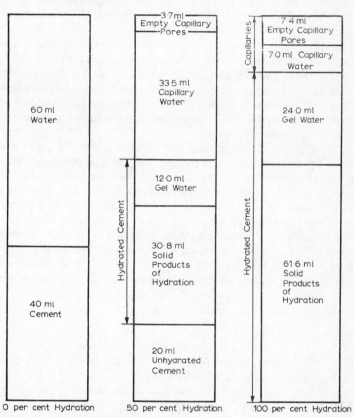

Fig. 1.10. *Diagrammatic representation of the volumetric proportions of cement paste at different stages of hydration*

Capillary Pores

We can thus see that at any stage of hydration the capillary pores represent that part of the gross volume which has not been filled by the products of hydration. Since these products occupy more than twice the volume of the original solid phase (i.e. cement) alone, the volume of the capillary system is reduced with the progress of hydration.

Thus the capillary porosity of the paste depends on both the water/cement ratio of the mix and the degree of hydration. The rate of hardening of the cement is of no importance *per se*, but the type of cement influences the degree of hydration achieved at a given age. As mentioned before, at water/cement ratios higher than about 0·38 the volume of the gel is not sufficient to fill all the space available to it so that there will be some volume of capillary pores left even after the process of hydration has been completed.

Capillary pores cannot be viewed directly but their size has been estimated from vapour pressure measurement to be of the order of 1·3 μm (5 × 10^{-5} in.). They vary in shape but, as shown by measurement of permeability, form an interconnected system randomly distributed throughout the cement paste.[1.25] These interconnected capillary pores are mainly responsible for the permeability of the hardened cement paste and for its vulnerability to frost.

However, hydration increases the solid content of the paste and in mature and dense pastes the capillaries may become blocked by gel and segmented so that they turn into capillary pores interconnected solely by the gel pores. The absence of continuous capillaries is due to a combination of a suitable water/cement ratio and a sufficiently long period of moist curing; the degree of maturity required for different water/cement ratios for ordinary Portland cements is indicated in Fig. 1.11. The actual time to achieve the required maturity depends on the characteristics of the cement used, but approximate values of the time required can be gauged from the data of Table 1.6. For water/cement ratios above about

Table 1.6: *ApproximateTime Required to Produce Maturity at which Capillaries become Segmented*[1.26]

Water/cement ratio by weight	Time required
0·40	3 days
0·45	7 days
0·50	14 days
0·60	6 months
0·70	1 year
over 0·70	impossible

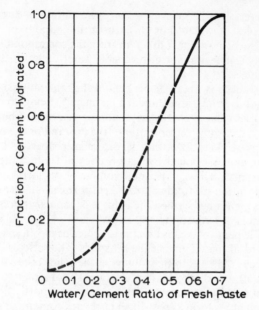

Fig. 1.11. *Relation between the water/cement ratio and the degree of hydration at which the capillaries cease to be continuous*[1.26]

0·7 even complete hydration would not produce enough gel to block all the capillaries. For extremely fine cement the maximum water/cement ratio would be higher, possibly up to 1·0; conversely, for coarse cements it would be below 0·7. The importance of eliminating continuous capillaries is such that this might be regarded a necessary condition for a concrete to be classified as "good."

Gel Pores

Let us now consider the gel itself. From the fact that it can hold large quantities of evaporable water it follows that the gel is porous, but the gel pores are really interconnected interstitial spaces between the gel particles. The gel pores are much smaller than the capillary pores: between 15 and

20 Å in diameter. This is only one order of magnitude greater than the size of molecules of water. For this reason the vapour pressure and mobility of adsorbed water are different from the corresponding properties of free water. The amount of reversible water indicates directly the porosity of the gel.[1.24]

The gel pores occupy about 28 per cent of the total volume of gel, the material left after drying in a standard manner[1.48] being considered as solids. The actual value is characteristic for a given cement but is largely independent of the water/cement ratio of the mix and of the progress of hydration. This would indicate that gel of similar properties is formed at all stages and that continued hydration does not affect the products already in existence. Thus as the total volume of gel increases with the progress of hydration the total volume of gel pores also increases. On the other hand, as mentioned earlier, the volume of capillary pores decreases with the progress of hydration.

Porosity of 28 per cent means that the gel pores occupy a space equal to about one-third of the volume of the gel solids. The ratio of the surface of the solid part of the gel to the volume of the solids is equal to that of spheres about 90 Å in diameter. This must not be construed to mean that gel consists of spherical elements; the particles are mostly fibrous, and bundles of such fibres form a cross-linked network containing some more or less amorphous interstitial material.[1.27]

Another way of expressing the porosity of the gel is to say that the volume of the pores is about three times the volume of the water providing a layer one molecule thick over the entire solid surface in the gel.

From measurements of water adsorption, the specific surface of gel has been estimated to be of the order of 5.5×10^8 m² per m³, or approximately 200,000 m²/kg.[1.27] By contrast, unhydrated cement has a specific surface of some 200 to 500 m²/kg.

In connexion with the pore structure it may be relevant to note that high-pressure steam-cured cement has a specific surface of some 7000 m²/kg only. This indicates an entirely different particle size of the products of hydration at a high pressure and temperature and, in fact, steam curing seems to result in an almost entirely micro-crystalline material.

The specific surface of normally cured cement depends on the curing temperature and on the chemical composition of cement. It has been suggested[1.27] that the ratio of the specific surface to the weight of non-evaporable water (which in turn is proportional to the porosity of the hydrated cement paste) is proportional to

$$0.230(C_3S) + 0.320(C_2S) + 0.317(C_3A) + 0.368(C_4AF)$$

where the symbols in brackets represent the percentages of the compounds present in the cement. There seems to be little variation between the numerical coefficients of the last three compounds, and this indicates that the specific surface of the paste varies little with a change in the composition of cement. The rather lower coefficient of C_3S is due to the fact that it produces a large quantity of micro-crystalline $Ca(OH)_2$, which has a very much lower specific surface than the gel.

The proportionality between the weight of water forming a mono-molecular layer over the surface of the gel and the weight of non-evaporable water in the paste (for a given cement) means that gel of nearly the same specific surface is formed throughout the progress of hydration. In other words, particles of the same size are formed all the time and the already existing gel particles do not grow in size. This is not, however, the case in cement with a high C_2S content.[1.28]

Mechanical Strength of Cement Gel

There are two classical theories of hardening or gain of strength of cement. That put forward by H. Le Chatelier in 1882 states that the products of hydration of cement have a lower solubility than the original compounds, so that the hydrates precipitate from a supersaturated solution. The precipitate is in the form of interlaced elongated crystals with high cohesive and adhesive properties.

The colloidal theory propounded by W. Michaëlis in 1893 states that the crystalline aluminate, sulpho-aluminate and hydroxide of calcium give the initial strength. The lime-saturated water then attacks the silicates and forms a hydrated calcium silicate which, being almost insoluble, forms a gelatinous mass. This mass hardens gradually owing to the loss of water either by external drying or by hydration of the inner unhydrated core of the cement grains: in this manner cohesion is obtained.

In the light of modern knowledge it appears that both theories contain elements of truth and are in fact by no means irreconcilable. In particular, colloidal chemists have found that many if not most colloids consist of crystalline particles but these, being extremely small, have a large surface area which gives them what appear to be different properties from the usual solids. Thus colloidal behaviour is essentially a function of the size of the surface area rather than of the non-regularity of internal structure of the particles involved.[1.42]

In the case of Portland cement it has been found that, when mixed with a large quantity of water, cement produces within a few hours a solution supersaturated with $Ca(OH)_2$ and containing concentrations of calcium silicate hydrate in a metastable condition.[1.2] This hydrate rapidly precipitates in agreement with Le Chatelier's theory; the subsequent hardening may be due to the withdrawal of water from the hydrated material as postulated by Michaëlis.

Further experimental work has shown that the calcium silicate hydrates are in fact in the form of extremely small (sub-microscopic) interlocking crystals[1.20] which, because of their size, could be equally well described as gel. When cement is mixed with a small quantity of water the degree of crystallization is probably even poorer, the crystals being ill-formed. Thus the Le Chatelier–Michaëlis controversy is largely reduced to a matter of terminology as we are dealing with a gel consisting of crystals.

The term cement gel is considered, for convenience, to include the crystalline calcium hydroxide. Gel is thus taken to mean the cohesive mass of hydrated cement in its densest paste, i.e. inclusive of gel pores, the characteristic porosity being about 28 per cent.

The actual source of strength of the gel is not fully understood but it proably arises from two kinds of cohesive bonds.[1.27] The first type is the physical attraction between solid surfaces, separated only by the small (15 to 20 Å) gel pores; this attraction is usually referred to as van der Waals' forces.

The source of the second type of cohesion is the chemical bonds. Since cement gel is of the limited swelling type (i.e. the particles cannot be dispersed by addition of water) it seems that the gel particles are cross-linked by chemical forces. These are much stronger than van der Waals' forces but the chemical bonds cover only a small fraction of the boundary of the gel particles. On the other hand, a surface area as high as that of cement gel is not a necessary condition for high strength development as high-pressure steam-cured cement paste, which has a low surface area, exhibits extremely good hydraulic properties.[1.14]

We cannot thus estimate the relative importance of the physical and chemical bonds but there is no doubt that both contribute to the very considerable strength of the hardened paste.

Water Held in Hydrated Cement Paste

The presence of water in hydrated cement has been repeatedly mentioned. The cement paste is indeed hygroscopic owing to the hydrophilic character of cement coupled with the presence of sub-microscopic pores. The actual water content of the paste depends on the ambient humidity. In particular, capillary pores, because of their comparatively large size, empty when the ambient relative humidity falls below about 45 per cent,[1.25] but water is adsorbed in the gel pores even at very low ambient humidities.

We can thus see that water in hydrated cement is held with varying degrees of firmness. At one extreme there is free water; at the other, chemically combined water forming a definite part of the hydrated compounds. Between these two categories there is gel water held in a variety of other ways.

The water held by the surface forces of the gel particles is called adsorbed water, and that part of it which is held between the surfaces of certain planes in a crystal is called interlayer or zeolitic water. Lattice water is that part of the water of crystallization which is not chemically associated with the principal constituents of the lattice. The diagrammatic representation of Fig. 1.12 may be of interest.

Free water is held in capillaries and is beyond the range of the surface forces of the solid phase.

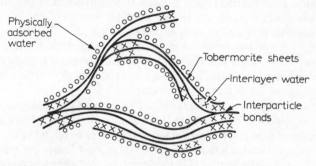

Fig. 1.12. *Probable structure of hydrated silicates*[1.53]

There is no technique available for determining how water is distributed between these different states, nor is it easy to predict these divisions from theoretical considerations as the energy of binding of combined water in the hydrate is of the same order of magnitude as the energy of binding of the adsorbed water. However, recent investigations using the nuclear magnetic resonance technique suggest that gel water has the same energy of binding as interlayer water in some swelling clays; thus the gel water may well be in interlayer form.[1.54]

A convenient division of water in the hydrated cement, necessary for investigation purposes, though rather arbitrary, is into two categories: evaporable and non-evaporable. This is achieved by drying the cement paste to equilibrium (i.e. to a constant weight) at a given vapour pressure. The usual value is 8×10^{-3} mm of mercury, obtained over $Mg(ClO_4)_2.2H_2O$. More recently, drying in an evacuated space which is connected to a moisture trap held at $-79°C$ has been used.[1.48] Alternatively, the evaporable water can be determined by the loss upon drying at a higher temperature, usually $105°C$, or by freezing out, or by removing with a solvent.

All these methods essentially divide water according to whether or not it can be removed at a certain reduced vapour pressure. Such a division is perforce arbitrary because the relation between vapour pressure and water content of cement is continuous. By contrast with crystalline hydrates, no breaks occur in this relationship. However, in general terms, the non-evaporable water contains nearly all chemically combined water and also some water not held by chemical bonds. This water has a vapour pressure lower than that of the ambient atmosphere and the quantity of such water is in fact a continuous function of the ambient vapour pressure.

The amount of non-evaporable water increases as hydration proceeds, but in a saturated paste non-evaporable water can never become more than one-half of the total water present. In well-hydrated cement the non-evaporable water is about 18 per cent by weight of the anhydrous material; this proportion rises to about 23 per cent in fully hydrated cement.[1.1] It follows from the proportionality between the amount of non-evaporable water and the solid volume of the cement paste that the former volume can be used as a measure of the quantity of the cement gel present, i.e. of the degree of hydration.

The manner in which water is held in a cement paste determines the energy of binding. For instance, 1670 Joules (400 calories) are used in establishing the bond of 1 gram of non-evaporable water, while the energy of the water of crystallization of $Ca(OH)_2$ is 3560 Joules per gram (850 cal/g). Likewise, the density of the water varies; it is approximately 1·2 for non-evaporable, 1·1 for gel, and 1·0 for free water.[1.24] It has been suggested that the increase in the density of the adsorbed water at low surface concentrations is not the result of compression but is caused by the orientation of the molecules in the adsorbed phase due to the action of the surface forces.[1.12]

Heat of Hydration of Cement

In common with many chemical reactions, the hydration of cement compounds is exothermic, up to 500 Joules per gram (120 cal/g) of cement being liberated. Since the conductivity of concrete is comparatively low, it acts as an insulator, and in the interior of a large concrete mass, hydration can result in a large rise in temperature. At the same time the exterior of the concrete mass loses some heat so that a steep temperature gradient may be established, and during subsequent cooling of the interior serious cracking may result. This behaviour is, however, modified by the creep of concrete.

At the other extreme, the heat produced by the hydration of cement may prevent freezing of the water in the capillaries of freshly placed concrete in cold weather, and a high evolution of heat is therefore advantageous. It is clear then that it is advisable to know the heat-producing properties of different cements in order to choose the most suitable cement for a given purpose.

The heat of hydration is the quantity of heat, in calories per gram of unhydrated cement, evolved upon complete hydration at a given temperature. The most common method of determining the heat of hydration is by measuring the heats of solution of unhydrated and hydrated cement in a mixture of nitric and hydrofluoric acids: the difference between the two values represents the heat of hydration. This method is described in B.S. 1370:1958, and is similar to the method of A.S.T.M. Standard C 186–68. While there are no particular difficulties in this test, care should be taken to prevent carbonation of the unhydrated cement as the absorption of one per cent of CO_2 results in an apparent decrease in the heat of hydration of 24·3 Joules per gram (5·8 cal/g) out of a total of between 250 and over 420 Joules per gram (60 and 100 cal/g).[1.29]

The temperature at which hydration takes place greatly affects the rate of heat development, as shown by the data of Table 1.7 which gives the heat developed in 72 hours at different temperatures.[1.30]

Table 1.7: *Heat of Hydration Developed After 72 Hours at Different Temperatures*[1.30]

Cement type	Heat of hydration developed at—							
	4°C (40°F)		24°C (75°F)		32°C (90°F)		41°C (105°F)	
	J/g	cal/g	J/g	cal/g	J/g	cal/g	J/g	cal/g
I	154	36·9	285	68·0	309	73·9	335	80·0
III	221	52·9	348	83·2	357	85·3	390	93·2
IV	108	25·7	195	46·6	192	45·8	214	51·2

Strictly speaking the heat of hydration, as measured, consists of the chemical heat of the reactions of hydration and the heat of adsorption of water on the surface of the gel formed by the processes of hydration. The latter heat accounts for about a quarter of the total heat of hydration. Thus the heat of hydration is really a composite quantity.[1.24]

For practical purposes it is not necessarily the total heat of hydration that matters but the rate of heat evolution. The same total heat produced over a longer period can be dissipated to a greater degree with a consequent smaller rise in temperature. The rate of heat development can be easily measured in an adiabatic calorimeter, and typical time–temperature curves obtained under adiabatic conditions are shown in Fig. 1.13.

For the usual range of Portland cements, Bogue[1.2] observed that about one-half of the total heat is liberated between 1 and 3 days, about three-quarters in 7 days, and 83 to 91 per cent of the total heat in six months. In fact, the heat of hydration depends on the chemical composition of the cement, and the heat of hydration of cement is very nearly a sum of the heats of hydration of the individual compounds when hydrated separately. It follows that, given the compound composition of cement, its heat of hydration can be calculated with a fair degree of accuracy. Typical values of the heat of hydration of pure compounds are given in Table 1.8.

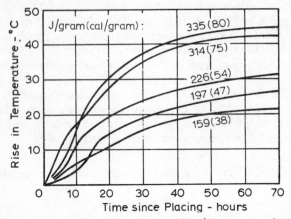

Fig. 1.13. *Temperature rise in 1:2:4 concrete (water/cement ratio of 0·60) made with different cements and cured adiabatically*[1.31]

The total heat of hydration of each cement at three days is shown. (*Crown copyright*)

Table 1.8: *Heat of Hydration of Pure Compounds*[1.32]

Compound	Heat of hydration J/g	cal/g
C_3S	502	120
C_2S	260	62
C_3A	867	207
C_4AF	419	100

It may be noted that there is no relation between the heat of hydration and the cementing properties of the individual compounds. Woods, Steinour and Starke[1.33] tested a number of commercial cements and, using the method of least squares, calculated the contribution of individual compounds to the total heat of hydration of cement. They obtained equations of the type—

heat of hydration of 1 gram of cement =

$$136(C_3S) + 62(C_2S) + 200(C_3A) + 30(C_4AF)$$

where the figures in brackets denote the percentage by weight of the individual compounds present in cement.

Since in the early stages of hydration the different compounds hydrate at different rates, the *rate* of heat evolution, as well as the total heat, depends on the compound composition of the cement. It follows that by reducing the proportions of the compounds that hydrate most rapidly (C_3A and C_3S) the high rate of heat liberation in the early life of concrete can be checked. The fineness of the cement also influences the rate of heat development, an increase in fineness speeding up the reactions of hydration and therefore the heat evolved but the total amount of heat liberated is not affected by the fineness of cement.

The influence of C_3A and C_3S can be gauged from Figs. 1.14 and 1.15.

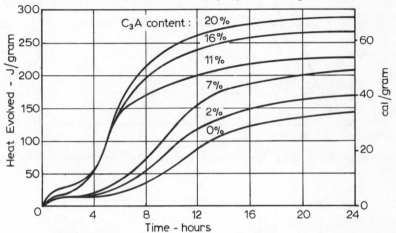

Fig. 1.14. *Influence of C_3A content on heat evolution*[1.32] (*C_3S content approximately constant*)

As mentioned before, for many uses of concrete a controlled heat evolution is advantageous and suitable cements have been developed. One such cement is low heat Portland cement discussed in more detail in the next chapter. The rate of heat development of this and other cements is shown in Fig. 1.16.

The quantity of cement in the mix will also affect the total heat development: thus the richness of the mix can be varied in order to help the control of heat development.

Influence of the Compound Composition on Properties of Cement
In the preceding section it was shown that the heat of hydration of cement is a simple additive function of the compound composition of cement. It would seem, therefore, that the various hydrates retain their identity in the cement gel, which can be considered thus to be a fine physical mixture or to consist of copolymers of the hydrates. A further corroboration of this is obtained from the measurement of specific surface of hydrated cements containing different amounts of C_3S and

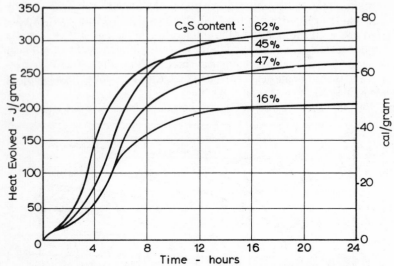

Fig. 1.15. *Influence of C₃S content on heat evolution*[1.32] (*C₃A content approximately constant*)

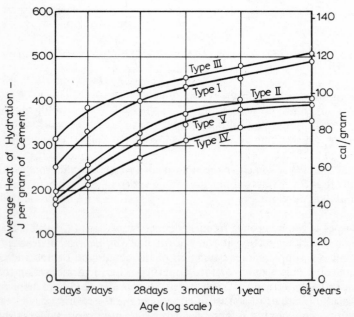

Fig. 1.16. *Development of heat of hydration of different cements cured at 21°C (70°F)* (*water/cement ratio of 0·40*)[1.34]

C_2S: the results agree with the specific surface areas of hydrated neat C_3S and C_2S. Likewise, the water of hydration agrees with the additivity of the individual compounds.

This argument does not, however, extend to all properties of hardened cement paste, notably to shrinkage, creep, and strength; nevertheless, the compound composition gives some indication of the properties to be expected. In particular, the composition controls the rate of evolution of heat of hydration and the resistance of cement to sulphate attack, so that limiting values of oxide or compound composition of different types of cement are prescribed by some specifications. The limitations of A.S.T.M. Standard C 150–72 are less restrictive than they used to be (see Table 1.9).

Table 1.9: *A.S.T.M. Specification C 150–72: Compound Composition Limits for Cement*

Compound	Cement type				
	I	II	III	IV	V
C_3S maximum				35	
C_2S minimum				40	
C_3A maximum		8	15	7	5

The difference in the early rates of hydration of C_3S and C_2S—the two silicates primarily responsible for the strength of cement paste—has been mentioned earlier. A convenient approximate rule assumes that C_3S contributes most to the strength development during the first four weeks and C_2S influences the gain in strength from four weeks onwards.[1.35] At the age of about one year the two compounds, weight for weight, contribute approximately equally to the ultimate strength.[1.36] Neat C_3S and neat C_2S have been found to have a strength of the order of 70 MN/m² (10,000 lb/in²) at the age of 18 months, but at the age of 7 days C_2S had no strength while the strength of C_3S was about 40 MN/m² (6,000 lb/in²). The development of strength of neat compounds is shown in Fig. 1.17.

As mentioned on page 13, the calcium silicates appear in commercial cements in "impure" form. These impurities may strongly affect the rate of reaction and of strength development of the hydrates. For instance, the addition of 1 per cent of Al_2O_3 to pure C_3S increases the early strength of the paste, as shown in Fig. 1.18. According to Verbeck,[1.55] this increase in strength probably results from activation of the silicate crystal lattice due to introduction of the alumina (or magnesia) into the crystal lattice with resultant activating structural distortions.

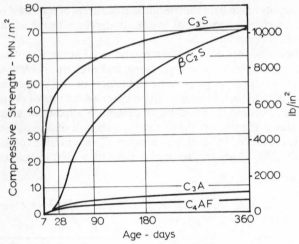

Fig. 1.17. *Development of strength of pure compounds*[1.2]

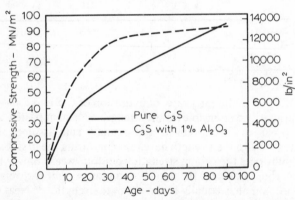

Fig. 1.18. *Development of strength of pure C_3S and C_3S with 1 per cent of Al_2O_3*[1.55]

The influence of the other major compounds on the strength development of cement has been established less clearly. C_3A contributes to the strength of the cement paste at one to three days, and possibly longer, but causes retrogression at an advanced age, particularly in cements with a high C_3A or $(C_3A + C_4AF)$ content. The rôle of C_3A is still controversial.

The rôle of C_4AF in the development of strength of cement is also debatable but there certainly is no appreciable positive contribution. It is likely that colloidal hydrated $CaO.Fe_2O_3$ is deposited on the cement grains, thus delaying the progress of hydration of other compounds.[1.7]

From knowledge of the contribution to strength of the individual compounds present it might be possible to predict the strength of cement on the basis of its compound composition. This would be in the form of a formula of the type—

$$\text{strength} = a.(C_3S) + b.(C_2S) + c.(C_3A) + d.(C_4AF),$$

where the symbols in brackets represent the percentage by weight of the compound, and a, b, etc. are constant parameters representing, the contribution of one per cent of the corresponding compound to the strength of the cement paste.

The use of such a formula would make it easy to forecast at the time of manufacture the strength of cement and would reduce the need for conventional testing. In practice, however, the influence of different compounds is not always significant and has been found to depend on age and on the curing conditions. In general terms, an increase in the C_3S content increases strength up to 28 days;[1.56] Fig. 1.19 shows the 7-day

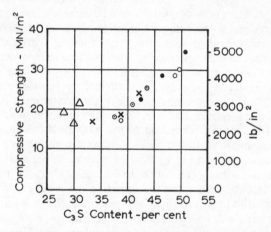

Fig. 1.19. *Relation between 7-day strength of cement paste and the C_3S content in cement*[1.37] *(Cement from each works shown by a different mark)*

strength of standard mortar made with cements of different composition and obtained from different works.[1.37] The C_2S content has a positive influence on strength at 5 and 10 years only, and C_3A a positive influence up to 7 or 28 days but a negative influence later on.[1.56,1.57] The influence of alkalis is considered on page 45.

Prediction of the effects of compounds other than silicates on strength is unreliable. According to Lea[1.38] these discrepancies may be due to the presence of glass in clinker, discussed more fully in the succeeding section.

In other words, the relations observed are statistical in nature, and deviations arise from the fact that we are ignoring some of the variables involved.[1.14] It can be argued, in any case, that all constituents of hydrated Portland cement contribute in some measure to strength in so far as all products of hydration fill space and thus reduce porosity.

Furthermore, there are some indications that the additive behaviour cannot be fully realized. In particular, Powers[1.22] suggested that the same products are formed at all stages of hydration of the paste; this follows from the fact that for a given cement the surface area of hydrated cement is proportional to the amount of water of hydration, whatever the water/cement ratio and age. Thus the fractional rates of hydration of all compounds in a given cement would be the same. This probably is not entirely correct and the composition of products of hydration of cement varies with time. More important still, the composition is not the same at different points in space. This arises from the fact that for diffusion to take place from the face of the still unhydrated part of the cement grain to the space outside (*see* p. 11), there must be a difference in ion concentration: the space outside is saturated but that inside is supersaturated. This diffusion varies the rate of hydration. It is likely therefore that neither the suggestion of equal fractional rates of hydration nor the assumption that each compound hydrates at a rate independent of other compounds is valid. Probably, the exact composition of the individual compounds in the anhydrous cement is affected by the oxide composition of the cement as a whole and therefore the compounds in different cements are not identical; likewise, the oxide composition influences to some extent the rate of formation of the individual hydrates, but our understanding of the hydration rates is still unsatisfactory.

For instance, the amount of heat of hydration per unit weight of hydrated material has been found to be constant at all ages[1.34] (*see* Fig. 1.20), thus suggesting that the nature of the products of hydration does not vary with time. It is therefore reasonable to use the assumption of equal fractional rates of hydration within the limited range of composition of ordinary and rapid hardening Portland cements. However, other cements which have a higher C_2S content than ordinary or rapid hardening cements do not conform to this behaviour. Measurements of heat of hydration indicate that C_3S hydrates earlier, and some C_2S is left to hydrate later.

Furthermore, the initial framework of the paste established at the time of setting affects to a large degree the subsequent structure of the products of hydration. This framework influences especially the shrinkage and development of strength.[1.14] It is not surprising, therefore, that there is a definite relation between the degree of hydration and strength. Fig. 1.21 shows, for instance, an experimental relation between the compressive strength of concrete and the combined water in a cement

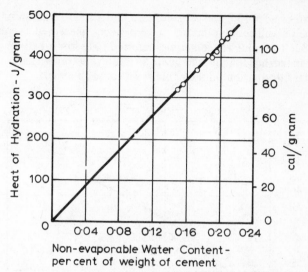

Fig. 1.20. *Relation between the heat of hydration and the amount of non-evaporable water for Type I Portland cement*[1.22]

paste with a water/cement ratio of 0·25.[1.39] These data agree with Powers' observations on the gel/space ratio, according to which the increase in strength of a cement paste is a function of the increase in the relative volume of gel, regardless of age, water/cement ratio, or compound composition of cement. However, the total surface area of the solid phase is related to the compound composition, which does affect the actual value of the ultimate strength.[1.22]

The effects of the minor compounds on the strength of cement paste have not been thoroughly investigated as these compounds were not thought to be of importance as far as strength is concerned. K_2O is believed to replace one molecule of CaO in C_2S with a consequent rise in the C_3S content above that calculated.[1.6] Tests[1.3] on the influence of alkalis have shown that the increase in strength beyond the age of 28 days is strongly affected by the alkali content: the greater the amount of alkali present the lower the gain in strength. This has been confirmed by two statistical evaluations of strength of several hundred commercial cements.[1.56,1.57] The poor gain in strength between 3 and 28 days can be attributed more specifically to water-soluble K_2O present in the cement.[1.58] On the other hand, in the total absence of alkalis, the early strength of cement paste can be abnormally low.[1.58] It seems thus that the rôle of alkalis in strength development of cement paste is not yet clearly understood.

The alkalis are known to react with the so-called alkali-reactive aggregates (*see* p. 139) and cements used under such circumstances often have their alkali content limited to 0·6 per cent (measured as equivalent soda). Such cements are sometimes referred to as low-alkali cements.

We can see then that the alkalis are an important constituent of cement, but fuller information on their rôle is yet to be obtained.

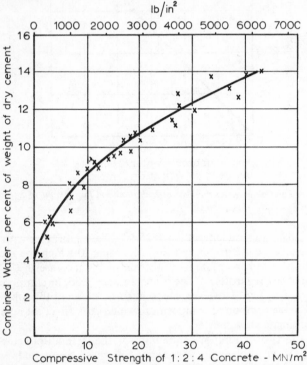

Fig. 1.21. *Relation between compressive strength and combined water content*[1.1]

Effects of Glass in Clinker

It may be recalled that during the formation of cement clinker in the kiln some 20 to 30 per cent of the material becomes liquid; on subsequent cooling crystallization takes place but there is always some material which undercools to glass. In fact, the rate of cooling of clinker greatly affects the properties of cement: if cooling were so slow that full crystallization could be achieved (e.g. in a laboratory), β-C_2S might become converted to γ-C_2S, this conversion being accompanied by expansion and powdering, known as dusting. Furthermore, γ-C_2S hydrates too slowly to be a useful cementitious material. However, Al_2O_8, MgO and the alkalis may stabilize β-C_2S, even on very slow cooling in all practical cases.

Another reason why some glass is desirable is the effect of glass on the crystalline phases. Alumina and ferric oxide are completely liquefied at clinkering temperatures, and on cooling produce C_3A and C_4AF. The extent of glass formation would thus affect these compounds to a large degree while the silicates, which are formed mainly as solids, would be relatively unaffected. It may be noted too that glass may also hold a large proportion of minor compounds such as the alkalis and MgO, which are thus not available for expansive hydration.[1.40] Thus a rapid cooling of high-magnesia clinkers is advantageous. Since the aluminates are attacked by sulphates their presence in glass would also be an advantage. C_3A and C_4AF in glass form can hydrate to a solid solution of C_3AH_6 and C_3FH_6 which is resistant to sulphates.

On the other hand, there are some advantages of a lower glass content. In some cements a greater degree of crystallization leads to an increase in the amount of C_3S produced. Also, a high glass content adversely affects the grindability of clinker.

It can be seen, then, that a strict control of the rate of cooling of clinker so as to produce a desired degree of crystallization is very important. The range of glass content in commercial clinkers, determined by the heat of solution method, is between 2 and 21 per cent.[1.41] An optical microscope indicates much lower values.

It may be recalled that the Bogue compound composition assumes that the clinker has crystallized completely to yield its equilibrium products, and, as we have seen, the reactivity of glass is different from that of crystals of similar composition.

It can be seen then that the rate of cooling of clinker, as well as, possibly, other characteristics of the process of cement manufacture, affects the strength of cement and defies attempts to develop a formula of the type mentioned in the preceding section. Nevertheless, if one process of manufacture is used and the rate of cooling of clinker is kept constant there is a definite relation between compound composition and strength.

Tests on Physical Properties of Cement

The manufacture of cement requires stringent control, and a number of tests are performed in the cement works laboratory to ensure that the cement is of the desired quality and that it conforms to the requirements of the relevant national standards. It is desirable none the less for the purchaser or for an independent laboratory to make acceptance tests or, more frequently, to examine the properties of a cement to be used for some special purpose. Tests on the chemical composition and fineness have already been described; further tests prescribed by B.S. 12:1958 for ordinary and rapid hardening Portland cements are given below. Other relevant standards are mentioned when other types of cement are discussed in Chapter 2.

Consistence of Standard Paste

For the determination of the initial and final setting times and for the
Le Chatelier soundness test, neat cement paste of a standard consistence
has to be used. It is, therefore, necessary to determine for any given
cement the water content of the paste which will produce the desired
consistence.

The consistence is measured by the Vicat apparatus shown in Fig. 1.22,
using a 10 mm diameter plunger fitted into the needle holder. A trial paste

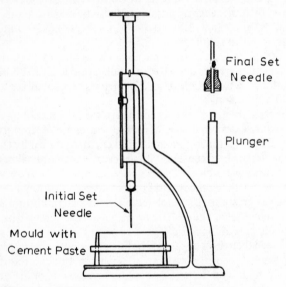

Fig. 1.22. *Vicat apparatus*

of cement and water is mixed in a prescribed manner and placed in the
mould. The plunger is then brought into contact with the top surface of
the paste and released. Under the action of its weight the plunger will
penetrate the paste, the depth of penetration depending on the consistence. This is considered to be standard, in the meaning of B.S. 12 : 1958,
when the plunger penetrates the paste to a point 5 to 7 mm from the bottom of the mould. The water content of the standard paste is expressed as
a percentage by weight of the dry cement, the usual range of values being
between 26 and 33 per cent.

Setting Time

The physical processes of setting were discussed on page 16; here, the
actual determination of setting times will be briefly dealt with. The setting

times are measured using the Vicat apparatus (Fig. 1.22) with different penetrating attachments.

For the determination of the initial set a round or square needle with a cross-sectional area of 1 mm^2 is used. This needle, acting under a prescribed weight, is used to penetrate a paste of standard consistence placed in a special mould. When the paste stiffens sufficiently for the needle to penetrate only to a point about 5 mm from the bottom, initial set is said to have taken place. Initial set is expressed as the time elapsed since the mixing water was added to the cement. A minimum time of 45 minutes is prescribed by B.S. 12:1958 for ordinary and rapid hardening Portland cements and for Portland blast-furnace cement; for low heat Portland cement (B.S. 1370:1958) the minimum setting time is 60 minutes. The initial setting time of aluminous cement is prescribed by B.S. 915:1947 as between 2 and 6 hours.

Final set is determined by a 1 mm square needle fitted with a metal attachment hollowed out so as to leave a circular cutting edge 5 mm in diameter and set 0·5 mm behind the tip of the needle. Final set is said to have taken place when the needle, gently lowered to the surface of the paste, makes an impression on it but the circular cutting edge fails to do so. The final setting time is reckoned from the moment when mixing water was added to the cement, and is required by the relevant British Standards to be not more than 10 hours for ordinary, rapid hardening, low heat and blast-furnace Portland cements. For aluminous cement B.S. 915:1947 specifies the final setting time as not more than 2 hours after the initial set.

It may sometimes be useful to take advantage of the observation that for the majority of American commercial ordinary and rapid hardening Portland cements at room temperature, the initial and final setting times are related. An approximate (within $\pm$ 15 min) simple formula has been suggested: final setting time (min) $= 90 + 1·2 \times$ initial setting time (min).

Since the setting of cement is affected by the temperature and the humidity of the surrounding medium, these are specified by B.S. 12:1958: temperature of between 14 and 18°C (58 and 64°F) and relative humidity of air of not less than 90 per cent.

It should be remembered that the speed of setting and the rapidity of hardening, i.e. of gain of strength, are entirely independent of one another. For instance, the prescribed setting times of rapid hardening cement are no different from those for ordinary Portland cement, although the two cements harden at different rates.

Tests[1.59] have shown that setting of cement paste is accompanied by a change in the ultrasonic pulse velocity through it (cf. page 504). This could lead to an alternative method of measurement of setting time of cement.

Soundness

It is essential that a cement paste, once it has set, does not undergo a large change in volume. In particular, there must be no appreciable expansion, which under conditions of restraint could result in a disruption of the hardened cement paste. Such expansion may take place due to the delayed or slow hydration or other reaction of some compounds present in the hardened cement, namely free lime, magnesia, and calcium sulphate.

If the raw materials fed into the kiln contain more lime than can combine with the acidic oxides, the excess will remain in a free condition. This hard burnt lime hydrates only very slowly, and since slaked lime occupies a larger volume than the original free calcium oxide, expansion takes place. Cements which exhibit this expansion are known as unsound.

Lime added to cement does not produce unsoundness because it hydrates rapidly before the paste has set. On the other hand, free lime present in clinker is intercrystallized with other compounds and is only partially exposed to water during the time before the paste has set.

Free lime cannot be determined by chemical analysis of cement since it is not possible to distinguish between unreacted CaO and $Ca(OH)_2$ produced by a partial hydration of the silicates when cement is exposed to the atmosphere. On the other hand, a test on clinker, immediately after it has left the kiln, would show the free lime content since no hydrated cement is then present.

A cement can also be unsound due to the presence of MgO, which reacts with water in a manner similar to CaO. However, only periclase (crystalline MgO) is deleteriously reactive, and MgO present in glass is harmless.

Calcium sulphate is the third compound liable to cause expansion: in this case calcium sulphoaluminate is formed. It may be recalled that a hydrate of calcium sulphate—gypsum—is added to cement clinker in order to prevent flash set, but if gypsum is present in excess of the amount that can react with C_3A during setting, unsoundness in the form of a slow expansion will result. For this reason, B.S. 12:1958 limits very strictly the amount of gypsum that can be added to clinker, but the limits are well on the safe side as far as the danger of unsoundness is concerned.[1.46]

Since unsoundness of cement is not apparent until after a period of months or years it is essential to test the soundness of cement in an accelerated manner: a test devised by Le Chatelier is prescribed by B.S. 12:1958. The Le Chatelier apparatus, shown in Fig. 1.23, consists of a small brass cylinder split along its generatrix. Two indicators with pointed ends are attached to the cylinder on either side of the split; in this manner the widening of the split, caused by the expansion of cement, is greatly magnified and can be easily measured. The cylinder is placed on a glass plate, filled with cement paste of standard consistence, and covered with another glass plate. The whole assembly is then immersed in water at 18 to 20°C (64 to 68°F) for 24 hours. At the end of that period the distance between the indicators is measured, and the mould is immersed in water again and brought to the boil in 30 minutes. After boiling for one hour the mould is removed, and after cooling the distance between the indicators

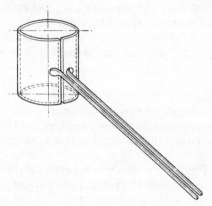

Fig. 1.23. *The Le Chatelier apparatus*

is again measured. The increase in this distance represents the expansion of the cement, and for Portland cements is limited to 10 mm. If the expansion exceeds this value a further test is made after the cement has been spread and aerated for 7 days. During this time some of the lime may hydrate or even carbonate, and a physical breakdown in size may also take place. At the end of the 7-day period, the Le Chatelier test is repeated and the expansion of aerated cement must not exceed 5 mm. A cement not satisfying at least one of these tests should not be used.

The Le Chatelier test detects unsoundness due to free lime only. Magnesia is rarely present in large quantities in the raw materials from which cement is manufactured in England, but is encountered in other countries.* For this reason, in the United States for instance, soundness of cement is checked by the autoclave test, which is sensitive to both free magnesia and free lime. In this test, prescribed by A.S.T.M. Standard C 151–71, a neat cement bar 25 mm (or 1 in.) square in cross-section and with a 250 mm (or 10 in.) gauge length is cured in humid air for 24 hours. The bar is then placed in an autoclave (a high pressure steam boiler), which is raised to a temperature of 216°C (420°F) (steam pressure of

* An example is India, where low-magnesia limestone occurs only to a limited extent. The resulting cement has therefore a high MgO content but expansion can be significantly reduced by the addition of active siliceous material such as pulverized fuel ash or finely ground burnt clay.

$2 \pm 0.07\,MN/m^2\,(295\,lb/in^2))$ in one hour, and maintained at this temperature for 3 hours. The expansion of the bar due to autoclaving must not exceed 0·5 per cent. The high steam pressure accelerates the hydration of both magnesia and lime.

The results of the autoclave test are affected by, in addition to the compounds causing expansion, the C_3A content, and are also subject to other anomalies. The test gives, therefore, no more than a broad indication of the risk of long-term expansion in practice.[1.1]

No test is available for the detection of unsoundness due to an excess of calcium sulphate, but its content can be easily determined by chemical analysis.

Strength of Cement

The mechanical strength of hardened cement is the property of the material that is perhaps most obviously required for structural use. It is not surprising, therefore, that strength tests are prescribed by all specifications for cement.

The strength of mortar or concrete depends on the cohesion of the cement paste, on its adhesion to the aggregate particles, and to a certain extent on the strength of the aggregate itself. This last is not considered at this stage, and is eliminated in tests on the quality of cement by the use of standard aggregates.

Strength tests are not made on a neat cement paste because of difficulties of moulding and testing with a consequent large variability of test results. Cement-sand mortar and, in some cases, concrete of prescribed proportions and made with specified materials under strictly controlled conditions are used for the purpose of determining the strength of cement.

There are several forms of strength tests: direct tension, direct compression, and flexure. The latter determines in reality the tensile strength in bending because, as is well known, cement paste is considerably stronger in compression than in tension. Since the flexure test is not used in Great Britain and little used elsewhere it will not be further discussed.

The direct tension test used to be commonly employed but pure tension is rather difficult to apply so that the results of such a test show a fairly large scatter. Furthermore, since structural techniques are designed mainly to exploit the good strength of concrete in compression, the tensile strength of cement is often of lesser interest than its compressive strength. For these reasons the tension test has gradually given way to compression tests.

However, the tension test has been retained in B.S. 12:1958 as a permitted test for a one-day strength of rapid hardening Portland cement. In this test a 1:3 cement-sand mortar with a water content of

8 per cent of the weight of the solids is mixed and moulded into a briquette of the shape shown in Fig 1.24. The sand is the standard Leighton Buzzard sand obtained from a quarry near the village of that name in Bedfordshire. This sand consists of pure siliceous material and is practically all of one size; all particles are nearly spherical and are smaller than a 850 μm (No. 18 B.S.) sieve and at least 90 per cent of the sand is retained on a 600 μm (No. 25 B.S.) sieve.

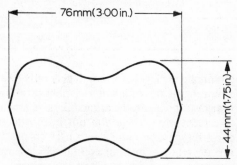

Fig. 1.24. *Briquette for the tension test of mortar*

The briquettes are moulded in a standard manner, cured for 24 hours at a temperature between 18 and 20°C (64 and 68°F) in an atmosphere of at least 90 per cent relative humidity, and tested in direct tension, the pull being applied through special jaws engaging the wide ends of the briquette. B.S. 12: 1958 prescribes the minimum one-day strength of rapid hardening Portland cement as 2·1 MN/m$_2$ (300 lb/in^2), taken as the average value for six briquettes.

There are two standard methods of testing the compressive strength of cement: one uses mortar, the other concrete.

In the mortar test, a 1 : 3 cement-sand mortar is used. The sand is again the standard Leighton Buzzard sand, and the weight of water in the mix is 10 per cent of the weight of the dry materials. Expressed as a water/cement ratio this corresponds to 0·40 by weight. A standard procedure, prescribed by B.S. 12: 1958, is followed in mixing, and 70·6 mm (2·78 in.) cubes are made using a vibrating table with a frequency of 200 Hz applied for two minutes. The cubes are demoulded after 24 hours and further cured in water until tested in a wet-surface condition. The B.S. 12: 1958 requirements for minimum strengths (average values for three cubes) are given in Table 1.10.

The vibrated mortar test gives fairly reliable results but it has been suggested that mortar made with one-size aggregate leads to a greater scatter of strength values than would be obtained with concrete made under similar conditions. Moreover, the values of strength obtained in a

Table 1.10: *B.S. 12:1958 Requirements for Strength of Cement*

Age, days	Minimum compressive strength							
	Mortar test				Concrete test			
	Ordinary Portland		Rapid-hardening Portland		Ordinary Portland		Rapid-hardening Portland	
	MN/m²	lb/in²	MN/m²	lb/in²	MN/m²	lb/in²	MN/m²	lb/in²
3	15	2,200	21	3,000	8	1,200	12	1,700
7	23	3,400	28	4,000	14	2,000	17	2,500

test should approximate the level of strength generally found in concrete, and this would require the use of a water/cement ratio higher than that used in the mortar test. It can also be argued that we are interested in the performance of cement in concrete and not in mortar, especially one made with a one-size aggregate and never used in practice. For these reasons a test on concrete was introduced in the 1958 revision of B.S. 12 as an alternative to the vibrated mortar test.

In the concrete test, fixed weights of cement and water (corresponding to a water/cement ratio of 0·60) are mixed with such amounts of coarse and fine aggregate as will produce a workable concrete with a slump of between 10 and 50 mm (or ½ and 2 in.). There are some limitations on the mineral character and grading of the aggregates. Three 100 mm (or 4 in.) cubes are made by hand, one at a time, in a prescribed manner. After demoulding, the cubes are stored in water until tested at 18 to 20°C (64 to 68°F). The B.S. 12:1958 requirements for the minimum values of the average strength of three cubes at each age are given in Table 1.10. Apart from satisfying these minima the strength at later ages is to be higher than at an earlier age, as strength retrogression might be a sign of unsoundness or other faults in the cement. The condition of strength increase with age applies also to the vibrated mortar cubes.

The hand-made concrete cubes are supposed to show a smaller cube-to-cube variation than the vibrated mortar cubes but this claim is yet to be fully verified.

The influence of cement on the properties of mortar and concrete is qualitatively the same, and the relation between the strengths of corresponding specimens of the two materials is linear. This is shown, for instance, in Fig. 1.25: mortar and concrete of fixed proportions, each with a water/cement ratio of 0·65 were used. The strengths are not the same for the specimens of each pair, at least in part because specimens of different shape and size were used, but there may also be an inherent quantitative difference between the strengths of mortar and concrete due to the greater amount of entrapped air in mortar.

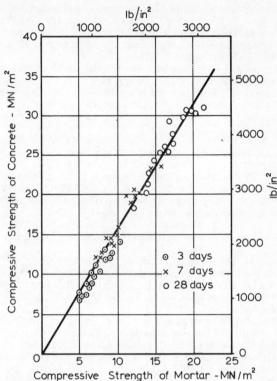

Fig. 1.25. *Relation between the strengths of concrete and mortar of the same water/cement ratio*[1.37]

REFERENCES

1.1. F. M. LEA, *The Chemistry of Cement and Concrete* (London, Arnold, 1970).

1.2. R. H. BOGUE, *Chemistry of Portland Cement* (New York, Reinhold, 1955).

1.3. A. M. NEVILLE, Role of cement in creep of mortar, *J. Amer. Concr. Inst.*, **55**, pp. 963–84 (March 1959).

1.4. M. A. SWAYZE, The quaternary system CaO–C_5A_3–C_2F–C_2S as modified by saturation with magnesia, *Amer. J. Sci.*, **244**, pp. 65–94 (1946).

1.5. W. CZERNIN, *Cement Chemistry and Physics for Civil Engineers* (London, Crosby Lockwood, 1962).

1.6. H. H. STEINOUR, The reactions and thermochemistry of cement hydration at ordinary temperature, *Proc. 3rd Int. Symp. on the Chemistry of Cement*; London, 1952; pp. 261–89.

1.7. R. H. BOGUE and W. LERCH, Hydration of Portland cement compounds, *Industrial and Engineering Chemistry*, **26**, No. 8, pp. 837–47 (Easton, Pa., 1934).

1.8. E. P. FLINT and L. S. WELLS, Study of the system CaO–SiO_2–H_2O at 30°C and the reaction of water on the anhydrous calcium silicates, *J. Res. Nat. Bur. Stand.*, **12**, No. 687, pp. 751–83 (1934).

1.9. S. GIERTZ-HEDSTROM, The physical structure of hydrated cements, *Proc. 2nd Int. Symp. on the Chemistry of Cements*, Stockholm, 1938, pp. 505–34.

1.10. T. C. POWERS, The non-evaporable water content of hardened Portland cement paste: its significance for concrete research and its method of determination, *A.S.T.M. Bul. No.* 158, pp. 68–76 (May 1949).

1.11. L. S. BROWN and R. W. CARLSON, Petrographic studies of hydrated cements, *Proc. A.S.T.M.*, **36**, Part II, pp. 332–50 (1936).

1.12. L. E. COPELAND, Specific volume of evaporable water in hardened Portland cement pastes, *J. Amer. Concr. Inst.*, **52**, pp. 863–74 (1956).

1.13. S. BRUNAUER, J. C. HAYES and W. E. HASS, The heats of hydration of trical-cium silicate and beta-dicalcium silicate, *J. Phys. Chem.*, **58**, pp. 279–87 (Ithaca, N.Y., 1954).

1.14. F. M. LEA, Cement research: retrospect and prospect, *Proc. 4th Int. Symp. on the Chemistry of Cement*, Washington D.C., 1960, pp. 5–8.

1.15. H. F. W. TAYLOR, Hydrated calcium silicates, Part I: Compound formation at ordinary temperatures, *J. Chem. Soc.*, pp. 3682–90 (London, 1950).

1.16. H. H. STEINOUR, The system $CaO-SiO_2-H_2O$ and the hydration of the calcium silicates, *Chemical Reviews*, **40**, pp. 391–460 (U.S., 1947).

1.17. J. W. T. SPINKS, H. W. BALDWIN and T. THORVALDSON, Tracer studies of diffusion in set Portland cement, *Can. J. Technol.*, **30**, Nos. 2 and 3, pp. 20–8 (1952).

1.18. J. D. BERNAL, The structures of cement hydration compounds, *Proc. 3rd Int. Symp. on the Chemistry of Cement*, London, 1952, pp. 216–36.

1.19. T. C. POWERS, H. M. MANN and L. E. COPELAND, The flow of water in hardened Portland cement paste, *Highw. Res. Bd. Sp. Rep.* 40, pp. 308–23 (Washington D.C., 1959).

1.20. J. D. BERNAL, J. W. JEFFERY and H. F. W. TAYLOR, Crystallographic research on the hydration of Portland cement: A first report on investigations in progress, *Mag. Concr. Res.*, No. 11, pp. 49–54 (Oct. 1952).

1.21. W. C. HANSEN, Discussion on "Aeration cause of false set in Portland cement," *Proc. A.S.T.M.*, **58**, pp. 1053–4 (1958).

1.22. T. C. POWERS, The physical structure and engineering properties of concrete, *Portl. Cem. Assoc. Res. Dept. Bul.* 90, p. 39 (Chicago, July 1958).

1.23. T. C. POWERS, A discussion of cement hydration in relation to the curing of concrete, *Proc. Highw. Res. Bd.*, **27**, pp. 178–88 (Washington, 1947).

1.24. T. C. POWERS and T. L. BROWNYARD, Studies of the physical properties of hardened Portland cement paste (Nine parts), *J. Amer. Concr. Inst.*, **43** (Oct. 1946 to April 1947).

1.25. G. J. VERBECK, Hardened concrete—pore structure, *A.S.T.M. Sp. Tech. Publicn. No.* 169, pp. 136–42 (1955).

1.26. T. C. POWERS, L. E. COPELAND and H. M. MANN, Capillary continuity or discontinuity in cement pastes, *J. Portl. Cem. Assoc. Research and Development Laboratories*, **1**, No. 2, pp. 38–48 (May 1959).

1.27. T. C. POWERS, Structure and physical properties of hardened Portland cement paste, *J. Amer. Ceramic Soc.*, **41**, pp. 1–6 (Jan. 1958).

1.28. L. E. COPELAND and J. C. HAYES, Porosity of hardened Portland cement pastes, *J. Amer. Concr. Inst.*, **52**, pp. 633–40 (Feb. 1956).

1.29. R. W. CARLSON and L. R. FORBRICK, Correlation of methods for measuring heat of hydration of cement, *Industrial and Engineering Chemistry* (Analytical Edition), **10**, pp. 382–6 (Easton, Pa., 1938).

1.30. W. LERCH and C. L. FORD, Long-time study of cement performance in concrete, Chapter 3: Chemical and physical tests of the cements, *J. Amer. Concr. Inst.*, **44**, pp. 743–95 (April 1948).

1.31. N. DAVEY and E. N. FOX, Influence of temperature on the strength development of concrete, *Build. Res. Sta. Tech. Paper No.* 15 (London, H.M.S.O., 1933).

1.32. W. LERCH and R. H. BOGUE, Heat of hydration of Portland cement pastes, *J. Res. Nat. Bur. Stand.*, **12**, No. 5, pp. 645–64 (May 1934).

1.33. H. WOODS, H. H. STEINOUR and H. R. STARKE, Heat evolved by cement in relation to strength, *Engng. News Rec.*, **110**, pp. 431–3 (New York, 1933).

1.34. G. J. VERBECK, Long-time study of cement performance in concrete, Chapter 6: The heats of hydration of the cements, *Proc. A.S.T.M.*, **50**, pp. 1235–57 (1950).

1.35. H. WOODS, H. R. STARKE and H. H. STEINOUR, Effect of cement composition on mortar strength, *Engng. News Rec.*, **109**, No. 15, pp. 435–7 (New York, 1932).

1.36. R. E. DAVIS, R. W. CARLSON, G. E. TROXELL and J. W. KELLY, Cement investigations for the Hoover Dam, *J. Amer. Concr. Inst.*, **29**, pp. 413–31 (1933).

1.37. S. WALKER and D. L. BLOEM, Variations in Portland cement, *Proc. A.S.T.M.*, **58**, pp. 1009–32 (1958).

1.38. F. M. LEA, The relation between the composition and properties of Portland cement, *J. Soc. Chem. Ind.*, **54**, pp. 522–7 (London, 1935).

1.39. F. M. LEA and F. E. JONES, The rate of hydration of Portland cement and its relation to the rate of development of strength, *J. Soc. Chem. Ind.*, **54**, No. 10, pp. 63–70T (London, 1935).

1.40. L. S. BROWN, Long-time study of cement performance in concrete, Chapter 4: Microscopical study of clinkers, *J. Amer. Concr. Inst.*, **44**, pp. 877–923 (May 1948).

1.41. W. LERCH, Approximate glass content of commercial Portland cement clinker, *J. Res. Nat. Bur. Stand.*, **20**, pp. 77–81 (Jan. 1938).

1.42. F. M. LEA, *Cement and Concrete*, Lecture delivered before the Royal Institute of Chemistry, London, on 19th Dec. 1944 (Cambridge, W. Heffer and Sons, 1944).

1.43. W. H. PRICE, Factors influencing concrete strength, *J. Amer. Concr. Inst.*, **47**, pp. 417–32 (Feb. 1951).

1.44. U.S. BUREAU OF RECLAMATION, Investigation into the effects of cement fineness and alkali content on various properties of concrete and mortar, *Concrete Laboratory Report No. C*–814 (Denver, Colorado, 1956).

1.45. S. BRUNAUER, P. H. EMMETT and E. TELLER, Adsorption of gases in multimolecular layers, *J. Amer. Chem. Soc.*, **60**, pp. 309–19 (1938).

1.46. W. LERCH, The influence of gypsum on the hydration and properties of Portland cement pastes, *Proc. A.S.T.M.*, **46**, pp. 1252–92 (1946).

1.47. L. E. COPELAND and R. H. BRAGG, Determination of $Ca(OH)_2$ in hardened pastes with the X-ray spectrometer, *Portl. Cem. Assoc. Rep.* (Chicago, 14th May 1953).

1.48. L. E. COPELAND and J. C. HAYES, The determination of non-evaporable water in hardened Portland cement paste, *A.S.T.M. Bul. No.* 194, pp. 70–4 (Dec. 1953).

1.49. L. E. COPELAND, D. L. KANTRO and G. VERBECK, Chemistry of hydration of Portland cement, *Proc. 4th Int. Symp. on the Chemistry of Cement*, Washington D.C., 1960, pp. 429–65.

1.50. P. SELIGMANN and N. R. GREENING, Studies of early hydration reactions of Portland cement by X-ray diffraction, *Highway Research Record*, No. 62, pp. 80–105 (Highway Research Board, Washington D.C., 1964).

1.51. W. G. HIME and E. G. LaBONDE, Particle size distribution of Portland cement from Wagner turbidimeter data, *J. Portl. Cem. Assoc. Research and Development Laboratories*, **7**, No. 2, pp. 66–75 (May 1965).

1.52. C. D. LAWRENCE, The properties of cement paste compacted under high pressure, *Cement Concr. Assoc. Res. Rep. No.* 19 (London, June 1969).

1.53. R. F. FELDMAN and P. J. SEREDA, A model for hydrated Portland cement paste as deduced from sorption-length change and mechanical properties, *Materials and Structures*, No. 1, pp. 509–19 (Paris, Jan.–Feb. 1968).

1.54. P. SELIGMANN, Nuclear magnetic resonance studies of the water in hardened cement paste, *J. Portl. Cem. Assoc. Research and Development Laboratories*, **10**, No. 1, pp. 52–65 (Jan. 1968).

1.55. G. VERBECK, Cement hydration reactions at early ages, *J. Portl. Cem. Assoc. Research and Development Laboratories*, **7**, No. 3, pp. 57–63 (Sept. 1965).

1.56. R. L. BLAINE, H. T. ARNI and M. R. DeFORE, Interrelations between cement and concrete properties, Part 3, *Nat. Bur. Stand. Bldg. Sc. Series* 8 (Washington D.C., April 1968).

1.57. M. VON EUW and P. GOURDIN, Le calcul prévisionnel des résistances des ciments Portland, *Materials and Structures*, **3**, No. 17, pp. 299–311 (Paris, Sept.–Oct. 1970).

1.58. W. J. McCOY and D. L. ESHENOUR, Significance of total and water soluble alkali contents of cement, *Proc. 5th Int. Symp. on the Chemistry of Cement, Tokyo*, 1968, **2**, pp. 437–43.

1.59. M. DOHNALIK and K. FLAGA, Nowe spostrzezenia w problemie czasu wiazania cementu, *Archiwum Inzynierii Ladowej*, **16**, No. 4, pp. 745–52 (1970).

2. Cements of Different Types

THE previous chapter dealt with the properties of Portland cement in general, and we have seen that cements differing in chemical composition and physical characteristics may exhibit different properties when hydrated. It should thus be possible to select mixtures of raw materials for the production of cements with various desired properties. In fact, several types of Portland cement are available commercially and additional special cements can be produced for specific uses. The various types of Portland cement will now be described. Several non-Portland cements will also be discussed, notably aluminous cement.

Types of Portland Cement

In order to facilitate the discussion, a list of different Portland cements, together with the American description where available, is given in Table 2.1. The A.S.T.M. composition limits for some of these cements have already been listed (Table 1.9), and typical values of compound composition are given in Table 2.2.

Table 2.1: *Main Types of Portland Cement*

English description	A.S.T.M. description
Ordinary Portland	Type I
Rapid Hardening Portland	Type III
Extra Rapid Hardening Portland	
Ultra High Early Strength Portland	
Low Heat Portland	Type IV
Modified cement	Type II
Sulphate Resisting Portland	Type V
Portland Blast-furnace	Type IS
White Portland	
Portland-Pozzolana	Type IP

Note: Cements Type I, IS, IP, II and III are also available with an interground air-entraining agent, and are then denoted by letter A, e.g. Type IA.

Many of the cements have been developed to ensure good durability of concrete under a variety of conditions. It has not been possible, however, to find in the constitution of cement a complete answer to the problem of

Table 2 2: *Typical Values of Compound Composition of Portland Cements of Different Types*[2.34]

Cement	Value	Compound composition, per cent								Number of samples
		C_3S	C_2S	C_3A	C_4AF	$CaSO_4$	Free CaO	MgO	Ignition Loss	
	Max.	67	31	14	12	3·4	1·5	3·8	2·3	
Type I	Min.	42	8	5	6	2·6	0·0	0·7	0·6	
	Mean	49	25	12	8	2·9	0·8	2·4	1·2	21
	Max.	55	39	8	16	3·4	1·8	4·4	2·0	
Type II	Min.	37	19	4	6	2·1	0·1	1·5	0·5	
	Mean	46	29	6	12	2·8	0·6	3·0	1·0	28
	Max.	70	38	17	10	4·6	4·2	4·8	2·7	
Type III	Min.	34	0	7	6	2·2	0·1	1·0	1·1	
	Mean	56	15	12	8	3·9	1·3	2·6	1·9	5
	Max.	44	57	7	18	3·5	0·9	4·1	1·9	
Type IV	Min.	21	34	3	6	2·6	0·0	1·0	0·6	
	Mean	30	46	5	13	2·9	0·3	2·7	1·0	16
	Max.	54	49	5	15	3·9	0·6	2·3	1·2	
Type V	Min.	35	24	1	6	2·4	0·1	0·7	0·8	
	Mean	43	36	4	12	2·7	0·4	1·6	1·0	22

durability of concrete: the principal mechanical properties of hardened concrete, such as strength, shrinkage, permeability, resistance to weathering, and creep, are affected also by factors other than cement constitution, although this determines to a large degree the rate of gain of strength.[2.2] Fig. 2.1 shows the rate of development of strength of concretes made with cements of different types: while the rates vary considerably, there is little difference* in the 90-day strength of cements of all types. The general tendency is for the cements with a low rate of hardening to have a slightly higher ultimate strength. For instance, Fig. 2.1 shows that Type IV cement has the lowest strength at 28 days but develops the second highest strength at the age of 5 years. A comparison of Fig. 2.1 and Fig. 2.2 illustrates the fact that differences between cement types are not readily quantified. Also, the retrogression of strength of the concrete made with Type II cement is not characteristic of this type of cement. The pattern of low early and high late strength agrees with the influence of the initial framework of hardened cement on the ultimate development of strength: the more slowly the framework is established the denser the

* In some cases, e.g. Fig. 2.2, the differences are greater.

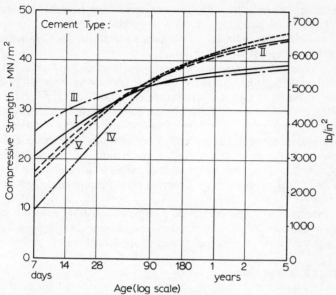

Fig. 2.1. *Strength development of concretes containing 335 kg of cement per cubic metre (565 lb/yd³) and made with cements of different types*[2.1]

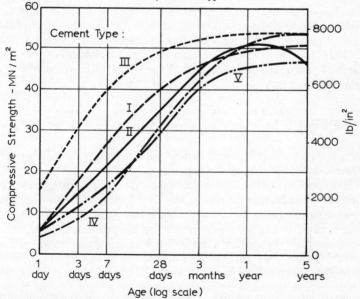

Fig. 2.2. *Strength development of concretes with a water/cement ratio of 0·49 made with cements of different types*[2.4]

gel and the higher the ultimate strength. Nevertheless, significant differences in the important physical properties of cements of different types are found only in the earlier stages of hydration:[2,3] in well-hydrated pastes the differences are only minor.

The division of cements into different types is necessarily no more than a broad classification and there may sometimes be wide differences between cements of nominally the same type. On the other hand, there are often no sharp discontinuities in the properties of different types of cement, and many cements can be classified as more than one type.

Obtaining some special property of cement may lead to undesirable features in another respect. For this reason a balance of requirements may be necessary, and the economic aspect of manufacture must also be considered. Type II cement is an example of a "compromise" all-round cement.

The methods of manufacture have improved steadily over the years, and there has been a continual development of cements to serve different purposes with a corresponding change in specifications.

Ordinary Portland Cement
This is by far the most common cement in use: nearly 90 per cent of all cement used in the United Kingdom (about 17 million tonnes per annum) is of the ordinary type.*

Ordinary Portland cement (Type I) is admirably suitable for use in general concrete construction when there is no exposure to sulphates in the soil or in ground water. The specification for this cement is given in B.S. 12:1958. The limitations of chemical composition are: the lime saturation factor is to be not greater than 1·02 and not less than 0·66. The factor is defined as—

$$\frac{(CaO) - 0.7(SO_3)}{2.8(SiO_2) + 1.2(Al_2O_3) + 0.65(Fe_2O_3)}$$

where the symbols in brackets denote the percentage by weight of the given compound present in the cement.

The upper limit of the lime saturation factor ensures that the amount of lime is not so high as to result in free lime appearing at the clinkering temperature in equilibrium with the liquid present. The unsoundness of cement caused by free lime was discussed in the previous chapter, and is indeed controlled by the Le Chatelier test. But the importance of avoiding unsound cement is so great that in England the safeguard of controlled compound composition is considered desirable. Nevertheless, the

---

* It may be interesting to note that in 1975 the consumption of cement in the United Kingdom was equivalent to a little more than 300 kg per head of population; the corresponding figure for Austria was 730 kg.

A.S.T.M. Standard and the majority of European specifications for cement prescribe no limits of the lime content.

Further requirements of B.S. 12:1958 for the the chemical composition of ordinary Portland cement are that the magnesia content does not exceed 4·0 per cent and that the ratio Al_2O_3/Fe_2O_3 be not less than 0·66. In addition, the insoluble residue must not exceed 1·5 per cent and the loss on ignition is limited to 3 per cent in temperate climates and 4 per cent in the tropics. The maximum gypsum content is also specified (*see* p. 16).

Over the years there have been some changes in the characteristics of ordinary Portland cement.[2.5] In particular, modern cements have a higher C_3S content and a greater fineness than 40 years ago but there has been no significant change in the quality of cement for a number of years. As a consequence, cements have nowadays a higher 28-day strength, but the gain in strength between 28 days and 10 years is unaltered: approximately 20 MN/m^2 (3,000 lb/in^2) for continuously water-cured concrete with a water/cement ratio of about 0·53.[2.4] (*See* Fig. 2.3.)

Rapid Hardening Portland Cement

This cement is very similar to ordinary Portland cement, and is also covered by B.S. 12:1958. Rapid hardening Portland cement (Type III), as its name implies, develops strength more rapidly, and should therefore be correctly described as high early strength cement. The rate of hardening must not be confused with the rate of setting: in fact, the two cements have similar setting times.

The strength developed at the age of three days is of the same order as the 7-day strength of ordinary Portland cement with the same water/cement ratio. This expected rate of hardening is reflected in the minimum strengths specified by B.S. 12:1958, listed in Table 1.10. The increased rate of gain of strength of the rapid hardening cement is achieved by a higher C_3S content and by finer grinding of the cement clinker. B.S. 12:1958 prescribes a minimim fineness of 325 m^2/kg, but as a rule a higher fineness is encountered.

Ordinary Portland cement manufactured in Great Britain is invariably finer than the 225 m^2/kg prescribed by B.S. 12:1958—usually, above 300 m^2/kg. Many works also produce cement with a high C_3S content so that sometimes in practice there is little difference between rapid hardening and *some* ordinary cements; this cannot, however, be assumed to be the rule (*see* p. 289).

The requirements of soundness and chemical composition are the same for rapid hardening as for ordinary Portland cement and need not therefore be repeated.

The use of rapid hardening cement is indicated where a rapid strength development is desired, e.g. when formwork is to be removed early for re-use, or where sufficient strength for further construction is wanted as quickly as practicable. Rapid hardening cement is only about £0·50 ($1·20) per tonne dearer than ordinary cement, and it is not surprising that rapid hardening cement is used extensively, accounting for about 10 per cent of all cement manufactured in the United Kingdom. Since,

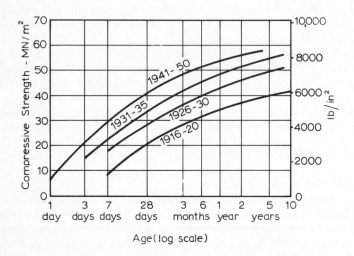

Fig. 2.3. *The rate of gain of strength of cements manufactured between 1916 and 1950 measured on standard concrete cylinders with a water/cement ratio of 0·53*[2.4]

however, the rapid gain of strength means a high rate of heat development, rapid hardening Portland cement should not be used in mass construction or in large structural sections. On the other hand, for construction at low temperatures the use of cement with a high rate of heat liberation may prove a satisfactory safeguard against early frost damage.

Special Rapid Hardening Portland Cements
There exist several specially manufactured cements which are particularly rapid hardening. One of these, a so-called *extra rapid hardening Portland cement*, is obtained by intergrinding calcium chloride with rapid hardening Portland cement. The quantity of calcium chloride should not exceed 2 per cent. Because calcium chloride is deliquescent it is vital to store extra rapid hardening cement under dry conditions, and it should generally be used within one month of despatch from the cement works.

Extra rapid hardening cement is suitable for cold weather concreting, or when a very high early strength is required but when it is inadvisable to use aluminous cement. The strength of extra rapid hardening cement is about 25 per cent higher than that of rapid hardening cement at 1 or 2 days and 10 to 20 per cent higher at 7 days. The setting time of extra rapid hardening cement is short: depending on temperature it can be 5 to 30 minutes so that early placing is essential. Shrinkage is rather higher than when rapid hardening Portland cement is used.

If extra rapid hardening cement is not available it is possible to use rapid hardening Portland cement whose speed of hardening is increased by means of an addition of calcium chloride immediately prior to mixing the concrete. In this case, calcium chloride can be classified as an accelerator; its effects are discussed on page 94.

Another type of cement with very rapid hardening properties is the so-called *ultra high early strength Portland cement*, marketed in Great Britain. This cement contains no admixture; the rapid strength development is due to the very high fineness of the cement: 700 to 900 m^2/kg. Because of this, the gypsum content has to be higher (4 per cent expressed as SO_3) than in cements complying with B.S. 12:1958 but in all other respects the ultra high early strength cement satisfies the requirement of that standard. We may note that the high gypsum content has no adverse effect on long-term soundness as the gypsum is used up in the early reactions of hydration.

The cement is manufactured by separating fines from rapid hardening Portland cement by a cyclone air elutriator. Because of its high fineness, the ultra high early strength cement has a low bulk density and deteriorates rapidly on exposure. High fineness leads to rapid hydration, and therefore to a high rate of heat generation at early ages and to a rapid strength development; for instance, the 3-day strength of rapid hardening Portland cement is reached at 16 hours, and the 7-day strength at 24 hours.[2.35] There is, however, little gain in strength beyond 28 days. Typical strengths of 1:3 concretes made with the ultra high early strength cement are given in Table 2.3.

Table 2.3: *Typical Values of Strength of a 1:3 Concrete made with Ultra High Early Strength Portland Cement*[2.35]

| Age | Compressive strength at water/cement ratio of— | | | | | |
| | 0·40 | | 0·45 | | 0·50 | |
	MN/m^2	lb/in^2	MN/m^2	lb/in^2	MN/m^2	lb/in^2
8 hours	12	1,800	10	1,400	7	1,000
16 hours	33	4,800	26	3,800	22	3,200
24 hours	39	5,700	34	5,000	30	4,300
28 days	59	8,600	57	8,200	52	7,600
1 year	62	9,000	59	8,600	57	8,200

The cement has been used successfully in a number of structures where early prestressing or putting in service is of importance. Shrinkage and creep are not significantly different from those obtained with other Portland cements when the mix proportions are the same:[2.36] in the case of creep, the comparison has to be made on the basis of the same stress/strength ratio (*see* p. 350). We should note, however, that for the same mix proportions, the use of ultra high early strength cement results in lower workability.

The ultra high early strength Portland cement is marketed as *Swift-crete*. A somewhat less fine cement is *Speed* cement, developed in Belgium but manufactured also in England. It contains no accelerator and has a specific surface of 450 to 500 m²/kg. The standard vibrated mortar cube test gives strengths of about 28 MN/m² (4000 lb/in²) at 1 day, 48 MN/m² (7000 lb/in²) at 3 days, and 68 MN/m² (9800 lb/in²) at 28 days. The Speed cement is suitable for winter concreting or for urgent jobs such as road repair, well sealing, etc.

In some countries, e.g. Italy and Sweden, extremely high early strength cement is manufactured by double burning in the kiln.

One more cement of the very high early strength cement variety should be mentioned. This is the so-called *regulated-set cement*, recently introduced in the U.S. The cement consists essentially of a mixture of Portland cement and calcium fluoro-aluminate ($C_{11}A_7.CaF_2$) with an appropriate retarder. The setting time of the cement can vary between 1 and 30 minutes and is controlled in the manufacture of the cement.

The early strength development is controlled by the content of calcium fluoro-aluminate: when this is 5 per cent, about 6 MN/m² (900 lb/in²) can be achieved at 1 hour; a 50 per cent mixture will produce 20 MN/m² (3000 lb/in²) at the same time. These values are based on a mix with a cement content of 330 kg/m³ (560 lb/yd³). The later strength development is the same as for the parent Portland cement.

Low Heat Portland Cement
The rise in temperature in the interior of a large concrete mass due to the heat developed by the hydration of cement can lead to serious cracking (*see* p. 375). For this reason it is necessary to limit the rate of heat evolution of the cement used in this type of structure: a greater proportion of the heat can then be dissipated and a lower rise in temperature results.

Cement having such a low rate of heat development was first produced for use in large gravity dams in the United States, and is known as low heat Portland cement (Type IV).

B.S. 1370:1958 limits the heat of hydration of this cement to 251 J per gram (60 cal/g) at the age of 7 days, and 293 J per gram (70 cal/g) at 28 days.

The limits of lime content, after correction for the lime combined with SO_3, are—

$$\frac{CaO}{2\cdot4(SiO_2) + 1\cdot2(Al_2O_3) + 0\cdot65(Fe_2O_3)} \leqq 1$$

and

$$\frac{CaO}{1\cdot9(SiO_2) + 1\cdot2(Al_2O_3) + 0\cdot65(Fe_2O_3)} \geqq 1$$

The rather lower content of the more rapidly hydrating compounds, C_3S and C_3A, results in a slower development of strength of low heat cement as compared with ordinary Portland cement, but the ultimate strength is unaffected. In any case, to ensure a sufficient rate of gain of strength the specific surface of the cement must be not less than 320 m^2/kg.

A low heat Portland blast-furnace cement is also available; this is covered by B.S. 4246:1968.

In some applications a very low early strength may be a disadvantage, and for this reason a so-called modified (Type II) cement was developed in the United States. This modified cement successfully combines a somewhat higher rate of heat development than that of low heat cement with a rate of gain of strength similar to that of ordinary Portland cement. Modified cement is recommended for use in structures where a moderately low heat generation is desirable or where moderate sulphate attack may occur. This cement is extensively used in the United States.

Modified cement, referred to as Type II cement, and low heat cement (Type IV) are covered by A.S.T.M. Specification C 150–72.

Sulphate-resisting Cement

In discussing the reactions of hydration of cement, and in particular the setting process, mention was made of the reaction between C_3A and gypsum ($CaSO_4.2H_2O$) and of the consequent formation of calcium sulphoaluminate. In hardened cement, calcium aluminate hydrate can react with a sulphate salt from outside the concrete in a similar manner: the product of addition is calcium sulphoaluminate, forming within the framework of the hydrated cement paste. Since the increase in the volume of the solid phase is 227 per cent, gradual disintegration of concrete results. A second type of reaction is that of base exchange between calcium hydroxide and the sulphates, resulting in the formation of gypsum with an increase in the volume of the solid phase of 124 per cent.

These reactions are known as sulphate attack. The salts particularly active are magnesium and sodium sulphate. Sulphate attack is greatly accelerated if accompanied by alternate wetting and drying, as is the case, for instance, in a marine structure in the zone between the tides (*see* Chapter.7).

The remedy lies in the use of cement with a low C_3A content, and such cement is known as sulphate-resisting Portland cement. The British Standard for this cement, B.S. 4027: 1966, stipulates a C_3A content of 3·5 per cent. The minimum fineness is 250 m²/kg. In other respects, sulphate-resisting cement is expected to conform to B.S. 12: 1958 for ordinary Portland cement. In the United States sulphate-resisting cement is known as Type V cement and is covered by A.S.T.M. Standard C 150–72. This specification limits the C_3A content to 5 per cent, and also restricts the total content of C_4AF plus twice the C_3A content to 20 per cent. The magnesia content is limited to 5 per cent.

The rôle of C_4AF is not quite clear. From the chemical standpoint C_4AF would be expected to form calcium sulphoaluminate, as well as calcium sulphoferrite, and thus cause expansion. It seems, however, that the action of calcium sulphate on *hydrated* cement is smaller the lower the $Al_2O_3:Fe_2O_3$ ratio. Some solid solutions are formed and they are liable to comparatively little attack. The tetracalcium ferrite is even more resistant, and it may form a protective film over any free calcium aluminate.[2.6]

Since it is often not feasible to reduce the Al_2O_3 content of the raw material, Fe_2O_3 may be added to the mix so that the C_4AF content increases at the expense of C_3A.[2.7]

An example of a cement with a very low $Al_2O_3:Fe_2O_3$ ratio is the *Ferrari* cement, in whose manufacture iron oxide is substituted for some of the clay. A similar cement is produced in Germany under the name of *Erz* cement. The name of iron ore cement is also used for this type of cement.

The low C_3A and comparatively low C_4AF contents of sulphate-resisting cement mean that it has a high silicate content and this gives the cement a high strength but, because C_2S represents a high proportion of the silicates, the early strength is low. The heat developed by sulphate-resisting cement is not much higher than that of low heat cement. It seems thus that sulphate-resisting cement is theoretically an ideal cement,[2.1] but because of the special requirements for the composition of the raw materials used in its manufacture, sulphate-resisting cement cannot be generally and cheaply made.

Portland Blast-furnace Cement
This type of cement is made by intergrinding Portland cement clinker and granulated blast-furnace slag, the proportion of the latter not exceeding 65 per cent of the weight of the mixture, as prescribed by B.S. 146: 1958.

Slag is a waste product in the manufacture of pig iron, the amounts of iron and slag obtained being of the same order. The slag is a mixture of lime, silica, and alumina, that is, the same oxides that make up Portland cement, but not in the same proportions. While it is not possible to give ranges of values it may be noted that a slag known to be satisfactory had the following composition: 42 per cent lime, 30 per cent silica, 19 per cent alumina, 5 per cent magnesia, and 1 per cent alkalis.

Blast-furnace slag varies greatly in composition and physical structure depending on the processes used and on the method of cooling of the slag. For use in the manufacture of blast-furnace cement the slag has to be quenched so that it solidifies as a glass, crystallization being largely prevented. This rapid cooling by water results also in a fragmentation of the material into a granulated form.

Slag can make a cementitious material in different ways. Firstly, it can be used together with limestone as a raw material for the conventional manufacture of Portland cement. Clinker made from these materials is often used (together with slag) in the manufacture of Portland blast-furnace cement.

The latter cement represents the second major use of slag. Dry granulated slag is fed with Portland cement clinker into a grinding mill, gypsum being added in order to control setting. Portland blast-furnace cement has been manufactured in Scotland for a number of years, and is also made in the United States where it is known as Type IS cement, and is covered by A.S.T.M. Standard C 595–72. Portland blast-furnace cement is used also in Germany, under the name of *Eisenportland* (up to 30 per cent slag) and *Hochofen* cements (31 to 85 per cent slag), and in France where the most common ones are *ciment métallurgique mixte* (50 per cent slag) and *ciment de haut fourneau* (65 to 75 per cent slag). A Belgian development is the *Trief process* in which wet-ground granulated slag is fed in the form of a slurry direct into the concrete mixer, together with Portland cement and aggregate. The cost of drying the slag is thus avoided, and grinding in the wet state results in a greater fineness than would be obtained with dry grinding for the same power input.

A variant used in England under the name of *Cemsave* and in South Africa known as *Slagment* is a process where dry-ground granulated slag of the same fineness as cement is added at the mixer as a partial replacement of Portland cement; Portland blast-furnace cement concrete is thus manufactured *in situ*. Like Portland blast-furnace cement concrete, Cemsave concrete has a lower early strength than when Portland cement only is used, but at later ages at least equal strengths are reached. However, with Cemsave the workability is somewhat higher so that some reduction in water/(cement plus slag) ratio is possible compared with the water/cement ratio of a Portland cement mix with the same aggregate content.

The exact nature of hydration of Portland blast-furnace cement is not quite clear. The Portland cement component hydrates in the normal manner and it appears that the calcium hydroxide thus liberated provides a "starter" for the hydration of the granulated slag. However, the further hydration of the slag is direct and does not depend on combination with lime.

Portland blast-furnace cement is rather similar to ordinary Portland cement, and B.S. 146: 1958 requirements for fineness, setting times and soundness are the same for both cements. In actual fact, the fineness of

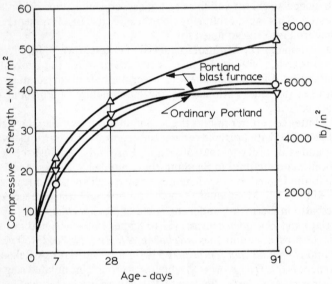

Fig. 2.4. *Strength development of concretes made with Portland blast-furnace cement (water/cement ratio = 0·6)*[2.8]

Portland blast-furnace cement tends to be higher, but even so the rate of hardening of Portland blast-furnace cement is somewhat slower during the first 28 days, and adequate curing is therefore of importance; the strength requirements of B.S. 146: 1958 are therefore lower than for ordinary Portland cement. However, at later ages there is little difference between the strengths of Portland blast-furnace and ordinary Portland cements. Fig. 2.4 shows typical strength–time curves. The heat of hydration of Portland blast-furnace cement is lower than that of ordinary Portland cement so that the former can be used in mass concrete structures. (The cement should then comply with B.S. 4246: 1968.) However, in cold weather the low heat of hydration of Portland blast-furnace cement, coupled with a moderately low rate of strength development, can lead to frost damage. Because of its fairly high sulphate resistance Portland blast-furnace cement is frequently used in sea-water construction.

Supersulphated Cement

Because it is made from granulated blast-furnace slag supersulphated cement will be considered at this stage, even though it is not a Portland cement.

Supersulphated cement is made by intergrinding a mixture of 80 to 85 per cent of granulated slag with 10 to 15 per cent of calcium sulphate (in the form of dead-burnt gypsum or anhydrite) and about 5 per cent of Portland cement clinker. A fineness of 400 to 500 m²/kg is usual.

Supersulphated cement is used extensively in Belgium, where it is known as *ciment métallurgique sursulfaté*, in France, and in Germany (under the name of *Sulfathüttenzement*). The increased use in U.K. has led to the preparation of a British Standard: B.S. 4248: 1973. The cement is highly resistant to sea water and can withstand the highest concentrations of sulphates normally found in soil or ground water, and is also resistant to peaty acids and oils. Concrete with a water/cement ratio not greater than 0·45 has been found not to deteriorate in contact with weak solutions of mineral acids of pH down to 3·5. For these reasons supersulphated cement is used in the construction of sewers and in contaminated ground, although it has been suggested that this cement is less resistant than sulphate-resisting Portland cement when the sulphate concentration exceeds 1 per cent.[2.31]

The heat of hydration of supersulphated cement is low: about 170 to 190 J per gram (40 to 45 cal/g) at 7 days and 190 to 210 J/g (45 to 50 cal/g) at 28 days.[2.6] The cement is therefore suitable for mass concrete construction but care must be taken if used in cold weather as the rate of strength development is considerably reduced at low temperatures. The rate of hardening of supersulphated cement increases with temperature up to about 50°C (122°F), but at higher temperatures anomalous behaviour has been encountered. For this reason steam curing above 50°C (122°F) should not be used without prior tests. It may also be noted that supersulphated cement should not be mixed with other cements.

Wet curing for not less than four days after casting is essential as premature drying out results in a friable or powdery surface layer, especially in hot weather, but the depth of this layer does not increase with time.

Supersulphated cement combines chemically with more water than is required for the hydration of Portland cement, so that concrete with a water/cement ratio of less than 0·4 should not be made. Mixes leaner than about 1:6 are not recommended. The decrease in strength with an increase in the water/cement ratio has been reported to be smaller than in other cements but, since the early strength development depends on the type of slag used in the manufacture of the cement, it is advisable to determine the actual strength characteristics prior to use. Typical strengths attainable are given in Table 2.4. It should be noted that for the concrete test B.S. 4248:1973 prescribes a water/cement ratio of 0·55 instead of 0·60 used with other cements.

Table 2.4: *Typical Values of Strength of Supersulphated Cement*[2.6]

Age, days	Compressive strength			
	Standard vibrated mortar test		Standard concrete test	
	MN/m²	lb/in²	MN/m²	lb/in²
1	7	1,000	5 to 10	700 to 1,500
3	28	4,000	17 to 28	2,500 to 4,000
7	35 to 48	5,000 to 7,000	28 to 35	4,000 to 5,000
28	38 to 66	5,500 to 9,500	38 to 45	5,500 to 6,500
6 months	—	—	52	7,500

Portland-pozzolana Cements and Pozzolanas

The first of these is the name given to interground or blended mixtures of Portland cement and pozzolana.

Pozzolana is a natural or artificial material containing silica in a reactive form. A more formal definition of A.S.T.M. Specification C 618–72 describes pozzolana as a siliceous or siliceous and aluminous material which in itself possesses little or no cementitious value but will, in finely divided form and in the presence of moisture, chemically react with calcium hydroxide at ordinary temperatures to form compounds possessing cementitious properties. It is essential that pozzolana be in a finely divided state as it is only then that silica can combine with lime (liberated by the hydrating Portland cement) in the presence of water to form stable calcium silicates which have cementitious properties. Pozzolanic materials most commonly met with are: volcanic ash—the original pozzolana—pumicite, opaline shales and cherts, calcined diatomaceous earth, burnt clay, fly ash, etc. The latter, known also as pulverized-fuel ash (*see* p. 526), is probably the most common artificial pozzolana. The fly ash particles are spherical (which is advantageous from the water requirement point of view) and are approximately of the same fineness as cement so that the silica is readily available for reaction. In considering pozzolanas in general, we should note that the silica has to be amorphous as crystalline silica has very low reactivity.

It is not possible to make a generalized statement on the Portland-pozzolana cements because the rate of strength development depends on the activity of the pozzolanas and on the proportion of Portland cement in the mixture. As a rule, however, Portland-pozzolana cements gain strength very slowly and require, therefore, curing over a comparatively long period, but their ultimate strength is approximately the same as that of òrdinary Portland cement alone. A typical strength curve is shown in Fig. 2.5.

A.S.T.M. Standard C 595–72 describes Portland-pozzolana cement as Type IP, and limits the pozzolana content to between 15 and 40 per cent of the weight of the Portland-pozzolana cement. Tests on the chemical activity of pozzolanas are given in A.S.T.M. Standard C 618–72.

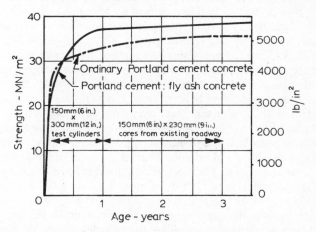

Fig. 2.5. *Strength development of concrete made with Portland cement and fly ash*[2.9]

Pozzolanas may often be cheaper than the Portland cement that they replace but their chief advantage lies in slow hydration and, therefore, low rate of heat development: this is of great importance in mass construction, and it is there that Portland-pozzolana cement or a partial replacement of Portland cement by the pozzolana is mostly used.

Portland-pozzolana cements show also good resistance to sulphate attack and some other destructive agents. It should be remembered, however, that they vary very considerably in their effects, both good and bad, and it is advisable to test any untried pozzolanic material in combination with the cement and the aggregate which are to be used in the actual construction.

When pozzolanas are used as a partial replacement for cement, the cement and the pozzolanas are batched separately and mixed with the other ingredients in the concrete mixer. The required properties of pozzolanas for such a purpose are prescribed by A.S.T.M. Standard C 618–72. For pulverized-fuel ash, the required main properties are: a minimum content of silica, alumina and ferric oxide of 70 per cent, a maximum SO_3 content of 5 per cent,* a maximum loss on ignition of 12 per cent, and a maximum alkali content (expressed as Na_2O) of 1·5 per cent. The latter value is applicable only when the fly ash is to be used with reactive aggregate.

For an assessment of pozzolanic activity with cement, the specification prescribes the measurement of a pozzolanic activity index. This is established by the determination of strength of mixtures with a specified replacement of cement by pozzolana. There is also a pozzolanic activity index with lime, which determines the total activity of a pozzolana.

Partial replacement of Portland cement by pozzolana has to be carefully defined, as the specific gravity of pozzolanas is much lower than that of cement; for instance, the specific gravity of fly ash is 2·1 to 2·4, compared with 3·15 for cement. Thus replacement by weight results in a considerably greater volume of the cementitious material in the mix. With replacement, concrete mixes have a lower early strength than when Portland cement is used but beyond about three months there is no loss of strength. With lean mixes, there may even be a long-term gain of strength due to the replacement (*see* Fig. 2.6). If equal early strength is required and pozzolana is to be used (e.g. because of alkali-aggregate reactivity) then addition of pozzolana rather than partial replacement of cement is necessary. For instance, when pulverized-fuel ash is used in lean mixes, almost 100 kg (220 lb) of pozzolana may be necessary to replace 50 kg (110 lb) of cement, but the amount of pozzolana is lower in rich mixes. Because the continuing formation of hydrates fills the pores and also because of the absence of free lime which could be leached out, partial replacement of Portland cement by pozzolana reduces the permeability of concrete: a 7- to 10-fold reduction has been reported.[2.37]

It must be remembered that, although pozzolana may be cheaper than Portland cement, the use of an additional material on site (and especially of an extremely fine one) results in additional cost. Thus pozzolanas are generally used for technical rather than economic reasons.

White Cement

For architectural purposes white concrete or, particularly in tropical countries, a pastel colour paint finish are sometimes required. To achieve best results it is advisable to use white cement with, of course, a suitable fine aggregate and, if the surface is to be treated, also coarse aggregate. This type of cement has also the advantage that it is not liable to cause staining, since it has a low content of soluble alkalis.

White Portland cement is made from raw materials containing very little iron oxide and manganese oxide. China clay is generally used, to-

---

* Four per cent according to the British Code of Practice for the Structural Use of Concrete CP 110: 1972.

gether with chalk or limestone, free from from specified impurities. Oil is used as fuel for the kiln in order to avoid contamination by coal ash. Since iron acts as a flux in clinkering its absence necessitates higher kiln temperatures but sometimes cryolite (sodium aluminium fluoride) is added as a flux.

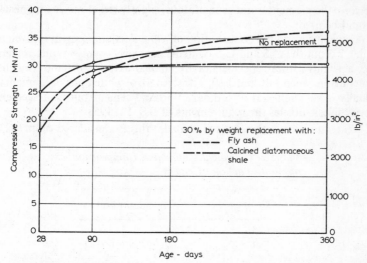

Fig. 2.6. *Effect of partial replacement of Portland cement by pozzolana on the strength development of concrete made with 167 kg of "cement" material per cubic metre of concrete (282 lb/yd³)*[2.37]

Contamination of the cement with iron during grinding has also to be avoided. For this reason, the rather inefficient pebble grinding is used instead of the usual ball mill, although recently nickel and molybdenum alloy balls have been introduced. The cost of grinding is thus higher, and this, coupled with the more expensive raw materials, makes white cement rather expensive (about double the price of ordinary Portland cement).

Because of this, white cement concrete is often used in the form of a facing placed against ordinary concrete backing, but great care is necessary to ensure full bond between the two concretes. To obtain good colour, white concrete of rich mix proportions is generally used, the water/cement ratio being not higher than about 0·4.

When a pastel colour is required white concrete can be used as a base for painting. Alternatively, pigments can be added to the mixer, but it is essential that the pigments do not affect adversely the development of strength of the cement or affect air entrainment. Mixing of concrete with pigments is not common because it is rather difficult to maintain a uniform colour of the resulting concrete.

A better way to obtain a uniform and durable coloured concrete is to use coloured cement. This consists of white cement interground with 2 to 10 per cent of pigment. Specifications for the use of this type of cement are given by the individual manufacturers of this rather specialized product. The specification for pigments is given in B.S. 1014:1961. Because the pigment is not cementitious, slightly richer mixes than usual should be used.

A typical compound composition of white Portland cement is given in Table 2.5 but the C_3S and C_2S contents may vary widely. White cement has a slightly lower specific gravity than ordinary Portland cement, generally between 3·05 and 3·10. The strength of white Portland cement is usually lower than that of ordinary Portland cement but white cement nevertheless satisfies the requirements of B.S. 12: 1958.

White aluminous cement is also made; this is considered on p. 91.

Table 2.5: *Typical Compound Composition of White Portland Cement*

Compound	Content, per cent
C_3S	51
C_2S	26
C_3A	11
C_4AF	1
SO_3	2·6
alkalis	0·25

Other Portland Cements

Among the numerous cements developed for special uses, anti-bacterial cement is of interest. It is a Portland cement interground with an anti-bacterial agent which prevents microbiological fermentation. This bacterial action is encountered in concrete floors of food processing plants where the leaching out of cement by acids is followed by fermentation caused by bacteria in the presence of moisture. Anti-bacterial cement can also be successfully used in swimming pools, public baths and similar places where bacteria or fungi are present.

Another special cement is the so-called hydrophobic cement, which deteriorates very little during prolonged storage under unfavourable conditions. This cement is obtained by intergrinding Portland cement with 0·1 to 0·4 per cent of oleic acid. Stearic acid or pentachlorophenol can also be used.[2.10] These additions increase the grindability of clinker, probably due to electrostatic forces resulting from a polar orientation of the acid molecules on the surface of the cement particles.

Oleic acid reacts with alkalis in cement to form calcium and sodium oleates which foam, so that air-entraining results. When this is not desired a detraining agent, such as tri-n-butyl phosphate, has to be added during grinding.[2.11]

The hydrophobic properties are due to the formation of a water-repellent film around each particle of cement. This film is broken during the mixing of the concrete, and normal hydration takes place but early strength is rather low.

Hydrophobic cement is similar in appearance to ordinary Portland cement but has a characteristic musty smell. In handling, the cement seems more fluid than other Portland cements.

Hydrophobic cement should not be confused with waterproofed cements, which are claimed to make a more impermeable concrete than ordinary Portland cement. There is considerable controversy about the effectiveness of these waterproofed cements.

Masonry cement, used in mortar in brickwork, is made by inter-grinding very finely ground Portland cement, limestone and an air-entraining agent, or alternatively Portland cement and hydrated lime, granulated slag or an inert filler, and an air-entraining agent. Masonry cements make a more plastic mortar than ordinary Portland cement, they also have a greater water-retaining power and lead to lower shrinkage. The strength of masonry cements is lower than that of ordinary Portland cement, particularly since a high air content is introduced, but this low strength is generally an advantage in brick construction. Masonry cement must not be used in structural concrete.

Natural Cements

This is the name given to a cement obtained by calcining and grinding a so-called cement rock, which is a clayey limestone containing up to 25 per cent of argillaceous material. The resulting cement is similar to Portland cement, and is really intermediate between Portland cement and hydraulic lime.

Since natural cement is calcined at temperatures too low for sintering, it contains practically no C_3S and is therefore slow hardening. Natural cements are rather variable in quality as adjustment of composition by blending is not possible. Because of this, as well as for economic reasons, natural cements are nowadays little used. In the United States, natural cements represent no more than one per cent of the production of all Portland cements.

Expanding Cements*

For many purposes it would be advantageous to use a cement which does not change in volume owing to drying shrinkage or, in special cases, even expands on hardening.

---

* Known in the U.S. as expansive cements.

Cements of this type have been developed by H. Lossier[2.12] in France, who used a mixture of Portland cement, an expanding agent, and a stabilizer. The expanding agent is obtained by burning a mixture of gypsum, bauxite, and chalk, which form calcium sulphate and calcium aluminate (mainly C_5A_3). In the presence of water, these compounds react to form calcium sulphoaluminate hydrate (ettringite), with an accompanying expansion of the paste. The stabilizer, which is blast-furnace slag, slowly takes up the excess calcium sulphate and brings expansion to an end. Very careful proportioning of the cement ingredients is necessary in order to obtain the desired expansion. Generally, about 8 to 20 parts of the "sulphoaluminate" clinker are mixed with 100 parts of Portland cement and 15 parts of the stabilizer.[2.6]

Since expansion takes place only as long as the concrete is moist, curing must be carefully controlled, and the use of expanding cement requires skill and experience.

Strictly speaking, the use of expanding cement cannot produce "shrinkless" concrete, as shrinkage takes place after the moist curing has ceased, but the magnitude of expansion can be adjusted so that the expansion and the subsequent shrinkage are numerically equal.

Another type of expanding cement, called high-energy expanding cement, is made by intergrinding Portland cement clinker, aluminous cement clinker, and gypsum, approximately in the proportions 65 : 20 : 15. Expansion is due to the formation of calcium sulphoaluminate, as in Lossier's cement, and takes place within 2 or 3 days of casting.

High-energy expanding cement is quick-setting and rapid-hardening, reaching a strength of about 7 MN/m^2 (1,000 lb/in^2) in 6 hours, and 50 MN/m^2 (7,000 lb/in^2) in 28 days. The cement has a high resistance to sulphate attack.

A more recent development, used to some extent commercially, is expanding cement, known as Type K, developed in California.[2.38] The ingredients of this cement are similar to those used by Lossier but the material selection and clinkering conditions (maximum temperature of about 1300°C) of the expanding agent probably lead to the formation of an anhydrous calcium sulphoaluminate ($C_4A_3.SO_3$). The expansion is due to the formation of hydrated calcium sulphoaluminate (ettringite), as in Lossier's cement, but the rate and magnitude of expansion appear to be more reliable. Similar cement is manufactured in Japan.

There are, however, some practical difficulties still to be resolved in order to ensure the required expansion under variable site conditions. The important requirement is that CaO, SO_3 and especially Al_2O_3 become available for ettringite formation at the right time.[2.39] Specifically, a major part of it must form after a certain strength has been attained; otherwise, the expansive force will be dissipated in the deformation of the still plastic concrete and no stress against the restraint will result. On the other hand, if the ettringite continues to form rapidly for too long, disruptive expansion may occur.[2.39] The controlling factors are the presence of $CaSO_4$ and the ratio of sulphate to aluminate in the paste. A complicating factor arises from the fineness of the cement: for a given sulphate content, the greater the fineness the smaller the expansion.[2.39]

Concretes based on expanding cements have been classified by the American Concrete Institute:[2.39] two basic types are recognized. Shrinkage-compensating concrete is one in which expansion, if restrained, induces compressive stresses which approximately offset the tensile stresses induced by shrinkage. Self-stressing concrete is concrete in which the induced compressive stresses are large enough to result in a significant compressive stress after drying shrinkage has occurred. It is clear that both these definitions involve a restraint of the expansion, usually in the form of reinforcement, preferably trixial. Indeed, it is not expansion that is of interest but an induced compressive stress which can compensate for tensile stresses that would otherwise manifest themselves as tensile strains and possibly as cracks.

The performance of these concretes on a limited scale has been found to be encouraging and there is a good prospect of prestressing concrete by the use of expanding cement. Nevertheless, further improvements in the cements are necessary before they can be more widely employed.

Aluminous Cement
The search for a solution to the problem of attack of gypsum-bearing waters on Portland cement concrete structures in France led Jules Bied to the development of a high-alumina cement, at the beginning of this century. This cement is very different in its composition and also in some properties from Portland cements, but the concreting techniques are similar.

Manufacture
From the name of the cement—aluminous or high alumina—it can be inferred that it contains a large proportion of alumina: the cement consists, in fact, of approximately equal parts, about 40 per cent each, of alumina and lime, with some ferrous and ferric oxides, and up to about 8 per cent of silica.

The raw materials are limestone or chalk, and bauxite. Bauxite is a residual deposit formed by the weathering under tropical conditions of rocks containing aluminium, and consists of hydrated alumina, oxides of iron and titanium, with small amounts of silica. There are no bauxite deposits in Great Britain and it is imported from Greece and France.

In the British process of manufacture of aluminous cement, bauxite is crushed into lumps not larger than 100 mm (or 4 in.). Dust and small particles formed during this fragmentation are cemented into briquettes of similar size as dust would tend to damp the furnace. The second main raw material is usually limestone, also crushed to lumps of about 100 mm (or 4 in.).

Limestone and bauxite in the required proportions are fed into the top of a furnace which is a combination of the cupola (vertical stack) and reverberatory (horizontal) types. Pulverized coal is used for firing, its quantity being about 22 per cent of the weight of the cement produced. In the furnace, the moisture and carbon dioxide are driven off and the materials are heated by the furnace gases to the point of fusion at about 1,600°C. The fusion takes place at the lower end of the stack so that the molten material falls into the reverberatory furnace and thence through a spout into steel pans. The melt is now solidified into pigs, fragmented in a rotary cooler, and then ground in a tube mill. A very dark grey powder with a fineness of 250 to 320 m²/kg is produced.

Because of the high hardness of aluminous cement clinker, the power consumption and the wear of tube mills are considerable. This, coupled with the high prime cost of bauxite and the high temperature of firing, leads to a high price of aluminous cement, compared with, say, rapid hardening Portland cement. The price is, however, compensated for by some valuable properties of aluminous cement.

It may be noted that, unlike the case of Portland cement, the materials used in the manufacture of aluminous cement are completely fused in the kiln. This fact gave rise to the French name *Ciment Fondu*, which is now used in England as a trade name. Because trade names are so widely used the other names should also be mentioned: *Lightning* cement (in England) and *Lumnite* (in the United States).

Composition

Table 2.6 gives typical values of oxide composition of aluminous cement. A minimum alumina content of 32 per cent is prescribed by B.S. 915:

Table 2.6: *Typical Oxide Composition of Aluminous Cement*

Oxide	Content, per cent
SiO_2	3 to 8
Al_2O_3	37 to 41
CaO	36 to 40
Fe_2O_3	9 to 10
FeO	5 to 6
TiO_2	1·5 to 2
MgO	1
Insoluble residue	1

1947, which requires also the alumina/lime ratio to be between 0·85 and 1·3.

Considerably less is known about the compound composition of aluminous cement than of Portland cement, and no simple method of calculation is available. The main cementitious compounds are calcium aluminates of low basicity: CA and C_5A_3. The latter is now believed to be really $C_{12}A_7$.[2.32] Other phases are also present: $C_6A_4.FeO.S$ and an isomorphous $C_6A_4.MgO.S$.[2.13] C_2S or C_2AS does not account for more than a few per cent, and there are, of course, minor compounds present but no free lime can exist. Thus unsoundness is never a problem in aluminous cement although B.S. 915: 1947 prescribes the conventional Le Chatelier test.

Hydration
The hydration of CA, which has the highest rate of strength development, results in the formation of CAH_{10}, a small quantity of C_2AH_8, and alumina gel $(Al_2O_3.aq)$. With time, these hexagonal hydrates, which are unstable both at normal and at higher temperatures, become transformed into cubic crystals of C_3AH_6 and alumina gel. This transformation is encouraged by a higher temperature and a higher concentration of lime or a rise in alkalinity.[2.14]

C_5A_3 is believed to hydrate to C_2AH_8. C_2S forms CSH_x, the lime liberated by hydrolysis reacting with excess alumina; no $Ca(OH)_2$ exists. The reactions of hydration of the other compounds, particularly those containing iron, have not been determined with any degree of certainty, but the iron held in glass is known to be inert.[2.15]

The water of hydration of aluminous cement is calculated to be up to 50 per cent of the weight of the dry cement,[2.6] which is about twice as much as the water required for the hydration of Portland cement. For this reason, mixes with a water/cement ratio lower than about 0·5 were not usually recommended in the past. However, recently, possibly because of the effects of conversion (*see* p. 86), the use of high water/cement ratios has gone out of favour, and 1:7 or even 1:9 mixes with a water/cement ratio of 0·35 are considered preferable, and 0·5 is a recommended maximum for structural use. Compaction by vibration is, of course, necessary. From the fact that aluminous cement combines with more water than Portland cement it follows that for the same mix proportions aluminous cement results in a concrete with a lower porosity and therefore higher impermeability.

Resistance to Chemical Attack
As mentioned earlier, aluminous cement was first developed to resist sulphate attack, and it is indeed extremely satisfactory in this respect. This resistance to sulphates is due to the absence of $Ca(OH)_2$ in hydrated

aluminous cement and also to the protective influence of the relatively inert alumina gel formed during hydration.[2.16] However, mixes leaner than about 1:8 are very much less resistant to sulphates.[2.6]

Aluminous cement is not attacked by CO_2 dissolved in pure water and is therefore suitable for the manufacture of pipes. The cement is not acid-resisting but it can withstand tolerably well very dilute solutions of acids (pH greater than 3·5 to 4·0) found in industrial effluents but not of hydro-chloric, hydrofluoric or nitric acids. On the other hand, caustic alkalis, even in dilute solutions, attack aluminous cement with great vigour by dissolving the alumina gel. The behaviour of this cement in the presence of many agents has been studied by Hussey and Robson.[2.16]

It may be noted that, although aluminous cement stands up extremely well to sea water, this water should not be used as mixing water; the setting and hardening of the cement are adversely affected, possibly be-cause of the formation of chloroaluminates. Likewise, calcium chloride must never be added to aluminous cement.

Physical Properties of Aluminous Cement Concrete

Another outstanding feature of aluminous cement is its very high rate of strength development. About 80 per cent of its ultimate strength is achieved at the age of 24 hours, and even at 6 to 8 hours the concrete is strong enough for the side formwork to be struck and for the preparation for further concreting to take place. Fig. 2.7 shows strength–time curves for concrete cylinders with different water/cement ratios, cured at room temperature; the relation between 1-day strength and water/cement ratio is shown in Fig. 5.7.

Concrete made with aluminous cement and aluminous cement clinker as aggregate, with a water/cement ratio of 0·5, can reach a strength of about 100 MN/m² (14,000 lb/in²) in 24 hours, and 120 MN/m² (18,000 lb/in²) in 28 days. This extremely high strength development is due to the cementitious character of the aggregate but this aggregate is, of course, very expensive.

It should be stressed that the rapidity of hardening is not accompanied by rapid setting. In fact, aluminous cement is slow setting but the final set follows the initial set more rapidly than is the case in Portland cement. Typical values for aluminous cement are: initial set—4 to 5 hours, final set—30 minutes later.

B.S. 915: 1947 requires the initial set to take place between two and six hours after mixing, and the final set not more than two hours after the initial set. Of the compounds present in the aluminous cement C_5A_3 sets in a few minutes while CA is considerably more slow-setting, so that the higher the C:A ratio in the cement the more rapid the set. On the other hand, the higher the glass content of the cement the slower the setting.

The setting time of aluminous cement is greatly affected by the addition

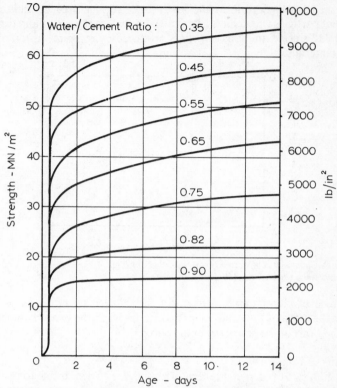

Fig. 2.7. *Strength development of concrete cylinders with different water/cement ratios made with aluminous cement and cured at 18°C (64°F) and 95 per cent relative humidity*[2.17]

of plaster, lime, Portland cement and organic matter and for this reason no additives should be used. The use of aluminous–Portland cement mixtures for special purposes is discussed on p. 92.

It is likely that, because of its rapid setting properties, C_5A_3 is responsible for the loss of workability of many aluminous cement concretes, which takes place within 15 or 20 minutes of mixing.* Fig. 2.8 shows the change in the compacting factor with time for a $1:2:4$ mix with a water/cement ratio of 0.55. The use of the slump test is not recommended because, unlike Portland cement, aluminous cement does not produce a "fatty" lubricant.

It may be noted that for equal mix proportions aluminous cement produces a somewhat more workable mix than when Portland cement is

---

* Cf. behaviour of concrete made with Portland cement (*see* p. 197).

used. This may be due to the lower total surface area of aluminous cement particles which have a "smoother" surface than Portland cement particles, since aluminous cement is produced by complete fusion of the raw materials.

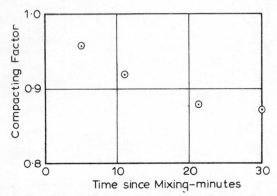

Fig. 2.8. *Change in compacting factor with time for a 1:2:4 aluminous cement concrete with a water/cement ratio of 0·55*[2.17]

Creep of aluminous cement concrete has been found to differ little from the creep of Portland cement concretes when the two are compared on the basis of the stress/strength ratio.[2.22]

Effects of Temperature
The high rate of gain of strength of aluminous cement is due to its rapid hydration, which in turn means a high rate of heat development. This can be as high as 38 J per gram per hour (9 cal/g/h) whereas for rapid hardening Portland cement the rate is never higher than 15 J per gram per hour (3·5 cal/g/h). However, the *total* heat of hydration is of the same order for both types of cement.

The high rate of heat liberation makes it necessary for aluminous cement concrete to be placed in thin sections and never in a large mass. This requirement is of the utmost importance as apart from cracking caused by the temperature cycle, encountered also in Portland cement concretes, the rise in temperature *per se* affects adversely the strength of aluminous cement paste. The influence of temperature is illustrated in Table 2.7, listing the strength of concretes cured for the first 24 hours at 21°C and 38°C (70°F and 100°F) but subsequently at 21°C (70°F). A considerable loss in strength due to the higher temperature and regression at later ages can be seen.

Efficient wet curing during the first 24 hours can help in limiting the temperature rise but wet curing as such is not essential.[2.33] Nevertheless, the rapid gain of strength between about 6 and 14 hours after casting

Table 2.7: Influence of Curing Temperature During the First 24 Hours on the Strength of Aluminous Cement Concrete Cylinders (W/C = 0.53) Subsequently Cured at 21°C (70°F)[2.6]

Cement	Curing temperature during first 24 hours °C	°F	Compressive strength at the age of							
			1 day		3 days		7 days		28 days	
			MN/m²	lb/in²	MN/m²	lb/in²	MN/m²	lb/in²	MN/m²	lb/in²
A	21	70	28·1	4,080	29·6	4,290	28·9	4,200	25·2	3,660
	38	100	18·8	2,730	19·3	2,800	19·3	2,800	12·5	1,820
B	21	70	24·2	3,510	28·1	4,070	29·5	4,280	31·4	4,560
	38	100	15·7	2,280	17·6	2,550	18·9	2,750	11·9	1,730
C	21	70	20·3	2,950	28·3	4,110	28·9	4,190	27·8	4,040
	38	100	16·2	2,350	17·2	2,500	19·0	2,760	15·6	2,260
D	21	70	25·6	3,710	27·9	4,050	30·6	4,440	27·8	4,030
	38	100	18·9	2,750	22·7	3,290	18·5	2,680	11·2	1,620
E	21	70	24·3	3,520	30·3	4,390	32·4	4,700	34·4	4,990
	38	100	19·8	2,870	18·9	2,740	19·9	2,890	13·2	1,920
F	21	70	30·9	4,480	39·2	5,680	40·3	5,850	42·7	6,200
	38	100	21·9	3,180	21·9	3,180	22·0	3,190	18·5	2,680

N.B.—The generally low level of strength is due to the use of cements made in the 1920s.

means that the rate of hydration at that time is very high, and a large amount of water in the mix is being combined. Satisfactory hydration cannot be achieved if loss of water from the concrete occurs within about 18 hours of casting, even if the concrete is subsequently stored in water. Thus evaporation from the concrete should be prevented from the time it has been placed, and after the final set has taken place active wet curing should be applied up to the age of 18 to 24 hours.

The strength of aluminous cement concrete is affected adversely also by a rise in temperature at later ages when the concrete is stored under moist conditions. This means that concrete properly placed and cured and which has developed a high strength will, upon exposure to hot and moist conditions, lose a considerable proportion of its strength.

This loss of strength is associated with the conversion of the unstable hexagonal aluminate hydrates, CAH_{10} and C_2AH_8, to the cubic hydrate C_3AH_6, the reaction being of the type—

$$3CAH_{10} \rightarrow C_3AH_6 + 2AH_3 + 18H$$

The cubic hydrate contains less water of crystallization than the hexagonal plates and yet the reaction does not appear to take place in a fully desiccated paste. It is, however, also possible that a hydrate with more water of crystallization, C_4AH_{19}, is formed. The specific gravity of the converted hydrate C_3AH_6 is 2·52 while that of CAH_{10} is 1·72. Thus, on conversion, the porosity of the cement paste increases substantially and this must be reflected in a serious loss of strength.[2.33] Scanning microscope studies have confirmed the formation of voids on conversion[2.40] and mercury pressure porosimeter measurements have shown an increase both in the total porosity and in the mean pore size.[2.42] However, very slow conversion of mixes with low water/cement ratios need not lead to a loss of strength as there is a concurrent hydration of the still unhydrated cement and a consequent filling of the newly formed voids.

The magnitude of the changes in strength can be gauged from Fig. 2.9 which shows the loss of strength of a 1 : 5·6 cement-sand mortar with a water/cement ratio of 0·65, as a result of continuous storage in water at 40°C (104°F). The strength of the same mortar cured in water at room temperature was 51 MN/m² (7,400 lb/in²) at 3 days and 57 MN/m² (8,200 lb/in²) at 14 days. The same figure shows that a loss of strength occurs also at temperatures which can be described as no more than warm: 25 and 30°C (77 and 86°F). In these tests, the specimens were exposed to the temperature in question from the age of 6 hours onwards. Similar effects were found for storage above water. In the first day or two the higher temperature speeds up the chemical reactions of hardening, thus increasing strength, but this is of little practical consequence. Later, the strength falls off, and the rate of drop in strength is greater the higher the temperature. The loss of strength occurs whatever the age at which the rise in temperature takes place[2.20] (*see* Fig. 2.10) although a delay of 24 hours from the time of casting reduces the rate of loss of strength.[2.20]

Regardless of the rate of loss of strength, concrete of given mix proportions appears to reach with time a characteristic residual strength. This residual strength is the strength of the concrete when all hexagonal

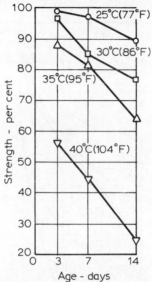

Fig. 2.9. *Strength of an aluminous cement 1:5·6 mortar stored at different temperatures as a percentage of strength of similar mortar cured in water at room temperature*[2.18]

calcium aluminate hydrate has changed to the stable cubic C_3AH_6. The loss of strength is more severe in weaker mixes. For instance, concrete with a water/cement ratio of 0·29 has been found to have its strength reduced from 91 to 54 MN/m^2 (13,200 to 7,800 lb/in^2) but for concrete with a water/cement ratio of 0·65 the loss was from 43 to 5 MN/m^2 (6,200 to 750 lb/in^2).[2.20] Fig. 2.11 shows the strength of different concretes cured for 100 days at 18°C and 40°C (64°F and 104°F), following setting at room temperature and Fig. 2.12 gives the relation between strength and water/cement ratio at the two temperatures.

The difference in the loss of strength of weak and strong mixes may be associated with the porosity of the cement paste and the amount of water which can find ingress: the lower the water/cement ratio of the original mix the lower the porosity of the resulting concrete and, therefore, the more difficult it is for water to be available throughout the cement paste. This explanation is indirectly confirmed by the observation that for a given water/cement ratio a leaner mix shows a lower loss, and such a mix is known to have a lower porosity.

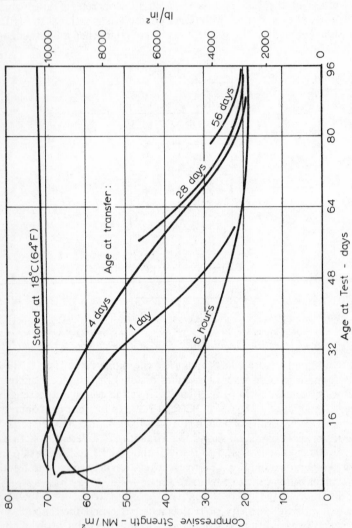

Fig. 2.10. *Influence of the delay in the exposure of aluminous cement mortar (water/cement ratio of 0·45) to water at 40°C (104°F)*[2.20]

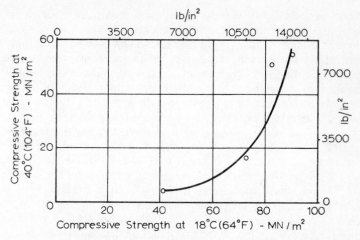

Fig. 2.11. *Relation between strengths of aluminous cement concrete cubes cured for 100 days at 40°C and 18°C (104°F and 64°F), following setting at 18°C (64°F)*[2.20]

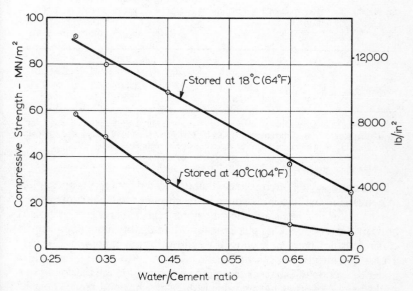

Fig. 2.12. *Relation between strength and water/cement ratio for concrete made with aluminous cement and cured at 18 and 40°C (64 and 104°F) for 100 days*

Short periods of exposure to hot-wet conditions cause only a small loss of strength but their effects are cumulative.

Recent studies[2.33] have shown that conversion does not occur only at temperatures above some critical value but also at ordinary temperatures prevailing in England. The rate of conversion is lower the lower the temperature, and conversion is indeed very slow at ordinary temperatures, so that the loss of strength is generally apparent only from the age of about 5 years onward. Over a long period of time the loss is, nevertheless, serious, as indicated by the data of Table 2.8. For this reason, the British Code of Practice for the Structural Use of Concrete CP 110: 1972 required aluminous cement concrete to have a higher 24-hour strength than the 28-day strength of Portland cement concrete for the same purpose: for reinforced concrete 50 MN/m² (7,200 lb/in²) compared with 20 MN/m² (2,900 lb/in²), and for prestressed concrete 60 MN/m² (8,700 lb/in²) compared with 30 or 40 MN/m² (4,300 or 5,800 lb/in²). There is also a limitation on the maximum water/cement ratio: 0·40; for prestressed concrete a limit of 0·35 is desirable. In 1974, all reference to high-alumina cement was deleted from the code and the cement is no longer used structurally in the United Kingdom.

Table 2.8: *Long-term Strength of Aluminous Cement Concrete*[2.33]

Storage*	Water/cement ratio†	Workability	Strength at the age of 21 years		Strength at 21 years as a percentage of maximum strength‡
			MN/m²	lb/in²	
Laboratory	0·40	low	37·9	5,500	55
	0·64	high	26·7	3,870	41
Outdoors	0·40	low	19·9	2,880	31
	0·64	high	13·2	1,920	23

* 203 mm (8 in.) cubes were stripped at 24 hours, cured under moist hessian for 7 days, then stored in the laboratory at 18 to 20°C (64 to 68°F) for 28 days, and thereafter either continuing in the laboratory or stored in open air in Budapest.

† Cement content of 267 kg/m³ (450 lb/yd³).

‡ Maximum strength was achieved at the age of about 1 year in most cases.

As mentioned earlier, it is likely that desiccated concrete is not subject to conversion but, if heat is applied to a specimen still containing some mixing water in a free state, the rate of desiccation is not fast enough to prevent conversion taking place because good quality aluminous cement concrete is not permeable enough to allow a rapid expulsion of water. Thus some airport slabs in Italy have been found to deteriorate rapidly under the action of exhaust gases from jet engines though the conditions could be described as hot and dry rather than hot and moist. Loss of strength of aluminous cement concrete has been found to take place also when radiant or infra-red heat is applied to moist concrete, e.g. immediately after curing.[2.19]

It can be seen that, except for rich and strong mixes, aluminous cement concrete should not be generally used in structural members and the Comité Européen du Béton recommend that the use of aluminous cement be subject to special justification. Furthermore, the conversion of the aluminate hydrates increases the porosity of the paste so that the loss of strength due to a rise in temperature is accompanied by a marked decrease in the resistance of the concrete to sulphate attack, and to freezing and thawing,[2.42] but the resistance to acids seems unaffected.[2.6] There is also an increase in permeability.[2.42] Because of the relatively low loss of strength of very rich mixes no risk is involved in using aluminous cement grout in jointing precast post-tensioned members, but many countries forbid the use of aluminous cement concrete in various structures.[2.33]* In Britain, structural use of the material is no longer acceptable.

An interesting means of preventing the harmful effects of conversion of the aluminate hydrates has been suggested by Budnikov.[2.21] Gypsum or anhydrite ($CaSO_4$) is added to aluminous cement in a weight ratio of about 1 : 4. The aluminates (CA and C_5A_3) then react with gypsum to form $C_3A . 3CaSO_4 . H_{31}$ but, since the reaction takes place before the mass has set and its volume has been stabilized, the formation of calcium sulphoaluminate has no harmful effects (cf. sulphate attack, p. 392). The hardened cement gains strength rapidly and develops high strength, which is improved by a rise in temperature. The cement has also a high resistance to chemical attack by sulphates and chlorides. Although the addition of gypsum is effective in overcoming the ill-effects of conversion it really alters the nature of the cement: we are no longer dealing with aluminous cement but with an anhydrite-aluminous cement.

Refractory Properties

Desiccated aluminous cement concrete heated to a high temperature has a satisfactory strength, and the resistance of aluminous cement to dry heat is, in fact, so high that this cement is one of the foremost refractory materials. This is largely due to a good ceramic bond which replaces the hydraulic bond. Since conversion due to an early rise in temperature must be avoided, it is still necessary for the concrete to be moist cured at room temperature during the first 24 hours after casting.

Concrete made with aluminous cement and refractory aggregate, such as crushed firebrick, is stable at temperatures up to about 1,300°C. For temperatures up to 1,600°C special aggregates, such as fused alumina or carborundum, have to be used. A temperature as high as 1,800°C can be

---

* See A. M. Neville, *High-Alumina Cement Concrete* (Lancaster, Construction Press, 1975).

withstood by concrete made from special white calcium aluminate cement with a fused alumina aggregate. This cement consists of about 72 per cent of alumina, 26 per cent lime, about 1 per cent of iron and silica, and has a composition approaching C_3A_5. By comparison, Portland cement concrete cannot withstand prolonged exposure to temperatures much over 500°C. Refractory concrete made with aluminous cement has a good resistance to acid attack (e.g. acids in flue gases), and firing at 900 to 1,000°C in fact appreciably increases this resistance.[2.16]

Refractory concrete is superior to refractory brickwork, which expands on heating and therefore needs expansion joints. On the other hand, aluminous cement concrete can be cast monolithically or with butt joints only to exactly the required shape and size. The loss of water on first firing results in a contraction approximately equal and opposite to the thermal expansion. Upon subsequent cooling, for instance during the shut-down of works, butt joints would open slightly due to the thermal contraction but they would close up again on reheating.

For insulating purposes, when temperatures up to about 950°C are expected, lightweight concrete can be made with aluminous cement and lightweight aggregate. Such concrete has a density of 500 to 1,000 kg/m³ (30 to 60 lb/ft³) and a thermal conductivity of 0·21 to 0·29 J m/m²s °C (0·12 to 0·17 Btu/ft²h°F/ft).

Setting of Portland-Aluminous Cement Mixtures

As mentioned earlier, the setting of mixtures of aluminous and Portland cements is greatly accelerated, and, when either cement constitutes between 20 and 80 per cent of the mixture, flash set may occur. Typical data are shown in Fig. 2.13 but actual values vary for different cements, and trial tests should be made with any given cements. The accelerated setting is due to the formation of a hydrate of C_4A by the addition of lime from the Portland cement to calcium aluminate from the aluminous cement. Also, gypsum contained in the Portland cement may react with hydrated calcium aluminates, and as a consequence the now non-retarded Portland cement may exhibit a flash set.

Mixtures of the two cements in suitable proportions are used when rapid setting is of vital importance, e.g. for stopping the ingress of water, or for construction between the tides, but the ultimate strength of such pastes is quite low.

Because of the rapid setting just described, in normal concrete construction it is essential to make sure that the two cements do not come in contact with one another. Thus, placing concrete made with one type of cement against concrete made with the other must be delayed by at least 24 hours if aluminous cement was cast first, or 3 to 7 days if the earlier concrete was made with Portland cement. Contamination through plant or tools must also be avoided.

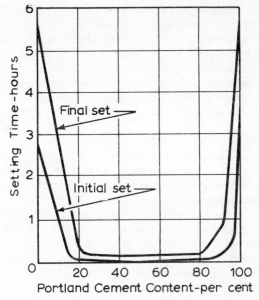

Fig. 2.13. *Setting time of Portland-aluminous cement mixtures*[2.6]

Admixtures

Often, instead of using a special cement, it is possible to change some of the properties of the cement in hand by the use of a suitable additive. A great number of proprietary products is available: their effects are described by the manufacturers but the full details of the action of many of these additives, known as admixtures, are yet to be determined, and the performance of any one admixture should be carefully checked before it is used.

Admixtures may be classified according to the purpose for which they are used in concrete; the approach of A.S.T.M. Standard C 494–71 can be used. In this chapter, we shall consider only one well tried *accelerating admixture* (Type C, according to the A.S.T.M. classification) (calcium chloride), *retarding admixtures* or retarders (Type B), and *water-reducing admixtures* (Type A). The nomenclature is somewhat confusing in that retardation refers to the setting of concrete while acceleration primarily to the early strength development, i.e. to hardening (*see* p. 63), more rapid setting being generally only coincidental.

Additives inducing air entrainment are considered on p. 417. There exist also additives for other purposes, such as air detrainment, fungicidal action, water-proofing, etc., but these are not sufficiently standardized or developed for systematic treatment here. Pigments can also be considered as an additive; they are mentioned on p. 75.

An important feature of the majority of admixtures for concrete is that they are used primarily on the basis of experience or *ad hoc* tests: theoretical information is generally not available to permit a reliable quantitative prediction of behaviour in concrete under the various possible circumstances. However, more studies of admixtures are being undertaken.

Calcium Chloride
The addition of calcium chloride to the mix increases the rate of development of strength, and this accelerator is, therefore, used when concrete is to be placed at low temperatures (in the region of 2 to 4°C (35 to 40°F)) or when urgent repair work is to be done.

Calcium chloride increases the rate of heat liberation during the first few hours after mixing, the action of $CaCl_2$ being probably that of a catalyst in the reactions of hydration of C_3S and C_2S; it is possible that the reduction in the alkalinity of the solution promotes the hydration of the silicates. The hydration of C_3A is delayed somewhat, but the normal process of hydration of cement is not changed.

Calcium chloride may be added to rapid hardening as well as to ordinary Portland cement, and the more rapid the natural rate of hardening of the cement the earlier becomes apparent the action of the accelerator. Calcium chloride must not, however, be used with aluminous cement. With rapid hardening Portland cement the increase in strength due to $CaCl_2$ can be as much as 7 MN/m^2 (1,000 lb/in^2) at 1 day while with ordinary Portland cement this increase would be achieved only after 3 to 7 days. By the age of 28 days there is no difference between the the strengths of rapid hardening cements with and without $CaCl_2$, but in the case of ordinary Portland cement the addition of $CaCl_2$ would still show an improved strength.

Hickey's[2.23] results for cements of different types are shown in Fig. 2.14. The long-term strength of concrete is believed to be unaffected by $CaCl_2$. Calcium chloride is generally more effective in increasing the early strength of rich mixes with a low water/cement ratio than of lean ones.

The quantity of $CaCl_2$ added to the mix must be carefully controlled. To calculate the quantity required it can be assumed that the addition of 1 per cent of $CaCl_2$ (as a fraction of the weight of cement) affects the rate of hardening as much as a rise in temperature of 6°C (11°F). A calcium chloride content of 1 to 2 per cent is generally sufficient; the latter figure should not be exceeded unless a test with the cement to be actually used in construction is made, as the effects of calcium chloride depend to a certain

---

* When marine aggregate is used its chloride content should be taken into account (*see* p. 135).

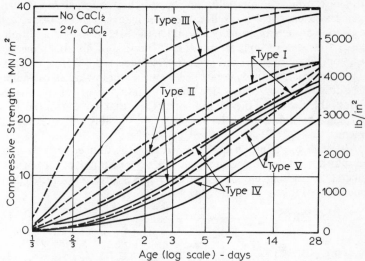

Fig. 2.14. *Influence of CaCl₂ on the strength of concretes made with different types of cement*[2.23] (For type V curve reference is 2.1).

degree on the composition of the cement. This applies in particular to the setting time. Calcium chloride generally accelerates setting and an excess of CaCl₂ can cause a flash set. Typical figures showing the influence of CaCl₂ on the setting time are given in Table 2.9. This acceleration of setting makes the addition of CaCl₂ useful in repair work, for instance, when the ingress of water can be prevented for a short time only.

Table 2.9: *Influence of Calcium Chloride on Setting Time*[2.24]

Weight of calcium chloride as percentage of weight of cement	Acceleration of set, minutes
0·1	25
0·3	15
0·5	45
1·0	85

It is important that calcium chloride be uniformly distributed throughout the mix, and this is best achieved by dissolving the additive in the mixing water before it enters the mixer. It is convenient to prepare a concentrated aqueous solution of CaCl₂, and this is more easily obtained using commercial calcium chloride flakes rather than granular calcium chloride which dissolves very slowly.

In cases when the durability of concrete may be impaired by outside agencies the use of calcium chloride may be inadvisable. For instance, the resistance of cement to sulphate attack is reduced by the addition of $CaCl_2$, particularly in lean mixes and the risk of an alkali-aggregate reaction, when the aggregate is reactive, is increased.[2.24] However, when this reaction is effectively controlled by the use of low-alkali cement and the addition of pozzolanas, the effect of $CaCl_2$ is very small. Another undesirable feature of the addition of $CaCl_2$ is that it increases the drying shrinkage by about 10 to 15 per cent, and possibly increases also the creep.

Although the addition of $CaCl_2$ reduces the danger of frost attack during the first few days after placing, the resistance of air-entrained concrete to freezing and thawing at later ages is adversely affected. Some indication of this is given in Fig. 2.15.

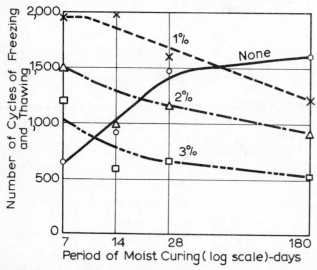

Fig. 2.15. *Resistance to freezing and thawing of concrete cured moist at $4°C$ $(40°F)$ for different contents of $CaCl_2$*[2.24]

On the credit side, $CaCl_2$ has been found to raise the resistance of concrete to erosion and abrasion, and this improvement persists at all ages.[2.24]

The possibility of corrosion of reinforcing steel by calcium chloride is still a subject of controversy: the U.S. Bureau of Reclamation found no corrosion of reinforcement when calcium chloride is used in correct proportions.[2.1] At high water/cement ratios some corrosion takes place but the attack is not progressive.[2.24] However, the chloride ions in concrete are often non-uniformly distributed. In practice, the British Code CP 110: 1972 limits the amount of calcium chloride to 1·5 per cent of weight of cement. Calcium chloride has been found to lead to corrosion of prestressing wire[2.30] and must not be used in the manufacture of prestressed concrete unless the tendons are permanently protected by an impermeable barrier. When steam-curing concrete with calcium chloride, there is a serious risk of severe corrosion of reinforcement and the use of calcium chloride when the curing temperature exceeds 60°C (140°F) is not recommended. When plain concrete is steam cured, however, $CaCl_2$ increases the strength of concrete and permits the use of a more rapid temperature rise during the curing cycle[2.25] (*see* p. 284).

The action of sodium chloride is similar to that of calcium chloride but is of lower intensity. The effects of NaCl are also more variable and a depression in the heat of hydration, with a consequent loss of strength at 7 days and later has been observed. For this reason the use of NaCl is definitely undesirable.

Retarders

A delay in the setting of the cement paste can be achieved by the addition to the mix of a retarding admixture. These admixtures generally slow down also the hardening of the paste although some salts may speed up the setting but inhibit the development of strength.

Retarders are useful in concreting in hot weather when the normal setting time is shortened by the higher temperature and in preventing the formation of cold joints. The delay in hardening caused by the retarders can be used to obtain an architectural finish of exposed aggregate: the retarder is applied to the interior surface of the formwork so that the hardening of the adjacent cement is delayed. This cement can be brushed off after the formwork has been struck so that an exposed aggregate surface is obtained.

Retarding action is exhibited by sugar, carbohydrate derivatives, soluble zinc salts, soluble borates, and others. In practice, retarders which are also water-reducing are more commonly used; these are described in the next section. Great care is necessary in using retarders as in incorrect quantities they can totally inhibit the setting and hardening of concrete. Cases are known of seemingly inexplicable results of strength tests when sugar bags have been used for the shipment of aggregate samples to the laboratory or when molasses bags have been used to transport freshly mixed concrete. The effects of sugar depend greatly on the quantity used, and conflicting results were reported in the past.[2.6] It seems now that, used in a carefully controlled manner, a small quantity of sugar (about 0·05 per cent of the weight of cement) will act as an acceptable retarder; the delay in setting of concrete (determined according to A.S.T.M. Standard C403–70) is about 4 hours.[2.41] However, the exact effects of sugar depend greatly on the chemical composition of cement. For this reason, the performance of sugar, and indeed of any retarder, should be determined by trial mixes with the actual cement which is to be used in construction.

A large quantity of sugar, say 0·2 to 1 per cent of the weight of cement, will virtually prevent the setting of cement. Such quantities of sugar can therefore be used as a "kill", for instance when a mixer or an agitator has broken down and cannot be discharged.

When sugar is used as a controlled set retarder, the early strength of concrete is severely reduced[2.26] but beyond about 7 days there is an increase in strength of several per cent compared with a non-retarded mix.[2.41] This is probably due to the fact that delayed setting produces a denser gel (cf. p. 276).

It is interesting to note that the effectiveness of an admixture depends on the time when it is added to the mix: a delay of even 2 minutes after water has come into contact with the cement increases the retardation. This is especially so with cements which have a high C_3A content because, once some C_3A has hydrated, it does not absorb the admixture so that more of it is left to retard the hydration of the calcium silicates.

Water-reducing Admixtures
According to A.S.T.M. Standard C494–71, admixtures which are water-reducing only are called Type A, but if the water-reducing properties are associated with set retardation, the admixture is classified as Type D. There exist also water-reducing and accelerating admixtures (Type E) but these are of little interest. The two main groups of admixtures of Type D are—

(*a*) lignosulphonic acids and their salts, (known as Class 1 in the A.S.T.M. nomenclature[2.27]) and

(*b*) hydroxylated carboxylic acids and their salts (known as Class 3).

The modifications and derivatives of these, known as Class 2 and 4 respectively, do not act as retarders, and may even behave as accelerators (*see* Fig. 2.16): they are therefore of Type A or E (*see* p. 93).

The principal active component of the admixtures are surface-active agents.[2.27] These are substances which are concentrated at the interface between two immiscible phases and which alter the physico-chemical forces acting at this interface. The substances are adsorbed on the cement particles, giving them a negative charge which leads to repulsion between the particles and results in stabilizing their dispersion. In addition, the charge causes the development around each particle of a sheath of orientated water molecules which prevent a close approach of the particles to one another. The particles have, therefore, a greater mobility, and water freed from the restraining influence of the flocculated system becomes available to lubricate the mix so that the workability is increased.[2.27]

One effect of dispersion is to expose a greater surface area of cement to hydration, which progresses therefore at a higher rate in the early stages. For this reason, there is an increase in the strength of concrete, compared with a mix of the *same* water/cement ratio but without the admixture. A more uniform distribution of the dispersed cement throughout the concrete may also contribute to the improved strength.[2.27] The increase in strength is particularly noticeable in very young concretes[2.29] but under certain conditions persists for a long time.

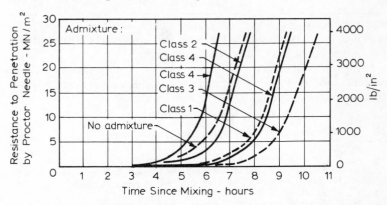

Fig. 2.16. *Effect of various water-reducing admixtures on the setting time of concrete*[2.28]

The influence of admixtures on strength varies considerably with the composition of cement, the greatest increase in strength occurring when used with cements of low alkali or low C_3A content. With some cements, the influence of admixtures is very small, but in general terms admixtures are effective with all types of Portland cement and also with aluminous cement. Some water-reducing admixtures are more effective when used in mixes containing pozzolanas than in plain mixes.[2.43]

The reduction in the quantity of mixing water that is possible owing to the use of admixtures varies between 5 and 15 per cent (Table 2.10). A part of this is in many cases (especially with Class 1 admixtures) due to entrained air introduced by the admixture.[2.29] The actual decrease in the mixing water depends on the cement content, type of aggregate used, presence of air-entraining agents or pozzolanas. It is therefore apparent that trial mixes, containing the actual materials to be used on the job, are essential in order to determine the type and quantity of admixture to achieve optimum properties. Data given by the manufacturers of

Table 2.10: *Water Reduction Obtained with Different Water-reducing and Set-retarding Admixtures*[2.28]

Admixture	Air-entraining agent (neutralized vinsol resin)	Water-reducing and set-retarding admixture			
		Class 1	Class 2	Class 3	Class 4
Slump, mm (in.)	70 (2¾)	70 (2¾)	65 (2½)	75 (3)	75 (3)
Air content, per cent	4·8	4·3	4·5	4·5	4·7
Water reduction compared with non-air entrained concrete, per cent	8	13	11	13	12

admixtures cannot generally be unquestionably accepted. It should also be noted that even though set is retarded, admixtures do not always reduce the rate of loss of workability with time.[2.29]

The quantity of admixture represents generally only a fraction of one per cent of the weight of cement in the mix so that use of reliable dispensing equipment is essential.

REFERENCES

2.1.   U.S. BUREAU OF RECLAMATION, *Concrete Manual*, 7th Edit. (Denver, Colorado, 1966).

2.2.   H. WOODS, Rational development of Cement Specifications, *J. Portl. Cem. Assoc. Research and Development Laboratories*, **1**, No. 1, pp. 4–11 (Jan. 1959).

2.3.   W. H. PRICE, Factors influencing concrete strength, *J. Amer. Concr. Inst.*, **47**, pp. 417–32 (Feb. 1951).

2.4.   H. F. GONNERMAN and W. LERCH, Changes in characteristics of Portland cement as exhibited by laboratory tests over the period 1904 to 1950, *A.S.T.M. Sp. Publicn. No.* 127 (1951).

2.5.   H. F. THOMSON, Ready mix: Quality control in cement manufacture, *Modern Concrete*, **22**, No. 11, pp. 32–6 (Chicago, March 1959).

2.6.   F. M. LEA, *The Chemistry of Cement and Concrete* (London, Arnold, 1970).

2.7.   R. H. BOGUE, Portland cement, *Portl. Cem. Assoc. Fellowship Paper No.* 53, pp. 411–31 (Washington D.C., August 1949).

2.8.   D. L. BLOEM, Comparison of strength development between Portland cement and Portland Blast-furnace slag cement, *Nat. Ready-mixed Concr. Assoc. Publicn. No.* 90, pp. 11 (Washington D.C., Oct. 1959).

2.9.   A. A. FULTON and W. T. MARSHALL, The use of fly ash and similar materials in concrete, *Proc. Inst. C. E.*, Part I, **5**, No. 6, pp. 714–30 (London, Nov. 1956).

2.10. R. W. NURSE, Hydrophobic cement, *Cement and Lime Manufacture*, **26**, No. 4, pp. 47–51 (London, July 1953).

2.11. U. W. STOLL, Hydrophobic cement, *A.S.T.M. Sp. Tech. Publicn., No.* 205, pp. 7–15 (1958).

2.12. H. LOSSIER, Cements with controlled expansions and their applications to prestressed concrete, *The Struct. E.*, **24**, No. 10, pp. 505–34 (Oct. 1946).

2.13. T. W. PARKER, La recherche sur la chimie des ciments au Royaume-Uni pendant les années d'après-guerre, *Revue Génerale des Sciences Appliquées*, **1**, No. 3, pp. 74–83 (1952).

2.14. H. LAFUMA, Quelques aspects de la physico-chimie des ciments alumineux, *ibid*, pp. 66–74.

2.15. F. M. LEA, *Cement and concrete*, Lecture delivered before the Royal Institute of Chemistry, London, on the 19th Dec., 1944 (Cambridge, W. Heffer and Sons, 1944).

2.16. A. V. HUSSEY and T. D. ROBSON, High-alumina cement as a constructional material in the chemical industry, *Symposium on Materials of Construction in the Chemical Industry*, Birmingham, Soc. Chem. Ind., 1950.

2.17. A. M. NEVILLE, Tests on the strength of high-alumina cement concrete, *J. New Zealand Inst. E.*, **14**, No. 3, pp. 73–6 (March 1959).

2.18. A. M. NEVILLE, The effect of warm storage conditions on the strength of concrete made with high-alumina cement, *Proc. Inst. C. E.*, **10**, pp. 185–92 (London, June 1958).

2.19. A. M. NEVILLE and I. E. ZEKARIA, The effect on concrete strength of drying during fixing electrical resistance strain gauges, *R.I.L.E.M.*, *Bul. No.* 38, pp. 95–6 (Paris, 1958).

2.20. A. M. NEVILLE, Further tests on the strength of high-alumina cement concrete under hot-wet conditions, *R.I.L.E.M. Int. Symp. on Concrete and Reinforced Concrete in Hot Countries*, Haifa, July 1960.

2.21. P. P. BUDNIKOV, Role of gypsum in the hardening of hydraulic cements, *Proc. 4th Int. Symp. on the Chemistry of Cement*, Washington D.C., 1960, pp. 469–77.

2.22. A. M. NEVILLE and H. KENINGTON, Creep of aluminous cement concrete, *ibid*, pp. 703–8.

2.23. U.S. BUREAU OF RECLAMATION, Strength and properties of concrete at early ages, *Materials Laboratory Report No. C–444* (Denver, Colorado, Sept. 1949).

2.24. J. J. SHIDELER, Calcium chloride in concrete, *J. Amer. Concr. Inst.*, **48**, pp. 537–59 (March 1952).

2.25. A. G. A. SAUL, Steam curing and its effect upon mix design, *Proc. of a Symposium on Mix Design and Quality Control of Concrete*, pp. 132–42 (London, Cement and Concrete Assoc., 1954).

2.26. D. L. BLOEM, Preliminary tests of effect of sugar on strength of mortar, *Nat. Ready-mixed Concr. Assoc. Publcn.* (Washington D.C., August 1959).

2.27. M. E. PRIOR and A. B. ADAMS, Introduction to Producers' papers on water-reducing admixtures and set-retarding admixtures for concrete, *A.S.T.M. Sp. Tech. Publcn. No.* 266, pp. 170–9 (1960).

2.28. C. A. VOLLICK, Effect of water-reducing admixtures and set-retarding admixtures on the properties of plastic concrete, *ibid*, pp. 180–200.

2.29. B. FOSTER, Summary: Symposium on effect of water-reducing admixtures and set-retarding admixtures on properties of concrete, *ibid*, pp. 240–6.

2.30. G. E. MONFORE and G. J. VERBECK, Corrosion of prestressed wire in concrete, *J. Amer. Concr. Inst.* **57**, pp. 491–515 (Nov. 1960).

2.31. E. BURKE, Discussion on Comparison of chemical resistance of super-sulphated and special purpose cements, *Proc. 4th Int. Symp. on the Chemistry of Cement*, Washington D.C., 1960, pp. 877–9.

2.32. P. LHOPITALLIER, Calcium aluminates and high-alumina cement, *ibid*, pp. 1007–33.

2.33. A. M. NEVILLE, Deterioration of high-alumina cement concrete, *Proc. Inst. C.E.*, **25**, pp. 287–324 (London, July 1963).

2.34. U.S. BUREAU OF RECLAMATION, *Concrete Manual*, 5th Edit. (Denver, Colorado, 1949).

2.35. AGRÉMENT BOARD, Certificate No. 69/32 of 14th April 1969 for Swiftcrete ultra high early strength Portland cement.

2.36. E. W. BENNETT and D. R. LOAT, Shrinkage and creep of concrete as affected by the fineness of Portland cement, *Mag. Concr. Res.*, **22,** No. 71, pp. 69–78 (June 1970).

2.37. E. C. HIGGINSON, Mineral admixtures, *A.S.T.M. Sp. Tech. Publicn. No. 169-A*, pp. 543–55 (1966).

2.38. A. KLEIN and G. E. TROXELL, Studies of calcium sulfoaluminate admixtures for expansive cements, *Proc. A.S.T.M.*, **58,** pp. 986–1008 (1958).

2.39. A.C.I. COMMITTEE 223, Expansive cement concretes—present state of knowledge, *J. Amer. Concr. Inst.*, **67,** pp. 583–610 (Aug. 1970).

2.40. P. K. MEHTA and G. LESNIKOFF, Conversion of $CaO.Al_2O_3.10H_2O$ to $3CaO.Al_2O_3.6H_2O$, *J. Amer. Ceramic Soc.*, **54,** pp. 210–12 (April 1971).

2.41. R. ASHWORTH, Some investigations into the use of sugar as an admixture to concrete, *Proc. Inst. C.E.*, **31,** pp. 129–45 (London, June 1965).

2.42. R. TSUKAYAMA, Effect of conversion on properties of concrete using high-aluminous cement, *Proc. 5th Int. Symp. on the Chemistry of Cement, Part 3, Tokyo*, pp. 316–327 (1968).

2.43. R. E. DAVIS, A review of pozzolanic materials and their use in concretes, *Symposium on Use of Pozzolanic Materials in Mortars and Concretes, A.S.T.M. Sp. Tech. Publicn. No. 99*, pp. 3–15 (1950).

3. Properties of Aggregate

SINCE at least three-quarters of the volume of concrete is occupied by aggregate, it is not surprising that its quality is of considerable importance. Not only may the aggregate limit the strength of concrete, as weak aggregate cannot produce strong concrete, but the properties of aggregate greatly affect the durability and structural performance of concrete.

Aggregate was originally viewed as an inert material dispersed throughout the cement paste largely for economic reasons. It is possible, however, to take an opposite view and to look on aggregate as a building material connected into a cohesive whole by means of the cement paste, in a manner similar to masonry construction. In fact, aggregate is not truly inert and its physical, thermal, and sometimes also chemical properties influence the performance of concrete.

Aggregate is cheaper than cement and it is, therefore, economical to put into the mix as much of the former and as little of the latter as possible. But economy is not the only reason for using aggregate: it confers considerable technical advantages on concrete, which has a higher volume stability and better durability than the cement paste alone.

General Classification of Aggregate

The size of aggregate used in concrete ranges from several inches down to particles of a few thousandths of an inch in cross-section. The maximum size actually used varies but in any mix particles of different sizes are incorporated, the particle size distribution being referred to as grading. In making low-grade concrete, aggregate from deposits containing a whole range of sizes, from the largest to the smallest, is sometimes used; this is referred to as all-in or pit-run aggregate. The alternative, very much more common, and always used in the manufacture of good quality concrete, is to obtain the aggregate in at least two size groups, the main division being between fine aggregate, often called sand, not larger than 5 mm or $\frac{3}{16}$ in., and coarse aggregate, which comprises material at least 5 mm or $\frac{3}{16}$ in. in size. In the United States the division is made at No. 4 sieve, which is 4·76 mm ($\frac{3}{16}$ in.) in size (*see* Table 3.14). More will be said about grading later but this basic division makes it possible to distinguish in the ensuing description between fine and coarse aggregate. It should be noted that the use of the term aggregate (to mean coarse aggregate) in contradistinction to sand is not correct, although comparatively common.

Sand is generally considered to have a lower size limit of about 0·07 mm or a little less. Material between 0·06 mm and 0·002 mm is classified as silt, and particles smaller still are termed clay. Loam is a soft deposit consisting of sand, silt, and clay in about equal proportions.

All aggregate particles originally formed a part of a larger parent mass. This may have been fragmented by natural processes of weathering and abrasion or artificially by crushing. Thus many properties of the aggregate depend entirely on the properties of the parent rock, e.g. chemical and mineral composition, petrographic description, specific gravity, hardness, strength, physical and chemical stability, pore structure, colour, etc. On the other hand, there are some properties possessed by the aggregate but absent in the parent rock: particle shape and size, surface texture, and absorption. All these properties may have a considerable influence on the quality of the concrete either fresh or in the hardened state.

It is only reasonable to add, however, that, although these different properties of aggregate *per se* can be examined, it is difficult to define a good aggregate other than by saying that it is an aggregate from which good concrete (for the given conditions) can be made. While aggregate whose properties all appear satisfactory will always make good concrete, the converse is not necessarily true and this is why the criterion of performance in concrete has to be used. In particular, it has been found that aggregate may appear to be unsatisfactory on some count but no trouble need be experienced when it is used in concrete. For instance, a rock sample may disrupt on freezing but need not do so when embedded in concrete, especially when the aggregate particles are well covered by a paste of low permeability. However, aggregate considered poor in more than one respect is unlikely to make a satisfactory concrete, so that tests on aggregate alone are of help in assessing its suitability for use in concrete.

Classification of Natural Aggregates

So far we have considered only aggregate formed from naturally occurring materials, and the present chapter deals almost exclusively with this type of aggregate. Aggregate can, however, also be manufactured from industrial products: since these artificial aggregates are generally either heavier or lighter than ordinary aggregate they are considered in Chapter 9.

A further distinction can be made between aggregate reduced to its present size by natural agents and crushed aggregate obtained by a deliberate fragmentation of rock.

From the petrological standpoint the aggregates, whether crushed or naturally reduced in size, can be divided into several groups of rocks having common characteristics. The classification of B.S. 812: 1967 is most convenient and is given in Table 3.1. The group classification does not imply suitability of any aggregate for concrete-making: unsuitable material can be found in any group, although some groups tend to have a better record than others. It should also be remembered that many

Table 3.1: *B.S. 812: 1967 Classification of Natural Aggregates According to Rock Type*

Basalt Group	*Flint Group*	*Gabbro Group*
Andesite	Chert	Basic diorite
Basalt	Flint	Basic gneiss
Basic Porphyrites		Gabbro
Diabase		Hornblende-rock
Dolerites of all kinds including theralite and teschenite		Norite
		Peridotite
Epidiorite		Picrite
Hornblende-schist		Serpentine
Lamprophyre		
Quartz-dolerite		
Spilite		
Granite Group	*Gritstone Group*	*Hornfels group*
Gneiss	Agglomerate	Contact-altered rocks
Granite	Arkose	of all kinds except
Granodiorite	Breccia	marble
Granulite	Conglomerate	
Pegmatite	Greywacke	
Quartz-diorite	Grit	
Syenite	Sandstone	
	Tuff	
Limestone Group	*Porphyry Group*	*Quartzite Group*
Dolomite	Aplite	Ganister
Limestone	Dacite	Quartzitic sandstones
Marble	Felsite	Re-crystallized
	Granophyre	quartzite
	Keratophyre	
	Microgranite	
	Porphyry	
	Quartz-porphyrite	
	Rhyolite	
	Trachyte	

Schist Group
Phyllite
Schist
Slate
All severely sheared rocks

trade and customary·names of aggregates are in use, and these often do not correspond to the correct petrographic classification.

A.S.T.M. Standard C 294–69 gives a description of some of the more common or important minerals found in aggregates. Mineralogical classification is of help in recognizing properties of aggregate but cannot provide a basis for predicting its performance in concrete as there are no minerals universally desirable and few invariably undesirable ones. The A.S.T.M. classification is summarized below:

Silica minerals—(quartz, opal, chalcedony, tridymite, cristobalite)
Feldspars
Micaceous minerals
Carbonate minerals
Sulphate minerals
Iron sulphide minerals
Ferromagnesian minerals
Zeolites
Iron oxides
Clay minerals

The details of petrological and mineralogical methods are outside the scope of this book, but it is important to realize that geological examination of aggregate is a useful aid in assessing its quality, and, in particular, in comparing a new aggregate with one for which service records are available. Furthermore, adverse properties, such as the presence of some unstable forms of silica, can be detected. In the case of artificial aggregates the influence of manufacturing methods and of processing can also be studied.

Sampling

Tests of various properties of aggregate are perforce performed on samples of the material and, therefore, the results of the tests apply, strictly speaking, to the aggregate in the sample only. Since, however, we are interested in the bulk of the aggregate as supplied or as available for supply we should ensure that the sample is typical of the average properties of the aggregate. Such a sample is said to be representative, and to obtain it certain precautions in procuring the sample have to be observed.

No detailed procedures can, however, be laid down because the conditions and situations involved in taking samples in the field can vary widely from case to case. Nevertheless, an intelligent experimenter can obtain reliable results if he bears in mind at all times that the sample taken is to be representative of the bulk of the material considered. An instance of such care would be to use a scoop rather than a shovel so as to prevent rolling off of particles of some sizes when the shovel is lifted. This became recognized in the 1967 revision of B.S. 812.

The main sample is made up of a number of portions drawn from different parts of the whole. The minimum number of these portions, called increments, is ten, and they should add up to a weight not less than that given in Table 3.2 for particles of different sizes, as prescribed by B.S. 812: 1967. If, however, the source from which the sample is being obtained is variable or segregated, a larger number of increments should be taken and a larger sample ought to be dispatched for testing. This is particularly the case in stockpiles when increments have to be taken from all parts of the pile, not only near its surface but also from the centre.

Table 3.2: *Minimum Weights of Samples for Testing (B.S. 812: 1967)*

Maximum particle size present in substantial proportion		Minimum weight of sample despatched for testing	
mm	in.	kg	lb
25 or larger	1 or larger	50	112
between 5 and 25	between $\frac{3}{16}$ and 1	25	56
5 or smaller	$\frac{3}{16}$ or smaller	13	28

It is clear from Table 3.2 that the main sample may be rather large, particularly when large-size aggregate is used, and so the sample has to be reduced before testing. At all stages of reduction it is necessary to ensure that the representative character of the sample is retained so that the actual test sample has the same properties as the main sample and *ipso facto* as the bulk of the aggregate.

There are two ways of reducing the size of a sample, each essentially dividing it into two similar parts: quartering and riffling. For quartering, the main sample is thoroughly mixed and in the case of fine aggregate dampened in order to avoid segregation. The material is heaped into a cone and then turned over to form a new cone. This is repeated twice, the material always being deposited at the apex of the cone so that the fall of particles is evenly distributed round the circumference. The final cone is flattened and divided into quarters. One pair of diagonally opposite quarters is discarded, and the remainder forms the sample for testing or, if still too large, can be reduced by further quartering. Care must be taken to include all fine material in the appropriate quarter.

As an alternative, the sample can be split into halves using a riffler. This is a box with a number of parallel vertical divisions, alternate ones discharging to the left and to the right. The sample is discharged into the riffler over its full width, and the two halves are collected into two boxes at the bottom of the chutes on each side. One half is discarded, and riffling of the other half is repeated until the sample is reduced to the desired size. B.S. 812: 1967 describes a typical riffler.

Particle Shape and Texture

In addition to the petrological character of aggregate, its external characteristics are of importance, in particular the particle shape and surface texture. The shape of three-dimensional bodies is rather difficult to describe, and it is, therefore, convenient to define certain geometrical characteristics of such bodies.

Roundness measures the relative sharpness or angularity of the edges and corners of a particle. Roundness is controlled largely by the strength and abrasion resistance of the parent rock and by the amount of wear to which the particle has been subjected. In the case of crushed aggregate, the particle shape depends on the nature of the parent material and on the type of crusher and its reduction ratio, i.e. the ratio of the size of material fed into the crusher to the size of the finished product. A convenient broad classification of roundness is that of B.S. 812: 1967, given in Table 3.3.

Table 3.3: *Particle Shape Classification of B.S. 812: 1967*

Classification	Description	Examples
Rounded	Fully water-worn or completely shaped by attrition	River or seashore gravel; desert, seashore and wind-blown sand
Irregular	Naturally irregular, or partly shaped by attrition and having rounded edges	Other gravels; land or dug flint
Flaky	Material of which the thickness is small relative to the other two dimensions	Laminated rock
Angular	Possessing well-defined edges formed at the intersection of roughly planar faces	Crushed rocks of all types; talus; crushed slag
Elongated	Material, usually angular, in which the length is considerably larger than the other two dimensions	—
Flaky and Elongated	Material having the length considerably larger than the width, and the width considerably larger than the thickness	—

A classification sometimes used in the United States is as follows—

> *Well-rounded*—no original faces left
> *Rounded*—faces almost gone
> *Subrounded*—considerable wear, faces reduced in area
> *Subangular*—some wear but faces untouched
> *Angular*—little evidence of wear

Since the degree of packing of particles of one size depends on their shape, the angularity of aggregate can be estimated from the proportion of voids in a sample compacted in a prescribed way. B.S. 812: 1967 defines the angularity number as 67 minus the percentage of solid volume in a vessel filled with aggregate in a standard manner. The size of particles used in the test must be controlled within narrow limits, and should preferably lie between any of the following—

19·0 and 12·7 mm ($\frac{3}{4}$ in. and $\frac{1}{2}$ in.)
12·7 and 9·5 mm ($\frac{1}{2}$ in. and $\frac{3}{8}$ in.)
9·5 and 6·3 mm ($\frac{3}{8}$ in. and $\frac{1}{4}$ in.)
6·3 and 4·8 mm ($\frac{1}{4}$ in. and $\frac{3}{16}$ in.)

The figure 67 in the expression for the angularity number represents the solid volume of the most rounded gravel, so that the angularity number measures the percentage of voids in excess of that in the rounded gravel (i.e. 33). The higher the number the more angular the aggregate, the range for practical aggregate being between 0 and 11. A recent development in measurement of angularity of aggregate, both coarse and fine but of single size, is an angularity factor defined as the ratio of the solid volume of loose aggregate to the solid volume of glass spheres of specified grading;[3.41] thus no packing is involved and the attendant error is avoided. The usefulness of the test is yet to be determined.

The void content of aggregate can be calculated from the change in the the volume of air when a known decrease in pressure is applied; hence, the volume of air, i.e. the volume of interstitial space, can be calculated.[3.52]

An indirect proof of the dependence of the percentage of voids on the shape of particles is obtained from Fig. 3.1, based on Shergold's[3.1] data. The sample consisted of a mixture of two aggregates, one angular, the other rounded, in varying proportions, and it can be seen how increasing the proportion of rounded particles decreases the percentage of voids.

Another aspect of the shape of coarse aggregate is its *sphericity*, defined as a function of the ratio of the surface area of the particle to its volume. Sphericity is related to the bedding and cleavage of the parent rock, and is also influenced by the type of crushing equipment when the size of particles has been artificially reduced. Particles with a high ratio of surface area to volume are of particular interest as they lower the workability of the mix. Elongated and flaky particles are of this type. The latter can also affect adversely the durability of concrete as they tend to be oriented in one plane, with water and air voids forming underneath.

The presence of elongated or flaky particles in excess of 10 to 15 per cent of the weight of coarse aggregate is generally considered undesirable, but no recognized limits are laid down.

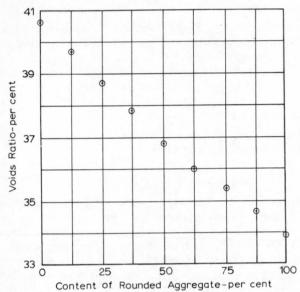

Fig. 3.1. *Influence of angularity of aggregate on voids ratio*[3.1]
(*Crown copyright*)

The weight of flaky particles expressed as a percentage of the weight of the sample is called the flakiness index. Elongation index is similarly defined. Some particles are both flaky and elongated, and are, therefore, counted in both categories.

The classification is made by means of simple gauges described in B.S. 812: 1967. The division is based on the rather arbitrary assumption that a particle is flaky if its thickness (least dimension) is less than 0·6 times the mean sieve size of the size fraction to which the particle belongs. Similarly, a particle whose length (largest dimension) is more than 1·8 times the mean sieve size of the size fraction is said to be elongated. The mean size is defined as the arithmetic mean of the sieve size on which the particle is just retained and the sieve size through which the particle just passes. As closer size control is necessary, the sieves considered are not those of the standard concrete aggregate series but: 63·5, 50·8, 38·1, 25·4, 19·0, 12·7, 9·52 and 6·35 mm (2½, 2, 1½, 1, ¾, ½, ⅜ and ¼ in.) sieves.

The flakiness and elongation tests are useful for general assessment of aggregates but they do not adequately describe the particle shape.

The classification of the surface texture is based on the degree to which the particle surfaces are polished or dull, smooth or rough; the type of roughness has also to be described. Surface texture depends on the hardness, grain size, and pore characteristics of the parent material (hard, dense and fine-grained rocks generally having smooth fracture surfaces), as well as on the degree to which forces acting on the particle surface have smoothed or roughened it. Visual estimate of roughness is quite reliable, but in order to reduce misunderstanding the classification of B.S. 812: 1967, given in Table 3.4, should be followed. There is no recognized

Table 3.4: *Surface Texture of Aggregates* (*B.S. 812: 1967*)

Group	Surface Texture	Characteristics	Examples
1	Glassy	Conchoidal fracture	Black flint, vitreous slag
2	Smooth	Water-worn, or smooth due to fracture of laminated or fine-grained rock	Gravels, chert, slate, marble, some rhyolites
3	Granular	Fracture showing more or less uniform rounded grains	Sandstone, oolite
4	Rough	Rough fracture of fine- or medium-grained rock containing no easily visible crystalline constituents	Basalt, felsite, porphyry, limestone
5	Crystalline	Containing easily visible crystalline constituents	Granite, gabbro, gneiss
6	Honeycombed	With visible pores and cavities	Brick, pumice, foamed slag, clinker, expanded clay

method of measuring the surface roughness but Wright's approach[3.2] is of interest: the interface between the particle and a resin in which it is set is magnified, and the difference between the length of the profile and the length of an unevenness line drawn as a series of chords is determined. This is taken as a measure of roughness. Reproducible results are obtained, but the method is laborious and is not widely used.

It seems that the shape and surface texture of aggregate influence considerably the strength of concrete. The flexural strength is more affected than the compressive strength, and the effects of shape and texture are particularly significant in the case of high strength concrete. Some data of Kaplan's[3.3] are reproduced in Table 3.5 but this gives no more than an indication of the type of influence, as some other factors may not have been taken into account. The full rôle of shape and texture of aggregate in the development of concrete strength is not known, but possibly a rougher texture results in a greater adhesive force between the particles and the cement matrix. Likewise, the larger surface area of angular aggregate means that a larger adhesive force can be developed.

The shape and texture of fine aggregate have a significant effect on the water requirement of the mix made with the given aggregate. If these properties of fine aggregate are expressed indirectly by its packing, i.e. by the percentage voids in a loose condition (*see* p. 124), then the influence on the water requirement is quite definite[3.42] (*see* Fig. 3.2). The influence of the voids in coarse aggregate is less definite.[3.42]

Table 3.5: *Average Relative Importance of the Aggregate Properties Affecting the Strength of Concrete*[3.3]

Property of concrete	Relative effect of aggregate properties per cent		
	Shape	Surface texture	Modulus of elasticity
Flexural Strength	31	26	43
Compressive Strength	22	44	34

N.B. Values represent the ratio of variance due to each property to the total variance accounted for by the three characteristics of aggregate in tests on three mixes made with 13 aggregates.

Flakiness and the shape of coarse aggregate in general have an appreciable effect on the workability of concrete. Fig. 3.3, reproduced from Kaplan's[3.4] paper, shows the pattern of the relation between the angularity of coarse aggregate and the compacting factor of concrete made with it. An increase in angularity from minimum to maximum would reduce the compacting factor by about 0·09 but in practice there can clearly be no unique relation between the two factors as other properties of aggregate also affect the workability. Kaplan's experimental results,[3.4] however, do not confirm that the surface texture is a factor.

Bond of Aggregate
Bond between aggregate and cement paste is an important factor in the strength of concrete, especially the flexural strength, the full rôle of bond

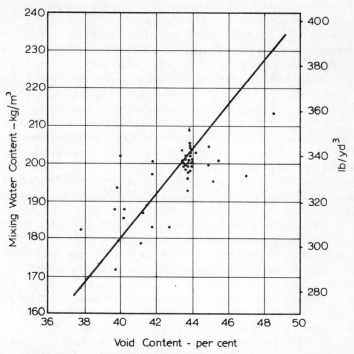

Fig. 3.2. *Relation between void content of sand in a loose condition and the water requirement of concrete made with the given sand*[3.42]

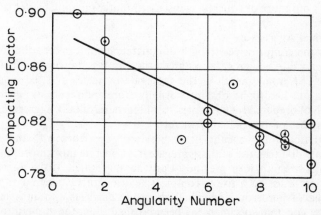

Fig. 3.3. *The relation between the angularity number of aggregate and the compacting factor of concrete made with the given aggregate*[3.4]

being only now realized. Bond is due, in part, to the interlocking of the aggregate and the paste owing to the roughness of the surface of the former. A rougher surface, such as that of crushed particles, results in a better bond; better bond is also usually obtained with softer, porous, and mineralogically heterogeneous particles. Generally, texture characteristics which permit no penetration of the surface of the particles are not conducive to good bond. In addition, bond is affected by other physical and chemical properties of aggregate, related to its mineralogical and chemical composition, and to the electrostatic condition of the particle surface. For instance, some chemical bond may exist in the case of limestone and possibly siliceous aggregates, and at the surface of polished particles some capillary forces may develop. However, little is known about these phenomena, and relying on experience is still necessary in predicting the bond between the aggregate and the surrounding cement paste.

The determination of the quality of bond of aggregate is rather difficult and no accepted tests exist. Generally, when bond is good, a crushed concrete specimen should contain some aggregate particles broken right through, in addition to the more numerous ones pulled out from their sockets. An excess of fractured particles, however, might suggest that the aggregate is too weak. Because it depends on the paste strength as well as on the properties of aggregate surface, bond strength increases with the age of concrete; it seems that the ratio of bond strength to the strength of the paste increases with age.[3.43] Thus, providing it is adequate, the bond strength *per se* may not be a controlling factor in the strength of concrete. However, in high strength concrete there is probably a tendency for the bond strength to be lower than the tensile strength of the cement paste so that preferential failure in bond takes place. The problem of failure of concrete is discussed more fully in Chapter 5.

Strength of Aggregate

Clearly the compressive strength of concrete cannot exceed that of the major part of the aggregate contained therein. It is, however, difficult to test the crushing strength of the aggregate by itself, and the required information has to be obtained usually from indirect tests: crushing strength of prepared rock samples, crushing value of bulk aggregate, and performance of aggregate in concrete.

The latter simply means either previous experience with the given aggregate or a trial use of the aggregate in a concrete mix known to have a certain strength with previously proven aggregates. If the aggregate under test leads to a lower compressive strength of concrete, and in particular if numerous individual aggregate particles appear fractured after the concrete specimen has been crushed, then the strength of the aggregate is lower than the nominal compressive strength of the concrete mix in which the aggregate was incorporated. Clearly, such aggregate can be used only in a concrete of lower strength.

Inadequate strength of aggregate represents a limiting case as the properties of aggregate have some influence on the strength of concrete even when the aggregate by itself is strong enough not to fracture prematurely. If we compare concretes made with different aggregates we can observe that the influence of aggregate on the strength of concrete is qualitatively the same whatever the mix proportions, and is the same regardless of whether the concrete is tested in compression or in tension.[3.5] It is possible that the influence of aggregate on the strength of concrete is due not only to the mechanical strength of the aggregate but also, to a considerable degree, to its absorption and bond characteristics.

In general, the strength and elasticity of aggregate depends on its composition, texture and structure. Thus a low strength may be due to the weakness of constituent grains or the grains may be strong but not well knit or cemented together.

The modulus of elasticity of aggregate is rarely determined; this is, however, not unimportant as the modulus of elasticity of concrete is generally higher the higher the modulus of the constituent aggregate, but depends on other factors as well. The modulus of elasticity of aggregate affects also the magnitude of creep and shrinkage that can be realized by the concrete.

Table 3.6: *Compressive Strength of American Rocks Commonly Used as Concrete Aggregates*[3.6]

Type of rock	Number of samples*	Compressive strength					
		Average †		After deletion of extremes‡			
				Maximum		Minimum	
		MN/m²	lb/in²	MN/m²	lb/in²	MN/m²	lb/in²
Granite	278	181	26,200	257	37,300	114	16,600
Felsite	12	324	47,000	526	76,300	120	17,400
Trap	59	283	41,100	377	54,700	201	29,200
Limestone	241	159	23,000	241	34,900	93	13,500
Sandstone	79	131	19,000	240	34,800	44	6,400
Marble	34	117	16,900	244	35,400	51	7,400
Quartzite	26	252	36,500	423	61,300	124	18,000
Gneiss	36	147	21,300	235	34,100	94	13,600
Schist	31	170	24,600	297	43,100	91	13,200

* For most samples, the compressive strength is an average of 3 to 15 specimens. † Average of all samples. ‡ 10 per cent of all samples tested with highest or lowest values have been deleted as not typical of the material.

A good average value of the crushing strength of aggregate is about 200 MN/m² (30,000 lb/in²) but many excellent aggregates range in strength down to 80 MN/m² (12,000 lb/in²). One of the highest values recorded is 530 MN/m² (77,000 lb/in²) for a certain quartzite. Values for other rocks are given in Table 3.6. It should be noted that the required strength of aggregate is considerably higher than the normal range of concrete strengths because the actual stresses at the points of contact of individual particles within the concrete may be far in excess of the nominal compressive stress applied.

On the other hand, aggregate of moderate or low strength and modulus of elasticity can be valuable in preserving the durability of concrete. Volume changes of concrete, arising from hygral or thermal reasons, lead to a lower stress in the cement paste when the aggregate is compressible. Thus compressibility of aggregate would reduce distress in concrete while a strong and rigid aggregate might lead to cracking of the surrounding cement paste.

It may be noted that no general relation exists between the strength and modulus of elasticity of different aggregates.[3.3] Some granites, for instance, have been found to have a modulus of elasticity of 45 GN/m² (6·5 × 10⁶ lb/in²), and gabbro and diabase a modulus of 85·5 GN/m² (12·4 × 10⁶ lb/in²), the strength of all these rocks ranging between 145 and 170 MN/m² (21,000 to 25,000 lb/in²). Values of the modulus in excess of 160 GN/m² (23 × 10⁶ lb/in²) have been encountered.

A test to measure the compressive strength of prepared rock cylinders is prescribed by B.S. 812: 1967. A 25·4 mm (1 in.) diameter cylinder 25·4 mm (1 in.) high, is used, and the nominal crushing strength of an oven-dry specimen is determined to the nearest 0·5 MN/m² (or 100 lb/in²). The preparation of the sample involves drilling, sawing and grinding—all rather laborious operations. The results of the crushing test are affected by the presence of planes of weakness in the rock, and there is, therefore, some doubt about the value of this test, particularly as the structural weakness in the rock may not be significant once the rock has been comminuted to the size used in concrete. In essence, the crushing strength test measures the quality of the parent rock rather than the quality of the aggregate as used in concrete. For this reason tests on prepared specimens are nowadays less used than tests on bulk aggregate, but are nevertheless useful when dealing with a potential new source of crushed aggregate.

Sometimes, the strength of a wet as well as of a dry specimen is determined. The ratio of wet to dry strengths measures the softening effect, and when this is high, poor durability of the rock may be suspected.

A test on the crushing properties of bulk aggregate is the so-called *crushing value test* of B.S. 812: 1967. There is no mathematical relation between this crushing value and the compressive strength, but qualitatively the results of the two tests are in agreement. The crushing value is a useful guide when dealing with aggregates of unknown performance, particularly when lower strength may be suspected, as for instance with limestone and some granites and basalts.

The material to be tested should pass a 12·7 mm ($\frac{1}{2}$ in.) B.S. test sieve and be retained on a 9·5 mm ($\frac{3}{8}$ in.) sieve. When, however, this size is not available, particles of other sizes may be used, but those larger than standard will in general give a higher crushing value, and the smaller ones a lower value than would be obtained with the same rock of standard size. The sample to be tested should be dried in an oven at 100 to 110°C (212 to 230°F) for four hours, and then placed in a cylindrical mould and tamped in a prescribed manner. A plunger is put on top of the aggregate and the whole assembly is placed in a compression testing machine and subjected to a load of 40·6 tonne (40 ton) (pressure of 21·8 MN/m² (3,170 lb/in²)) over the gross area of the plunger, the load being increased gradually over a period of 10 minutes. After the load has been released the aggregate is removed and sieved on a 2·4 mm (No. 7) B.S. test sieve in the case of a sample of the 12·7 to 9·5 mm ($\frac{1}{2}$ to $\frac{3}{8}$ in.) standard size; for samples of other sizes the sieve size is prescribed in B.S. 812: 1967. The ratio of the weight of the material passing this sieve to the total weight of the sample is called the aggregate crushing value.

In the United States, where large quantities of artificial lightweight aggregates are used, attempts were made to develop a strength test for these aggregates, rather similar to the crushing value test described above, but no test has been standardized.

The crushing value test is rather insensitive to the variation in strength of weaker aggregates, i.e. those with a crushing value of over 25 to 30. This is so because, having been crushed before the full load of 40·6 tonne (40 ton) has been applied, these weaker materials become compacted so that the amount of crushing during later stages of the test is reduced. For this reason a *ten per cent fines value* test has been introduced and is included in the B.S. 812: 1967. In this test the apparatus of the standard crushing test is used to determine the load required to produce 10 per cent fines from the 12·7 to 9·5 mm ($\frac{1}{2}$ to $\frac{3}{8}$ in.) particles. This is achieved by applying a progressively increasing load on the plunger so as to cause its penetration in 10 minutes of about—

15 mm (0·60 in.) for rounded aggregate,
20 mm (0·80 in.) for crushed aggregate, and
24 mm (0·95 in.) for honeycombed aggregate (such as expanded shale or foamed slag).

These penetrations should result in a percentage of fines passing a 2·4 mm (No. 7) B.S. sieve of between 7·5 and 12·5 per cent. If y is the actual percentage of fines due to a maximum load of x ton, then the load required to give 10 per cent fines is given by

$$\frac{14x}{y+4}.$$

It should be noted that in this test, unlike the standard crushing value test, a higher numerical result denotes a higher strength of the aggregate. B.S. 882: 1965 prescribes a minimum value of 10 tons for aggregate to be used in concrete wearing surfaces, and 5 tons when used in other concretes.

The ten per cent fines value test shows a good correlation with the standard crushing value test for strong aggregates, while for weaker aggregates the ten per cent fines value test is more sensitive and gives a truer picture of differences between more or less weak samples. For this reason, the test is of use in assessing lightweight aggregates but there is no simple relation between the test result and the upper limit of strength of concrete made with the given aggregate.

Other Mechanical Properties of Aggregate

Several mechanical properties of aggregate are of interest, especially when the aggregate is to be used in road construction or is to be subjected to high wear.

The first of these is toughness, which can be defined as the resistance of aggregate to failure by impact. Toughness can be determined on prepared cylindrical samples of rock: the minimum height from which a standard weight has to be dropped so as to cause failure of the specimen represents the toughness of the material. This test was devised in the days of horse-drawn and steel-tyred traffic, and is now little used,* although it would disclose adverse effects of weathering of the rock under test.

It is possible also to determine the impact value of bulk aggregate, and toughness determined in this manner is related to the crushing value, and can, in fact, be used as an alternative test. The size of the particles tested is the same as in the crushing value test and the permissible values of the crushed fraction smaller than a 2·4 mm (No. 7) B.S. test sieve are also the same. The impact is applied by a standard hammer falling 15 times under its own weight upon the aggregate in a cylindrical container. This results in fragmentation in a manner similar to that produced by the pressure of the plunger in the aggregate crushing value test. Full details of the test are prescribed in B.S. 812: 1967 and B.S. 882: 1965 prescribes the following maximum values of the average of duplicate samples—

30 per cent when the aggregate is to be used in concrete for wearing surfaces, and 45 per cent when to be used in other concretes.

---

* The test had been covered by A.S.T.M. Standard D3-18, which was discontinued in 1965.

These figures serve as useful guides, but it is clear that a direct correlation between the crushing value and the performance of aggregate in concrete or the strength of the concrete is not possible.

One advantage of the impact test is that it can be performed in the field with some modifications, such as the measurement of quantities by volume rather than by weight.

In addition to strength and toughness, hardness or resistance to wear is an important property of concrete used in roads and in floor surfaces subjected to heavy traffic. Several tests are available, and it is possible to cause wear by abrasion, i.e. by rubbing of a foreign material against the stone under test, or by attrition of stone particles against one another.

In the abrasion (Dorry) test a cylindrical specimen, similar to those used in the crushing strength test, is subjected to wear by quartz (Leighton Buzzard) sand pressed against the cylinder by a rotating metal disc. The abrasion value is expressed as 20 minus one-third of the loss of weight of the cylinder in grams. Good stone has an abrasion value of not less than 17; stone with a value of less than 14 would be considered poor.

The details of the abrasion test were included in B.S. 812 up to the 1951 edition. Nowadays the test has gone out of favour both in Britain and in the United States, and, in keeping with the tendency to test aggregate in bulk, a new abrasion value test was introduced in B.S. 812: 1960. Aggregate particles between 12·7 and 9·5 mm ($\frac{1}{2}$ and $\frac{3}{8}$ in.) whose total volume is 33,000 mm^3 are made up in a tray in a single layer, using a setting compound. The sample is subjected to abrasion in a standard machine, the grinding lap being turned 500 revolutions with Leighton Buzzard sand being fed continuously at a prescribed rate. The aggregate abrasion value is defined as the percentage loss in weight on abrasion, so that a high value denotes low resistance to abrasion.

The attrition (Deval) test also uses aggregate in bulk. Particles of known total weight are subjected to wear in an iron cylinder rotated 10,000 times at 30 to 33 revolutions per minute. The proportion of broken material expressed as a percentage represents the attrition value. The test can be performed on dry or wet aggregate, and the difference in results indicates the influence of the condition of the aggregate on its resistance to attrition.

An attrition value of about 7 to 8 is usually considered as the maximum permissible, but a shortcoming of the test is that it gives only small numerical differences between widely differing aggregates. The test was covered by A.S.T.M. Standard D2-33 (re-approved in 1968) but was discontinued in 1971.

An American test combining attrition and abrasion is the Los Angeles test; it is quite frequently used in other countries, too, because its results show extremely good correlation not only with the actual wear of aggregate when used in concrete but also with the compressive and flexural strengths of concrete made with the given aggregate. In this test, aggregate of specified grading is placed in a cylindrical drum, mounted horizontally, with a shelf inside. A charge of steel balls is added, and the drum is rotated a specified number of revolutions. The tumbling and dropping of the aggregate and the balls results in abrasion and attrition of the aggregate, and this is measured in the same way as in the attrition test.

The Los Angeles test can be performed on aggregates of different sizes, the same wear being obtained by an appropriate weight of the sample, and of the charge of steel balls, and by a suitable number of revolutions. The various quantities are prescribed by A.S.T.M. Standard C 131–69.

Table 3.7: *Average Test Values for British Rocks of Different Groups**

Rock group	Crushing strength MN/m^2	Crushing strength lb/in^2	Aggregate crushing value*	Abrasion value	Impact value	Attrition value† Dry	Attrition value† Wet	Specific gravity
Basalt	200	29,000	12	17·6	16	3·3	5·5	2·85
Flint	205	30,000	17	19·2	17	3·1	2·5	2·55
Gabbro	195	28,500	—	18·7	19	2·5	3·2	2·95
Granite	185	27,000	20	18·7	13	2·9	3·2	2·69
Gritstone	220	32,000	12	18·1	15	3·0	5·3	2·67
Hornfels	340	49,500	11	18·8	17	2·7	3·8	2·88
Limestone	165	24,000	24	16·5	9	4·3	7·8	2·69
Porphyry	230	33,500	12	19·0	20	2·6	2·6	2·66
Quartzite	330	47,500	16	18·9	16	2·5	3·0	2·62
Schist	245	35,500	—	18·7	13	3·7	4·3	2·76

* Courtesy of Dr. J. F. Kirkaldy.　† Lower value denotes a better quality.

Table 3.7 gives average values of crushing strength, aggregate crushing value, abrasion, impact, and attrition for the different rock groups of B.S. 812: 1967. It should be noted that the values for hornfels and schists are based on a few specimens only; these groups would appear to be better than they really are, presumably because only good quality hornfels and schists were tested. As a rule they are not suitable for use in concrete. Likewise, chalk is not included in the limestone group data as it is not generally suitable as a concrete aggregate.

As far as the crushing strength is concerned, basalt is extremely variable, fresh basalts with little olivine reaching some 400 MN/m^2 (60,000 lb/in^2), while decomposed basalt at the other end of the scale may have a strength of no more than 100 MN/m^2 (15,000 lb/in^2). Limestone and porphyry show much less variation in strength, and in Britain porphyry has a good general performance—rather better than that of granites, which tend to be variable.

Table 3.8 *Reproducibility of Test Results on Aggregate*[3.40]

Test	Coefficient of variation, per cent	Number of samples to be tested to ensure 0·9 probability that mean will be—	
		within ± 3 per cent of true mean	within ± 10 per cent of true mean
Dry attrition	5·7	10	1
Wet attrition	5·6	9	1
Abrasion	9·7	28	3
Impact of prepared specimen	17·1	90	8
Impact of bulk aggregate	3·0	—	—
Crushing strength	14·3	60	6
Aggregate crushing value	1·8	1	—
Los Angeles test	1·6	1	—

(*Crown Copyright*)

An indication of the accuracy of the results of the different tests is given by Table 3.8, listing the number of samples to be tested in order to ensure a 0·9 probability that the mean value for the samples is within ±3 and also within ±10 per cent of the true mean.[3.7] The aggregate crushing value shows up as particularly consistent. On the other hand, the prepared specimens show a greater scatter of results than the bulk samples, which is of course to be expected. While the various tests described in this and succeeding sections give an indication of the quality of the aggregate, it is not possible to predict from the properties of aggregate the potential strength development of concrete made with the given aggregate, and indeed it is not yet possible to translate physical properties of aggregate into its concrete-making properties.

Specific Gravity
Since aggregate generally contains pores, both permeable and impermeable (*see* p. 125), the meaning of the term specific gravity has to be carefully defined, and there are indeed several types of specific gravity.

The *absolute* specific gravity refers to the volume of the solid material excluding all pores, and can, therefore, be defined as the ratio of the weight of the solid, referred to vacuum, to the weight of an equal volume of gas-free distilled water, both taken at a stated temperature. Thus, in order to eliminate the effect of totally enclosed impermeable pores the material has to be pulverized, and the test is both laborious and sensitive. Fortunately, it is not normally required in concrete technology work.

If the volume of the solid is deemed to include the impermeable pores, but not the capillary ones, the resulting specific gravity is prefixed by the word *apparent*. The apparent specific gravity is then the ratio of the weight of the aggregate dried in an oven at 100 to 110°C (212 to 230°F) for 24 hours to the weight of water occupying a volume equal to that of the solid including the impermeable pores. The latter weight is determined using a vessel which can be accurately filled with water to a specified volume. Thus, if the weight of the oven-dried sample is D, the weight of the vessel full of water is B, and the weight of the vessel with the sample and topped up with water is A, then the weight of the water occupying the same volume as the solid is $B - (A-D)$. The apparent specific gravity is then

$$\frac{D}{B - A + D}.$$

The vessel referred to earlier, and known as a pycnometer, is usually a one-litre jar with a watertight metal conical screwtop having a small hole at the apex. The pycnometer can thus be filled with water so as to contain precisely the same volume every time.

Calculations with reference to concrete are generally based on the saturated surface-dry condition of the aggregate (*see* p. 128) because the water contained in *all* the pores in the aggregate does not take part in the chemical reactions of cement and can, therefore, be considered as part of the aggregate. Thus, if a sample of the saturated and surface-dry aggregate weighs C, the gross apparent specific gravity is

$$\frac{C}{B - A + C}.$$

This is the specific gravity most frequently and easily determined and necessary for calculations of yield of concrete or of the quantity of aggregate required for a given volume of concrete.

The apparent specific gravity of aggregate depends on the specific gravity of the minerals of which the aggregate is composed and also on the amount of voids. The majority of natural aggregates have a specific gravity of between 2·6 and 2·7, and the range of values is given in Table 3.9. The values for artificial aggregates extend from considerably below to very much above this range (*see* Chapter 9).

As mentioned earlier, specific gravity of aggregate is used in the calculation of quantities but the actual value of the specific gravity of aggregate is not a measure of its quality. Thus the value of specific gravity should not be specified unless we are dealing with a material of a given petrological character when a variation in specific gravity would reflect the porosity of the particles. An exception to this is the case of mass construction, such as a gravity dam, where a minimum density of concrete is essential for the stability of the structure.

Table 3.9: *Apparent Specific Gravities of Different Rock Groups*[3.7]

Rock group	Average specific gravity	Range of specific gravities
Basalt	2·80	2·6–3·0
Flint	2·54	2·4–2·6
Granite	2·69	2·6–3·0
Gritstone	2·69	2·6–2·9
Hornfels	2·82	2·7–3·0
Limestone	2·66	2·5–2·8
Porphyry	2·73	2·6–2·9
Quartzite	2·62	2·6–2·7

(*Crown Copyright*)

Bulk Density

It is well known that in the metric system the density of a material is numerically equal to its specific gravity although, of course, the latter is a ratio while density is expressed in kilogrammes per litre. However, in concrete practice, expressing the density in kilogrammes per cubic metre is more common. In the Imperial system, specific gravity has to be multiplied by the unit weight of water (approximately 62·4 lb/ft³) in order to be converted into absolute density (specific weight) expressed in pounds per cubic foot.

This absolute density, it must be remembered, refers to the volume of the individual particles only, and of course it is not physically possible to pack these particles so that there are no voids between them. When aggregate is to be actually batched by volume it is necessary to know the weight of aggregate that would fill a container of unit volume. This is known as the *bulk density* of aggregate, and this density is used to convert quantities by weight to quantities by volume.

The bulk density clearly depends on how densely the aggregate is packed, and it follows that for a material of a given specific gravity the bulk density depends on the size distribution and shape of the particles: particles all of one size can be packed to a limited extent but smaller particles can be added in the voids between the larger ones, thus increasing the bulk density of the packed material. The shape of the particles greatly affects the closeness of packing that can be achieved.

For a coarse aggregate of given specific gravity, a higher bulk density means that there are fewer voids to be filled by sand and cement, and the bulk density test has at one time been used as a basis of proportioning of mixes.

The actual bulk density of aggregate depends not only on the various characteristics of the material which determine the potential degree of packing, but also on the actual compaction achieved in a given case. For instance, using spherical particles all of the same size, the densest packing is achieved when their centres lie at the apexes of imaginary tetrahedra. The bulk density is then 0·74 of the specific weight of the material. For the loosest packing, the centres of spheres are at the corners of imaginary cubes and the bulk density is only 0·52 of the specific weight of the solid.

Thus for test purposes the degree of compaction has to be specified. B.S. 812: 1967 recognizes two degrees: loose (or uncompacted) and compacted. The test is performed in a metal cylinder of prescribed diameter and depth, depending on the maximum size of the aggregate and also on whether compacted or uncompacted bulk density is being determined.

For the determination of loose bulk density, the dried aggregate is gently placed in the container to overflowing and then levelled by rolling a rod across the top. In order to find the compacted or rodded bulk density, the container is filled in three stages, each third of the volume being tamped a prescribed number of times with a 16 mm ($\frac{5}{8}$ in.) diameter round-nosed rod. Again, the overflow is removed. The nett weight of the aggregate in the container divided by its volume then represents the bulk density for either degree of compaction.

Knowing the apparent specific gravity for the saturated and surface-dry condition, ρ, the voids ratio can be calculated from the expression

$$\text{voids ratio} = 1 - \frac{\text{bulk density}}{\rho \times \text{unit weight of water}}.$$

If the aggregate contains surface water it will pack less densely owing to the bulking effect. This is discussed on p. 130. Moreover, the bulk density as determined in the laboratory may not be directly suitable for conversion of weight to volume of aggregate for purposes of volume batching as the degree of compaction in the laboratory and on the site may not be the same.

The bulk density of aggregate is of interest in connexion with the use of lightweight and heavy aggregates (*see* Chapter 9).

Porosity and Absorption of Aggregate

The presence of internal pores in the aggregate particles was mentioned in connexion with the specific gravity of aggregate, and indeed the characteristics of these pores are very important in the study of its properties. The porosity of aggregate, its permeability, and absorption, influence such properties of aggregate as the bond between it and the cement paste, the resistance of concrete to freezing and thawing, as well as its chemical stability and resistance to abrasion. As stated earlier, the apparent specific gravity of aggregate also depends on its porosity and, as a consequence, the yield of concrete for a given weight of aggregate is affected.

The pores in aggregate vary in size over a wide range, the largest being large enough to be seen under a microscope or even with the naked eye, but even the smallest aggregate pores are generally larger than the gel pores in the cement paste. Pores smaller than 4 μm are of special interest as they are generally believed to affect the durability of aggregates subjected to alternating freezing and thawing (*see* p. 407).

Some of the aggregate pores are wholly within the solid; others open on to the surface of the particle. The cement paste, because of its viscosity, cannot penetrate to a great depth any but the largest of the aggregate pores, so that it is the gross volume of the particle that is considered solid for the purpose of calculating the aggregate content in concrete. However, water can enter the pores, the amount and rate of penetration depending on their size, continuity and total volume. The order of porosity of some common rocks is given in Table 3.10, and since aggregate represents some three-quarters of the volume of concrete it is

Table 3.10: *Porosity of Some Common Rocks*

Rock group	Porosity, per cent
Gritstone	0·0–48·0
Quartzite	1·9–15·1
Limestone	0·0–37·6
Granite	0·4–3·8

clear that the porosity of aggregate materially contributes to the overall porosity of concrete.

When all the pores in the aggregate are full it is said to be saturated and surface-dry. If aggregate in this condition is allowed to stand free in dry air, e.g. in the laboratory, some of the water contained in the pores will evaporate and the aggregate will be less than saturated, i.e. air-dry. Prolonged drying in an oven would reduce the moisture content of the aggregate still further until, when no moisture whatever is left, the aggregate is said to be bone-dry. These various stages are shown diagrammatically in Fig. 3.4, and some typical values of absorption are given in Table 3.11.

The water absorption of aggregate is determined by measuring the increase in weight of an oven-dried sample when immersed in water for 24 hours (the surface water being removed). The ratio of the increase in weight to the weight of the dry sample, expressed as a percentage, is termed absorption. Standard procedures are prescribed in B.S. 812: 1967.

Table 3.11: *Typical Values of Absorption of Different Aggregates*[3.8]

Aggregate size and type	Shape	Moisture contained in air-dry aggregate as a percentage of dry weight per cent	Absorption (moisture contained in saturated and surface-dry aggregate as a percentage of dry weight) per cent
19·0–9·5 mm ($\frac{3}{4}$–$\frac{3}{8}$ in.) Thames Valley river gravel	Irregular	0·47	2·07
9·5–4·8 mm ($\frac{3}{8}$–$\frac{3}{16}$ in.) Thames Valley river gravel	,,	0·84	3·44
4·8–2·4 mm ($\frac{3}{16}$ in.—No. 7)	,,	0·50	3·15
2·4–1·2 mm (No. 7–14) Thames	,,	0·30	2·90
1·2 mm–600 μm (No. 14–25) Valley	,,	0·30	1·70
600–300 μm (No. 25–52) river	,,	0·40	1·10
300–150 μm (No. 52–100) sand	,,	0·50	1·25
150–75 μm (No. 100–200)	,,	0·60	1·60
4·8 mm–150 μm ($\frac{3}{16}$ in.—No. 100) Thames Valley river sand zone 2	,,	0·80	1·80
19·0–9·5 mm ($\frac{3}{4}$–$\frac{3}{8}$ in.) Test river gravel	,,	1·13	3·30
9·5–4·8 mm ($\frac{3}{8}$–$\frac{3}{16}$ in.) Test river gravel	,,	0·53	4·53
19·0–9·5 mm ($\frac{3}{4}$–$\frac{3}{8}$ in.) Bridport gravel	Rounded	0·40	0·93
9·5–4·8 mm ($\frac{3}{8}$–$\frac{3}{16}$ in.) Bridport gravel	,,	0·50	1·17
19·0–9·5 mm ($\frac{3}{4}$–$\frac{3}{8}$ in.) Mountsorrel granite	Angular	0·30	0·57
9·5–4·8 mm ($\frac{3}{8}$–$\frac{3}{16}$ in.) Mountsorrel granite	,,	0·45	0·80
19·0–9·5 mm ($\frac{3}{4}$–$\frac{3}{8}$ in.) crushed limestone	,,	0·15	0·50
9·5–4·8 mm ($\frac{3}{8}$–$\frac{3}{16}$ in.) crushed limestone	,,	0·20	0·73
850–600 μm (No. 18–25) Leighton Buzzard standard sand	Rounded	0·05	0·20

Some typical values of absorption of different aggregates are given in Table 3.11, based on Newman's[3.8] data. The moisture content in the air-dry condition is also tabulated. It may be noted that gravel has generally a higher absorption than crushed rock of the same petrological character since weathering results in the outer layer of the gravel particles being more porous and absorbent.

Although there is no clear-cut relation between the strength of concrete and the water absorption of aggregate used, the pores at the surface of the particle affect the bond between the aggregate and the cement paste, and may thus exert some influence on the strength of concrete.

Normally, it is assumed that at the time of setting of concrete the aggregate is in a saturated and surface-dry condition. If the aggregate is batched in a dry condition it is assumed that sufficient water will be absorbed from the mix to bring the aggregate to a saturated condition, and this absorbed water is not included in the nett or effective mixing water. It is possible, however, that when dry aggregate is used the particles become quickly coated with cement paste which prevents further ingress of water necessary for saturation. This is particularly so with coarse aggregate, where water has further to travel from the surface of the

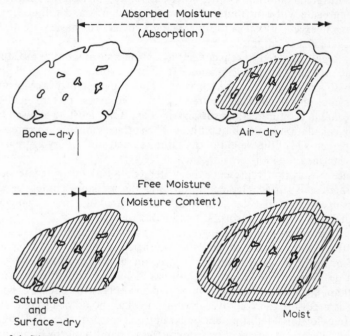

Fig. 3.4. *Diagrammatic representation of moisture in aggregate*

particle. As a result, the effective water/cement ratio is higher than would be the case had full absorption of water by the aggregate been possible. This effect is significant mainly in rich mixes where rapid coating of aggregate can take place; in lean, wet mixes the saturation of aggregate proceeds undisturbed. In practical cases the actual behaviour of the mix is affected also by the order of feeding the ingredients into the mixer.

The absorption of water by aggregate results also in some loss of workability with time, but beyond about 15 minutes the loss becomes small.

Since absorption of water by dry aggregate slows down or is stopped owing to the coating of particles with cement paste, it is often useful to determine the quantity of water absorbed in 10 to 30 minutes instead of the total water absorption, which may never be achieved in practice.

Moisture Content of Aggregate

It was mentioned in connexion with the specific gravity that in fresh concrete the volume occupied by the aggregate is the volume of the particles including all the pores. If no water movement into the aggregate is to take place the pores must be full of water, i.e. the aggregate must be in a saturated condition. On the other hand, any water on the surface of the aggregate will contribute to the water in the mix and will occupy a volume in excess of that of the aggregate particles. The basic state of the aggregate is then saturated and surface-dry.

Aggregate exposed to rain collects a considerable amount of moisture on the surface of the particles, and, except at the surface of the stockpile, keeps this moisture over long periods. This is particularly true of fine aggregate, and the surface or free moisture (in excess of that held by aggregate in a saturated and surface-dry condition) must be allowed for in the calculation of batch quantities. The surface moisture is expressed as a percentage of the weight of the saturated and surface-dry aggregate, and is termed the moisture content.

Since absorption represents the water contained in aggregate in a saturated and surface-dry condition, and the moisture content is the water in excess of that state, the total water content of a moist aggregate is equal to the sum of absorption and moisture content.

As the moisture content of aggregate changes with weather and changes also from one part of a stockpile to another, the value of the moisture content has to be determined frequently and a number of methods has been developed. The oldest one consists simply of finding the loss in weight of an aggregate sample when dried on a tray over a source of heat. Care is necessary to avoid over-drying: the sand should be brought to a just free-flowing condition, and must not be heated further. This stage can be determined by feel or by forming the sand into a pile by means of a conical mould; when the mould has been removed the material should slump freely. If the sand has acquired a brownish tinge, this is a sure sign that excessive drying has taken place. This method of determining the moisture content of aggregate, colloquially referred to as the "frying-pan method," is simple, can be used in the field and is quite reliable.

In the laboratory the moisture content of aggregate can be determined by means of a pycnometer. The apparent specific gravity of the aggregate on a saturated and surface-dry basis, ρ, must be known. Then, if B is the weight of the pycnometer full of water, C the weight of the moist sample and A the weight of the pycnometer with the sample and topped up with water, the moisture content of the aggregate is

$$\left[\frac{C}{A - B}\left(\frac{\rho - 1}{\rho}\right) - 1\right] \times 100.$$

The test is slow and requires great care in execution (e.g. all air must be expelled from the sample) but can yield accurate results.

In the siphon can test[3.9] the volume of water displaced by a known weight of moist aggregate is measured, the siphon making this determination more accurate. Preliminary calibration for each aggregate is required as the results depend on its specific gravity but once this has been done the test is rapid and accurate.

The moisture content of aggregate can also be found using a steelyard moisture meter: the moist aggregate is added to a vessel containing a fixed amount of water and suspended at one end of a steelyard until it balances. We measure thus the quantity of water that has to be replaced by the moist aggregate for a constant weight and *total* volume. For this condition it can be shown that the amount of displaced water is proportional to the moisture content of the aggregate. A calibration curve for any aggregate used has to be obtained. The moisture content can be determined with an accuracy of $\frac{1}{2}$ per cent.

In the buoyancy meter test[3.10] the moisture content of the aggregate of known specific gravity is determined from the apparent loss in weight on immersion in water. The balance can read the moisture content directly if the size of the sample is adjusted, according to the specific gravity of the aggregate, to such a value that a saturated and surface-dry sample has a standard weight when immersed. The test is rapid and gives the moisture content to the nearest $\frac{1}{2}$ per cent. A simple version of the test is prescribed by A.S.T.M. Standard C 70–72.

Numerous other methods have been developed. For instance moisture can be removed by burning the aggregate with methyl alcohol, the resulting loss in weight of the sample being measured. There are also proprietary meters based on the measurement of pressure of gas formed in a closed vessel by the reaction of calcium carbide with the moisture in the sample.

Electrical devices which give instantaneous or continuous reading of the moisture content of aggregate in a storage bin, on the basis of the variation of resistance or capacitance with a change in the moisture content of the aggregate, have been developed. In some batching plants, meters of this type are used in automatic devices which regulate the quantity of water to be added to the mixer but an accuracy greater than 1 per cent of moisture cannot in practice be achieved. Microwave meters have recently been developed but these are very expensive.

It can be seen that a great variety of tests is available but, however accurate the test, its result is significant only if a representative sample has been used. Furthermore, if the moisture content of aggregate varies between adjacent parts of a stockpile, the adjustment of mix proportions becomes laborious. Since the variation in moisture content occurs mainly in the vertical direction from a water-logged bottom of a pile to its drying or dry surface, care in laying out of stockpiles is necessary: storing in horizontal layers, having at least two stockpiles and allowing each pile to drain before use, and not using the bottom 300 mm (12 in.) or so, all help to keep the variation in moisture content to a minimum. Coarse aggregate holds very much less water than sand, has a less variable moisture content, and generally causes fewer difficulties.

Bulking of Sand

The presence of moisture in aggregate necessitates correction of the actual mix proportions: the weight of water added to the mix has to be decreased by the weight of the free moisture in the aggregate, and the weight of the aggregate must be increased by a like amount. In the case of sand there is a second effect of the presence of moisture: bulking. This is the increase in the volume of a given weight of sand caused by the films of water pushing the sand particles apart. While bulking *per se* does not affect the proportioning of materials by weight, in the case of volume batching bulking results in a smaller weight of sand occupying the fixed volume of the measuring box. For this reason the mix becomes deficient in sand and appears "stony", and the concrete may be prone to segregation and honeycombing. Also, the yield of concrete is reduced. The remedy, of course, lies in increasing the apparent volume of sand to allow for bulking.

The extent of bulking depends on the percentage of moisture present in the sand and on its fineness. The increase in volume relative to that occupied by a saturated and surface-dry sand increases with an increase in the moisture content of the sand up to a value of some 5 to 8 per cent, when bulking of 20 to 30 per cent occurs. Upon further addition of water, the films merge and the water moves into the voids between the particles so that the total volume of sand decreases until when fully saturated (flooded) its volume is approximately the same as the volume of dry sand for the same method of filling the container. This is apparent from Fig. 3.5, which also shows that finer sand bulks considerably more and reaches maximum bulking at a higher water content than does coarse sand. Extremely fine sand has been known to bulk as much as 40 per cent at a moisture content of 10 per cent, but such a sand is in any case unsuitable for the manufacture of good quality concrete.

Coarse aggregate shows only a negligible increase in volume due to the presence of free water, as the thickness of moisture films is very small compared with the particle size.

Since the volume of saturated sand is the same as that of dry sand, the most convenient way of determining bulking is by measuring the decrease in volume of the given sand when inundated. A container of known volume is filled with loosely packed moist sand. The sand is then tipped out, the container is partially filled with water and the sand is

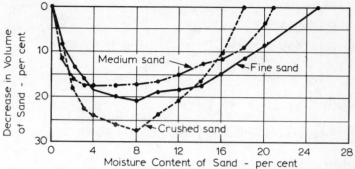

Fig. 3.5. *Decrease in true volume of sand due to bulking (for a constant volume of moist sand)*

gradually fed back, with stirring and rodding to expel all air bubbles. The volume of sand in the saturated state, V_s, is now measured. If V_m is the initial volume of the sand (i.e. the volume of the container), then bulking is given by—

$$\frac{V_m - V_s}{V_s}$$

With volume batching, bulking has to be allowed for by increasing the total volume of (moist) sand used. Thus volume V_s is multiplied by a factor—

$$1 + \frac{V_m - V_s}{V_s} = \frac{V_m}{V_s}$$

sometimes known as the bulking factor, and a graph of bulking factor against moisture of three typical sands is shown in Fig. 3.6.

The bulking factor can also be found from the bulk densities of dry and moist sand, D_d and D_m respectively, and the moisture content per unit volume of sand, m/V_m. The bulking factor is then—

$$\frac{D_d}{D_m - \dfrac{m}{V_m}}.$$

Since D_d represents a ratio of the weight of dry sand, w, to its bulk volume V_s (the volumes of dry and inundated sand being the same)—

$$\frac{D_d}{D_m - \dfrac{V_m}{m}} = \frac{\dfrac{w}{V_s}}{\dfrac{w + m}{V_m} - \dfrac{m}{V_m}} = \frac{V_m}{V_s}$$

i.e. the two factors are identical.

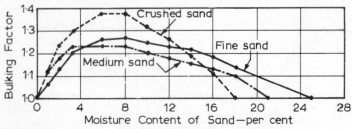

Fig. 3.6. *Bulking factor for sands with different moisture contents*

Deleterious Substances in Aggregate

There are three broad categories of deleterious substances that may be found in aggregates: *impurities* which interfere with the processes of hydration of cement; *coatings* preventing the development of good bond between aggregate and the cement paste; and certain individual particles which are *weak* or *unsound* in themselves. All or part of an aggregate can also be harmful through the development of chemical reactions between the aggregate and the cement paste: these chemical reactions are discussed on page 139.

Organic Impurities

Natural aggregates may be sufficiently strong and resistant to wear and yet they may not be satisfactory for concrete-making if they contain organic impurities which interfere with the chemical reactions of hydration. The organic matter found in aggregate consists usually of products of decay of vegetable matter (mainly tannic acid and its derivatives) and appears in the form of humus or organic loam. Such materials are more likely to be present in sand than in coarse aggregate, which is easily washed.

Not all organic matter is harmful and it is best to check its effects by making actual test cubes. Generally, however, it saves time to ascertain first whether the amount of organic matter is sufficient to warrant further tests. This is done by the so-called colorimetric test of A.S.T.M. Standard C 40–66. The acids in the sample are neutralized by a 3 per cent solution of NaOH, prescribed quantities of aggregate and of solution being placed in a bottle. The mixture is vigorously shaken to allow the intimate contact necessary for chemical action, and then left to stand for 24 hours, when the organic content can be judged by the colour of the solution: the greater the organic content, the darker the colour. If the colour of the liquid above the test sample is not darker than the standard yellow colour defined by the standard, the sample can be assumed to contain only a harmless amount of organic impurities.

If the observed colour is darker than the standard, i.e. if the solution appears brownish or brown, the aggregate has a rather high organic content, but this does not necessarily mean that the aggregate is not fit for use in concrete. The organic matter present may not be harmful to concrete or the colour may be due to some iron-bearing minerals. For this reason, further tests are necessary: concrete cubes are made using the suspected aggregate and their strength is compared with concrete of the same mix proportions but made with aggregate of known quality.

In earlier editions, B.S. 812 contained the colorimetric test, but B.S. 812: 1967 specifies the measurement of the pH values of cement mortars under standard conditions. The test is rather laborious and is suitable for laboratory work only. The change was made because the colorimetric test was judged to be unreliable but it is nevertheless useful as a preliminary assessment of suitability of aggregate: if no colour change in excess of that specified is observed, no further tests are necessary unless contamination with industrial effluents has occurred.

In some countries, the quantity of organic matter in aggregate is determined from the loss of weight of a sample on treating with hydrogen peroxide.

It is interesting to note that in some cases the effects of organic impurities may be only temporary. In one investigation[3.11] concrete made with a sand containing organic matter had a 24-hour strength equal to 53 per cent of the strength of similar concrete made with clean sand. At 3 days this ratio rose to 82 per cent, then to 92 per cent at 7 days, and at 28 days equal strengths were recorded.

Clay and Other Fine Material
Clay may be present in aggregate in the form of surface coatings which interfere with the bond between aggregate and the cement paste. Since good bond is essential to ensure a satisfactory strength and durability of concrete, the problem of clay coatings is an important one.

There are two more types of fine material which can be present in aggregate: silt and crusher dust. Silt is a material between 0·002 mm and 0·06 mm, reduced to this size by natural processes of weathering; silt may thus be found in aggregate won from natural deposits. On the other hand, crusher dust is a fine material formed during the process of comminution of rock into crushed stone or, less frequently, of gravel into crushed sand. In a properly laid out processing plant this dust should be removed by washing. Other soft or loosely adherent coatings can also be removed during the processing of the aggregate. Well-bonded coatings cannot be so removed, but if they are chemically stable and have no deleterious effect there is no objection to the use of aggregate with such a coating although shrinkage may be increased. However, aggregates with chemically re-active coatings, even if physically stable, can lead to serious trouble.

Silt and fine dust may form coatings similar to those of clay, or may be present in the form of loose particles not bonded to the coarse aggregate. Even when they are in the latter form, silt and fine dust should not be present in excessive quantities because, owing to their fineness and therefore large surface area, silt and fine dust increase the amount of water necessary to wet all the particles in the mix.

In view of the above, it is necessary to control the clay, silt and fine dust contents of aggregate. B.S. 882: 1965 limits the content of all three materials together to not more than—

15 per cent by weight in crushed stone sand,
3 per cent by weight in natural or crushed gravel sand, and
1 per cent by weight in coarse aggregate.

A.S.T.M. Standard C 33–71a lays down similar requirements, but distinguishes between concrete subject to abrasion and other concretes. In the former case, the amount of material passing a 75 μm (No. 200) B.S. test sieve is limited to 3 per cent of the weight of sand, instead of the 5 per cent value permitted for other concretes. The corresponding value for coarse aggregate is laid down as 1 per cent.

The clay content is specified separately as 1 per cent in fine and $\frac{1}{4}$ per cent in coarse aggregate. It may be noted that different test methods are prescribed in different specifications so that the results are not directly comparable.

The clay, silt and fine dust content of fine aggregate can be determined by the sedimentation method described in B.S. 812: 1967. The sand sample is placed in a sodium oxalate solution in a stoppered jar and rotated with the axis of the jar horizontal for 15 minutes at approximately 80 revolutions per minute. The fine solids become dispersed and the amount of suspended material is then measured by means of an Andreason pipette. A simple calculation gives the percentage of clay, fine silt and fine dust in the sand, the separation size being 20 μm.

A similar method, with suitable modifications, can be used for coarse aggregate, but it is simpler to wet-sieve the aggregate on a 75 μm (No. 200) B.S. test sieve, as prescribed in B.S. 812: 1967 and A.S.T.M. Standard C 117–69. This type of sieving is resorted to because fine dust or clay adhering to larger particles would not be separated in ordinary dry sieving. In wet sieving, on the other hand, the aggregate is placed in water and agitated sufficiently vigorously for the finer material to be brought into suspension. By decantation and sieving, all material smaller than a 75 μm (No. 200) B.S. test sieve can be removed. To protect this sieve from damage by large particles during decantation a 1·2 mm (No. 14) sieve is placed above the 75 μm (No. 200) sieve.

For natural sands and crushed gravel sands there is also a field test available which can be performed quite easily and rapidly, with very little laboratory equipment. 50 ml of an approximately 1 per cent solution of common salt in water is placed in a 250 ml B.S. measuring cylinder. Sand, as received, is added until its level reaches the 100 ml mark and more solution is then added until the total volume of the mixture in the cylinder is 150 ml. The cylinder is now covered with the palm of the hand, shaken vigorously, repeatedly turned upside down and then allowed to stand for 3 hours. The silt which became dispersed on shaking will now settle in a layer above the sand, and the height of this layer can be expressed as a percentage of the height of the sand below.

It should be remembered that this is a volumetric ratio, which cannot easily be converted to a ratio by weight since the conversion factor depends on the fineness of the material. It has been suggested that for natural sand the weight ratio is obtained by multiplying the volumetric ratio by a factor of $\frac{1}{4}$, the corresponding figure for crushed gravel sand being $\frac{1}{2}$ but with some aggregates an even wider variation is obtained. These conversions are not reliable, and B.S. 882: 1965 recommends that when the volumetric content exceeds 8 per cent tests by the more accurate methods, described earlier, should be made.

Salt Contamination

Sand won from the seashore or from a river estuary contains salt, and opinions differ as to the suitability of such a sand for use in concrete. The simplest course is to wash the sand in fresh water, but special care is required with deposits just above the high-water mark in which large quantities of salt, sometimes over 6 per cent of weight of sand, may be found. Generally, sand from the sea bed, washed even in sea water, does not contain harmful quantities of salt.

The British Code of Practice for the Structural Use of Concrete CP 110: 1972 specifies the maximum content of anhydrous calcium chloride (or equivalent) in marine aggregate as 1 per cent of the weight of cement to be used, but only 0·1 per cent with aluminous cement or in prestressed concrete unless the tendons are permanently protected.

If salt is not removed it will absorb moisture from the air and cause efflorescence—unsightly white deposits on the surface of the concrete (*see* also p. 400). A slight corrosion of reinforcement may also result, but this is not believed to progress to a dangerous degree, especially when the concrete is of good quality and adequate cover to reinforcement is provided. No trouble need be expected in mass concrete structures.

Sea sands are often extremely fine and the grading of any new sand should be carefully checked. Sea-dredged coarse aggregate may have a large shell content. This has no adverse effect on strength but workability of concrete made with aggregate having a large shell content is slightly reduced.[3.44]

Unsound Particles

Tests on aggregate sometimes reveal that the majority of the component particles are satisfactory but that a few are unsound: the quantity of such particles must clearly be limited.

There are two broad types of unsound particles: those that fail to maintain their integrity, and those that lead to disruptive expansion on freezing or even on exposure to water. The disruptive properties are characteristic of certain rock groups, and will therefore be discussed in relation to the durability of aggregate in general (mainly in the next section). In this section, non-durable impurities only will be considered.

Shale and other particles of low density are regarded as unsound, and so are soft inclusions, such as clay lumps, wood, and coal, as they lead to pitting and scaling. If present in large quantities (over 2 to 5 per cent of the weight of the aggregate) these particles may adversely affect the strength of concrete and should certainly not be permitted in concrete which is exposed to abrasion.

Coal, in addition to being a soft inclusion, is undesirable for other reasons: it can swell, causing disruption of concrete, and if present in large quantities in a finely divided form, it can disturb the process of hardening of the cement paste. However, discrete particles of hard coal amounting to no more than $\frac{1}{4}$ per cent of the weight of the aggregate have no adverse effect on the strength of concrete.

The presence of coal and other materials of low density can be determined by flotation in a liquid of suitable specific gravity, as, for instance, by the method of A.S.T.M. Standard C 123–69. If the danger of pitting and scaling is not thought important, and strength of concrete is the main consideration, a trial mix should be made.

Mica should be avoided because in the presence of active chemical agents produced during the hydration of cement, alteration of mica to other forms may result. Also, free mica in fine aggregate, even in quantities of a few per cent of the weight of the aggregate, affects adversely the water requirement and hence the strength of concrete.[3.45] These facts should be borne in mind when materials such as china clay sand are considered for use in concrete. Gypsum and other sulphates must not be present.

Iron pyrites and marcasite represent the most common expansive inclusions in aggregate. These sulphides react with water and oxygen in the air to form a ferrous sulphate which subsequently decomposes to form the hydroxide, while the sulphate ions react with calcium aluminates in the cement. Surface staining of the concrete and pop-outs may result, particularly under warm and humid conditions.

Not all pyrites are reactive but, since the decomposition of pyrites takes place only in lime water, it is possible to test a suspect aggregate for reactivity by placing the material in a saturated solution of lime.[3.12] If the aggregate is reactive a blue-green gelatinous precipitate of ferrous sulphate appears within a few minutes, and on exposure to air this changes to brown ferric hydroxide. The absence of this reaction means that no staining need be feared. Lack of reactivity was found by Midgley[3.12] to be associated with the presence of a number of metal cations, while their absence makes the pyrites active. Generally, particles of pyrites likely to cause trouble are those between 5 and 10 mm (or $\frac{3}{16}$ and $\frac{3}{8}$ in.) in size.

The permissible quantities of unsound particles laid down by A.S.T.M. Standard C 33–71a are summarized in Table 3.12.

Table 3.12: *Permissible Quantities of Unsound Particles Prescribed by A.S.T.M. Standard C 33–71a*

Type of particles	Maximum content—per cent of weight	
	In fine aggregate	In coarse aggregate
Friable particles	3·0	5·0
Coal	0·5 to 1·0*	0·5 to 1·0*
Soft particles	—	5·0
Chert that will readily disintegrate	—	1·0 to 5·0**

* Depending on importance of appearance.
** Depending on exposure.

The majority of impurities discussed in the present section are found in natural aggregate deposits and are much less frequently encountered in crushed aggregate. However, some processed aggregates, such as mine tailings, can contain harmful substances. For instance, small quantities of lead soluble in limewater (e.g. 0·1 per cent of PbO by weight of aggregate) greatly delay the set and reduce the early strength of concrete; the long-term strength is unaffected.[3.46]

Soundness of Aggregate

This is the name given to the ability of aggregate to resist excessive changes in volume as a result of changes in physical conditions. Lack of soundness is thus distinct from expansion caused by the chemical reactions between the aggregate and the alkalis in cement.

The physical causes of large or permanent volume changes of aggregate are freezing and thawing, thermal changes at temperatures above freezing, and alternating wetting and drying.

Aggregate is said to be unsound when volume changes, induced by the above causes, result in deterioration of the concrete. This may range from local scaling to extensive surface cracking and to disintegration over a considerable depth, and can thus vary from no more than impaired appearance to a structurally dangerous situation.

Unsoundness is exhibited by porous cherts, especially the lightweight ones with a fine-textured pore structure, by some shales, by limestones with laminae of expansive clay, and by other particles containing clay minerals, particularly of the montmorillonite or illite group. For instance, an altered dolerite has been found to move as much as 0·0006 with wetting and drying, and concrete containing this aggregate might fail under conditions of alternating wetting and drying, and will certainly do so on freezing and thawing.

A test for soundness of aggregate is prescribed by the A.S.T.M. Standard C 88–71a. A sample of graded aggregate is subjected alternately to immersion in a saturated solution of sodium or magnesium sulphate (generally the more severe of the two) and drying in an oven. The formation of salt crystals in the pores of the aggregate tends to disrupt the particles, probably in a manner similar to the action of ice. The reduction in size of the particles, as shown by a sieve analysis, after a number of cycles of exposure denotes the degree of unsoundness. The test is no more than qualitative in predicting the behaviour of the aggregate under actual site conditions, and cannot be used as a basis of acceptance or rejection of unknown aggregates. Specifically, there is no clear reason why soundness as tested by A.S.T.M. Standard C 88–71a should be related to performance in concrete subjected to freezing and thawing.

Other tests consist of subjecting the aggregate to cycles of alternating freezing and thawing, and sometimes this treatment is applied to mortar or concrete made with the suspect aggregate. Unfortunately, none of the tests gives an accurate indication of the behaviour of aggregate under actual conditions of moisture and temperature changes above the freezing point.

Likewise, there are no tests which could satisfactorily predict the durability of aggregate in the concrete under conditions of freezing and thawing. The main reason for this is that the behaviour of aggregate is related to the presence of the surrounding cement paste, so that only a service record can satisfactorily prove the durability of aggregate.

Nevertheless, certain aggregates are known to be susceptible to frost damage and it is on these that our attention is centred. These are: porous cherts, shales, some limestones, particularly laminated limestones, and some sandstones. A common characteristic of these rocks with a poor record is their high absorption, but it should be emphasized that many durable rocks also exhibit high absorption (*see* Fig. 3.7).

For frost damage to occur there must exist critical conditions of water content and lack of drainage. These are governed, *inter alia*, by the size, shape and continuity of pores in the aggregate, because these characteristics of the pores control the rate and amount of absorption and the rate at which water can escape from the aggregate particle. Indeed, these features of the pores are more important than merely their total volume as reflected by the magnitude of absorption.

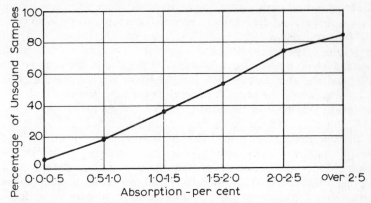

Fig. 3.7. *Distribution of sound and unsound aggregate samples as a function of absorption*[3.37]

It has been found that pores smaller than 4 to 5 μm are critical, for they are large enough to permit water to enter but not large enough to allow easy drainage under the pressure of ice. This pressure, in fully confined space at $-20°C$ ($-4°F$), may be as high as 200 MN/m² (29,000 lb/in²). Thus, if splitting of aggregate particles and disruption of the surrounding cement paste are to be avoided, flow of water towards unfilled pores within the aggregate particle or into the surrounding paste must be possible before the hydraulic pressure becomes high enough to cause disruption.

This argument illustrates the statement made earlier that the durability of aggregate cannot be fully determined other than when it is embedded in cement paste: the particle may be strong enough to resist the pressure of ice but expansion may cause disruption of the surrounding mortar.

It has been said that the pore size is an important factor in the durability of aggregate. In most aggregates, pores of different sizes are present so that we are really confronted with a pore size distribution. A means of expressing this quantitatively has been developed by Brunauer, Emmett and Teller.[3.13] The specific surface of the aggregate is determined from the amount of a gas sorbate required to form a layer one molecule thick over the entire internal surface of the aggregate pores. The total volume of the pores is measured by absorption, and the ratio of the volume of pores to their surface represents the hydraulic radius of the pores. This value, familiar from flow problems in hydraulics, gives an indication of the pressure required to produce flow.

Alkali-Aggregate Reaction

During the last thirty years some deleterious chemical reactions between the aggregate and the surrounding cement paste have been observed.

The most common reaction is that between the active silica constituents of the aggregate and the alkalis in cement. The reactive forms of silica are opal (amorphous), chalcedony (cryptocrystalline fibrous), and tridymite (crystalline). These reactive materials occur in: opaline or chalcedonic cherts, siliceous limestones, rhyolites and rhylotic tuffs, dacite and dacite tuffs, andesite and andesite tuffs, and phyllites.[3.29] The reaction starts with the attack on the siliceous minerals in the aggregate by the alkaline hydroxides derived from the alkalis (Na_2O and K_2O) in the cement. As a result, an alkali-silicate gel is formed, and alteration of the borders of the aggregate takes place. The gel is of the "unlimited swelling" type; it imbibes water with a consequent tendency to increase in volume. Since the gel is confined by the surrounding cement paste, internal pressures result and eventually lead to expansion, cracking and disruption of the cement paste. Thus expansion appears to be due to hydraulic pressure generated through osmosis, but expansion can also be caused by the swelling pressure of the still solid products of the alkali-silica reaction.[3.30] For this reason it is believed that it is the swelling of the hard aggregate particles that is most harmful to concrete. Some of the relatively soft gel is later leached out by water and deposited in the cracks already formed by the swelling of the aggregate.

While we can predict that with given materials an alkali-aggregate reaction will take place, it is not generally possible to estimate the deleterious effects from the knowledge of the quantities of the reactive materials alone. For instance, the actual reactivity of aggregate is affected by its particle size and porosity as these influence the area over which the reaction can take place. Since the quantity of alkalis depends on the cement only, their concentration at the reactive surface of aggregate will be governed by the magnitude of this surface. The minimum alkali content of cement at which expansive reaction may take place is 0·6 per cent of the soda equivalent. This is calculated from stoichiometry as the actual Na_2O content plus 0·658 times the K_2O content of the clinker. In exceptional cases, however, cements with an even lower alkali content have been known to cause expansion.[3.31] Within limits, the expansion of concrete made with a given reactive aggregate is greater the higher the alkali content of the cement and, for a given composition of cement, the greater its fineness.[3.32]

Other factors influencing the progress of the alkali-aggregate reaction include the availability of non-evaporable water in the paste and the permeability of the paste. Moisture is necessary and the reaction is accelerated under conditions of alternating wetting and drying. The temperature accelerates the reaction, at least in the range 10 to 38°C (50 to 100°F). It can thus be seen that various physical and chemical factors make the problem of alkali-aggregate reaction highly complex. In particular, the gel can change its constitution by absorption and thus exert a considerable pressure, while at other times diffusion of the gel out of the confined area takes place.[3.32]

It is not surprising, therefore, that although we know that certain types of aggregate tend to be reactive, there is no simple way of determining whether a given aggregate will cause excessive expansion due to reaction with alkalis in the cement. Service record has generally to be relied upon. If no record is available it is possible only to determine the potential reactivity of the aggregate but not to prove that reaction will take place. A quick chemical test is prescribed by A.S.T.M. Standard C 289–71: the reduction in the alkalinity of a normal solution of NaOH when placed in contact with pulverized aggregate at 80°C (176°F) is determined, and the amount of dissolved silica is measured. The interpretation of the result is in many cases not clear, but generally a potentially deleterious reaction is indicated if the plotted test result falls to the right of the boundary line of Fig. 3.8, reproduced from the A.S.T.M. Standard but based on Mielenz and Witte's paper.[3.33] However, potentially deleterious aggregates represented by points lying above the dashed line in Fig. 3.8 may be extremely reactive with alkalis so that a relatively low expansion may result. These aggregates should therefore be tested further to determine whether their reactivity is deleterious by the mortar bar test described below.

In the mortar bar test for the physical reactivity of aggregate, the suspected aggregate, crushed if need be and made up to a prescribed grading, is used in making special sand-cement mortar bars, using a cement with an equivalent alkali content of not less than 0·6 per cent. The bars are stored over water at 38°C (100°F), at which temperature the expansion is more rapid and usually higher than at higher or lower temperatures.[3.34] The reaction is also accelerated by the use of a fairly high water/cement ratio. The details of procedure are prescribed by A.S.T.M. Standard C 227–71.

The aggregate under test is considered harmful if it expands more than 0·05 per cent after 3 months or more than 0·1 per cent after 6 months

This test has shown a very good correlation with field experience, but a considerable time is required before judgement on the aggregate can be pronounced. On the other hand, as mentioned earlier, the results of the chemical test are often not conclusive. Likewise, petrographic examination, although a useful tool in identifying the mineral constituents, cannot establish that a given mineral will result in abnormal expansion. A rapid and conclusive test for aggregate reactivity is thus still to be developed.

It has been found that expansion due to the alkali-aggregate reaction can be reduced or eliminated by addition to the mix of reactive silica in a finely powdered form. This apparent paradox can be explained by reference to Fig. 3.9, showing the relation between the expansion of a

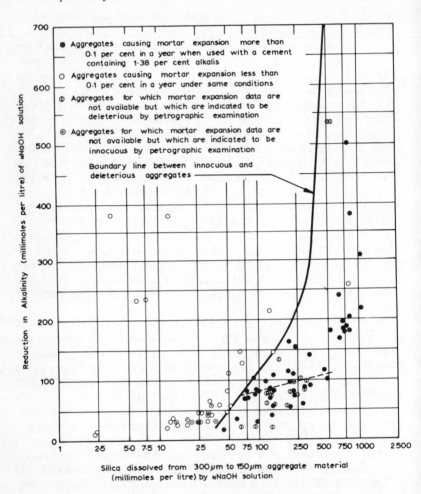

Fig. 3.8. *Results of chemical test of A.S.T.M. Standard C 289–71*

mortar bar and the content of reactive silica of size between 850 and 300 μm (No. 18 and No. 52) sieves, i.e. not in a powdered form. In the range of low silica contents, the greater quantity of silica for a given amount of alkalis increases expansion, but with higher values of silica content the situation is reversed: the greater the surface area of the reactive aggregate the lower the quantity of alkalis available per unit of this area, and the less alkali-silica gel can be formed. On the other hand, owing to the extremely low mobility of calcium hydroxide, only that adjacent to the surface of the aggregate is available for reaction, so that the quantity of calcium hydroxide per unit area of aggregate is independent of the magnitude of the total surface area of the aggregate. Thus, increasing the surface area increases the calcium hydroxide/alkali ratio of the solution at the boundary of the aggregate. Under such circumstances an innocuous (non-expanding) calcium alkali silicate product is formed.[3.36]

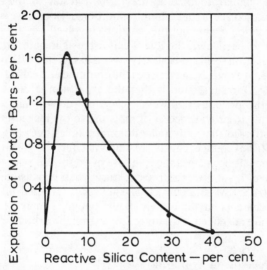

Fig. 3.9. *Relation between expansion after 224 days and reactive silica content in the aggregate*[3.35]

By a similar argument, finely divided siliceous material added to the coarse reactive particles already present would reduce expansion, although the reaction with alkalis still takes place. These pozzolanic additions, such as crushed pyrex glass or fly ash, have indeed been found effective in reducing the penetration of the coarser aggregate particles. The performance of any pozzolana in this respect should be tested following the A.S.T.M. Standard C 441–69. With a sufficient amount of added silica, the initial reaction lowers the alkali concentration to that obtained with a low alkali cement when no additive is present. It is essential, however, that a sufficient amount of pulverized silica be added; it is generally recommended that 20 g of reactive silica be added for each gram of alkali in excess of 0·5 per cent of the weight of the cement.[3.30] Thus the amount of pozzolana is quite large. Inadequate amounts can actually aggravate the situation and increase expansion if a particularly bad silica-alkali ratio is reached (cf. Fig. 3.9). A side effect of the addition of pozzolanas which has to be borne in mind is the increase in the water requirement of the mix.

The alkali-aggregate reaction of the type described has fortunately not been encountered in Great Britain, but is widespread in many other countries, notably North America, Scandinavia, India, Australia and New Zealand.

Another type of deleterious aggregate reaction is that between some dolomitic limestone aggregates and alkalis in the cement. Expansion of concrete, similar to that occurring as a result of the alkali-silica reaction, takes place under humid conditions. Tests have shown that de-dolomitization occurs but the reactions involved are still imperfectly understood; in particular, the rôle of clay in the aggregate is not clear but expansive reaction seems to be nearly always associated with the presence of clay. Also, in expansive aggregates the dolomite and calcite crystals are very fine.[3.47] One suggestion is that the expansion is due to moisture uptake by the previously unwetted clay, the de-dolomitization being necessary only to provide access of moisture to the locked-in clay.[3.48] It should be stressed that only *some* dolomitic limestones (e.g. in certain parts of Ontario) cause expansive reaction in concrete. No simple test to identify them has been developed; in case of doubt, help can be obtained from an investigation of the rock texture, rock expansion in sodium hydroxide (A.S.T.M. Standard C 586–69) or when used in test beams made with cements having a high alkali content. It may be noted that pozzolanas are not effective in controlling the alkali-carbonate expansion.

Thermal Properties of Aggregate

There are three thermal properties of aggregate that may be significant in the performance of concrete: coefficient of thermal expansion, specific heat, and conductivity. The last two are of importance in mass concrete or where insulation is required, but not in ordinary structural work, and are discussed in the section dealing with the thermal properties of concrete (*see* p. 432).

The coefficient of thermal expansion of aggregate influences the value of this coefficient for concrete containing the given aggregate: the higher the coefficient of the aggregate the higher the coefficient of the concrete, but the latter depends also on the aggregate content in the mix and on the mix proportions in general.

There is, however, another aspect of the problem. It has been suggested that if the coefficients of thermal expansion of the coarse aggregate and of the cement paste differ too much, a large change in temperature may introduce differential movement and a break in the bond between the aggregate particles and the surrounding paste. However, possibly because the differential movement is affected also by other forces, such as those due to shrinkage, a large difference between the coefficients is not necessarily detrimental when the temperature does not vary outside the range of, say, 4 to 60°C (40 to 140°F). Nevertheless, when the two co-efficients differ by more than 5.5×10^{-6} per °C (3×10^{-6} per °F) the durability of concrete subjected to freezing and thawing may be affected.

The coefficient of thermal expansion can be determined by means of a dilatometer devised by Verbeck and Hass[3.14] for use with both fine and coarse aggregate. The linear coefficient of thermal expansion varies with the type of parent rock, the range for the more common rocks being about 0.9×10^{-6} to 16×10^{-6} per °C (0.5×10^{-6} to 8.9×10^{-6} per °F), but the majority of aggregates lies between approximately 5×10^{-6} to 13×10^{-6} per °C (3×10^{-6} and 7×10^{-6} per °F) (see Table 3.13). For

Table 3.13: *Linear Coefficient of Thermal Expansion of Different Rock Types*[3.39]

Rock type	Thermal coefficient of linear expansion	
	10^{-6} per °C	10^{-6} per °F
Granite	1.8 to 11.9	1.0 to 6.6
Diorite, andesite	4.1 to 10.3	2.3 to 5.7
Gabbro, basalt, diabase	3.6 to 9.7	2.0 to 5.4
Sandstone	4.3 to 13.9	2.4 to 7.7
Dolomite	6.7 to 8.6	3.7 to 4.8
Limestone	0.9 to 12.2	0.5 to 6.8
Chert	7.4 to 13.1	4.1 to 7.3
Marble	1.1 to 16.0	0.6 to 8.9

hydrated Portland cement paste, the coefficient varies between 11×10^{-6} and 16×10^{-6} per °C (6×10^{-6} and 9×10^{-6} per °F), but values up to 20.7×10^{-6} per °C (11.5×10^{-6} per °F) have also been reported, the coefficient varying with the degree of saturation. Thus, a serious difference in coefficients occurs only with aggregates of a very low expansion; these are certain granites, limestones and marbles.

If extreme temperatures are expected, the detailed properties of any given aggregate have to be known. For instance, quartz undergoes inversion at 574°C and expands suddenly by 0.85 per cent. This would disrupt the concrete, and for this reason fire-resistant concrete is never made with quartz aggregate.

Sieve Analysis

This somewhat grandiose name is given to the simple operation of dividing a sample of aggregate into fractions, each consisting of particles of the same size. In practice, each fraction contains particles between specific limits, these being the openings of standard test sieves.

The test sieves used for concrete aggregate have square openings and their properties are fixed by B.S. 410: 1969. Sieves used to be described by the size of the opening (in inches) for larger sizes, and by the number of openings per lineal inch for sieves smaller than about $\frac{1}{8}$ in. Thus a No. 100 B.S. test sieve has 100 × 100 openings in each square inch, the size of the opening and the width of the wire of which the mesh is made being laid down in previous editions of B.S. 410. Nowadays, sieve sizes are designated by the nominal aperture size in millimetres or microns.

Sieves smaller than 4 mm (0·16 in.) are normally made of wire cloth although, if required, this can be used up to 16 mm (0·62 in.). The wire cloth is made of phosphor bronze but for some coarser sieves brass and mild steel can also be used. The screening area, i.e. the area of the openings as a percentage of the gross area of the sieve, varies between 34 and 53 per cent.

Coarse test sieves (4 mm (0·16 in.) and larger) are made of perforated mild steel plate, with a screening area of 44 to 65 per cent.

All sieves are mounted in frames which can nest. It is thus possible to place the sieves one above the other in order of size with the largest sieve at the top, and the material retained on each sieve after shaking represents the fraction of aggregate coarser than the sieve in question but finer than the sieve above. 200 mm (8 in.) diameter frames are used for 5 mm ($\frac{3}{16}$ in.) or smaller sizes, and 300 or 400 mm (12 or 18 in.) diameter frames for 5 mm ($\frac{3}{16}$ in.) and larger sizes. It may be remembered that 5 mm ($\frac{3}{16}$ in., No. 4 in the U.S.) is the dividing line between the fine and coarse aggregate.

The sieves used for concrete aggregate consist of a series in which the clear opening of any sieve is approximately one-half of the opening of the next larger sieve size. The B.S. test sieve sizes in Imperial units for this series were as follows: 3 in., $1\frac{1}{2}$ in., $\frac{3}{4}$ in., $\frac{3}{8}$ in., $\frac{3}{16}$ in., Nos. 7, 14, 25, 52, 100, and 200.

For determination of oversize and undersize aggregate, and especially for research work on aggregate grading, additional sieve sizes are required. The full sequence of test sieves is based theoretically on the ratio of $\sqrt[4]{2}$ for the openings of two consecutive sieves. However, recently both the British (B.S. 410: 1969) and American (A.S.T.M. E11–70) sieves have been standardized in accordance with the R40/3 sieve series of the International Standards Organization. Not all of these sizes form a true geometric series but follow "preferred numbers".

Table 3.14 gives the standard sieve sizes according to their fundamental description by aperture in millimetres or microns and also the previous British and A.S.T.M. designations and approximate apertures in inches. For aggregate grading purposes, however, in the United Kingdom, sieve sizes of 20, 10, and 5 mm are used instead of the internationally standardized values of 19·0, 9·5, and 4·75 mm respectively.

Table 3.14: *Standard British and American Sieve Sizes*

Aperture mm or μm	Approximate Imperial equivalent in.	Previous designation of nearest size	
		B.S.	A.S.T.M.
125 mm	5	—	5 in.
106 mm	4·24	4 in.	4·24 in.
90 mm	3·5	$3\frac{1}{2}$ in.	$3\frac{1}{2}$ in.
75 mm	3	3 in.	3 in.
63 mm	2·5	$2\frac{1}{2}$ in.	$2\frac{1}{2}$ in.
53 mm	2·12	2 in.	2·12 and 2 in.
45 mm	1·75	$1\frac{3}{4}$ in.	$1\frac{3}{4}$ in.
37·5 mm	1·50	$1\frac{1}{2}$ in.	$1\frac{1}{2}$ in.
31·5 mm	1·25	$1\frac{1}{4}$ in.	$1\frac{1}{4}$ in.
26·5 mm	1·06	1 in.	1·06 and 1 in.
22·4 mm	0·875	$\frac{7}{8}$ in.	$\frac{7}{8}$ in.
19·0 mm	0·750	$\frac{3}{4}$ in.	$\frac{3}{4}$ in.
16·0 mm	0·625	$\frac{5}{8}$ in.	$\frac{5}{8}$ in.
13·2 mm	0·530	$\frac{1}{2}$ in.	0·530 in.
11·2 mm	0·438	—	$\frac{7}{16}$ in.
9·5 mm	0·375	$\frac{3}{8}$ in.	$\frac{3}{8}$ in.
8·0 mm	0·312	$\frac{5}{16}$ in.	$\frac{5}{16}$ in.
6·7 mm	0·265	$\frac{1}{4}$ in.	0·265 in.
5·6 mm	0·223	—	No. $3\frac{1}{2}$
4·75 mm	0·187	$\frac{3}{16}$ in.	No. 4
4·00 mm	0·157	—	No. 5
3·35 mm	0·132	No. 5	No. 6
2·80 mm	0·111	No. 6	No. 7
2·36 mm	0·0937	No. 7	No. 8
2·00 mm	0·0787	No. 8	No. 10
1·70 mm	0·0661	No. 10	No. 12
1·40 mm	0·0555	No. 12	No. 14
1·18 mm	0·0469	No. 14	No. 16
1·00 mm	0·0394	No. 16	No. 18
850 μm	0·0331	No. 18	No. 20
710 μm	0·0278	No. 22	No. 25
600 μm	0·0234	No. 25	No. 30
500 μm	0·0197	No. 30	No. 35
425 μm	0·0165	No. 36	No. 40
355 μm	0·0139	No. 44	No. 45
300 μm	0·0117	No. 52	No. 50
250 μm	0·0098	No. 60	No. 60
212 μm	0·0083	No. 72	No. 70
180 μm	0·0070	No. 85	No. 80
150 μm	0·0059	No. 100	No. 100
125 μm	0·0049	No. 120	No. 120
106 μm	0·0041	No. 150	No. 140
90 μm	0·0035	No. 170	No. 170
75 μm	0·0029	No. 200	No. 200
63 μm	0·0025	No. 240	No. 230
53 μm	0·0021	No. 300	No. 270
45 μm	0·0017	No. 350	No. 325
38 μm	0·0015	—	No. 400

Table 3.15: *Usual Sieve Sizes for Concrete Aggregate Based on B.S. 410: 1969*

		millimetres								microns			
B.S. 410: 1969	sieve size	75	50	37·5	20	10	5	2·36	1·18	600	300	150	75
		millimetres								microns			
Equivalent B.S. Imperial sieve	size	76·2	50·8	38·1	19·05	9·52	4·76	2·40	1·20	600	300	150	75
		inches								sieve numbers			
	designation	3	2	$1\frac{1}{2}$	$\frac{3}{4}$	$\frac{3}{8}$	$\frac{3}{16}$	7	14	25	52	100	200

A further difficulty arises from the fact that the aggregate industry in the United Kingdom is still in the process of adjustment to the S.I. system of units; the present situation is thus that the usual sieve sizes for grading of aggregate for concrete are as given in Table 3.15. Of these, the 50 mm (2 in.) size does not form part of the normal concrete aggregate series.

We can see thus that in discussing aggregate grading we have to contend with two sets of sieve sizes. In this book, results of measurements made with Imperial size sieves will be reported by the exact metric equivalent but grading curves for design purposes (*see* Chapter 10) will, wherever available, be based on the new B.S. metric sieve sizes.

Before the sieve analysis is performed, the aggregate sample has to be air-dried in order to avoid lumps of fine particles being classified as large particles and also to prevent clogging of the finer sieves. The weights of the reduced samples for sieving, as recommended by B.S. 812: 1967, are given in Table 3.16, and Table 3.17 shows the maximum weight of

Table 3.16: *Minimum Weight of Sample for Sieve Analysis* (*B.S. 812: 1967*)

Maximum size present in substantial proportions		Minimum weight of sample to be taken for sieving	
mm	in.	kg	lb
63·5	$2\frac{1}{2}$	50	112
50·8	2	35	80
38·1 or 31·8	$1\frac{1}{2}$ or $1\frac{1}{4}$	15	35
25·4	1	5	10
19·0	$\frac{3}{4}$	2	5
12·7	$\frac{1}{2}$	1	$2\frac{1}{2}$
9·5	$\frac{3}{8}$	0·5	1
6·3 or 4·8	$\frac{1}{4}$ or $\frac{3}{16}$	0·2	$\frac{1}{2}$
passing 2·4	passing No. 7 mesh	0·1	$\frac{1}{4}$

material with which each sieve can cope. If this weight is exceeded on a sieve, material which is really finer than this sieve may be included in the portion retained. The material on the sieve in question should, therefore, be split into two parts and each should be sieved separately.

The actual sieving operation can be performed by hand, each sieve in turn being shaken until not more than a trace continues to pass. The movement should be backwards and forwards, sideways left and right, circular clockwise and anticlockwise, all these motions following one another so that every particle "has a chance" of passing through the sieve. In most modern laboratories a sieve shaker is available, usually fitted with a time switch so that uniformity of the sieving operation can be ensured. None the less, care is necessary in order to make sure that no sieve is overloaded. (*See* Table 3.17.)

Table 3.17: *Maximum Weight to be Retained at the Completion of Sieving* (*B.S. 812: 1967*)

B.S. sieve size mm	in.	Maximum weight			
		457 mm (18 in.) diameter sieves		305 mm (12 in.) diameter sieves	
		kg	lb	kg	lb
50·8	2	10	22	4·5	10
38·1	1½	8	18	3·5	8
31·8 or 25·4	1¼ or 1	6	13	2·5	5½
19·0	¾	4	9	2·0	4½
12·7	½	3	6½	1·5	3
9·5	⅜	2	4½	1·0	2
6·3	¼	1·5	3	0·75	1½
4·8	3⁄16	1·0	2	0·50	1
3·2	⅛	—	—	0·30	½

B.S. sieve size mm	μm	No.	Maximum weight 203 mm (8 in.) diameter sieves	
			gram	oz
2·4		7	200	7
1·7 or 1·2		10 or 14	100	3½
	850, 600 or 420	18, 25 or 36	75	2½
	300 or 210	52 or 72	50	2
	150	100	40	1½
	75	200	25	1

The results of a sieve analysis are best reported in tabular form, as shown in Table 3.18. Column (2) shows the weight retained on each sieve. This is expressed as a percentage of the total weight of the sample and is shown in column (3). Now, working from the finest size upwards the *cumulative* percentage (to the nearest one per cent) passing each sieve can be calculated [column (4)], and it is this percentage that is used in the plotting of grading curves.

Grading Curves

The results of a sieve analysis can be grasped much more easily if represented graphically, and for this reason grading charts are very extensively used. By using a chart it is possible to see at a glance whether the grading of a given sample conforms to that specified, or is too coarse or too fine, or deficient in a particular size.

In the grading chart commonly used, the ordinates represent the cumulative percentage passing and the abscissae the sieve opening plotted to a logarithmic scale. Since the openings of sieves in a standard series are in the ratio of $\frac{1}{2}$, a logarithmic plot shows these openings at a constant spacing. This is illustrated in Fig. 3.10 which represents the data of Table 3.18.

Table 3.18: *Example of Sieve Analysis*

B.S. sieve size (1)		Weight retained g (2)	Percentage retained (3)	Cumulative percentage passing (4)	Cumulative percentage retained (5)
9·52 mm	$\frac{3}{8}$ in.	0	0·0	100	0
4·76 mm	$\frac{3}{16}$ in.	6	2·0	98	2
2·40 mm	7	31	10·1	88	12
1·20 mm	14	30	9·8	78	22
600 μm	25	59	19·2	59	41
300 μm	52	107	34·9	24	76
150 μm	100	53	17·3	7	93
smaller than 150 μm	smaller than 100	21	6·8	—	—

Total = 307   Total = 246
Fineness modulus = 2·46

It is convenient to choose a scale such that the scale spacing between two adjacent sieve sizes is approximately equal to 20 per cent on the ordinate scale; a visual comparison of different grading curves can then be made from memory.

Fineness Modulus
A single factor computed from the sieve analysis is sometimes used, particularly in the United States. This is the fineness modulus, defined as the sum of the cumulative percentages retained on the sieves of the standard series: 150, 300, 600 μm, 1·20, 2·40, 4·76 mm (Nos. 100, 52, 25, 14, 7, $\frac{3}{16}$ in.) and up to the largest sieve size present. It should be remembered that when all the particles in a sample are coarser than, say, 600 μm (No. 25), the cumulative percentage retained on 300 μm (No. 52) should be entered as 100; the same value, of course, would be entered for 150 μm (No. 100). The value of the fineness modulus is higher the coarser the aggregate. (*See* column (5), Table 3.18.)

The fineness modulus can be looked upon as a weighted average* size

---

* Popovics[3,49] showed it to be a logarithmic average of the particle size distribution.

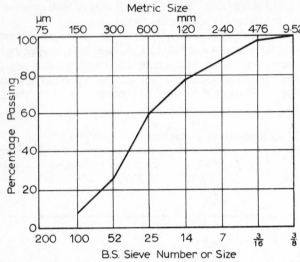

Fig. 3.10. *Example of a grading curve (see Table 3.17)*

of a sieve on which the material is retained, the sieves being counted from the finest. For instance, a fineness modulus of 4·00 can be interpreted to mean that the fourth sieve, 1·20 mm (No. 14), is the average size. However, it is clear that one parameter, the average, cannot be representative of a distribution: thus the same fineness modulus can represent an infinite number of totally different size distributions or grading curves. The fineness modulus cannot, therefore, be used as a description of the grading of an aggregate, but it is valuable for measuring slight variations in the aggregate from the same source, i.e. as a day-to-day check. Nevertheless, within certain limitations, the fineness modulus gives an indication of the probable behaviour of a concrete mix made with aggregate having a certain grading, and the use of the fineness modulus in assessment of aggregates and in mix design has many supporters.[3.49]

Grading Requirements

We have seen how to find the grading of a sample of aggregate but it still remains to determine whether or not a particular grading is suitable. A related problem is that of combining fine and coarse aggregates so as to produce a desired grading. What, then, are the properties of a "good" grading curve?

Since the strength of fully compacted concrete with a given water/ cement ratio is independent of the grading of the aggregate, grading is, in the first instance, of importance only in so far as it affects workability. As, however, the development of strength corresponding to a given water/

cement ratio requires full compaction, and this can be achieved only with a sufficiently workable mix, it is necessary to produce a mix that can be compacted to a maximum density with a reasonable amount of work.

It should be stated at the outset that there is no one *ideal* grading curve but a compromise is aimed at. Apart from the physical requirements, the economic problem must not be forgotten: concrete has to be made of materials which can be produced cheaply so that no narrow limits can be imposed on aggregate.

It has been suggested that the main factors governing the desired aggregate grading are: the surface area of the aggregate, which determines the amount of water necessary to wet all the solids; the relative volume occupied by the aggregate; the workability of the mix; and the tendency to segregation.

Segregation is discussed on p. 199, but it should be observed here that the requirements of workability and absence of segregation tend to be partially opposed to one another: the easier it is for the particles of different sizes to pack, smaller particles passing into the voids between the larger ones, the easier it is also for the small particles to be shaken out of the voids, i.e. to segregate in the dry state. In actual fact, it is the mortar (i.e. a mixture of sand, cement and water) that should be prevented from passing freely out of the voids in the coarse aggregate. It is also essential for the voids in the combined aggregate to be sufficiently small to prevent the cement paste from passing through and separating out.

The problem of segregation is thus rather similar to that of filters, although the requirements in the two cases are of course diametrically opposite: for the concrete to be satisfactory it is essential that segregation be avoided.

There is a further requirement for a mix to be satisfactorily workable: it must contain a sufficient amount of material smaller than a 300 μm (No. 52) sieve. Since the cement particles are included in this material, a richer mix requires a lower content of fine sand than a lean mix. If the grading of sand is such that it is deficient in finer particles, increasing the fine/coarse aggregate ratio may not prove a satisfactory remedy, as it may lead to an excess of middle sizes and possibly harshness. (A mix is said to be harsh when one size fraction is present in excess, as shown by a steep step in the middle of a grading curve, so that particle interference results.) This need for an adequate amount of fines (provided they are structurally sound) explains why minimum contents of particles passing 300 μm (No. 52) and sometimes also 150 μm (No. 100) sieves are laid down, as for instance in Tables 3·23 and 3.24.*

The requirement that the aggregate occupies as large a relative volume as possible is in the first instance an economic one, the aggregate being

---

* It is now thought that the U.S. Bureau of Reclamation requirements for the minimum percentage of particles passing the 300 and 150 μm (Nos. 52 and 100) sieves are too high.

cheaper than the cement paste, but there are also technical reasons why too rich a mix is undesirable. It has also been assumed that the greater the amount of solid particles that can be packed into a given volume of concrete the higher its strength. This maximum density theory has led to the advocacy of grading curves parabolic in shape, or in part parabolic and then straight (when plotted to a natural scale), as shown in Fig. 3.11.

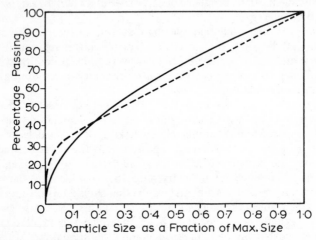

Fig. 3.11. *Fuller's grading curves*

It was found, however, that the aggregate graded to give maximum density makes a harsh and somewhat unworkable mix. The workability is improved when there is an excess of paste above that required to fill the voids in the sand, and also an excess of mortar (sand plus cement) above that required to fill the voids in the coarse aggregate.

Let us now consider the surface area of the aggregate particles. The water/cement ratio of the mix is generally fixed from strength considerations. At the same time, the amount of cement paste has to be sufficient to cover the surface of all the particles so that the lower the surface area of the aggregate the less paste, and therefore the less water, is required.

Taking for simplicity a sphere of diameter D as representative of the shape of the aggregate we have the ratio of the surface area to volume of $6/D$. This ratio of the surface of the particles to their volume (or, when the particles have a constant specific gravity, to their weight) is called specific surface. For particles of a different shape, a coefficient other than $6/D$ would be obtained but the surface area is still inversely proportional to the particle size, as shown in Fig. 3.12 reproduced from Shacklock and Walker's report.[3.15] It should be noted that a logarithmic scale is used for both the ordinates and the abcissae since the sieve sizes are in geometrical progression.

In the case of graded aggregate, the grading and the overall specific surface are related to one another, although of course there are many grading curves corresponding to the same specific surface. If the grading extends to a larger maximum aggregate size the overall specific surface is reduced and the water requirement decreases, but the relation is not

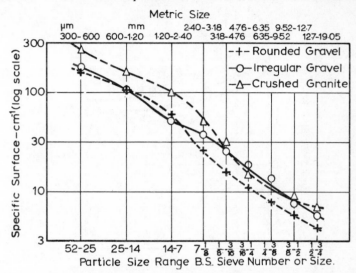

Fig. 3.12. *Relation between specific surface and particle size*[3.15]

linear. For instance increasing the maximum aggregate size from 9·5 mm to 63 mm ($\frac{3}{8}$ in. to $2\frac{1}{2}$ in.) can, under certain conditions, reduce the water requirement for a constant workability by as much as 50 kg per cubic metre (85 lb/yd³) of concrete. The corresponding decrease in the water/cement ratio may be as much as 0·15. Some typical values are shown in Fig. 3.13.

The practical limitations of the maximum size of aggregate that can be used under given circumstances and the problem of influence of the maximum size on strength in general are discussed on p. 176.

It can be seen that, having chosen the maximum size of aggregate and its grading, we can express the total surface area of the particles using the specific surface as a parameter, and it is the total surface of the aggregate that determines the water requirement or the workability of the mix. Mix design on the basis of the specific surface of the aggregate was first suggested by Edwards[3.50] as far back as 1918, and interest in this method has recently been renewed. Specific surface can be determined using the water permeability method,[3.17] but no simple field test is available, and a mathematical approach is made difficult by the variability in the shape of different aggregate particles.

This, however, is not the only reason why the design of mixes on the basis of the specific surface of aggregate is not universally recommended. The application of surface area calculations was found to break down for aggregate particles smaller than about 150 μm (No. 100 B.S. sieve), and for cement. These particles, and also some larger sand particles, appear in.

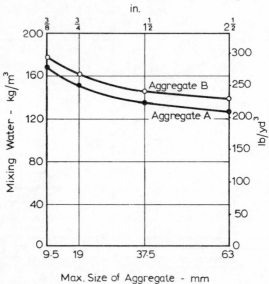

Fig. 3.13. *Influence of maximum size of aggregate on mixing water requirement for a constant slump*[3.16]

to act as a lubricant in the mix and do not seem to require wetting in quite the same way as coarse particles. An indication of this can be obtained from some results of tests by Glanville, Collins and Matthews,[3.18] reproduced in part in Table 3.19.

Because specific surface gives a somewhat misleading picture of the workability to be expected (largely owing to an overestimate of the effect of fine particles), an empirical surface index was suggested by Murdock[3.19] and its values as well as those of the specific surface are given in Table 3.20.

The overall effect of the surface area of an aggregate of given grading is obtained by multiplying the percentage weight of any size fraction by the coefficient corresponding to that fraction, and summing all the products. According to Murdock[3.19] the surface index (modified by an angularity index) should be used, and in fact the values of this index are based on empirical results. On the other hand, Davey[3.20] found that for the same total specific surface of the aggregate the water requirement

Table 3.19: *Water/Cement Ratio Required to Produce a Given Workability for Various Amounts of Crusher Dust (Smaller than 150 μm (No. 100) B.S. Sieve) in the Aggregate*[3.18]

Grading curve	Dust content as a percentage of total aggregate	Water/cement ratio for—		
		Low workability	Medium workability	High workability
1	0	0·612	—	—
	3·0	0·618	—	—
	6·0	0·634	—	—
	9·0	—	0·700	0·750
	12·0	—	0·730	0·760
2	0	0·630	—	—
	3·5	0·635	0·715	—
	7·0	0·648	0·715	0·750
	10·5	0·653	0·720	0·745
	14·0	—	0·720	0·750
3	0	0·665	0·735	0·780
	4·2	0·665	0·725	0·758
	8·4	0·682	0·735	0·766
	12·6	0·695	0·740	0·770
	16·8	0·740	0·775	0·790
4	0	0·713	0·780	0·820
	4·8	0·720	0·787	0·825
	9·6	0·732	0·787	0·825
	14·4	0·765	0·805	0·830
	19·2	0·807	0·835	0·850

(*Crown Copyright*)

Table 3.20: *Relative Values of Surface Area and Surface Index*

Particle size fraction		Relative surface area	Murdock's[3.19] surface index
76·2–38·1 mm	3–1$\frac{1}{2}$ in.	$\frac{1}{2}$	$\frac{1}{2}$
38·1–19·05 mm	1$\frac{1}{2}$–$\frac{3}{4}$ in.	1	1
19·05–9·52 mm	$\frac{3}{4}$–$\frac{3}{8}$ in.	2	2
9·52–4·76 mm	$\frac{3}{8}$–$\frac{3}{16}$ in.	4	4
4·76–2·40 mm	$\frac{3}{16}$ in.–7	8	8
2·40–1·20 mm	7–14	16	12
1·20 mm–600 μm	14–25	32	15
600–300 μm	25–52	64	12
300–150 μm	52–100	128	10
smaller than 150 μm	smaller than 100		1

Table 3.21: *Properties of Concretes Made with Aggregates of the Same Specific Surface*[3.20]

Size fraction	Aggregate grading—percent							Specific surface m²/kg	Water/cement ratio by weight	Compressive strength				Modulus of rupture			
	300–150 μm	600–300 μm	1·20 mm–600 μm	2·40–1·20 mm	4·76–2·40 mm	9·52–4·76 mm	19·05–9·52 mm			7 days		28 days		7 days		28 days	
Grading	52–100	25–52	14–25	7–14	$\frac{3}{16}$–7	$\frac{3}{8}$–$\frac{3}{16}$	$\frac{3}{4}$–$\frac{3}{8}$			MN/m²	lb/in²	MN/m²	lb/in²	MN/m²	lb/in²	MN/m²	lb/in²
A	11·2	11·2	11·2	11·2	11·2	22·0	22·0	3·2	0·575	23·7	3,440	32·9	4,770	3·72	539	4·38	636
B	12·9	12·9	12·9	0	0	30·6	30·7	3·2	0·575	24·2	3,510	32·3	4,690	3·74	543	4·48	651
C	15·4	15·4	0	0	0	34·6	34·6	3·2	0·575	24·6	3,570	32·8	4,760	3·84	557	4·54	659
D	25·4	0	0	0	0	0	74·6	3·2	0·575	23·3	3,380	32·1	4,650	3·46	502	4·16	603

and the compressive strength of the concrete are the same for very wide limits of aggregate grading. This applies both to continuously and gap-graded aggregate, and in fact three of the four gradings listed in Table 3.21, reproduced from Davey's paper, are of the gap type.

An increase in the specific surface of the aggregate for a constant water/cement ratio has been found to lead to a lower strength of concrete, as shown for instance in Table 3.22, reproducing Newman and

Table 3.22: *Specific Surface of Aggregate and Strength of Concrete for a 1:6 Mix with a Water/Cement Ratio of 0·60*[3.21]

Specific surface of aggregate m²/kg	28-day compressive strength of concrete		Density of fresh concrete	
	MN/m²	lb/in²	kg/m³	lb/ft³
2·24	36·1	5,240	2,330	145·5
2·80	34·9	5,060	2,325	145·1
4·37	30·3	4,390	2,305	144·0
5·71	27·5	3,990	2,260	141·0

(*Crown Copyright*)

Teychenné's[3.21] results. The reasons for this are not quite clear, but it is possible that a reduction in density of the concrete consequent upon an increase in fineness of the aggregate is instrumental in lowering the strength.[3.22]

It seems then that the surface area of the aggregate is an important factor in determining the workability of the mix, but the exact rôle played by the finer particles has by no means been ascertained.

The type gradings of Road Note No. 4[3.23] represent different values of overall specific surface. For instance, when river sand and gravel are used, the four grading curves, Nos. 1 to 4, of Fig. 3.14 correspond to the specific surface of 1·6, 2·0, 2·5 and 3·3 m²/kg, respectively.[3.21] In practice, when trying to approximate type gradings, the properties of the mix will remain largely unaltered when compensation of a small deficiency of fines by a somewhat larger excess of coarser particles is applied but the departure must not be too great. The deficiency and excess are, of course, mutually interchangeable in the above statement.

There is no doubt then that the grading of aggregate is a major factor in the workability of a concrete mix. Workability, in turn, affects the water and cement requirements, controls segregation, has some effect on bleeding, and influences the placing and finishing of the concrete. These factors represent the important characteristics of fresh concrete and affect also its properties in the hardened state: strength, shrinkage, and durability.

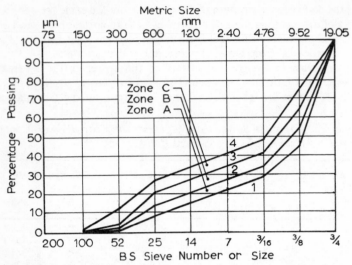

Fig. 3.14. *Road Note No. 4 type grading curves for 19·05 mm (¾ in.) aggregate*[3.23] (*Crown copyright*)

Grading is thus of vital importance in the proportioning of concrete mixes, but its exact rôle in mathematical terms is not fully known, and the behaviour of this type of semi-liquid mixture of granular materials is still imperfectly understood.

Finally, it must be remembered that far more important than devising a "good" grading is ensuring that the grading is kept constant; otherwise, variable workability results and, as this is usually corrected at the mixer by a variation in the water content, concrete of variable strength is obtained.

Practical Gradings

From the brief review in the previous section it can be seen how important it is to use aggregate with a grading such that a reasonable workability and a minimum segregation are obtained. The importance of the latter requirement cannot be over-emphasized: a workable mixture which *could* produce a strong and economical concrete will result in a honeycombed, weak, not durable and variable end product if segregation takes place.

The process of calculation of the proportions of aggregates of different size to achieve the desired grading comes within the scope of mix design, and is described in Chapter 10. Here, the properties of some "good" grading curves will be discussed. It should be remembered, however, that in practice the aggregate available locally or within an economic distance has to be used, and this can generally produce satisfactory concrete, given an intelligent approach and sufficient care. The curves most commonly referred to as a basis of comparison are those of the Road Research Note No. 4 on the Design of Concrete Mixes.[3.23] They have been prepared for aggregates of 19·05 and 38·1 mm ($\frac{3}{4}$ in. and $1\frac{1}{2}$ in.) maximum size, and are reproduced in Figs. 3.14 and 3.15 respectively. Similar curves for aggregate with a 9·52 mm ($\frac{3}{8}$ in.) maximum size have been prepared by McIntosh and Erntroy[3.24] of the Cement and Concrete Association, and are shown in Fig. 3.16.

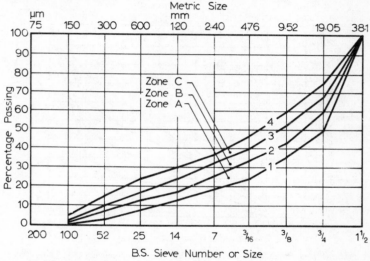

Fig. 3.15. *Road Note No. 4 type grading curves for 38·1 mm ($1\frac{1}{2}$ in.) aggregate*[3.23] (*Crown Copyright*)

Four curves are shown for each maximum size of aggregate, but due to the presence of over- and under-size aggregate and also because of variation *within* any fraction size, practical gradings are more likely to lie in the vicinity of these curves than to follow them exactly. It is therefore convenient to talk about grading zones, and these are marked on all the diagrams. In some specifications definite limits of grading, rather than a single curve, are laid down.

Curve No. 1 represents the coarsest grading in each of the Figs. 3.14 to 3.16. Such a grading is comparatively workable and can, therefore, be used for mixes with a low water/cement ratio or for rich mixes; it is, however, necessary to make sure that segregation does not take place. At the other extreme, curve No. 4 represents a fine grading: it will be cohesive but not very workable. In particular, an excess of material between 1·20 and 4·76 mm (No. 14 and $\frac{3}{16}$ in.) B.S. test sieves will produce a harsh concrete, which, although it may be suitable for compaction by vibration, is difficult to place by hand. If the same workability is to be obtained using aggregates with grading curves Nos. 1 and 4, the latter would require a considerably higher water content: this would mean a lower strength if both concretes are to have the same aggregate/cement ratio or, if the same strength is required, the concrete made with the fine aggregate would have to be considerably richer, i.e. each cubic metre would contain more cement than when the coarser grading is used.

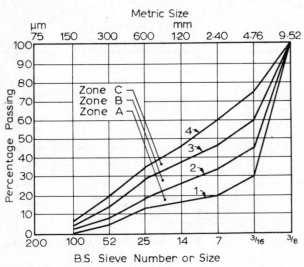

Fig. 3.16. *McIntosh and Erntroy's type grading curves for 9·52 mm (⅜ in.) aggregate*[3.24]

The change between the extreme gradings is progressive. In the case of gradings lying in part in one zone, in part in another, there is, however, a danger of segregation when too many intermediate sizes are missing (cf. gap grading). If, on the other hand, there is an excess of middle-sized aggregate the mix will be harsh and difficult to compact by hand and possibly even by vibration. For this reason, it is preferable to use aggregate with gradings similar to type rather than totally dissimilar ones.

Figs. 3.17 and 3.18 show the range of gradings used with 152·4 mm (6 in.) and 76·2 mm (3 in.) maximum aggregate size respectively, as given by McIntosh.[3.25] The actual gradings, as usual, run parallel with the limits rather than crossing over from one to the other.

In practice, the use of separate fine and coarse aggregate means that a grading can be made up to conform exactly with a type grading at one intermediate point, generally the 4·76 mm (3/16 in.) size. Good agreement can usually also be obtained at the ends of the curve (150 μm (No. 100)

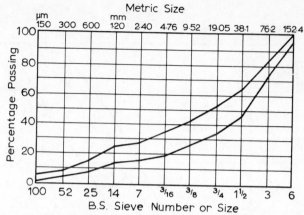

Fig. 3.17. *Range of gradings used with 152·4 mm (6 in.) aggregate*[3.25]

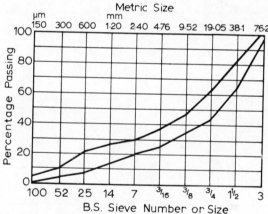

Fig. 3.18. *Range of gradings used with 76·2 mm (3 in.) aggregate* [3.25]

B.S. sieve and the maximum size used). If coarse aggregate is delivered in single-size fractions, as is usually the case, agreement at additional points above 4·76 mm ($\frac{3}{16}$ in.) can be obtained, but for sizes below 4·76 mm ($\frac{3}{16}$ in.) blending of two or more sands is necessary.

Grading of Fine and Coarse Aggregates
Since for any but unimportant work fine and coarse aggregates are batched separately, the grading of each type of aggregate should be known and controlled.

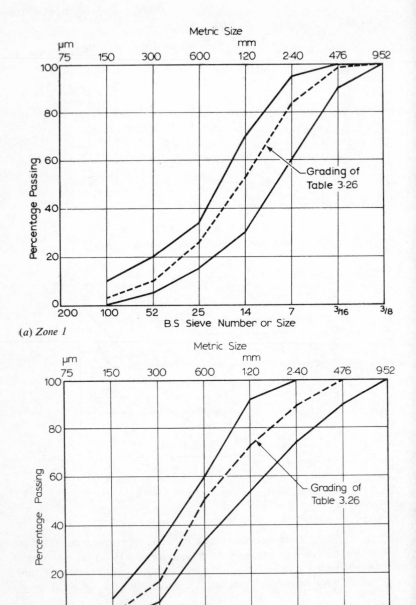

(a) Zone 1

(b) Zone 2

Fig. 3.19. *Grading limits for sand in zones 1 to 4 of B.S. 882: 1965*

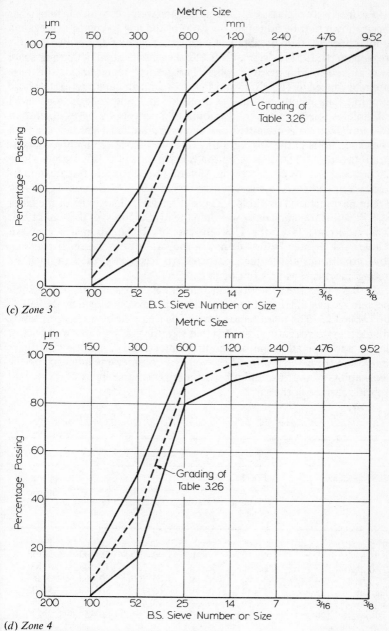

(c) Zone 3

(d) Zone 4

Fig. 3.19. (cont.)

Formerly, two classes of fine aggregate were recognized, but it has been shown that by adjusting the ratio of the fine to coarse aggregate a good concrete could be obtained with either class of aggregate. For this reason, in the 1954 revision of B.S. 882 the classification of fine aggregate was altered to four grading zones. The grading requirements for these are reproduced in Table 3.23 and Fig. 3.19, and any fine aggregate whose grading falls wholly within the limits of any one zone is considered suitable. A tolerance of a *total* amount of 5 per cent on sieves other than 600 μm (No. 25) is permitted, but the aggregate must not be finer than the exact limits of the finest grading (No. 4) or coarser than the coarsest grading (No. 1). The only exception is in the case of crushed stone where 20 per cent is allowed to pass the 150 μm (No. 100) B.S. test sieve in all zones. For comparison, the requirements of A.S.T.M. Standard C 33–71a are in part included in Table 3.23 (*see* Fig. 3.20). The limits of the latter specification are much narrower than the overall limits of B.S. 882: 1965. The requirements of the U.S. Bureau of Reclamation are given in Table 3.24. It may be noted that in the case of air-entrained concrete lower quantities of the finest particles are acceptable, the entrained air acting effectively as very fine aggregate.

In B.S. 882: 1965 the division into zones is based primarily on the percentage passing the 600 μm (No. 25) B.S. sieve, as shown by the values in Table 3.23. The main reason for this is that a large number of sands divide themselves naturally at just that size, the gradings above and below being approximately uniform. Furthermore, the content of particles finer than the 600 μm (No. 25) sieve has a considerable influence on the workability of the mix, and provides a fairly reliable index of the overall specific surface of the sand.

Table 3.23: *B.S. and A.S.T.M. Grading Requirements for Fine Aggregate*

B.S. sieve size		Percentage by weight passing B.S. sieves				
		B.S. 882: 1965				A.S.T.M. Standard C 33–71a
		Grading zone 1	Grading zone 2	Grading zone 3	Grading zone 4	
9·52 mm	$\frac{3}{8}$ in.	100	100	100	100	100
4·76 mm	$\frac{3}{16}$ in.	90–100	90–100	90–100	95–100	95–100
2·40 mm	7	60–95	75–100	85–100	95–100	80–100
1·20 mm	14	30–70	55–90	75–100	90–100	50–85
600 μm	25	15–34	35–59	60–79	80–100	25–60
300 μm	52	5–20	8–30	12–40	15–50	10–30
150 μm	100	0–10*	0–10*	0–10*	0–15*	2–10

* For crushed stone sands the permissible limit is increased to 20 per cent.

Table 3.24: *U.S. Bureau of Reclamation Grading Requirements for Fine Aggregate*

Sieve size		Individual percentage retained
4·76 mm	$\frac{3}{16}$ in.	0–5
2·40 mm	7	5–15⎫ or ⎧ 5–20
1·20 mm	14	10–25⎭ ⎩10–20
600 μm	25	10–30
300 μm	52	15–35
150 μm	100	12–20
smaller than 150 μm	smaller than 100	3–7

Sand falling into any zone can generally be used in concrete, although under some circumstances the suitability of a given sand may depend on the grading and shape of the coarse aggregate.

The suitability of the fine sand of zone 4 for use in reinforced concrete has to be tested. Since the greater part of this sand is smaller than the 600 μm (No. 25) B.S. sieve, a gap-graded or a nearly gap-graded aggregate is obtained, and special care in choosing the mix proportions must be exercised. The sand content of the mix should generally be low, and suggested values of coarse/fine aggregate ratio are given in Table 3.25.

Table 3.25: *Suggested Proportions by Weight of Coarse to Fine Aggregate for Sand of Different Zones*[3.38]

Maximum size of coarse aggregate		Coarse/fine aggregate ratio for sand of zone—			
mm	in.	1	2	3	4*
9·52	$\frac{3}{8}$	1	1$\frac{1}{2}$	2	3
19·05	$\frac{3}{4}$	1$\frac{1}{2}$	2	3	3$\frac{1}{2}$
38·1	1$\frac{1}{2}$	2	3	3$\frac{1}{2}$	—

(*Crown copyright*)

* The suitability of the mix for use in reinforced concrete should be ascertained by test.

Nevertheless, quite good concrete can be obtained with sand of zone 4, particularly using vibration.

At the other extreme, the coarse sand of zone 1 produces a harsh mix, and a high sand content may be necessary for higher workability. This sand is more suitable for rich mixes or for use in concrete of low workability.

Zone 2 represents a medium sand generally suitable for the "standard" 1:2 fine to coarse mix (when the maximum size of aggregate is 19·05 mm (¾ in.)).

In general terms, the ratio of coarse to fine aggregate should be higher the finer the grading of the fine aggregate: typical values are given in Table 3.25. When crushed rock coarse aggregate is used, a slightly higher proportion of sand is required than with gravel aggregate in order to

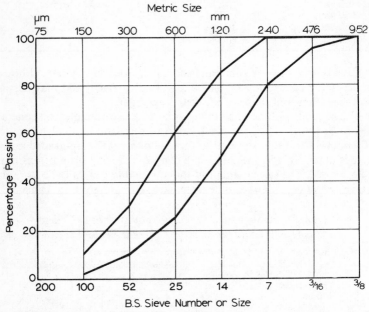

Fig. 3.20. *A.S.T.M. Standard C33–71a grading limits for fine aggregate*

compensate for the lowering of workability by the sharp, angular shape of the crushed particles.

The choice of correct proportions is particularly important as the grading of the sand approaches the fine outer limit of zone 4 or the coarse outer limit of zone 1. It is worth noting, however, that, proportioned correctly, fine sand can be utilized with success, and this is of considerable economic importance in the United Kingdom where there is a predominance of fine sand; in the past there has been a great deal of prejudice against this type of material.

An example of using sand from any of the four zones to produce equally good concrete is given in Table 3.26, based on results obtained at the Building Research Station.[3.21] The actual sand gradings are shown in

Fig. 3.19. An aggregate/cement ratio of 6·04 and a water/cement ratio of 0·60, both by weight, were used throughout. To keep the workability approximately constant the coarse/fine aggregate ratio was varied so that the overall specific surface of aggregate remained at 2·55 m²/kg. Table 3.26 shows that concrete of similar quality was obtained in all cases.

Table 3.26: *Properties of Concretes Made with Aggregates of Constant Overall Specific Surface*[3.21]

Properties of concrete	Sand-grading zone*			
	1	2	3	4
Overall specific surface, m²/kg	2·55	2·55	2·55	2·55
Percentage of material passing 4·76 mm ($\frac{3}{16}$ in.)	46	36	29	24
Approximate proportions by volume	1:2½:3½	1:2:4	1:1½:4½	1:1¼:4¾
Water/cement ratio by weight	0·60	0·60	0·60	0·60
Compacting factor	0·92	0·93	0·93	0·94
28-day crushing strength, MN/m²	27·1	28·1	29·2	29·0
lb/in²	3,930	4,080	4,230	4,200

(*Crown copyright*)

* The actual gradings are shown in Figure 3.19.

Table 3.27: *Properties of Concretes Made with Aggregates of Fixed Proportions*[3.21]

Properties of concrete	Sand grading zone*			
	1	2	3	4
Overall specific surface, m²/kg	2·08	2·55	3·04	3·55
Percentage of material passing 4·76 mm ($\frac{3}{16}$ in.)	36	36	36	36
Approximate proportions by volume	1:2:4	1:2:4	1:2:4	1:2:4
Water/cement ratio by weight	0·58	0·60	0·63	0·66
Compacting factor	0·92	0·95	0·95	0·96
28-day crushing strength, MN/m²	31·6	28·9	24·8	23·2
lb/in²	4,580	4,200	3,600	3,370

(*Crown copyright*)

* The actual gradings are shown in Figure 3.19.

By way of contrast, Table 3.27 shows test results for the case when the same materials were used but the coarse/fine aggregate ratio was

Table 3.28: *Grading Requirements for Coarse Aggregate (B.S. 882: 1965)*

B.S. sieve size		Percentage by weight passing B.S. sieves							
		Nominal size of graded aggregate			Nominal size of single-sized aggregate				
mm	in.	38·1 to 4·76 mm 1½ in. to 3/16 in.	19·0 to 4·76 mm ¾ in. to 3/16 in.	12·7 to 4·76 mm ½ in. to 3/16 in.	64·0 mm 2½ in.	38·1 mm 1½ in.	19·0 mm ¾ in.	12·7 mm ½ in.	9·52 mm ⅜ in.
76·2	3	100	—	—	100	—	—	—	—
64·0	2½	—	—	—	85–100	100	—	—	—
38·1	1½	95–100	100	—	0–30	85–100	100	—	—
19·0	¾	30–70	95–100	100	0–5	—	85–100	100	—
12·7	½	—	—	90–100	—	—	—	85–100	100
9·52	⅜	10–35	25–55	40–85	—	0–20	—	—	85–100
4·76	3/16	0–5	0–10	0–10	—	0–5	0–20	0–45	0–20
2·40	No. 7	—	—	—	—	—	0–5	0–10	0–5

kept constant. The use of finer sand led to a higher water requirement and consequently a lower strength of concrete.

The requirements of B.S. 882: 1965 for the grading of coarse aggregate are reproduced in Table 3.28: values are given both for graded aggregate and for nominal one-size fractions. For comparison, some of the limits of A.S.T.M. Standard C 33–71a are given in Table 3.29.

Table 3.29: *Grading Requirements for Coarse Aggregate (A.S.T.M. Standard C 33–71a)*

Sieve size		Percentage by weight passing sieves				
		Nominal size of graded aggregate			Nominal size of single-sized aggregate	
		38·1 to 4·76 mm	19·0 to 4·76 mm	12·7 to 4·76 mm	64 mm	38·1 mm
mm	in.	1½ in. to $\frac{3}{16}$ in.	$\frac{3}{4}$ in. to $\frac{3}{16}$ in.	½ in. to $\frac{3}{16}$ in.	2½ in.	1½ in.
76·2	3	—	—	—	100	—
64·0	2½	—	—	—	90–100	—
50·8	2	100	—	—	35–70	100
38·1	1½	95–100	—	—	0–15	90–100
25·4	1	—	100	—	—	20–55
19·0	$\frac{3}{4}$	35–70	90–100	100	0–5	0–15
12·7	½	—	—	90–100	—	—
9·52	$\frac{3}{8}$	10–30	20–55	40–70	—	0–5
4·76	$\frac{3}{16}$	0–5	0–10	0–15	—	—
2·40	No. 7	—	0–5	0–5	—	—

The actual grading requirements depend to some extent on the shape and surface characteristics of the particles. For instance, sharp, angular particles with rough surfaces should have a slightly finer grading in order to reduce the possibility of interlocking and to compensate for the high friction between the particles. The actual grading of crushed aggregate is affected primarily by the type of crushing plant employed. A roll granulator usually produces fewer fines than other types of crushers, but the grading depends also on the amount of material fed into the crusher.

The grading limits for all-in aggregate prescribed by B.S. 882: 1965 are reproduced in Table 3.30. It should be remembered that this type of aggregate is not used except for small and unimportant jobs, mainly because it is difficult to avoid segregation in stockpiling.

Table 3.30: *Grading Requirements for All-in Aggregate (B.S. 882: 1965)*

B.S. sieve size		Percentage by weight passing B.S. sieves: 38·1 mm (1½ in.) nominal size	19·0 mm (¾ in.) nominal size
76·2 mm	3 in.	100	—
38·1 mm	1½ in.	95–100	100
19·0 mm	¾ in.	45–75	95–100
4·76 mm	$\frac{3}{16}$ in.	25–45	30–50
600 μm	25	8–30	10–35
150 μm	100	0–6	0–6

Oversize and Undersize

Strict adherence to size limits of aggregate is not possible: breakage during handling will produce some undersize material, and wear of screens in the quarry or at the crusher will result in oversize particles being present.

Table 3.31: *Sizes of Over- and Under-size Screens of U.S. Bureau of Reclamation*

Nominal size fraction		Test screen for—			
		Undersize		Oversize	
mm	in.	mm	in.	mm	in.
4·76–9·52	$\frac{3}{16}$–$\frac{3}{8}$	4·00	No. 5*	13·2	$\frac{1}{2}$
9·52–19·0	$\frac{3}{8}$–$\frac{3}{4}$	8·0	$\frac{5}{16}$	22·4	$\frac{7}{8}$
19·0–38·1	$\frac{3}{4}$–1½	16·0	$\frac{5}{8}$	45	1¾
38·1–76·2	1½–3	31·5	1¼	90	3½
76·2–152·4	3–6	63	2½	178	7

* A.S.T.M. size.

In the United States it is usual to specify over- and undersize screen sizes as $\frac{7}{6}$ and $\frac{5}{6}$, respectively, of the nominal sieve size; actual values are given in Table 3.31. The quantity of aggregate smaller than the undersize or larger than the oversize is generally severely limited.

The grading requirements of B.S. 882: 1965 allow some under- and oversize both for coarse and fine aggregate. The figures for the former are given in Table 3.28, and it can be seen that between 5 and 15 per cent oversize is permitted. However, no aggregate must be retained on a sieve one size larger (in the standard series) than the nominal maximum size. 5 to 10 per cent undersize material is allowed. In the case of single-size aggregate, some undersize is also allowed, and the amount passing the sieve next smaller than the nominal size is also prescribed. It is important that this fine fraction of coarse aggregate be not neglected in the calculation of the actual grading.

For fine aggregate, a total departure of 5 per cent from zone limits is allowed but not beyond the coarser limit of zone 1 or the finer limit of zone 4 (*see* Table 3.23).

Gap-Graded Aggregate

As mentioned earlier, aggregate particles of a given size pack so as to form voids that can be penetrated only if the next smaller size of particles is sufficiently small. This means that there must be a minimum difference between the sizes of any two adjacent particle fractions. In other words, sizes differing but little cannot be used side by side, and this has led to advocacy of gap-graded aggregate.

Gap grading can then be defined as a grading in which one or more intermediate size fractions are omitted. The term continuously graded is used to describe conventional grading when it is necessary to distinguish it from gap grading. On a grading curve, gap grading is represented by a horizontal line over the range of sizes omitted. For instance, the top grading curve of Fig. 3.21 shows that no particles of size between 9·52 and 2·40 mm ($\frac{3}{8}$ in. and No. 7) B.S. sieve are present. In some cases, a gap between 9·52 and 1·20 mm ($\frac{3}{8}$ in. and No. 14) sieve is considered suitable. Omission of these sizes would reduce the number of stockpiles of aggregate required and lead to economy. In the case of aggregate of 19·0 mm ($\frac{3}{4}$ in.) maximum size there would be two piles only: 19·0 to 9·52 mm ($\frac{3}{4}$ to $\frac{3}{8}$ in.), and sand screened through a 1·20 mm (No. 14) screen. The particles smaller than 1·20 mm (No. 14) B.S. sieve size could easily enter the voids in the coarse aggregate so that the workability of the mix would be higher than that of a continuously graded mix of the same sand content.

Tests by Shacklock[3.26] have shown that for a given aggregate/cement ratio and water/cement ratio the highest workability is obtained with a lower sand content in the case of gap-graded aggregate than when continuously graded aggregate is used. However, in the more workable range of mixes, gap-graded aggregate showed a greater proneness to segregation. For this reason, gap grading is recommended mainly for mixes of relatively low workability that are to be compacted by vibration. Good control and, above all, care in handling so as to avoid segregation, are essential.

It may be observed that even when some "ordinary" aggregates are used gap grading exists; for instance, sand belonging to zone 4 of B.S.

882: 1965 is almost completely deficient in particles between the 4·76 and 2·40 or 1·20 mm ($\frac{3}{16}$ in. and No. 7 or No. 14) sieve sizes. Thus, whenever we use a zone 4 sand without blending with a coarser sand, we are in fact using a gap-graded aggregate. Gap-graded aggregate can be used in any concrete, but there are two cases of interest: preplaced aggregate con-

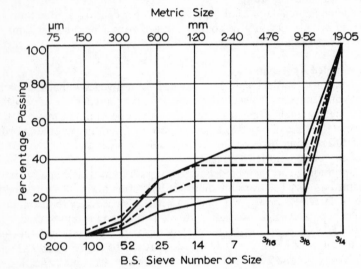

Fig. 3.21. *Typical gap gradings*

crete (*see* p. 223) and exposed aggregate concrete; in the latter, a pleasing finish is obtained since a large quantity of only one size of coarse aggregate becomes exposed after treatment.

From time to time various claims of superior properties have been made for concrete made with gap-graded aggregate, but these do not seem to have been substantiated. Strength, both compressive and tensile, does not appear to be affected. Likewise, Fig. 3.22, showing McIntosh's[3.27] results, confirms that using given materials with a fixed aggregate/cement ratio (but adjusting the sand content), approximately the same workability and strength are obtained with gap and continuous gradings.

Similarly, there is no difference in shrinkage of the concretes made with aggregate of either type of grading,[3.26] although it might be expected that a framework of coarse particles almost touching one another would result in a lower total change in dimensions on drying. The resistance of concrete to freezing and thawing is lower when gap-graded aggregate is used.[3.26]

It seems, therefore, that the rather extravagant claims made by

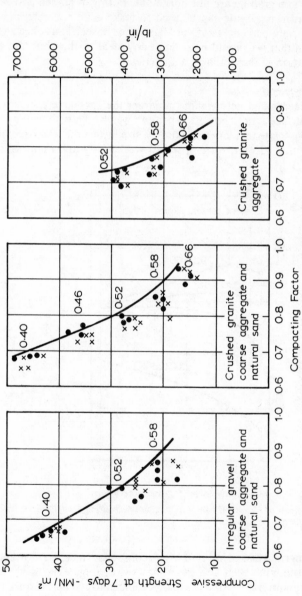

Fig. 3.22. *Workability and strength of 1:6 concretes made with gap- and continuously graded aggregates*[3.27]

Cross denotes gap-graded and circle continuously graded mixes. Each group of points represents mixes with the water/cement ratio indicated but with different sand contents.

advocates of gap grading are not borne out. Both gap-graded and continuously graded aggregate can be used to make good concrete, but in each case the right percentage of sand has to be chosen. Thus once again it can be seen that we should not aim at any ideal grading but find the best combination of the available aggregates.

Maximum Aggregate Size

It has been mentioned before that the larger the aggregate particle the smaller the surface area to be wetted per unit weight. Thus, extending the grading of aggregate to a larger maximum size lowers the water requirement of the mix, so that, for a specified workability and richness, the

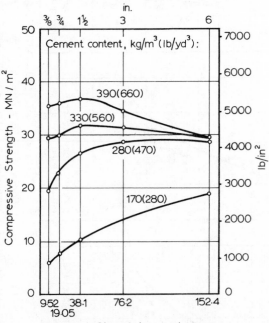

Fig. 3.23. *Influence of maximum size of aggregate on the 28-day compressive strength of concretes of different richness*[3.51]

water/cement ratio can be lowered with a consequent increase in strength.

This behaviour has been verified by tests with aggregates up to 38·1 mm (1½ in.) maximum size,[3.28] and is usually assumed to extend to larger sizes as well. Recent experimental results show, however, that above the 38·1 mm (1½ in.) maximum size the gain in strength due to the reduced water requirement is offset by the detrimental effects of lower bond area (so that volume changes in the paste cause larger stresses at interface) and of discontinuities introduced by the very large particles, particularly in rich mixes. Concrete becomes grossly heterogeneous and the resultant lowering of strength may possibly be similar to that caused by a rise in the crystal size and coarseness of texture of rocks.

This adverse effect of increase in the size of the largest aggregate particles in the mix exists, in fact, throughout the range of sizes, but below 38·1 mm ($1\frac{1}{2}$ in.) the effect of size on the decrease in the water requirement is dominant. For larger sizes, the balance of the two effects depends on richness of the mix,[3.42,3.51] as shown in Fig. 3.23. Thus the best maximum size of aggregate from the standpoint of strength is a function of the richness of the mix. Specifically, in lean concrete (165 kg of cement per cubic metre (280 lb/yd³)), the use of 150 mm (or 6 in.) aggregate is advantageous. However, in structural concrete of usual proportions, from the point of view of strength there is no advantage in using aggregate with a maximum size greater than about 25 or 40 mm (1 or $1\frac{1}{2}$ in.). Moreover, the use of larger aggregate would require the handling of a separate stockpile and might increase the risk of segregation. However, a practical decision would also be influenced by the availability and cost of different size fractions. There are, of course, structural limitations too: the maximum size of aggregate should be no more than $\frac{1}{5}$ to $\frac{1}{4}$ of the thickness of the concrete section and is related also to the spacing of reinforcement. The governing values are prescribed in codes of practice.

Use of "Plums"

The original idea of the use of aggregate as an inert filler can be extended to the inclusion of large stones in a normal concrete: thus the apparent yield for a given amount of cement is increased.

These stones are called "plums" and used in a large concrete mass they can be as big as a 300 mm (1 ft) cube but should not be greater than one-third of the least dimension to be concreted. The volume of plums should not exceed 20 to 30 per cent of the total volume of the finished concrete, and they have to be well dispersed throughout the mass. This is achieved by placing a layer of normal concrete, then spreading the plums, followed by another layer of concrete, and so on. Each layer should be of such thickness as to ensure at least 100 mm (4 in.) of concrete around each plum. Care must be taken to ensure that no air is trapped underneath the stones and that the concrete does not work away from their underside. The plums must have no adhering coating.

The placing of plums requires a large amount of labour and also breaks the continuity of concreting. It is, therefore, not surprising that with the current high ratio of the cost of labour to the cost of cement the use of plums is not economical except under special circumstances, but their use is covered by standards in some countries, e.g. South Africa.

Handling of Aggregate

Handling and stockpiling of coarse aggregate can easily lead to segregation. This is particularly so when discharging and tipping permits the aggregate to roll down a slope. A natural case of such segregation is a scree (talus): the size of particles is uniformly graded from largest at the bottom to smallest at the top.

A description of the precautions necessary in handling operations is outside the scope of this book, but one vital recommendation should be mentioned: coarse aggregate should be split into size fractions 5 to 10, 10 to 20, 20 to 40 mm (or $\frac{3}{16}$ to $\frac{3}{8}$, $\frac{3}{8}$ to $\frac{3}{4}$, $\frac{3}{4}$ to $1\frac{1}{2}$ in.), etc. These fractions should be handled and stockpiled separately and remixed only when being fed into the concrete mixer in the desired proportions. Thus segregation can occur only within the narrow size range of each fraction, and even this can be reduced by careful handling procedures.

Care is necessary to avoid breakage of the aggregate: particles greater than 40 mm (or $1\frac{1}{2}$ in.) should be lowered into bins by means of rock ladders and not dropped from a height.

On big and important jobs the results of segregation and breakage in handling (i.e. excess of undersize particles) are eliminated by "finish rescreening" immediately prior to feeding into the batching bins over the mixer. The proportions of different sizes are thus controlled much more effectively but the complexity and cost of the operations are correspondingly increased. This is, however, repaid by easier placing of uniformly workable concrete and by a possible saving in cement due to the uniformity of the concrete.

REFERENCES

3.1. F. A. SHERGOLD, The percentage voids in compacted gravel as a measure of its angularity, *Mag. Concr. Res.*, **5**, No. 13, pp. 3–10 (Aug. 1953).

3.2. P. J. F. WRIGHT, A method of measuring the surface texture of aggregate, *Mag. Concr. Res.*, **5**, No. 2, pp. 151–60 (Nov. 1955).

3.3. M. F. KAPLAN, Flexural and compressive strength of concrete as affected by the properties of coarse aggregates, *J. Amer. Concr. Inst.*, **55**, pp. 1193–208 (May 1959).

3.4. M. F. KAPLAN, The effects of the properties of coarse aggregates on the workability of concrete, *Mag. Concr. Res.*, **10**, No. 29, pp. 63–74 (Aug. 1958).

3.5. S. WALKER and D. L. BLOEM, Studies of flexural strength of concrete, Part 1: Effects of different gravels and cements, *Nat. Ready-mixed Concr. Assoc. Joint Research Laboratory Publicn. No.* 3 (Washington D.C., July 1956).

3.6. D. O. WOOLF, Toughness, hardness, abrasion, strength, and elastic properties, *A.S.T.M. Sp. Tech. Publicn. No.* 169, pp. 314–24 (1956).

3.7. ROAD RESEARCH, Roadstone test data presented in tabular form, *D.S.I.R. Road Note No.* 24 (London, H.M.S.O., 1959).

3.8. K. NEWMAN, The effect of water absorption by aggregates on the water-cement ratio of concrete, *Mag. Concr. Res.*, **11**, No. 33, pp. 135–42 (Nov. 1959).

3.9. J. D. McINTOSH, The siphon-can test for measuring the moisture content of aggregates, *Cement Concr. Assoc. Tech. Rep. TRA/198* (London, July 1955).

3.10. R. H. H. KIRKHAM, A buoyancy meter for rapidly estimating the moisture content of concrete aggregates, *Civil Engineering,* **50**, No. 591, pp. 979–80 (London, Sept. 1955).

3.11. NATIONAL READY-MIXED CONCRETE ASSOCIATION, *Technical Information Letter No.* 141 (Washington D.C., 15th Sept. 1959).

3.12. H. G. MIDGLEY, The staining of concrete by pyrite, *Mag. Concr. Res.*, **10**, No. 29, pp. 75–8 (Aug. 1958).

3.13. S. BRUNAUER, P. H. EMMETT and E. TELLER, Adsorption of gases in multi-molecular layers, *J. Amer. Chem. Soc.*, **60**, pp. 309–19 (1938).

3.14. G. J. VERBECK and W. E. HASS, Dilatometer method for determination of thermal coefficient of expansion of fine and coarse aggregate, *Proc. Highw. Res. Bd.*, **30**, pp. 187–93 (1951).

3.15. B. W. SHACKLOCK and W. R. WALKER, The specific surface of concrete aggregates and its relation to the workability of concrete, *Cement Concr. Assoc. Res. Rep. No.* 4 (London, July 1958).

3.16. S. WALKER, D. L. BLOEM and R. D. GAYNOR, Relationship of concrete strength to maximum size of aggregate, *Proc. Highw. Res. Bd.*, **38**, pp. 367–79 (Washington D.C., 1959).

3.17. A. G. LOUDON, The computation of permeability from simple soil tests, *Géotechnique*, **3**, No. 4, pp. 165–83 (Dec. 1952).

3.18. W. H. GLANVILLE, A. R. COLLINS and D. D. MATTHEWS, The grading of aggregates and workability of concrete, *Road Research Tech. Paper No.* 5 (London, H.M.S.O., 1947).

3.19. L. J. MURDOCK, The workability of concrete, *Mag. Concr. Res.*, **12**, No. 36, pp. 135–44 (Nov. 1960).

3.20. N. DAVEY, Concrete mixes for various building purposes, *Proc. of a Symposium on Mix Design and Quality Control of Concrete*, pp. 28–41 (London, Cement and Concrete Assoc., 1954).

3.21. A J. NEWMAN and D. C. TEYCHENNÉ, A classification of natural sands and its use in concrete mix design, *ibid.*, pp. 175–93.

3.22. B. W. SHACKLOCK, Discussion on reference 3.21, pp. 199–200.

3.23. ROAD RESEARCH, Design of concrete mixes, *D.S.I.R. Road Note No.* 4 (London, H.M.S.O., 1950).

3.24. J. D. McINTOSH and H. C. ERNTROY, The workability of concrete mixes with ⅜ in. aggregates, *Cement Concr. Assoc. Res. Rep. No.* 2 (London, June 1955).

3.25. J. D. McINTOSH, The use in mass concrete of aggregate of large maximum size, *Civil Engineering*, **52**, No. 615, pp. 1011–15 (London, Sept. 1957).

3.26. B. W. SHACKLOCK, Comparison of gap- and continuously graded concrete mixes, *Cement Concr. Assoc. Tech. Rep. TRA/*240 (London, Sept. 1959).

3.27. J. D. McINTOSH, The selection of natural aggregates for various types of concrete work, *Reinf. Concr. Rev.*, **4**, No. 5, pp. 281–305 (London, March 1957).

3.28. D. L. BLOEM, Effect of maximum size of aggregate on strength of concrete, *National Sand and Gravel Assoc. Circular No.* 74 (Washington D.C., Feb. 1959).

3.29. A. J. GOLDBECK, Needed research, *A.S.T.M. Sp. Tech. Publicn. No.* 169, pp. 26–34 (1956).

3.30. T. C. POWERS and H. H. STEINOUR, An interpretation of published researches on the alkali-aggregate reaction, *J. Amer. Concr. Inst.*, **51**, pp. 497–516 (Feb. 1955) and pp. 785–811 (April 1955).

3.31. W. C. HANNA, Additional information on inhibiting alkali-aggregate expansion, *J. Amer. Concr. Inst.*, **48**, p. 513 (Feb. 1952).

3.32. HIGHWAY RESEARCH BOARD, The alkali-aggregate reaction in concrete, *Research Report* 18–*C* (Washington D.C., 1958).

3.33. R. C. MIELENZ and L. P. WITTE, Tests used by Bureau of Reclamation for identifying reactive concrete aggregates, *Proc. A.S.T.M.*, **48**, pp. 1071–103 (1948).

3.34. W. LERCH, Concrete aggregates—chemical reactions, *A.S.T.M. Sp. Tech. Publicn. No.* 169, pp. 334–45 (1956).

3.35. H. E. VIVIAN, Studies in cement-aggregate reaction: X. The effect on mortar expansion of amount of reactive component, *Commonwealth Scientific and Industrial Research Organization Bul. No.* 256, pp. 13–20 (Melbourne, 1950).

3.36. G. J. VERBECK and C. GRAMLICH, Osmotic studies and hypothesis concerning alkali-aggregate reaction, *Proc. A.S.T.M.*, **55**, pp. 1110–28 (1955).

3.37. C. E. WUERPEL, *Aggregates for concrete* (Washington, National Sand and Gravel Assoc., 1944).

3.38. D.S.I.R. BUILDING RESEARCH STATION, *Principles of Modern Building* (London, H.M.S.O., 1959).

3.39. R. RHOADES and R. C. MIELENZ, Petrography of concrete aggregates, *J. Amer. Concr. Inst.*, **42**, pp. 581–600 (June 1946).

3.40. F. A. SHERGOLD, A review of available information on the significance of roadstone tests, *Road Research Tech. Paper No.* 10 (London, H.M.S.O., 1948).

3.41 B. P. HUGHES and B. BAHRAMIAN, A laboratory test for determining the angularity of aggregate, *Mag. Concr. Res.*, **18**, No. 56, pp. 147–52 (Sept. 1966).

3.42. D. L. BLOEM and R. D. GAYNOR, Effects of aggregate properties on strength of concrete, *J. Amer. Concr. Inst.*, **60**, pp. 1429–55 (Oct. 1963).

3.43. K. M. ALEXANDER, A study of concrete strength and mode of fracture in terms of matrix, bond and aggregate strengths, *Tewksbury Symp. on Fracture, University of Melbourne, August* 1963, 27 pp.

3.44. G. P. CHAPMAN and A. R. ROEDER, The effects of sea-shells in concrete aggregates, *Concrete*, **4**, No. 2, pp. 71–9 (London, Feb. 1970).

3.45. J. D. DEWAR, Effect of mica in the fine aggregate on the water requirement and strength of concrete, *Cement Concr. Assoc. Tech. Rep. TRA/370* (London, April 1963).

3.46. H. G. MIDGLEY, The effect of lead compounds in aggregate upon the setting of Portland cement, *Mag. Concr. Res.*, **22**, No. 70, pp. 42–4 (March 1970).

3.47. W. C. HANSEN, Chemical reactions, *A.S.T.M. Sp. Tech. Publicn. No.* 169–A, pp. 487–96 (1966).

3.48. E. G. SWENSON and J. E. GILLOTT, Alkali reactivity of dolomitic limestone aggregate, *Mag. Concr. Res.*, **19**, No. 59, pp. 95–104 (June 1967).

3.49 S. POPOVICS, The use of the fineness modulus for the grading evaluation of aggregates for concrete, *Mag. Concr. Res.*, **18**, No. 56, pp. 131–40 (Sept. 1966).

3.50. L. N. EDWARDS, Proportioning the materials of mortars and concretes by surface area of aggregates, *Proc. A.S.T.M.*, **18**, Part II, pp. 235–302 (1918).

3.51. E. C. HIGGINSON, G. B. WALLACE and E. L. ORE, Effect of maximum size of aggregate on compressive strength of mass concrete, *Symp. on Mass Concrete, Amer. Concr. Inst. Sp. Publicn. No.* 6, pp. 219–56 (1963).

3.52. E. KEMPSTER, Measuring void content: new apparatus for aggregates, sands and fillers, *Current Paper CP* 19/69 (Building Research Station, Garston, May 1969).

4. Fresh Concrete

THE strength of concrete of given mix proportions is very seriously affected by the degree of its compaction; it is vital, therefore, that the consistence of the mix be such that the concrete can be transported, placed and finished sufficiently easily and without segregation.

Definition of Workability

A concrete satisfying these conditions is said to be workable, but to say merely that workability determines the ease of placement and the resistance to segregation is too loose a description of this vital property of concrete. Furthermore, the desired workability in any particular case would depend on the means of compaction available; likewise, a workability suitable for mass concrete is not necessarily sufficient for thin, inaccessible, or heavily reinforced sections. For these reasons, workability should be defined as a physical property of concrete alone without reference to the circumstances of a particular type of construction.

To obtain such a definition it is necessary to consider what happens when concrete is being compacted. Whether compaction is achieved by ramming or by vibration, the process consists essentially of the elimination of entrapped air from the concrete until it has achieved as close a configuration as is possible for a given mix. Thus, the work done is used to overcome the friction between the individual particles in the concrete and also between the concrete and the surface of the mould or of the reinforcement. These two can be called internal friction and surface friction respectively. In addition, some of the work done is used in vibrating the mould or in shock and indeed in vibrating those parts of the concrete which have already been fully consolidated. Thus the work done consists of a "wasted" part and "useful" work, the latter, as mentioned before, comprising work done to overcome the internal friction and the surface friction. Since only the internal friction is an intrinsic property of the mix, workability can be best defined as the amount of useful internal work necessary to produce full compaction. This definition has been developed by Glanville, Collins and Matthews[4.1] of the Road Research Laboratory who have exhaustively examined the field of compaction and workability.

Another term used to describe the state of fresh concrete is *consistence*. In ordinary English usage this word refers to the firmness of form of a substance or to the ease with which it will flow. In the case of concrete, consistence is sometimes taken to mean the degree of wetness; within limits, wet concretes are more workable than dry concretes, but concretes of the same consistence may vary in workability.

The Need for Sufficient Workability

Workability has so far been discussed merely as a property of fresh concrete: it is, however, also a vital property as far as the finished product is concerned since concrete must have a workability such that compaction to maximum density is possible with a reasonable amount of work or with the amount that we are prepared to put in under given conditions.

The need for compaction becomes apparent from a study of the relation between the degree of compaction and the resulting strength. It is convenient to express the former as a density ratio, i.e. a ratio of the actual density of the given concrete to the density of the same mix if fully compacted. Likewise, the ratio of the strength of the concrete as actually (partially) compacted to the strength of the same mix when fully compacted can be called the strength ratio. Then the relation between the strength ratio and the density ratio is of the form shown in Fig. 4.1. The presence of voids in concrete greatly reduces its strength: 5 per cent of voids can lower strength by as much as 30 per cent and even 2 per cent voids can result in a drop of strength of more than 10 per cent.[4.1] This, of course, is in agreement with Feret's expression relating strength to the sum of the volumes of water and air in the hardened paste (*see* p. 233).

Voids in concrete are in fact either bubbles of entrapped air or spaces left after excess water has been removed. The volume of the latter depends solely on the water/cement ratio of the mix. The air bubbles, which represent "accidental" air, i.e. voids within an originally loose granular material, are governed by the grading of the finest particles in the mix and are more easily expelled from a wetter mix than from a dry one. It follows, therefore, that for any given method of compaction there may be an optimum water content of the mix at which the sum of the volumes of air bubbles and water space will be a minimum. At this optimum water content the highest density ratio of the concrete would be obtained. It can be seen, however, that the optimum water content may vary for different methods of compaction.

Factors Affecting Workability

The main factor is the water *content* of the mix, expressed in kilogrammes of water per cubic metre of concrete: it is convenient, though approximate, to assume that for a given type and grading of aggregate and workability of concrete, the water content is independent of the aggregate/cement ratio. On the basis of this assumption, the mix proportions of concretes of different richness can be estimated, and Table 4.1 gives typical values of water content for different slumps and maximum sizes of aggregate. These values are applicable to non-air-entrained concrete only. When air is entrained the water content can be reduced in accordance with the data of Fig. 4.2. This is indicative of the order of values only,

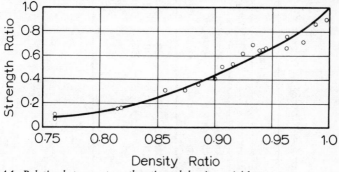

Fig. 4.1. *Relation between strength ratio and density ratio*[4.1]
(*Crown copyright*)

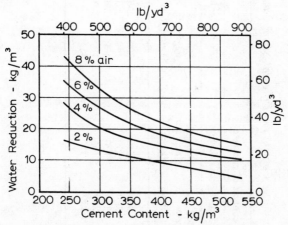

Fig. 4.2. *Reduction in mixing water requirement due to air-entrainment*[4.2]

since the effect of entrained air on workability depends on the mix proportions, as described in detail on p. 425.

If the water content and the other mix proportions are fixed, workability is governed by the maximum size of aggregate, its grading, shape and texture. The influence of these factors was discussed in Chapter 3. However, the grading and the water/cement ratio have to be considered

Table 4.1: *Approximate Water Content for Different Slumps and Maximum Sizes of Aggregate*[4.2]

Maximum size of aggregate		Water content of concrete										
		25–50 mm (1–2 in.) slump				75–100 mm (3–4 in.) slump				150–175 mm (6–7 in.) slump			
		Rounded aggregate		Angular aggregate		Rounded aggregate		Angular aggregate		Rounded aggregate		Angular aggregate	
mm	in.	kg/m³	lb/yd³	kg/m³	lb/yd³	kg/m³	lb/yd³	kg/m³	lb/yd³	kg/m³	lb/yd³	kg/m³	lb/yd³
9·5	$\frac{3}{8}$	190	320	210	360	200	340	225	380	230	390	255	430
19·0	$\frac{3}{4}$	170	290	195	330	190	320	210	350	210	350	225	380
38·1	$1\frac{1}{2}$	160	270	170	290	170	290	190	320	190	320	210	350
50·8	2	150	250	165	280	165	280	180	300	180	300	195	330
76·2	3	135	230	155	260	155	260	165	280	160	270	185	310

together, as a grading producing the most workable concrete for one particular value of water/cement ratio may not be the best for another value of the ratio. In particular, the higher the water/cement ratio the finer the grading required for the highest workability In actual fact, for a given value of water/cement ratio there is one value of the coarse/fine aggregate ratio (using given materials) that gives the highest workability.[4.1] Conversely, for a given workability, there is one value of coarse/fine aggregate ratio which needs the lowest water content. The influence of these factors was discussed in Chapter 3.

It should be remembered, however, that, although when discussing gradings of aggregate required for a satisfactory workability proportions by weight were laid down, these apply only to aggregate of a constant specific gravity. In actual fact, workability is governed by the volumetric proportions of particles of different sizes, so that when aggregates of varying specific gravity are used (e.g. in the case of some lightweight aggregates or mixtures of ordinary and lightweight aggregates) the mix proportions should be assessed on the basis of absolute volume of each size fraction. This applies also in the case of air-entrained concrete since the entrained air behaves like weightless fine particles. An example of calculation on absolute volume basis is given on p. 597. The influence of the properties of aggregate on workability decreases with an increase in the richness of the mix, and possibly disappears altogether when the aggregate/cement ratio is as low as $2\frac{1}{2}$ or 2.

In practice, predicting the influence of mix proportions on workability requires care since of the three factors water/cement ratio, aggregate/cement ratio, and water content, only two are independent. For instance, if the aggregate/cement ratio is reduced, but the water/cement ratio is kept constant, the water content increases, and consequently the workability also increases. If, on the other hand, the water content is kept constant when the aggregate/cement ratio is reduced, then the water/cement ratio decreases but workability is not seriously affected.

The last qualification is necessary because of some secondary effects: a lower aggregate/cement ratio means a higher total surface area of solids (aggregate and cement) so that the same amount of water results in a somewhat decreased workability. This could be offset by the use of a slightly coarser grading of aggregate. There are also other minor factors such as fineness of cement, but the influence of this is still controversial.

Measurement of Workability
Unfortunately no test is known that will measure directly the workability as defined earlier. Numerous attempts have been made, however, to correlate workability with some easily determinable physical measurement, but none of these is fully satisfactory although they may provide useful information within a range of variation in workability.

Slump Test

This is a test used extensively in site work all over the world. The slump test does not measure the workability of concrete but is very useful in detecting variations in the uniformity of a mix of given nominal proportions.

There are some slight differences in the details of procedure used in different countries, but these are not significant. The prescriptions of B.S. 1881: Part 2: 1970 are summarized below.

The mould for the slump test is a frustum of a cone, 300 mm (or 12 in.) high. It is placed on a smooth surface with the smaller opening at the top, and filled with concrete in four layers. Each layer is tamped 25 times with a standard 16 mm ($\frac{5}{8}$ in.) diameter steel rod, rounded at the end, and the top surface is struck with a trowel. The mould must be firmly held against its base during the entire operation; this is facilitated by handles or foot-rests brazed to the mould.

Immediately after filling, the cone is slowly lifted, and the unsupported concrete will now slump—hence the name of the test. The decrease in the height of the highest part* of the slumped concrete is called *slump*, and is measured to the nearest 5 mm (or $\frac{1}{4}$ in.). In order to reduce the influence on slump of the variation in the surface friction, the inside of the mould and its base should be moistened at the beginning of every test, and prior to lifting of the mould the area immediately around the base of the cone should be cleaned from concrete which may have dropped accidentally.

If instead of slumping evenly all round as in a true slump (Fig. 4.3) one half of the cone slides down an inclined plane, a shear slump is said to have taken place, and the test should be repeated. If shear slump persists, as may be the case with harsh mixes, this is an indication of lack of cohesion in the mix.

Mixes of stiff consistence have a zero slump, so that in the rather dry range no variation can be detected between mixes of different workability. Rich mixes behave satisfactorily, their slump being sensitive to variations in workability. However, in a lean mix with a tendency to harshness a true slump can easily change to the shear type, or even to collapse (Fig. 4.3), and widely different values of slump can be obtained in different samples from the same mix.

The order or magnitude of slump for different workabilities is given in Table 4.2. It should be remembered, however, that with different aggregates the same slump can be recorded for different workabilities, as indeed the slump bears no unique relation to the workability as defined earlier.

Despite these limitations the slump test is very useful on the site as a check on the day-to-day or hour-to-hour variation in the materials being fed into the mixer. An increase in slump may mean, for instance, that the

---

* The centre, according to some specifications, e.g. A.S.T.M. Standard C 143–71.

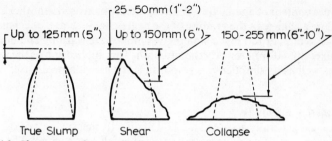

Fig. 4.3. *Slump: true, shear, and collapse*

Table 4.2 *Workability, Slump, and Compacting Factor of Concretes with 19 or 38* mm ($\frac{3}{4}$ *or* $1\frac{1}{2}$ in.) *Maximum Size of Aggregate*[4.3]

Degree of workability	Slump		Compacting factor		Use for which concrete is suitable
	mm	in.	Small apparatus	Large apparatus*	
Very low	0–25	0–1	0·78	0·80	Roads vibrated by power-operated machines. At the more workable end of this group, concrete may be compacted in certain cases with hand-operated machines.
Low	25–50	1–2	0·85	0·87	Roads vibrated by hand-operated machines. At the more workable end of this group, concrete may be manually compacted in roads using aggregate of rounded or irregular shape. Mass concrete foundations without vibration or lightly reinforced sections with vibration.
Medium	50–100	2–4	0·92	0·935	At the less workable end of this group, manually compacted flat slabs using crushed aggregates. Normal reinforced concrete manually compacted and heavily reinforced sections with vibration.
High	100–175	4–7	0·95	0·96	For sections with congested reinforcement. Not normally suitable for vibration.

* Not normally used.

moisture content of aggregate has unexpectedly increased; another cause would be a change in the grading of the aggregate, such as a deficiency of sand. Too high or too low a slump gives immediate warning and enables the mixer operator to remedy the situation. This application of the slump test, as well as its simplicity, is responsible for its widespread use.

Compacting Factor Test

There is no generally accepted method of directly measuring workability, i.e. the amount of work necessary to achieve full compaction. The most reliable test yet available uses the inverse approach: the degree of compaction achieved by a standard amount of work is determined. The work applied includes perforce the work done against the surface friction but this is reduced to a minimum, although probably the actual friction varies with the workability of the mix.

The degree of compaction, called the compacting factor, is measured by the density ratio, i.e. the ratio of the density actually achieved in the test to the density of the same concrete fully compacted.

The test, known as the compacting factor test, was developed at the Road Research Laboratory[4.1] and is now covered by B.S. 1881: Part 2: 1970. The apparatus consists essentially of two hoppers, each in the shape of a frustum of a cone, and one cylinder, the three being above one another. The hoppers have hinged doors at the bottom, as shown in Fig. 4.4. All inside surfaces are polished to reduce friction.

The upper hopper is filled with concrete, this being placed gently so that at this stage no work is done on the concrete to produce compaction. The bottom door of the hopper is then released and the concrete falls into the lower hopper. This is smaller than the upper one and is, therefore, filled to overflowing and thus always contains approximately the same amount of concrete in a standard state: the influence of the personal factor in filling the top hopper is greatly reduced. The bottom door of the lower hopper is released and the concrete falls into the cylinder. Excess concrete is cut by two floats slid across the top of the mould, and the net weight of concrete in the known volume of the cylinder is determined.

The density of the concrete in the cylinder is now calculated, and this density divided by the density of the fully compacted concrete is defined as the compacting factor. The latter density can be obtained by actually filling the cylinder with concrete in four layers, each tamped or vibrated, or alternatively calculated from the absolute volumes of the mix ingredients.

The compacting factor apparatus shown in Fig. 4.4 is about 1·2 m (4 ft) high. For concretes with a maximum aggregate size of over 19 mm and up to 38 mm ($\frac{3}{4}$ in. to $1\frac{1}{2}$ in.) a "large" apparatus should be employed. Its height is 1·8 m (6 ft) and for this reason the large apparatus is not used in practice. For the same concrete the large apparatus yields a somewhat higher value of the compacting factor than the small apparatus. Unfortunately, the compacting factor apparatus is not often used outside precast concrete works and large sites.

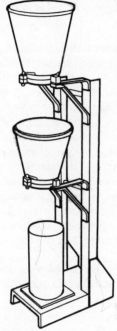

Fig. 4.4. *Compacting factor apparatus*

Table 4.2 lists the values of the compacting factor for different workabilities, as given in Road Note No. 4.[4.3] Unlike the slump test, variations in the workability of dry concrete are reflected in a large change in the compacting factor, i.e. the test is more sensitive at the low workability end of the scale than at high workability. However, very dry mixes tend to stick in one or both hoppers and the material has to be eased gently by poking with a steel rod. Moreover, it seems that for concrete of very low workability the actual amount of work required for full compaction depends on the richness of the mix while the compacting factor does not: leaner mixes need more work than richer ones.[4.4] This means that the implied assumption that all mixes with the same compacting factor require the same amount of useful work is not always justified. Likewise the assumption, mentioned earlier, that the wasted work represents a constant proportion of the total work done regardless of the properties of the mix is not quite correct Nevertheless, the compacting factor test undoubtedly provides a good measure of workability.

An automatic compacting factor test apparatus has also been developed. Here, the cylinder is supported by a spring balance which can be calibrated for a given mix so as to read workability directly, or even to indicate the excess or deficiency of water in kilogrammes per batch.

Flow Test

This laboratory test gives an indication of the consistence of concrete and its proneness to segregation by measuring the spread of a pile of concrete subjected to jolting. It is with regard to segregation that the flow test is of'greatest value, but it also gives a good assessment of consistence of stiff, rich, and rather cohesive mixes.

The test is covered by A.S.T.M. Standard C 124–39 (re-approved in 1966). The apparatus consists essentially of a brass-top table, 760 mm (30 in.) in diameter, mounted so that it can be jolted by a drop of 13 mm ($\frac{1}{2}$ in.). A mould in the shape of a frustum of a cone, much more squat than the slump cone, is placed at the centre of the table, filled with concrete in two layers and compacted in a manner similar to the slump test. The mould is now removed and the table is jolted 15 times in 15 seconds. This is done by a wheel operating an actuating cam. As a result, the concrete spreads over the table, and the average diameter of the spread concrete is measured.

The flow of concrete is defined as the percentage increase in the average diameter of the spread concrete (D in.) over the original diameter of the base (10 in.) i.e.

$$\text{Flow} = \frac{D - 10}{10} \times 100$$

Values from 0 to 150 per cent can be obtained.

The jolting applied during the test encourages segregation, and if the mix is not cohesive the larger particles of aggregate will separate out and move toward the edge of the table. Another form of segregation is possible: in a sloppy mix the cement paste tends to run away from the centre of the table leaving the coarser material behind.

It should be noted that the flow test does not measure workability, as concretes having the same flow may differ considerably in their workability.

Remoulding Test

Use is made of the flow table in another test, in which an assessment of workability is made on the basis of the effort involved in changing the shape of a sample of concrete. This is the remoulding test, developed by Powers.[4.5]

The apparatus is shown diagrammatically in Fig. 4.5. A standard slump cone is placed in a cylinder 305 mm (12 in.) in diameter and 203 mm (8 in.) high, the cylinder being mounted rigidly on a flow table, adjusted to give a 6·3 mm ($\frac{1}{4}$ in.) drop. Inside the main cylinder there is an

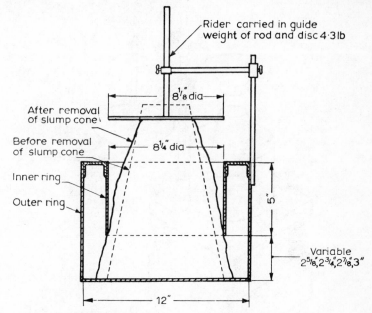

Fig. 4.5. *Remoulding test apparatus*

inner ring, 210 mm ($8\frac{1}{4}$ in.) in diameter and 127 mm (5 in.) high. The distance between the bottom of the inner ring and the bottom of the main cylinder can be set between 67 mm and 76 mm ($2\frac{5}{8}$ in. and 3 in.).

The slump cone is filled in the standard manner, removed, and a disc-shaped rider (weighing 1·9 kg (4·3 lb)) is placed on top of the concrete. The table is now jolted at the rate of one jolt per second until the bottom of the rider is 81 mm ($3\frac{3}{16}$ in.) above the base plate. At this stage the shape of the concrete has changed from a frustum of a cone to a cylinder. The effort required to achieve this remoulding is expressed as the number of jolts required. For very dry mixes a considerable effort may be necessary.

The test is purely a laboratory one but is valuable as the remoulding effort appears to be closely related to workability.

Vebe Test

This is a development of the remoulding test in which the inner ring of Powers' apparatus is omitted and compaction is achieved by vibration instead of jolting. The apparatus is shown diagrammatically in Fig. 4.6. The name *Vebe* is derived from the initials of V. Bährner of Sweden who developed the test. The test is now covered by B.S. 1881: Part 2: 1970, where it is referred to as the "V-B" consistometer test.

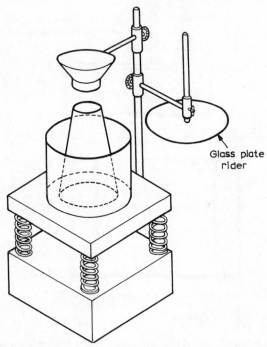

Fig. 4.6. *Vebe apparatus*

The remoulding is assumed to be complete when the glass plate rider is completely covered with concrete and all cavities in the surface of the concrete have disappeared. This is judged visually, and the difficulty of establishing the end point of the test may be a source of error. To overcome it an automatically operated device for recording the movement of the plate against time may be fitted.

Compaction is achieved using a vibrating table with an eccentric weight rotating at 3,000 rev/min and a maximum acceleration of $3g$ to $4g$. It is assumed that the input of energy required for compaction is a measure of workability of the mix, and this is expressed in Vebe seconds, i.e. the time required for the remoulding to be complete. Sometimes, a correction for the change in the volume of concrete from V_2 before to V_1 after vibration is applied, the time being multiplied by V_2/V_1.

Vebe is a good laboratory test, particularly for very dry mixes. This is in contrast to the compacting factor test where error may be introduced by the tendency of some dry mixes to stick in the hoppers. The Vebe test also has the additional advantage that the treatment of concrete during the test is comparatively closely related to the method of placing in practice.

Ball Penetration Test

This is a simple field test consisting of the determination of the depth to which a 152 mm (6 in.) diameter metal hemisphere, weighing 13·6 kg (30 lb), will sink under its own weight into fresh concrete. A sketch of the apparatus, devised by Kelly and known as the Kelly ball, is shown in Fig. 4.7.

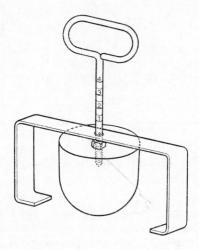

Fig. 4.7. *Kelly ball*

The use of this test is similar to that of the slump test, that is routine checking of consistence for control purposes. The test is essentially an American one, covered by A.S.T.M. Standard C 360–63 (re-approved in 1968), and is so far practically unknown in Britain. It is, however, worth considering the Kelly ball test as an alternative to the slump test, over which it has some advantages. In particular, the ball test is simpler and quicker to perform and, what is more important, it can be applied to concrete in a wheelbarrow or actually in the form. In order to avoid the effects of a boundary the depth of the concrete being tested should be not less than 200 mm (8 in.), and the least lateral dimension 460 mm (18 in.).

As would be expected, there is no simple correlation between penetration and slump, since neither test measures any basic property of concrete but only the response to specific conditions. On a site, when a particular mix is used, correlation can be found, as shown for instance in Fig. 4.8. In practice the ball test is essentially used to measure variations in the mix, such as those due to a variation in the moisture content of the aggregate.

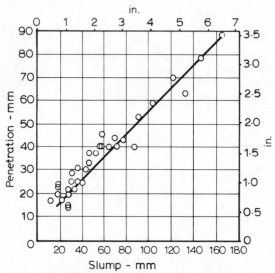

Fig. 4.8. *Relation between Kelly ball penetration and slump*[4.6]

Comparison of Tests

It should be said at the outset that no comparison is really possible as each test measures the behaviour of concrete under different conditions. The particular uses of each test have been mentioned.

The compacting factor test is closely related to the reciprocal of workability, and the remoulding and Vebe tests are direct functions of workability. The Vebe test measures the properties of concrete under vibration as compared with the free-fall conditions of the compacting factor test and the jolting in the remoulding test. All three tests are satisfactory in the laboratory, but the compacting factor apparatus is the most suitable for site use.

An indication of the relation between the compacting factor and the Vebe time is given by Fig. 4.9, but this applies only to the mixes used, and the relation must not be assumed to be generally applicable since it depends on factors such as the shape and texture of the aggregate or presence of entrained air, as well as on mix proportions. For specific mixes the relation between compacting factor and slump has been obtained, but such a relation is also a function of the properties of the mix. A general

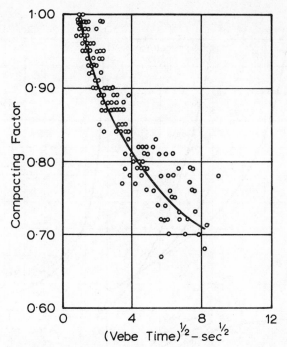

Fig. 4.9. *Relation between compacting factor and Vebe time*[4.4]

indication of the pattern of the relation between the compacting factor, Vebe time and slump is shown in Fig. 4.10. The influence of the richness of the mix in two of these relations is clear. The absence of influence in the case of the relation between slump and Vebe time is illusory because slump is insensitive at one end of the scale (low workability) and Vebe time at the other; thus two asymptotic lines with a small connecting part are present.

The flow test is valuable in assessing the cohesiveness of a laboratory mix.

The slump and penetration tests are purely comparative, and in that capacity both are very useful except that the slump test is unreliable with lean mixes, for which good control is often of considerable importance.

The ideal test for workability is yet to be devised. For this reason it is worth stressing the value of visual inspection of workability and of assessing it by patting with a trowel in order to see the ease of finishing. Experience is clearly necessary but once it has been acquired the "by eye" test, particularly for the purpose of checking uniformity, is both rapid and reliable.

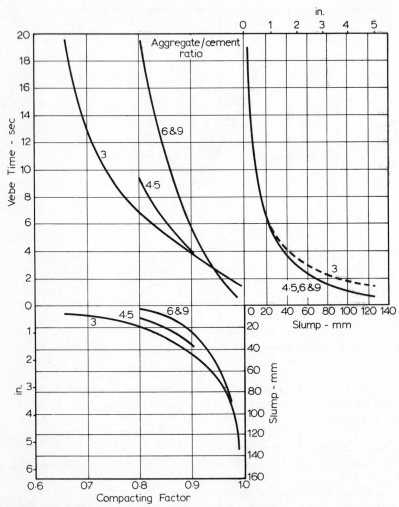

Fig. 4.10. *General pattern of relations between workability tests for mixes of varying aggregate/cement ratios*[4.14]

Effect of Time and Temperature on Workability

Freshly mixed concrete stiffens with time. This should not be confused with setting of cement. It is simply that some water from the mix is absorbed by the aggregate, some is lost by evaporation, particularly if the concrete is exposed to sun or wind, and some is removed by the initial chemical reactions. The compacting factor decreases by up to about 0·1 during a period of one hour from mixing; the exact value of the loss in workability varies with the richness of the mix, the type of cement, the temperature of the concrete, and the initial workability. An example of a slump–time curve, obtained by Evans,[4.25] is given in Fig. 4.11.

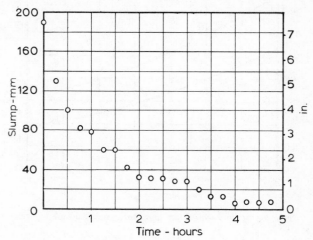

Fig. 4.11. *Relation between slump and time since completion of mixing for a 1:2:4 concrete with a water/cement ratio of 0·775*[4.25]

Because of this change in apparent consistence and also because we are really interested in the workability at the time of placing, i.e. some time after mixing, it is preferable to delay the appropriate test until, say, 15 minutes after mixing.

The workability of a mix is also affected by the ambient temperature, although, strictly speaking, we are concerned with the temperature of the concrete itself. Fig. 4.12 gives an example of the effect of temperature on slump of laboratory mixed concrete: it is apparent that on a hot day the water content of the mix would have to be increased for a constant workability to be maintained. Fig. 4.13 shows that as the concrete temperature increases the percentage increase in water required to effect a 25 mm (1 in.) change in slump also increases.

It is interesting that the loss of workability with increase in temperature has not been confirmed by recent site tests by Shalon[4.9] in a hot climate. Up to a temperature of 40°C (104°F) and with a relative humidity within the range 20 to 70 per cent, no effect of temperature on slump has been observed. Only above 50°C (122°F) or with humidities lower than 20 per cent does the slump fall off rapidly.

The opinion that higher temperature increases the mixing water demand, at least as far as reduction in slump affects the water requirement, has thus not been fully verified. These findings apply up to 40°C (104°F) and within 20 minutes of mixing. Over longer periods there is an

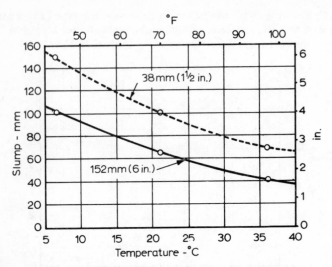

Fig. 4.12. *Influence of temperature on slump of concretes with different maximum aggregate size*[4.7]

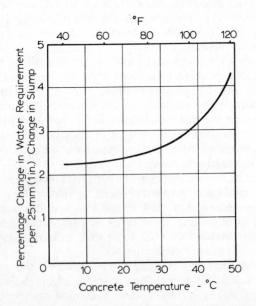

Fig. 4.13. *Influence of temperature on amount of water required to change slump*[4.8]

unmistakable loss of slump so that, for instance, with a long haul of ready-mixed concrete, high temperature would increase the water requirement for a given workability.

The discrepancy between the recent site tests and the laboratory tests of Fig. 4.12 may possibly be explained by the use of air-entrained concrete in the latter case: there may be some form of interaction between the entrained air, the cement, and the climate. The full answer is not known and it is recommended that for any new conditions actual site tests be made.

Segregation

In discussing workable concrete in general terms it was implied that such concrete should not easily segregate, i.e. it ought to be cohesive. However, strictly speaking the absence of a tendency to segregate is not included in the definition of a workable mix. Nevertheless, the absence of appreciable segregation is essential as full compaction of a segregated mix is impossible.

Segregation can be defined as separation of the constituents of a heterogeneous mixture so that their distribution is no longer uniform. In the case of concrete, it is the differences in the size of particles and in the specific gravity of the mix constituents that are the primary causes of segregation, but its extent can be controlled by the choice of suitable grading and by care in handling.

There are two forms of segregation. In the first, the coarser particles tend to separate out since they tend to travel further along a slope or to settle more than finer particles. The second form of segregation, occurring particularly in wet mixes, is manifested by the separation of grout (cement plus water) from the mix. With some gradings when a lean mix is used, the first type of segregation may occur if the mix is too dry; addition of water would improve the cohesion of the mix, but when the mix becomes too wet the second type of segregation would take place.

The influence of grading on segregation was discussed in detail in Chapter 3, but the actual extent of segregation depends on the method of handling and placing of concrete. If the concrete does not have far to travel and is transferred directly from the wheelbarrow to the final position in the form, the danger of segregation is small. On the other hand, dropping concrete from a considerable height, passing along a chute, particularly with changes of direction, and discharging against an obstacle—all these encourage segregation so that under such circumstances a particularly cohesive mix should be used. With a correct method of handling, transporting and placing, the likelihood of segregation can be greatly reduced: there are many practical rules but these are outside the scope of this book.

It should be stressed, however, that concrete should always be placed direct in the position in which it is to remain and must not be allowed to flow or be worked along the form. This prohibition includes the use of a vibrator to spread a heap of concrete over a larger area. Vibration provides a most valuable means of compacting concrete, but because a large amount of work is being done on the concrete the danger of segregation (in placing as distinct from handling) due to an improper use of a vibrator is increased. This is particularly so when vibration is allowed to continue too long: with many mixes separation of coarse aggregate toward the bottom of the form and of the cement paste toward the top may result. Such concrete would obviously be weak, and the laitance (scum) on its surface would be too rich and too wet so that a crazed surface with a tendency to dusting might result.

It may be noted that entrained air reduces the danger of segregation. On the other hand, the use of coarse aggregate whose specific gravity differs appreciably from that of the fine aggregate would lead to increased segregation.

Segregation is difficult to measure quantitatively, but is easily detected when concrete is handled on a site in any of the ways listed earlier as undesirable. A good picture of cohesion of the mix is obtained by the flow test. As far as proneness to segregation on over-vibration is concerned, a good test is to vibrate a concrete cube for about 10 min and then to strip it and observe the distribution of coarse aggregate: any segregation will be easily seen.

Bleeding

Bleeding, known also as water gain, is a form of segregation in which some of the water in the mix tends to rise to the surface of freshly placed concrete. This is caused by the inability of the solid constituents of the mix to hold all of the mixing water when they settle downwards. We are thus dealing with subsidence, and Powers[4.10] treats bleeding as a special case of sedimentation. Bleeding can be expressed quantitatively as the total settlement per unit height of concrete. The bleeding capacity as well as the rate of bleeding can be determined experimentally using the test of A.S.T.M. Standard C 232–71.

As a result of bleeding the top of every lift may become too wet and if the water is trapped by superimposed concrete, porous, weak, and non-durable concrete will result. If the bleeding water is remixed during finishing of the top surface a weak wearing surface will be formed. This can be avoided by delaying the finishing operations until the bleeding water has evaporated, and also by the use of wood floats and avoidance of over-working the surface. On the other hand, if evaporation of water from the surface of the concrete is faster than the bleeding rate plastic shrinkage cracking may result (*see* p. 321).

Some of the rising water becomes trapped on the underside of coarse aggregate particles or of reinforcement, thus creating zones of poor bond. This water leaves behind capillaries, and since all the voids are oriented in the same direction, the permeability of the concrete in a horizontal plane may be increased. A small amount of voids of this type is nearly always present, but appreciable bleeding must be avoided as the danger of frost damage may be increased. Bleeding is often pronounced in thin slabs, such as road slabs, and it is in these that frost generally constitutes a considerable danger.

Bleeding need not necessarily be harmful. If it is undisturbed (and the water evaporates) the effective water/cement ratio may be lowered with a resulting increase in strength. On the other hand, if the rising water carries with it a considerable amount of the finer cement particles a layer of laitance will be formed. If this is at the top of a slab a porous surface will result, with a permanently "dusty" surface. At the top of a lift a plane of weakness would form and the bond with the next lift would be inadequate. For this reason, laitance should always be removed by brushing and washing.

The tendency to bleeding depends largely on the properties of cement. Bleeding is decreased by increasing the fineness of cement and is also affected by certain chemical factors: there is less bleeding when the cement has a high alkali content, a high C_3A content, or when calcium chloride is added.[4.11] A higher temperature, within the normal range, increases the rate of bleeding, but the total bleeding capacity is probably unaffected. The physical properties of fine aggregate, especially that smaller than a 150 μm (No. 100) B.S. sieve, may also affect bleeding.[4.12] Rich mixes are less prone to bleeding than lean ones. Reduction in bleeding is obtained by the addition of pozzolanas or of aluminium powder. Air entrainment effectively reduces bleeding so that finishing can follow casting without delay.

Bleeding of concrete continues until the cement paste has stiffened sufficiently to put an end to the process of sedimentation.

The Mixing of Concrete

Mixing concrete by hand is expensive in labour and it is, therefore, not surprising that mechanical mixers have been in general use for a great many years.

Concrete Mixers

The object of mixing is to coat the surface of all aggregate particles with cement paste, and to blend all the ingredients of concrete into a uniform mass; this uniformity must furthermore not be disturbed by the process of discharging from the mixer. In fact, the method of discharging is one of the bases of classification of concrete mixers. Several types exist. In the tilting mixer the mixing chamber, known as the drum, is tilted for discharging. In the non-tilting type the axis of the mixer is always horizontal, and discharge is obtained either by inserting a chute into the drum or by reversing the direction of rotation of the drum (when the mixer is known as a reversing drum mixer), or rarely by splitting of the drum. There are also pan-type mixers rather similar in operation to an electric cake-mixer; these are called forced action mixers, as distinct from the tilting and non-tilting mixers which rely on the free fall of concrete in the drum.

Tilting mixers usually have a conical or bowl-shaped drum with vanes inside. The efficiency of the mixing operation depends on the details of design, but the discharge action is always good as all the concrete can be tipped out rapidly and in an unsegregated mass as soon as the drum is tilted. For this reason tilting-drum mixers are preferable for mixes of low workability and those containing large-size aggregate.

On the other hand, because of a rather slow rate of discharge from a non-tilting drum mixer, concrete is sometimes susceptible to segregation. In particular, the largest size of aggregate may tend to stay in the mixer so that the discharge starts as mortar and ends as a collection of coated stones. It is not suggested, of course, that this is a universal fault of non-tilting drum mixers but it may be advisable to check the performance of an untried type. This can be done by comparing the content of coarse aggregate in the second tenth of the batch with that in the ninth tenth. Limits for the difference between the two contents are discussed on p. 204.

Non-tilting mixers are always charged by means of a loading skip, which is also used with the larger tilting drum mixers. It is important that the whole charge from the skip be transferred into the mixer every time, i.e. no sticking must occur. Sometimes a shaker mounted on the skip assists in emptying it.

The pan mixer is generally not mobile and is therefore used either at a central mixing plant on a large concrete project, at a precast factory, or in a small version in the concrete laboratory. The mixer consists essentially of a circular pan rotating about its axis, with one or two stars of paddles rotating about a vertical axis not coincident with the axis of the pan. Sometimes the pan is static and the axis of the star travels along a circular path about the axis of the pan. In either case, the relative movement between the paddles and the concrete is the same, and concrete in every part of the pan is thoroughly mixed. Scraper blades prevent mortar sticking to the sides of the pan, and the height of the paddles can be adjusted so as to prevent a permanent coating of mortar forming on the bottom of the pan.

Pan mixers are particularly efficient with stiff and cohesive mixes and are, therefore, often used in the manufacture of precast concrete. They are also suitable, because of the scraping arrangements, for mixing very small quantities of concrete—hence their use in the laboratory. A bowl-and-stirrer of the cake-mixer type, working on the same principle as the pan mixer, is sometimes used for mixing of mortar.

It may be relevant to mention that in drum-type mixers no scraping of the sides takes place during mixing so that a certain amount of mortar adheres to the sides of the drum and stays there until the mixer has been cleaned at the end of the day's work. It follows that at the beginning of concreting the first mix would leave a large proportion of its mortar behind, and the discharge would consist largely of coated coarse particles. This initial batch should be discarded. As an alternative, a certain amount of mortar may be introduced into the mixer prior to the commencement of concreting, a procedure known as buttering the mixer. A convenient and simple way is to charge the mixer with the usual quantities of cement, water and fine aggregate, simply omitting the coarse material. The mix in excess of that stuck in the mixer can be used in construction and may in fact be particularly suitable for placing at a cold joint. The necessity of buttering should not be forgotten in laboratory work.

The size of a mixer is described by its volume, and two figures used to be given: the first denoted the volume of the unmixed ingredients in a loose state, and the second figure gave the volume of the concrete when mixed. For instance, mixers were described as 10/7, 14/10, 22/14, etc., the values being in cubic feet. Nowadays, the nominal size of the mixer is based on the volume of concrete after compaction (B.S. 1305: 1967). Mixers are made in a variety of sizes from 0.04 m^3 ($1\frac{1}{2}$ ft^3) for laboratory use up to 13 m^3 (17 yd^3). If the quantity mixed represents only a small fraction of the capacity of the mixer the resulting mix may not be uniform, and the operation would, of course, be uneconomical. Overload not exceeding 10 per cent is generally harmless.

One other type of mixer is of interest: the dual drum mixer used in road construction. In this there are two drums in series, concrete being mixed part of the time in one and then transferred to the other for the remainder of the mixing time, and finally discharged. In the meantime, the first drum is recharged and initial mixing takes place. The operations are synchronized so that there is no intermixing of batches. In this manner, the yield of concrete can be doubled compared with an ordinary mixer with the same batching equipment, and in road construction, where space and access are often limited, this is a considerable advantage. Triple-drum mixers are also used.

All the mixers considered so far are batch mixers, since one batch of concrete is mixed and discharged before any more materials are added. As opposed to this, a continuous mixer discharges mixed concrete steadily and is fed by a continuous weigh batching system. The mixer itself may be of a drum type or may rely on a screw moving in a stationary housing.

Specialized mixers are used in shotcreting and for mortar for preplaced aggregate concrete. The "colloid" mixer used for the latter may become of interest in concrete-making because the pre-mixing of cement and water allows better hydration and may lead to a higher strength at a given water/cement ratio than conventional mixing. For instance, at water/cement ratios of 0·45 to 0·50, a gain in strength of 10 per cent has been observed.[4.26] However, two-stage mixing undoubtedly represents a higher cost and is likely to be justifiable only in special cases.

Uniformity of Mixing

In any mixer it is essential that sufficient interchange of materials between different parts of the chamber takes place, so that uniform concrete is produced. The efficiency of the mixer can be measured by the variability of the mix discharged into a number of receptacles without interrupting the flow of concrete. For instance, in Belgium the mix is discharged in eight parts and these are compared for homogeneity.[4.13] The classification of the efficiency of mixing is then based on the data of Table 4.3. As is the case in some European countries, the Table uses mean deviation

Table 4.3: *Variability of Concrete in a "Satisfactory" Mixer*[4.13]

Test	Deviation—per cent	
	Maximum	Mean
Compressive strength	10–15	4–6
Percentage of coarse aggregate	15–20	6–8
Percentage of fine aggregate *or* cement	12–15	5–8

(arithmetic mean of the absolute values of deviation from the mean) and maximum deviation (the absolute value of deviation from the mean) instead of the standard deviation which is common in England and the U.S.

The values given in Table 4.3 would be typical of a satisfactory mixer, a smaller deviation denoting a very good mixer, and a greater deviation a poor one. It should be stressed, however, that the values listed give only the order of variation as the performance of a mixer depends on the consistence of the mix and on the maximum size of the aggregate used.

A rather rigid test of A.S.T.M. Standard C 94–72 (formally applicable only to mixers for ready-mixed concrete) lays down that samples of concrete should be taken from $\frac{1}{8}$ and $\frac{5}{8}$ points of a batch, and the differences in the properties of the two samples should not exceed any of the following—

density of concrete—16 kg/m³ (1 lb/ft³)
air content   —1 per cent
slump   —25 mm (1 in.) when the average is under 100 mm
                (4 in.), and 40 mm (1·5 in.) when the average is
                100 to 150 mm (4 to 6 in.)

percentage of aggregate retained on a 4·8 mm ($\frac{3}{16}$ in.) sieve—6 per cent density of air-free mortar—1·6 per cent compressive strength (average 7-day strength of 3 cylinders)—7·5 per cent.

The Bureau of Reclamation[4.7] lays down similar requirements.

In the United Kingdom, B.S. 3963: 1965 lays down a test of performance of mixers using a specified concrete mix. Maximum values of standard deviation of the contents of cement, fine aggregate, and water are prescribed; these values depend on the size of the mixer.

Mixing Time

On a site there is often a tendency to mix concrete as rapidly as possible, and it is, therefore, important to know what is the minimum mixing time necessary to produce a concrete uniform in composition and, as a result, of satisfactory strength. This time varies with the type of mixer, and, strictly speaking, it is not the mixing time but the number of revolutions of the mixer that is the criterion of adequate mixing. Generally, about 20 revolutions are sufficient. Since, however, there is an optimum speed of rotation recommended by the manufacturer of the mixer, the number of revolutions and the time of mixing are interdependent.

For a given mixer there exists a relation between mixing time and uniformity of the mix. Typical data are shown in Fig. 4.14, based on Shalon's[4.22] tests, the variability being represented as the range of strengths of specimens made from the given mix after a specified mixing time. Fig. 4.15 shows the results of the same tests plotted as a coefficient of variation against mixing time. It is apparent that mixing for less than 1 to 1¼ min produces an appreciably more variable concrete, but prolonging the mixing time beyond these values results in no significant improvement in uniformity.

The average strength of concrete increases also with an increase in mixing time, as shown for instance by Abrams' tests[4.23] (Fig. 4.16). The rate of increase falls rapidly beyond about one minute and is not significant beyond two minutes. Within the first minute, however, the influence of mixing time on strength is of considerable importance. For instance, Shalon[4.22] calculated that, for a given required strength, increasing the mixing time from 30 sec to 1 min permits a saving in the cement content of as much as 30 kilogrammes per cubic metre.

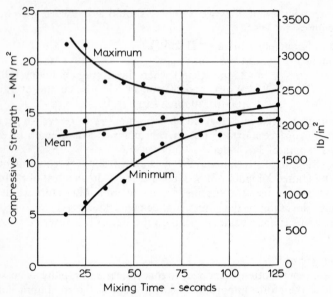

Fig. 4.14. *Relation between compressive strength and mixing time*[4.22]

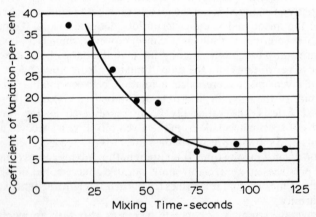

Fig. 4.15. *Relation between the coefficient of variation of strength and mixing time*[4.22]

As mentioned before, the exact value of the minimum mixing time varies with the type of mixer and depends also on its size. Table 4.4 gives typical values. The mixing time is reckoned from the time when all the solid materials have been put in the mixer, and it is usual to specify that

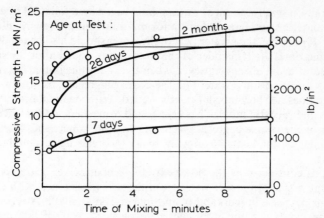

Fig. 4.16. *Effect of mixing time on strength of concrete*[4.23]

Table 4.4: *Recommended Minimum Mixing Times*

Capacity of mixer		Mixing time, min	
		Bureau of Reclamation[4.7]	American Concrete Institute[4.15] and A.S.T.M. Standard C 94–72
m³	yd³		
0·8	up to 1	1½	1
1·5	2	1½	1¼
2·3	3	2	1½
3·1	4	2½	1¾
3·8	5	2¾	2
4·6	6	3	2¼
7·6	10		3¼

all the water has to be added not later than after one quarter of the mixing time. The figures quoted refer to the usual mixers but there are many modern large mixers which perform satisfactorily with a mixing time of 1 to 1½ min. In high speed pan mixers the mixing time can be as short as 35 sec. On the other hand, when lightweight aggregate is used the mixing time should be not less than 5 min, sometimes divided into 2 min of mixing the aggregate with water, followed by 3 min with cement added. In general, the length of mixing time required for sufficient uniformity of the mix depends on the quality of blending of materials during charging of the mixer: simultaneous feed is beneficial.

No general rules on the order of feeding the ingredients into the mixer can be given as they depend on the properties of the mix and of the mixer.

Generally, a small amount of water should be fed first, followed by all the solid materials, preferably fed uniformly and simultaneously into the mixer. If possible, the greater part of the water should also be fed during the same time, the remainder of the water being added after the solids. With some drum mixers, however, when a very dry mix is used it is necessary to feed first some water and the coarse aggregate, as otherwise its surface does not become sufficiently wetted. With small laboratory pan mixers and very stiff mixes it has been found convenient to feed first sand, a part of the coarse aggregate, cement, then the water, and finally the remainder of the coarse aggregate so as to break up any nodules of mortar.

Let us consider now the other extreme—mixing over a long period. Generally, evaporation of water from the mix takes place, with a consequent decrease in workability and increase in strength. A secondary effect is that of grinding of the aggregate, particularly if soft: the grading of the aggregate thus becomes finer, and the workability lower. The friction effect also produces an increase in the temperature of the mix.

In the case of air-entrained concrete, prolonged mixing reduces the air content by about $\frac{1}{6}$ per hour (depending on the type of air-entraining agent), while a delay in placing without continuous mixing causes a drop in air content by only about $\frac{1}{10}$ per hour. On the other hand, a decrease in mixing time below 2 or 3 minutes may lead to inadequate entrainment of air.

Intermittent remixing up to about 3 hours, and in some cases up to 6 hours, is harmless as far as strength and durability are concerned, but the workability falls off with time unless loss of moisture from the mixer is prevented. Adding water to restore workability, known as re-tempering, will lower the strength of the concrete.

Some investigators[4.24] have reported that this loss of strength is smaller than would be expected from the consideration of the *total* water/cement ratio; others[4.27] have found the loss to be the same as if the retempering water were added during initial mixing. The explanation of this discrepancy probably lies in how the water was lost: water lost by evaporation should not be included in the *effective* water/cement ratio but if re-tempering replaces the water used up in hydration then the added water forms part of the effective water which governs the strength (*see* Fig. 4.19).

Re-tempering has been reported slightly to increase the shrinkage[4.27] but this probably occurs only if the effective water/cement ratio is increased by the added water.

Hand Mixing
There may be occasions when concrete has to be mixed by hand and, because in this case uniformity is more difficult to achieve, particular care

and effort are necessary. The aggregate should be spread in a uniform layer on a hard, clean and non-porous base; cement is then spread over the aggregate, and the dry materials are mixed by turning over from one end of the tray to another and "cutting" with a shovel until the mix appears uniform. Turning three times is usually required. Water is then gradually added so that neither water by itself nor with cement can escape. The mix is turned over again, usually three times, until it appears uniform in colour and consistence.

It is obvious that during hand mixing no soil or other extraneous material must be allowed to become included in the concrete.

Vibration of Concrete

The process of compacting the concrete consists essentially of the elimination of entrapped air. The oldest means of achieving this is by ramming or punning the surface of the concrete in order to dislodge the air and force the particles into a closer configuration. The more modern means is by vibration, by which the particles are momentarily separated, thus allowing them to be drawn into a compact mass.

The use of vibration as a means of compaction makes it possible to use drier mixes than can be compacted by hand (a compacting factor below about 0·75 or 0·80, down to 0·60 when pressure may also be required). In fact, extremely dry and stiff mixes can be vibrated satisfactorily so that for a given desired strength concrete can be made with a lower cement content. This means a saving in cost, but against that we have to offset the cost of the vibrating equipment, and of heavier and more sturdy formwork. In any case, the cost of labour would probably be the deciding factor if the choice is to be made on the basis of cost alone. As far as the quality of the concrete is concerned, both vibration and compaction by hand can, with the right mix and good workmanship, produce excellent concrete. Likewise, both means of compaction can produce poor concrete: in the case of hand-rammed concrete, inadequate compaction is the most common fault; when vibration is used it is possible that this has not been applied uniformly to the entire concrete mass so that some parts of it are not fully compacted while others are segregated owing to over-vibration. However, with a sufficiently stiff and well-graded mix the ill effects of over-vibration can be largely eliminated.

It has been mentioned that the two basic means of compaction require mixes of different workabilities: too dry a mix cannot be sufficiently worked by hand; and, conversely, too wet a mix should not be vibrated as segregation may result. This point has to be watched as, for instance, some mixes suitable for pumping may have too wet a consistence for vibration. Furthermore, different vibrators require different consistence of concrete for most efficient compaction so that the consistence of the concrete and the characteristics of the available vibrator have to be matched.

Internal Vibrators

Of the several types of vibrators this is perhaps the most common one. It consists essentially of a poker, housing an eccentric shaft driven through a flexible drive from a motor. The poker is immersed in concrete and thus applies approximately harmonic forces to it; hence, the alternative names of poker- or immersion-vibrator.

The frequency of vibration varies up to 12,000 cycles of vibration per minute: between 3,500 and 5,000 has been suggested as a desirable minimum with an acceleration of not less than $4g$.

The poker is easily moved from place to place, and is applied at 0·5 to 1 m (or 2 to 3 ft) centres for 5 to 30 sec, depending on the consistence of the mix, but with some mixes up to 2 min may be required. The actual completion of compaction can be judged by the appearance of the surface of the concrete, which should be neither honeycombed nor contain an excess of mortar. Gradual withdrawal of the poker at the rate of about 80 mm per sec (3 in./sec) is recommended[4.17] so that the hole left by the vibrator closes fully without any air being trapped. The vibrator should be immersed through the entire depth of the freshly deposited concrete and into the layer below if this is still plastic or can be brought again to a plastic condition. In this manner a plane of weakness at the junction of the two layers can be avoided and monolithic concrete is obtained. With a lift greater than about 0·5 m (2 ft) the vibrator may not be fully effective in expelling air from the lower part of the layer.

Internal vibrators are comparatively efficient since all the work is done directly on the concrete, unlike other types of vibrators. Pokers are made in sizes down to 20 mm ($\frac{3}{4}$ in.) diameter so that they can be used even with heavily reinforced and not easily accessible sections. An immersion vibrator will not expel air from the form boundary so that "slicing" along the form by means of a flat plate on edge is necessary. The use of absorptive linings to the form is helpful in this respect but expensive.

External Vibrators

This type of vibrator is rigidly clamped to the formwork resting on an elastic support, so that both the form and the concrete are vibrated. As a result, a considerable proportion of the work done is used in vibrating the formwork, which also has to be strong and tight so as to prevent distortion and leakage of grout.

The principle of an external vibrator is the same as that of an internal one, but the frequency is usually between 3,000 and 6,000 cycles of vibration per minute* although some vibrators reach 9,000 vibrations per minute. The power output varies between 80 and 1,100 W.

External vibrators are used for precast or thin *in situ* sections of such shape or thickness that an internal vibrator cannot be used.

When an external vibrator is used, concrete has to be placed in layers of suitable depth as air cannot be expelled through too great a thickness of concrete. The position of the vibrator may have to be changed as concreting progresses.

Portable, non-clamped external vibrators may be used at sections not otherwise accessible, but the range of compaction of this type of vibrator is very limited. One such vibrator is an electric hammer, sometimes used for compaction of concrete test cubes.

Vibrating Tables

This can be considered as a case of formwork clamped to the vibrator instead of the other way round, but the principle of vibrating the concrete and formwork together is unaltered.

The source of vibration, too, is similar. Generally a rapidly rotating eccentric weight makes the table vibrate with a circular motion. With two shafts rotating in opposite directions the horizontal component of vibration can be neutralized so that the table is subjected to a simple harmonic motion in the vertical direction only. There exist also some small good quality vibrating tables operated by an electro-magnet fed with alternating current. The range of frequencies used varies between 1,500 and about 7,000 cycles a minute. An acceleration of about $4g$ to $7g$ is desirable.[4.17] About $1 \cdot 5g$ and an amplitude of 40 μm (0·0015 in.) are believed to be the minima necessary for compaction,[4.18] but with these values a long period of vibration may be necessary. For simple harmonic motion the amplitude, a, and the frequency, f, are related by the equation

$$\text{acceleration} = a(2\pi f)^2.$$

When concrete sections of different sizes are to be vibrated, and in laboratory use, a table with a variable amplitude should be used. Variable frequency of vibration is an added advantage.

In practice, the frequency may rarely be varied during the actual compaction but, at least theoretically, there are considerable advantages in increasing the frequency and decreasing amplitude as consolidation progresses. The reason for this lies in the fact that initially the particles in the mix are far apart and the movement induced has to be of corresponding magnitude. On the other hand, once partial compaction has taken place the use of a higher frequency permits a greater number of adjusting

---

* Manufacturers' data have to be inspected carefully as sometimes the number of "impulses" is quoted, an impulse being *half* a cycle.

movements in a given time; a reduced amplitude means that the movement is not too large for the space available. Vibration at too large an amplitude relative to the inter-particle space results in the mix being in a constant state of flow so that full compaction is never achieved.

A vibrating table provides a reliable means of compaction of precast concrete and has the advantage of offering uniform treatment.

A variant of the vibrating table is a *shock table* used in some precast works in England and much more commonly in the Netherlands and in Denmark. The principle of this process of compaction is rather different from the high frequency vibration discussed earlier: in a shock table violent vertical shocks are imparted at the rate of about 60 to 200 per minute. The shocks are produced by a vertical drop of up to 13 mm ($\frac{1}{2}$ in.), this being achieved by means of cams. Concrete is placed in the form in shallow layers while the shock treatment progresses: extremely good results have been reported but the process is rather specialized and, although 50 years old, not widely used.

Other Vibrators

Various types have been developed for special purposes but only a very brief mention of these will be made.

A surface vibrator applies vibration through a flat plate direct to the top surface of the concrete. In this manner the concrete is restrained in all directions so that the tendency to segregate is limited; for this reason a more intense vibration can be used.

An electric hammer can be used as a surface vibrator when fitted with a bit having a large flat area, say 100 mm by 100 mm (4 in. by 4 in.); one of the main applications is compacting test cubes.

A vibrating roller is used for consolidating thin slabs. For road construction various vibrating screeds and finishers are available, but the details of these machines are outside the scope of this book. A power float is used mainly for granolithic floors in order to bind the granolithic layer to the main body of the concrete, and is more an aid in finishing than a means of compaction.

Revibration

It is usual to vibrate concrete immediately after placing so that consolidation is generally completed before the concrete has stiffened. All the preceding sections refer to this type of vibration.

It has been mentioned, however, that in order to ensure good bond between lifts the upper part of the underlying lift should be revibrated, provided the lower lift can still regain a plastic state; settlement cracks and the internal effects of bleeding can thus be eliminated.

This successful application of revibration raises the question whether revibration can be more generally used. On the basis of experimental results it appears that concrete can be successfully revibrated up to about 4 hours from the time of mixing.[4.19] Revibration at 1 to 2 hours after placing was found to result in an increase in the 28-day compressive strength of the form shown in Fig. 4.17. The comparison is on the basis of the same total period of vibration, applied either immediately after

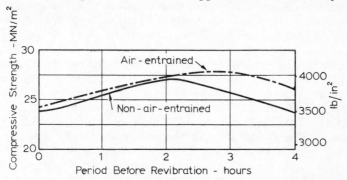

Fig. 4.17. *Relation between 28-day compressive strength and the time of revibration*[4.19]

placing or in part then and in part at a specified time later. An increase in strength of approximately 14 per cent has been reported, but actual values would depend on the workability of the mix and on details of the procedure. In general, the improvement in strength is more pronounced at earlier ages, and is greatest in concretes liable to high bleeding[4.20] since the trapped water is expelled on revibration. For the same reason, revibration greatly improves bond between concrete and reinforcement. It is possible also that some of the improvement in strength is due to a relief of the plastic shrinkage stresses around aggregate particles.

Despite these advantages revibration is not widely used as it involves an additional step in the production of concrete, and hence increased cost; also, if applied too late, revibration can damage the concrete.

Concreting in Hot Weather

There are some special problems involved in concreting in hot weather, arising both from a higher temperature of the concrete, and, in many cases, from an increased rate of evaporation from the fresh mix. These problems concern the mixing, placing and curing of the concrete.

A higher temperature of fresh concrete results in a more rapid hydration and leads therefore to accelerated setting and to a lower strength of hardened concrete since a less uniform framework of gel is established (*see* p. 276). Furthermore, rapid evaporation may cause plastic shrinkage and crazing, and subsequent cooling of the hardened concrete would introduce tensile stresses. It is generally believed that plastic shrinkage is likely to occur when the rate of evaporation exceeds the rate at which the bleeding water rises to the surface, but it has been observed that cracks also form under a layer of water and merely become apparent on drying.[4.16] This would suggest that cracking is associated with differential settlement of fresh concrete due to some obstruction to settlement, such as the coarse aggregate or the reinforcement. On the other hand, a drop in the ambient relative humidity encourages cracking[4.9] so that in fact the causes of it appear to be rather complex. In any case, one practical conclusion is that thin sections, such as shell roofs, should not be cast under hot and dry conditions.

Plastic shrinkage appears to be related to some physical characteristics of cement but the full extent of the problem is yet to be studied (*see* p. 321).

There are some further complications in hot weather concreting: air entraining is more difficult although this can be remedied by using larger quantities of the entraining agent. Curing also presents additional problems as the curing water tends to evaporate rapidly. The use of curing compounds is not entirely satisfactory since it leads to lower compressive strengths than when continuous water curing is applied. Some experimental values are given in Table 4.5.

There is a number of remedial measures that can be taken. In the first instance, the cement content should be kept as low as possible so that the heat of hydration does not unduly aggravate the effects of high ambient temperature. The temperature of the fresh concrete can be lowered by pre-cooling one or more of the ingredients of the mix. For instance, ice can be used instead of some of the mixing water but it is essential that the ice melts completely before the mixing has been completed. The cooling of the aggregate is more difficult and, because of the low specific heat of stone, less effective. All materials used should be protected from direct insolation.

The temperature of concrete delivered at site in hot weather should be as low as possible; an upper limit of 29°C(85°F) is often specified.

The temperature T of the freshly mixed concrete can be easily calculated from that of the ingredients, using the expression

$$T = \frac{0 \cdot 2(T_a W_a + T_c W_c) + T_w W_w}{0 \cdot 2(W_a + W_c) + W_w}$$

where T denotes temperature in °F, W weight of ingredient per unit volume of concrete and the suffixes a, c, w refer to aggregate, cement, and water (both added and in aggregate) respectively. $0 \cdot 2$ Btu/lb/°F is the approximate value of specific heat of the dry ingredients. In the S.I. system of units, the temperatures are in °C and the specific heat is 840 J/kg/°C.

The actual temperature of the concrete will be somewhat higher than indicated by the above expression owing to the mechanical work done in

Table 4.5 *Effect of Curing on the 56-day Compressive Strength of Concrete Cast in the Desert*[4.9]

Water/ cement ratio	Aggregate/ cement ratio	Means of curing	Relative compressive strength per cent for concrete cast during—		
			morning (28°C (82°F)), 40 per cent relative humidity	midday (38°C (100°F)), 18 per cent relative humidity	evening (33°C (91°F)), 25 per cent relative humidity
0·92	7·6	None	100	100	100
		Ordinary sealing compound	118	117	108
		White sealing compound	118	117	108
		7-day continuous watering	153	141	145
0·72	6·1	None	100	100	100
		Ordinary sealing compound	127	106	110
		White sealing compound	121	122	110
		7-day continuous watering	127	148	141

mixing, and will further rise due to the development of the heat of wetting and hydration of cement. To obtain a better picture we can say that if the water/cement ratio of a mix is 0·5 and the aggregate/cement ratio is 5·6, then a drop of 1°C (1°F) in the temperature of fresh concrete can be obtained by lowering the temperature either of the cement by 9°C (9°F) or of the water by 3·6°C (3·6°F) or of the aggregate by 1·6°C (1·6°F). It can be seen that because of its relatively small quantity in the mix the temperature of the cement is not important.

The use of hot cement *per se* is not detrimental to strength but it is preferable not to use cement at temperatures above about 75°C (170°F). This statement is of interest since hot cement is sometimes viewed with suspicion and various ill effects have at times been ascribed to its use. However, if hot cement is dampened by a small amount of water before it is well dispersed with other solids it may set quickly and form cement balls.

The influence of the temperature during setting on the strength at later ages is discussed on p. 276; here it suffices to say that a temperature not higher than about 16°C (60°F) should be aimed at, and, if possible, 32°C (90°F) should not be exceeded. After placing, concrete should be protected from the sun; otherwise, if a cold night follows, cracking is likely to occur, the extent of cracking being directly related to the temperature difference. In dry weather, wetting concrete and allowing evaporation to take place results in effective cooling; there is no cooling by this means when membrane curing is used so that a higher temperature may be reached. Large exposed areas such as roads and airfields are particularly vulnerable. Details of good practice for hot weather concreting have recently been published by the American Concrete Institute.[4.39]

Ready-Mixed Concrete

If instead of being batched and mixed on the site, concrete is delivered ready for placing from a central plant it is referred to as ready-mixed or pre-mixed concrete. This type of concrete is used extensively as it offers numerous advantages in comparison with the orthodox method of manufacture. In many countries, more than half the concrete used in *in situ* construction is ready-mixed.

Ready-mixed concrete is particularly useful on congested sites or in road construction where little space for the mixing plant and for extensive aggregate stockpiles is available, but perhaps the greatest single advantage of ready-mixed concrete is that it may be made under better conditions of control than are normally possible on any but large construction sites. Control has to be enforced but, since the central mixing plant operates under near-factory conditions, a really close control of all operations of manufacture of fresh concrete is possible. In a modern batching and mixing plant, interlocking prevents incorrect batching quantities, and sometimes a printed record of weights of ingredients of every batch is made. Proper care during transportation of the concrete is also ensured by the use of agitator trucks, but the placing and compaction remain, of course, the responsibility of the personnel on the site. Ready-mixed concrete can be considered to be more in the nature of a factory-made product, almost comparable with steel, so that a great deal of uncertainty and variability associated with the production of concrete on many a site is removed.

The use of ready-mixed concrete is also advantageous when only small quantities of concrete are required or when concrete is placed only at intervals. Usually, the price of ready-mixed concrete is somewhat higher than of site-mixed concrete, but this may often be offset by savings in the cement content, site organization, and supervisory staff.

There are two principal categories of ready-mixed concrete. In the first, the mixing is done at a central plant and the mixed concrete is then transported, usually in an agitator truck which revolves slowly so as to prevent segregation and undue stiffening of the mix. Such concrete is known as central-mixed as distinct from the second category—transit-mixed or truck-mixed concrete. Here, the materials are batched at a central plant but are mixed in a mixer truck either in transit to the site or immediately prior to the concrete being discharged. Transit-mixing permits a longer haul and is less vulnerable in case of delay, but the capacity of a truck used as a mixer is only about three-quarters of the same truck used solely to agitate pre-mixed concrete. Sometimes, the concrete is partially mixed at a central plant in order to increase the capacity of the agitator truck. The mixing is completed en route. Such concrete is known as shrink-mixed concrete.

It should be explained that agitating differs from mixing solely by the speed of rotation of the mixer: the agitating speed is between 2 and 6 rev/min compared with the mixing speed of 4 to about 16 rev/min; there is thus some overlap in the definitions. B.S. 1926: 1962 specifies a minimum mixing speed of 7 rev/min. It may be noted that the speed of mixing affects the rate of stiffening, while the total number of revolutions controls the uniformity of mixing. Usually, 70 revolutions at mixing speed are required as a minimum, and 100 revolutions at mixing speed is the maximum allowed. An overriding limit of 300 revolutions *in toto* is laid down by A.S.T.M. Standard C 94–72.

The details of batching plants are outside the scope of the present book and it will suffice to say that they are generally highly automated. As many as eight different ingredients can be weigh-batched into large capacity mixers, the quantities being accurately dispensed and automatically recorded for control purposes. For all this no more than one operator is required.

The main problem in the production of ready-mixed concrete is maintaining the workability of the mix right up to the time of placing. Concrete stiffens with time (Fig. 4.18) and handling ready-mixed concrete often takes quite a long while. The stiffening may also be aggravated by prolonged mixing and by a high temperature. In the case of transit-mixing water need not be added till nearer the commencement of mixing, but the time during which the cement and moist aggregate are allowed to remain in contact should be limited to about 90 minutes although B.S. 1926: 1962 allows 2 hours.

Many specifications impose the same limit on the time of haul of central-mixed concrete, and it is usual also to limit the total number of revolutions during both mixing and agitating to approximately 300. However, agitating up to 6 hours need not adversely affect the strength of concrete provided the mix remains sufficiently workable for full compaction. Unless, however, the initial workability is high, the stiffening

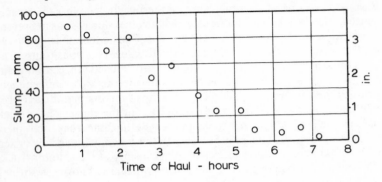

Fig. 4.18. *Loss in slump during agitating at 4 rev/min*[4.24]

caused by prolonged agitation would result in a concrete of very low workability, especially in hot weather, when a high loss of water by evaporation takes place in addition to the loss of free water by hydration of cement. For this reason concrete is sometimes re-tempered by the addition of water immediately before discharge; the workability is thus restored but it must be realized that the resultant compressive strength will be affected by the amount of water added to the mix (*see* Fig. 4.19).

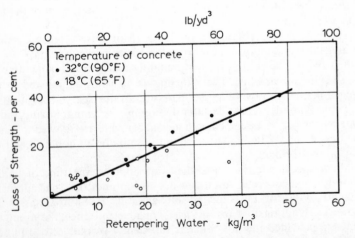

Fig. 4.19. *Effect of re-tempering water on the strength of concrete*[4.28]

Pumped Concrete

Since this book deals primarily with the properties of concrete the details of the means of transporting and placing are not considered. An exception should, however, be made in the case of pumping of concrete since this means of transportation requires the use of a mix having special properties.

The system consists essentially of a hopper into which concrete is discharged from the mixer, a concrete pump of the type shown in Fig. 4.20 or 4.21, and pipes through which the concrete is pumped.

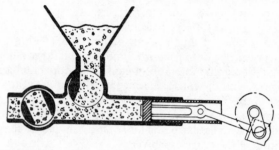

Fig. 4.20. *Direct-acting concrete pump*

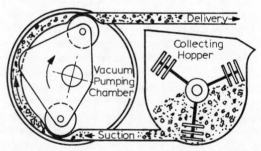

Fig. 4.21. *Squeeze-type concrete pump*

Many pumps are of the direct-acting, horizontal piston type with semi-rotary valves set so as to permit always the passage of the largest particles of aggregate being used: there is thus no full closure. Concrete is fed into the pump by gravity and is also partially sucked in during the suction stroke. The valves open and close with definite pauses so that concrete moves in a series of impulses but the pipe always remains full.

Recently, small portable pumps, sometimes called squeeze pumps, have been introduced for use with small (up to 75 or 100 mm (3 to 4 in.)) pipes; Fig. 4.21 shows such a pump. Concrete placed in a collecting hopper is fed by rotating blades into a pliable pipe located in the pumping chamber. The vacuum inside the chamber is about 660 mm (26 in.) of mercury. This ensures that, except when actually squeezed by a roller, the pipe has a normal (cylindrical) shape so that a continuous flow of concrete is ensured. Two rotating rollers progressively squeeze the tube and thus pump the concrete in the suction pipe toward the delivery pipe. Squeeze pumps are often lorry-mounted and may deliver concrete through a folding boom.

Squeeze pumps move concrete for distances up to 90 m (300 ft) horizontally or 30 m (100 ft) vertically. However, using piston pumps, concrete can be moved up to about 450 m (1,500 ft) horizontally or 40 m (140 ft) vertically or to proportionate combinations of distance and lift. We should note that the ratio of equivalent horizontal and vertical distances varies with the consistence of the mix and with the velocity of the concrete in the pipe: the greater the velocity the smaller the ratio.[4.29] Relay pumping is possible for greater distances. When bends are used, and these must never be sharp, the loss of head should be allowed for in the calculation of the range of delivery.

Pumps of different sizes are available and likewise pipes of various diameters are used but the pipe diameter must be at least three times the maximum aggregate size. Using squeeze pumps, an output of 20 m³ (25 yd³) of concrete per hour can be obtained but piston pumps with 220 mm (9 in.) pipes can deliver up to 55 m³ (73 yd³) per hour. Pumping is economical only if it can be used over long uninterrupted periods as at the beginning of each period of pumping the pipes have to be lubricated by mortar (at the rate of about 0·25 m³ per 100 m (1 yd³ per 1,000 ft) of 150 mm (6 in.) diameter pipe) and also because at the end of the operation a considerable effort is required in cleaning the pipes. However, alterations to the pipeline system can be made very quickly as special couplings are used. Aluminium pipes must not be used because aluminium reacts with the alkalis in cement and generates hydrogen. This gas introduces voids in the hardened concrete with a consequent loss of strength.

The main advantages of pumping concrete are that it can be delivered to points over a wide area otherwise not easily accessible, with the mixing plant clear of the site; this is especially valuable on congested sites or in special applications such as tunnel linings, etc. Pumping delivers the concrete direct from the mixer to the form and so avoids double handling. Placing can proceed at the rate of the output of the mixer and is not held back by the limitations of the transporting and placing equipment. A significant proportion of ready-mixed concrete is nowadays pumped.

Furthermore, pumped concrete is unsegregated but of course in order to be able to be pumped the mix must satisfy certain requirements. It might be added that unsatisfactory concrete cannot be pumped so that any pumped concrete is satisfactory as far as its properties in the fresh state are concerned.

First of all, the concrete must be well mixed before feeding into the pump and sometimes remixing in the hopper by means of a stirrer is carried out. Broadly speaking, the mix must not be harsh or sticky, too dry or too wet, i.e. its consistence is critical. A slump of between 40 and 100 mm ($1\frac{1}{2}$ and 4 in.) or a compacting factor of approximately 0·9 to 0·95 is generally recommended, but pumping produces a partial compaction so that at the point of delivery the slump may be decreased by 10 mm to 25 mm ($\frac{1}{2}$ in. to 1 in.). With a lower water content, the solid particles, instead of moving longitudinally in a coherent mass in suspension, would exert pressure on the walls of the pipe. When the water content is at the correct, or critical, value friction develops only at the surface of the pipe and in a thin, 1 to 2·5 mm (0·04 to 0·1 in.), layer of the lubricating mortar. It is possible that the formation of the lubricating film is aided by the fact that the dynamic action of the piston is transmitted to the pipe, but such a film is also caused by steel trowelling of a concrete surface. To allow for the film in the pipe a cement content slightly higher than otherwise would be used is desirable. The magnitude of the friction developed depends on the consistence of the mix, but there must be no excess water as segregation would result.

It may be useful to consider the problems of friction and segregation in more general terms. In a pipe through which a material is pumped, there is a pressure gradient in the direction of flow due to two effects: head of the material and friction. This is another way of saying that the material must be capable of transmitting a sufficient pressure to overcome all resistances in the pipeline. Of all the components of concrete, it is only water that is pumpable in its natural state, and it is the water, therefore, that transmits the pressure to the other mix components.

Two types of blockage can occur. In one, water escapes through the mix so that pressure is not transmitted to the solids, which therefore do not move. This occurs when the voids in the concrete are not small enough or intricate enough to provide sufficient internal friction within the mix to overcome the resistance of the pipeline. Therefore, an adequate amount of closely packed fines is essential to create a "blocked filter" effect, which allows the water phase to transmit the pressure but not to escape from the mix. In other words, the pressure at which segregation occurs must be greater than the pressure needed to pump the concrete.[4.30] It should be remembered of course that more fines mean a higher surface area of the solids and therefore a higher frictional resistance in the pipe.

We can see thus how the second type of blockage can occur. If the fines content is too high, the friction resistance of the mix can be so large that the pressure exerted by the piston through the water phase is not sufficient to move the mass of concrete, which becomes stuck. This type of failure is more common in high strength mixes or in mixes containing a high proportion of very fine material such as crusher dust or fly ash, while the segregation failure is more apt to occur in medium or low strength mixes with irregular or gap grading.

The optimum situation therefore is to produce maximum frictional resistance within the mix with minimum void sizes, and minimum frictional resistance against the pipe walls with a low surface area of the aggregate. This means that the coarse aggregate content should be high but the grading should be such that there is a low void content so that only little of the very fine material is required to produce the "blocked filter" effect. Thus, in general, except in very lean mixes, the content of material finer than 150 μm (No. 100 B.S. sieve) should be not less than 3 per cent and most of this can be finer than 75 μm (No. 200 B.S. sieve). This material may be the finer fraction of sand or a suitable additive, such as tuff or trass. This fine material gives continuity in grading right down to the cement fraction but still avoids a very high pipe friction.[4.30]

It may be noted that a sudden rise in pressure caused by a restriction or by a reduction in the diameter of the pipe may result in segregation of the aggregate which is left behind as the cement paste moves past the obstacle.[4.31]

The shape of the aggregate influences the suitability of a mix for pumping. In general, natural sand and rounded gravel are preferable to crushed aggregate but a good pumping mix can be made by suitable blending of crushed aggregate fractions. Recently, pumping of lightweight aggregate concrete has been introduced. If the aggregate surface is sealed, then it pumps as well as normal aggregate. But if the aggregate surface is porous then the internal voids may not become fully saturated even on thorough wetting. As a result, when pressure is applied in the pipeline, the air in these voids contracts, and water is forced into the pores with the result that the mix becomes too dry and too stiff. If pumping is stopped and pressure removed, water is discharged from the aggregate; this water may carry with it the fine material so that a plug forms on resumption of pumping. Some of the aggregate may also become crushed by pumping.

Concrete with a high content of entrained air behaves in a similar manner. Under a high pumping pressure, the air becomes compressed and no longer aids the mix by its "ball-bearing" effect. The friction rises and so does the pressure: the air becomes compressed further so that the workability drops even more. If the pipeline is long enough, the reduction in volume of the air under pressure can absorb the entire movement of the piston so that no concrete will come out at the delivery end. For this reason, air-entrained concrete is usually pumped only over short distances: about 45 m (150 ft).

Preplaced Aggregate Concrete

This type of concrete is produced in two stages. In the first operation coarse aggregate is placed and compacted in the forms. The voids between the particles, forming some 30 to 35 per cent of the overall volume to be concreted, are filled with mortar in the second stage. Preplaced aggregate concrete is also known as prepacked concrete, intrusion concrete or grouted concrete.

It is clear that the aggregate in the resulting concrete is of the gap-graded type. Typical fine and coarse aggregate gradings are shown in Tables 4.6 and 4.7.

Table 4.6: *Typical Gradings of Coarse Aggregate for Preplaced Aggregate Concrete*[4.32]

Sieve size	mm	150	75	38	19	13
	in.	6	3	$1\frac{1}{2}$	$\frac{3}{4}$	$\frac{1}{2}$
Cumulative percentage passing		100	67	40	6	1
			100	62	4	1
			100	97	9	1

Table 4.7: *Typical Grading of Fine Aggregate for Preplaced Aggregate Concrete*[4.32]

Sieve size	metric	2·36 mm	1·18 mm	600 μm	300 μm	150 μm
	B.S. No.	7	14	25	52	100
Cumulative percentage passing		100	98	72	34	11

The coarse aggregate must be thoroughly wetted or inundated before the mortar is intruded. The mortar is pumped under pressure through slotted pipes, starting from the bottom of the mass, the pipes being gradually withdrawn.

A typical mortar consists (by weight) of two parts of Portland cement, one part of a very finely divided and highly active pozzolana (for instance, fly-ash), and three to four parts of fine sand, with sufficient water to form a fluid mixture. The pozzolana reduces bleeding and segregation and improves fluidity of the mortar. An intrusion aid (representing about 1 per cent of the weight of the cement plus the pozzolana) is added in

order to improve the fluidity of the mortar and to hold the solid constituents in suspension. The intrusion aid also delays somewhat the
stiffening of the mortar and contains a small amount of aluminium
powder, which causes a slight expansion before setting takes place.

As an alternative, a mortar consisting of cement and fine sand can be
mixed in a special "colloid" mixer which disperses the cement to such a
degree that it remains in suspension until the pumping has been completed. This type of preplaced aggregate concrete is sometimes called
colloidal concrete.

The consistence of the mortar (about that of a thick cream) is expressed
as the time taken by a fixed quantity of mortar to discharge from a special
cone; this is known as the flow factor. Alternatively, a flowmeter is used;
this simply shows how far a given quantity of material discharged from
a funnel will travel along a horizontal channel.

Preplaced aggregate concrete is economical in cement, as little as 120
to 150 kg per cubic metre (200 to 250 lb/yd^3) of concrete being used, but
the strength of the resultant concrete is limited by the high water/cement
ratio necessary for a sufficient plasticity of mortar. However, for the usual
applications of preplaced aggregate concrete this strength is generally
adequate and a concrete of more uniform properties is obtained than is
the case with conventional methods of placing, as segregation is practically eliminated. As a result, a dense, impermeable and durable concrete
is produced. No internal vibration is used but external vibration at the
level of the grout surface may improve the exposed surfaces.

Preplaced aggregate concrete can be placed in locations not easily
accessible by ordinary concreting techniques; it can also be placed in
sections containing a large number of embedded items that have to be
precisely located: this arises, for instance, in nuclear shields. Likewise,
because the coarse and fine aggregate are placed separately, the danger of
segregation of heavy coarse aggregate, especially of steel aggregate used
in nuclear shields, is eliminated. Because of the reduced segregation preplaced aggregate concrete is also suitable for underwater construction,
and indeed the technique there differs little from placing under normal
conditions but for depths over 30 m (100 ft) a tremie should be used.

The drying shrinkage of preplaced aggregate concrete is lower than
that of ordinary concrete, usually 200×10^{-6} to 400×10^{-6}. The
reduced shrinkage is due to the point-to-point contact of the coarse
aggregate particles, without a clearance for the cement paste necessary in
ordinary concrete. This contact restrains the amount of shrinkage that
can actually be realized. Because of the reduced shrinkage preplaced
aggregate concrete is suitable for the construction of water-retaining and
large monolithic structures and for repair work. The low permeability of
preplaced aggregate concrete gives it a high resistance to freezing and
thawing.

Preplaced aggregate concrete may be used in mass construction where the temperature rise has to be controlled: cooling can be achieved by circulating refrigerated water round the aggregate and thus chilling it; the water is later displaced by the rising mortar. At the other extreme, in cold weather when frost damage is feared, steam can be circulated in order to pre-heat the aggregate.

Preplaced aggregate concrete is used also to provide an exposed aggregate finish: special aggregates are placed against the surfaces and become subsequently exposed by sandblasting or by acid wash.

Preplaced aggregate concrete appears thus to have many useful features, but because of numerous practical difficulties (e.g. the need for an extremely clean coarse aggregate) considerable skill and experience in application of the process are necessary for good results to be obtained.

Vacuum-processed Concrete

One solution to the problem of combining a sufficiently high workability with a minimum water/cement ratio is offered by vacuum-processing of freshly placed concrete. Such concrete is usually referred to as vacuum concrete—a rather misleading term but one in common usage.

The procedure is briefly as follows. A mix with a medium workability is placed in the forms in the usual manner. Since fresh concrete contains a continuous system of water-filled channels, the application of a vacuum to the surface of the concrete results in a large amount of water being extracted from a certain depth of the concrete. In other words, what might be termed "water of workability" is removed when no longer needed. It may be noted that air bubbles are removed only from the surface since they do not form a continuous system.

The final water/cement ratio before setting is thus reduced, and as this ratio largely controls the strength, vacuum-processed concrete has a higher strength and also a higher density, a lower permeability and a greater durability than would otherwise be obtained. Garnett's[4.21] data on strength are plotted in Fig. 4.22; the comparison should be made on the basis of the initial water/cement ratio. The magnitude of the decrease in the water/cement ratio due to vacuum-processing is given in Table 4.8.

The vacuum is applied through porous mats connected to a vacuum pump. The mats consist of an airtight cover, usually made of plywood, with a vacuum chamber formed by expanded metal. This is faced with a fine wire gauze covered by muslin which prevents the removal of cement together with the water. A diagrammatic representation of a mat is shown in Fig. 4.23. The mats can be placed on top of the concrete immediately after screeding, and can also be incorporated in the inside faces of forms.

Vacuum is created by a vacuum pump; its capacity is governed by the perimeter of the mat and not its area. The magnitude of the applied vacuum is usually in the range of 400 to 650 mm (15 to 25 in.) of mercury.

This vacuum reduces the water content by up to 20 per cent over a depth of 150 to 300 mm (6 to 12 in.). The reduction is greater nearer to the mat and it is usual to assume the suction to be fully effective over a depth of 150 mm (6 in.) only. Thus a concrete section 300 mm (12 in.) thick should have a vacuum applied from two opposite faces. The withdrawal of water

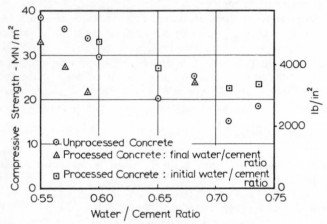

Fig. 4.22. *Strength of vacuum-processed and unprocessed concrete*[4.21]

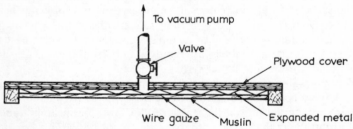

Fig. 4.23. *Cross-section of a vacuum mat*

produces settlement of the concrete to the extent of about 3 per cent of the depth over which the suction acts. The rate of withdrawal of water falls off with time, and it has been found that processing during 15 to 25 minutes is usually most economical. Little reduction in water content occurs beyond 30 minutes.

It has been said that, strictly speaking, no suction of water takes place during vacuum-processing but merely a fall of pressure below atmospheric is communicated to the interstitial fluid of the fresh concrete. This would mean that compaction by atmospheric pressure is taking place. Thus the amount of water removed would be equal to the contraction in the total volume of concrete and no voids would be produced. However, in

Table 4.8: *Water/Cement Ratio and Strength of Vacuum-Processed Concrete*[4.21]

Water/cement ratio		Compressive strength			
Before processing	After processing	Before processing MN/m^2	lb/in^2	After processing MN/m^2	lb/in^2
0·74	0·68	17·7	2,570	23·0	3,330
0·71	0·59	15·0	2,180	22·3	3,230
0·65	0·57	20·6	2,990	27·0	3,920
0·60	0·55	29·7	4,300	32·8	4,760

practice some voids are formed and for the same final water/cement ratio ordinary concrete has been found to have a somewhat higher strength than vacuum-processed concrete. This can be seen from Fig. 4.22.

The formation of voids can be prevented if in addition to vacuum-processing intermittent vibration is applied; under those circumstances a higher degree of consolidation is achieved and the amount of water withdrawn can be nearly doubled. In Garnett's[4.21] tests good results were obtained with vacuum-processing for 20 minutes accompanied by vibration between the 4th and 8th minutes, and again between the 14th and 18th minutes.

Vacuum-processing can be used over a fairly wide range of aggregate/cement ratios and aggregate gradings, but a coarser grading yields more water than a finer one. Furthermore, some of the finest material is removed by the processing, and fine additions, such as pozzolanas, should not be incorporated in the mix.

Vacuum-processed concrete stiffens very rapidly so that formwork can be removed within about 30 minutes of casting, even on columns 4·5 m (15 ft) high. This is of considerable economic value, particularly in a precast factory as the forms can be re-used at frequent intervals.

The surface of vacuum-processed concrete is entirely free from pitting and the uppermost 1 mm ($\frac{1}{32}$ in.) is highly resistant to abrasion. These characteristics are of special importance in the construction of concrete which is to be in contact with water flowing at a high velocity. Another useful characteristic of vacuum-processed concrete is that it bonds well to old concrete and can, therefore, be used for re-surfacing road slabs and other repair work. Vacuum treatment thus appears to be a valuable process; it is, however, rather an expensive one.

Shotcrete

This is the name given to mortar or concrete conveyed through a hose and pneumatically projected at high velocity onto a backup surface. The force of the jet impacting on the surface compacts the material so that it can support itself without sagging or sloughing even on a vertical face or overhead.

Since the essence of the process is that the material is projected pneumatically, shotcrete is more formally called *pneumatically applied mortar* or *concrete*; it is also known as *gunite*, although, strictly speaking, this name applies only to shotcrete placed by the dry mix process.[4.33]

The properties of shotcrete are no different from the properties of conventionally placed mortar or concrete of similar proportions: it is the method of placing that bestows on shotcrete significant advantages in many applications. At the same time, considerable skill and experience are required in the application of shotcrete so that its quality depends to a large extent on the performance of the operators involved, especially in control of the actual placing by the nozzle.

Since shotcrete is pneumatically projected on a backup surface and then gradually built-up, only one side of formwork is needed. This represents economy, especially when account is taken of the absence of form ties, etc. On the other hand, the cement content of shotcrete is high. Also, the necessary equipment and mode of placing are more expensive than in the case of conventional concrete. For these reasons, shotcrete is used primarily in certain types of construction: thin, lightly reinforced sections, such as roofs, especially shell or folded plate, tunnel linings, and prestressed tanks. Shotcrete is also used in repair of deteriorated concrete, in stabilizing rock slopes, in encasing steel for fireproofing, and as a thin overlay on concrete, masonry or steel. If shotcrete is applied to a surface covered by running water, an accelerator producing flash set, such as washing soda, is used. This adversely affects strength but makes repair work possible. Generally, shotcrete is applied in a thickness up to 100 mm (4 in.).

There are two basic processes by which shotcrete is applied. In the *dry mix process* (which is the more common of the two) cement and damp aggregate are intimately mixed and fed into a mechanical feeder or gun. The mixture is then transferred by a feed wheel or distributor (at a known rate) into a stream of compressed air in a hose, and carried up to the delivery nozzle. The nozzle is fitted inside with a perforated manifold through which water is introduced under pressure and intimately mixed with the other ingredients. The mixture is then projected at high velocity onto the surface to be shotcreted.

The fundamental feature of the *wet mix process* is that all the ingredients, including the mixing water, are mixed together to begin with. The mixture is then introduced into the chamber of the delivery equipment and from there conveyed pneumatically or by positive displacement. A pump similar to that of Fig. 4.21 can be used. Compressed air (or in the case of pneumatically conveyed air, additional air) is injected at the nozzle, and the material is projected at high velocity onto the surface to be shotcreted.

Either process can produce excellent shotcrete, but the dry mix process is better suited for use with porous lightweight aggregate and with flash set accelerators, and is also capable of greater delivery lengths.[4.34] On the other hand, the wet mix process gives a better control of the quantity of mixing water (which is metered, as opposed to judgement by the nozzle operator) and of any admixture used. Also, the wet mix process leads to less dust being produced.

Not all the shotcrete projected on a surface remains in position. Because of the high velocity of the impacting jet, some material rebounds. This consists of the coarsest particles in the mix, so that the shotcrete *in situ* is richer than would be expected from the mix proportions as batched. This may lead to increased shrinkage. The rebound is greatest in the initial layers and becomes smaller as a plastic cushion of shotcrete is built-up. Typical percentages of material rebounded are:[4.33]

in floors and slabs	5 to 15
on sloping or vertical surfaces	15 to 30
on soffits	25 to 50.

The significance of rebound is not so much in the waste of the material as in the danger from accumulation of rebounded sand in a position where it will become incorporated in the subsequent layers of shotcrete. This can occur if the rebound collects in inside corners, at the base of walls, behind reinforcement or embedded pipes, or on horizontal surfaces. Great care in placing of shotcrete is therefore necessary, and the use of large reinforcement is undesirable. The latter also leads to the risk of unfilled pockets behind the obstacle to the jet.

The projected shotcrete has to have a relatively dry consistence so that the material can support itself in any position; at the same time, the mix has to be wet enough to obtain compaction without excessive rebound. The usual range of water/cement ratios is 0·35 to 0·50; there is generally very little bleeding. In the case of mortar, the usual mix is 1:3·5 to 1:4·5, and 28-day strengths ranging between 20 and 50 MN/m² (3,000 and 7,000 lb/in²) are obtained. Sand with the same grading as for use in conventional mortar is satisfactory.

Concrete can also be shotcreted, the maximum aggregate size being 25 mm (1 in.) but the need for this material is small and its advantages are limited. The coarse aggregate content is lower than in conventionally placed concrete.

Curing of shotcrete is particularly important because the large surface–volume ratio can lead to rapid drying.

Analysis of Fresh Concrete

In considering the ingredients of a concrete mix we have so far assumed that the actual proportions correspond to those specified. If this were invariably so there would be little need for testing the strength of hardened concrete. However, in practice, mistakes, errors and even deliberate actions can lead to incorrect mix proportions, and it is sometimes useful to determine the composition of concrete at an early stage; the two values of greatest interest are the cement content and water/cement ratio.

Unfortunately, no simple and rapid analysis of fresh concrete is possible but there exists a test prescribed by B.S. 1881: Part 2: 1970.

The test has to be commenced virtually as soon as the concrete has been discharged from the mixer because loss of water can occur; even if evaporation is prevented, an unknown amount of hydration will take place during any period of delay. The analysis of the concrete has to be supplemented by tests on the fine and coarse fractions of the aggregate: specific gravity, absorption, and grading. This information is essential in the calculation of the quantity of the solid constituents of the mix. The weights of fine and coarse aggregate are determined by weighing in water, and the weight of cement is obtained as a difference between the weight in water of the concrete and of the aggregates. The water content is determined by drying the sample over a heater. Hence, the mix proportions can be calculated. The air content is determined independently but the analysis of air-entrained concrete is less accurate than of non-air-entrained concrete.

An accurate determination of the cement content is of major importance and several approaches have been proposed. Turton[4.35] suggested a modification of the B.S. 1881: Part 2: 1970 procedure: he collects the washed very fine material into a container, and obtains the weight of cement direct rather than as a difference. The calculation of the weight of cement has to include a correction for the silt and dust in the aggregate as determined by a sieve analysis on the aggregate samples. In another variant, the material smaller than 150 μm (No. 100 B.S. sieve) is separated out by filtering and pressing dry;[4.36] the weight of cement is taken as the weight of this fraction corrected for aggregate finer than 150 μm (No. 100 B.S. sieve) in the material as batched.

Another method[4.37] of determining the cement content of fresh concrete consists of the measurement of electrical conductivity of an aqueous suspension of the material. The test is based on the fact that hydration of cement produces salts whose concentration is proportional to the quantity of cement so that there is a definite relation between the cement content and electrical conductivity. The test must be performed within an hour of mixing but the limitations of the test conditions (e.g. temperature) are such that the determination cannot be made on site.

A totally different approach in the determination of cement content of fresh concrete is based on the separation of cement using a heavy liquid and a centrifuge.[4.38] This has not been very successful, especially when the finest aggregate particles have a specific gravity not significantly lower than that of cement.

It seems thus that no test for the composition of fresh concrete that is convenient and reliable enough to be used as a preplacement acceptance test is as yet available.[4.40]

REFERENCES

4.1. W. H. GLANVILLE, A. R. COLLINS and D. D. MATTHEWS, The grading of aggregates and workability of concrete, *Road Research Tech. Paper No.* 5, (London, H.M.S.O., 1947).

4.2. NATIONAL READY-MIXED CONCRETE ASSOCIATION, Control of quality of ready-mixed concrete, *Publicn. No.* 44 (Washington D.C., June 1957).

4.3. ROAD RESEARCH: Design of concrete mixes, *D.S.I.R. Road Note No.* 4 (London, H.M.S.O., 1950).

4.4. A. R. CUSENS, The measurement of the workability of dry concrete mixes, *Mag. Concr. Res.*, **8**, No. 22, pp. 23–30 (March 1956).

4.5. T. C. POWERS, Studies of workability of concrete, *J. Amer. Concr. Inst.*, **28**, pp. 419–48 (1932).

4.6. J. W. KELLY and M. POLIVKA, Ball test for field control of concrete consistency, *J. Amer. Concr. Inst.*, **51**, pp. 881–8 (May 1955).

4.7. U.S. BUREAU OF RECLAMATION, *Concrete Manual*, 7th Ed. (Denver, 1966).

4.8. P. KLIEGER, Effect of mixing and curing temperature on concrete strength, *J. Amer. Concr. Inst.*, **54**, pp. 1063–81 (June 1958).

4.9. R. SHALON and D. RAVINA, Studies in concreting in hot countries, *R.I.L.E.M. Int. Symp. on Concrete and Reinforced Concrete in Hot Countries* (Haifa, July 1960).

4.10. T. C. POWERS, The bleeding of Portland cement paste, mortar and concrete, *Portl. Cem. Assoc. Bul. No.* 2 (Chicago, July 1939).

4.11. H. H. STEINOUR, Further studies of the bleeding of Portland cement paste, *Portl. Cem. Assoc. Bul. No.* 4 (Chicago, Dec. 1945).

4.12. I. L. TYLER, Uniformity, segregation and bleeding, *A.S.T.M. Sp. Tech. Publicn. No.* 169, pp. 37–41 (1956).

4.13. CENTRE D'INFORMATION DE L'INDUSTRIE CIMENTIÈRE BELGE, L'execution du béton: le malaxage du béton, *Bul. No.* 69 (May 1958).

4.14. J. D. DEWAR, Relations between various workability control tests for ready-mixed concrete, *Cement Concr. Assoc. Tech. Report TRA/375* (London, Feb. 1964).

4.15. AMERICAN CONCRETE INSTITUTE, *Standard 614–59: Recommended Practice for Measuring, Mixing, and Placing Concrete* (Detroit, 1959).

4.16. F. D. BERESFORD and F. A. BLAKEY, Discussion on paper by W. Lerch: Plastic shrinkage, *J. Amer. Concr. Inst.*, **56**, Part II, pp. 1342–3 (Dec. 1957).

4.17. JOINT COMMITTEE OF THE I.C.E. AND THE I. STRUCT. E., *The Vibration of Concrete* (London, 1956).

4.18. J. KOLEK, The external vibration of concrete, *Civil Engineering*, **54**, No. 633, pp. 321–25 (London, March 1959).

4.19. C. A. VOLLICK, Effects of revibrating concrete, *J. Amer. Concr. Inst.*, **54**, pp. 721–32 (March 1958).

232   *Properties of Concrete*

4.20. E. N. MATTISON, Delayed screeding of concrete, *Constructional Review*, **32**, No. 7, p. 30 (Sydney, July 1959).

4.21. J. B. GARNETT, The effect of vacuum processing on some properties of concrete, *Cement Concr. Assoc. Tech. Report TRA/326* (London, Oct. 1959).

4.22. R. SHALON and R. C. REINITZ, Mixing time of concrete—technological and economic aspects, *Research Paper No.* 7 (Building Research Station, Technion, Haifa, 1958).

4.23. D. A. ABRAMS, Effect of time of mixing on the strength of concrete, *The Canadian Engineer* (25th July, 1st August, 8th August 1918, Reprinted by Lewis Institute, Chicago).

4.24. G. C. COOK, Effect of time of haul on strength and consistency of ready-mixed concrete, *J. Amer. Concr. Inst.*, **39**, pp. 413–26 (April 1943).

4.25. V. S. WIGMORE, Ready-mixed concrete, *Reinf. Concr. Rev.*, **5**, No. 12, pp. 793–816 (London, Dec. 1961).

4.26. W. JURECKA, Neuere Entwicklungen und Entwicklungstendenzen von Betonmischern und Mischanlagen, *Österreichische Ingenieur-Zeitschrift*, **10**, No. 2, pp. 27–43 (1967).

4.27. J. D. DEWAR, Some effects of prolonged agitation of concrete, *Cement, Lime and Gravel*, **38**, No. 4, pp. 121–8 (London, April 1963).

4.28. R. C. MEININGER, Study of A.S.T.M. limits on delivery time, *Nat. Ready-mixed Concr. Assoc. Publicn. No.* 131, pp. 17 (Washington D.C., Feb. 1969).

4.29. R. WEBER, *Rohrförderung von Beton*, Düsseldorf Beton-Verlag GmbH (1963), The transport of concrete by pipeline (London, *Cement and Concrete Assoc. Translation No.* 129, 1968).

4.30. E. KEMPSTER, Pumpable concrete, *Current Paper* 26/69, 8 pp. (Building Research Station, Garston, 1968).

4.31. E. KEMPSTER, Pumpability of mortars, *Contract Journal*, **217**, pp. 28–30 (May 4, 1967).

4.32. A.C.I. COMMITTEE 304, Preplaced aggregate concrete for structural and mass concrete, *J. Amer. Concr. Inst.*, **66**, pp. 785–97 (Oct. 1969).

4.33. Shotcreting, *Amer. Concr. Inst. Sp. Publicn. No.* 14, pp. 223 (1966).

4.34. A.C.I. COMMITTEE 506, Proposed A.C.I. Standard Recommended practice for shotcreting, *J. Amer. Concr. Inst.*, **63**, pp. 219–46 (Feb. 1966).

4.35. C. D. TURTON, Rapid analysis of freshly mixed concrete, *Engineering*, **191**, No. 4960, p. 659 (London, May 12, 1961).

4.36. R. BAVELJA, A rapid method for the wet analysis of fresh concrete, *Concrete*, **4**, No. 9, pp. 351–3 (London, Sept. 1970).

4.37. L. R. CHADDA, The rapid determination of cement content in concrete and mortar, *Indian Concrete J.*, **29**, No. 8, pp. 258–60 (Aug. 1955).

4.38. W. G. HIME and R. A. WILLIS, A method for the determination of the cement content of plastic concrete, *A.S.T.M. Bul. No.* 209, pp. 37–43 (Oct. 1955).

4.39. A.C.I. COMMITTEE 305, Recommended practice for hot weather concreting, *J. Amer. Concr. Inst.*, **68**, pp. 489–501 (July 1971).

4.40. A. M. NEVILLE, Analysis of fresh concrete, *Concrete*, **7**, No. 3, p. 37 (London, Mar. 1973).

5. Strength of Concrete

STRENGTH of concrete is commonly considered its most valuable property, although in many practical cases other characteristics, such as durability and impermeability, may in fact be more important. Nevertheless, strength usually gives an overall picture of the quality of concrete, as strength is directly related to the structure of the hardened cement paste.

The mechanical strength of cement gel was discussed on page 33; below, some empirical relations concerning the strength of concrete will be considered.

Water/Cement Ratio

In engineering practice, the strength of concrete at a given age and cured at a prescribed temperature is assumed to depend primarily on two factors only: the water/cement ratio and the degree of compaction. The influence of air voids on strength was discussed on page 182, and at this stage we shall consider fully-compacted concrete only: in practice this is taken to mean that the hardened concrete contains about 1 per cent of air voids.

When concrete is fully compacted its strength is taken to be inversely proportional to the water/cement ratio according to the "law" established by Duff Abrams in 1919. He found strength to be equal to

$$S = \frac{K_1}{K_2^{w/c}}$$

where w/c represents the water/cement ratio of the mix (originally taken by volume), and K_1 and K_2 are empirical constants. A typical strength versus water/cement ratio curve is shown in Fig. 5.1.

Abrams' "law," although established independently, is a special case of a general rule formulated by Feret in 1896. This was in the form

$$S = K \left(\frac{c}{c + e + a}\right)^2$$

where S is the strength of concrete, c, e, and a are the absolute volumes of cement, water, and air respectively, and K is a constant.

It may be recalled that the water/cement ratio determines the porosity of the hardened cement paste at any stage of hydration (*see* p. 30). Thus the water/cement ratio and the degree of compaction both affect the volume of voids in concrete, and this is why the volume of air in concrete is included in Feret's expression.

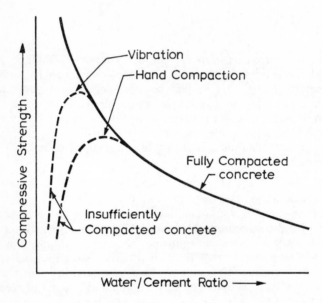

Fig. 5.1. *The relation between strength and water/cement ratio of concrete*

The relation between strength and the total volume of voids is not a unique property of concrete but is found also in other brittle materials in which water leaves behind pores: for instance, the strength of plaster is also a direct function of its void content[5.1] (*see* Fig. 5.2). Moreover, if the strength of different materials is expressed as a fraction of the strength at a zero porosity, a wide range of materials conform to the same relation between relative strength and porosity, as shown in Fig. 5.3 for plaster, steel, iron,[5.72] alumina and zirconia.[5.73] This general pattern is of interest in understanding the role of voids in the strength of concrete. Moreover, the relation of Fig. 5.3 makes it clear why cement compacts (*see* p. 27), which have a very low porosity, have a very high strength.

Strictly speaking, strength of concrete is probably influenced by the volume of all voids in concrete: entrapped air, capillary pores, gel pores, and entrained air if present.[5.104] An example of the calculation of the total air content may be of interest and is given in a footnote.*

Fig. 5.1 shows that the range of validity of the water/cement ratio rule is limited. At the lower end of the scale the curve ceases to be followed when full compaction is no longer possible; the actual position of the point of departure depends on the means of compaction available. It seems also that mixes with a very low water/cement ratio and an extremely high cement content (470 to 530 kg/m³ (800 to 900 lb/yd³)) exhibit retrogression of strength, particularly when large size aggregate is used. Thus at later ages, in this type of mix, a lower water/cement ratio would not lead to a higher strength. This behaviour may be due to stresses induced by shrinkage, whose restraint by aggregate particles causes cracking of the cement paste or a loss of the cement-aggregate bond.[5.2]

From time to time the water/cement ratio rule has been criticized as not being sufficiently fundamental (*see* the following section). Nevertheless, *in practice* the water/cement ratio is the largest single factor in the

---

*Let the given mix have proportions 1:3·4:4·2, with a water/cement ratio of 0·80. The entrapped air content has been measured to be 2·3 per cent. Given that the specific gravity of the fine and coarse aggregates is respectively 2·60 and 2·65, and on the assumption that the specific gravity of cement is 3·15, the volumetric ratio of cement:fine aggregate:coarse aggregate:water = $1/3·15:3·4/2·60:4·2/2·65:0·80 = 0·318:1·31:1·58:0·80$.

Since the air content is 2·3 per cent, the volume of the remaining materials must add up to 97·7 per cent of the total volume of concrete. Thus, on a percentage basis, the volumes are as follows:

cement (dry)	= 7·8
fine aggregate	= 32·0
coarse aggregate	= 38·5
water	= 19·4
Total	= 97·7 per cent.

We know that in the given case 0·7 of the cement has hydrated after 7 days of curing in water (see, for instance, ref. 5.105). Therefore, continuing in percentage volume units, we find the volume of cement which has hydrated to be 5·5 and the volume of unhydrated cement 2·3.

The volume of combined water is 0·23 of the weight of cement which has hydrated (*see* p. 24), i.e. $0·23 \times 5·5 \times 3·15 = 4·0$. On hydration, the volume of the solid products of hydration becomes smaller than the sum of volumes of the constituent cement and water by 0·254 of the volume of combined water (*see* p. 24). Hence, the volume of the solid products of hydration is $5·5 + (1 − 0·254) \times 4·0 = 8·5$.

Since the gel has a characteristic porosity of 28 per cent (*see* p. 24), the volume of gel pores is w_g such that $w_g/(8·5 + w_g) = 0·28$, whence the volume of gel pores is 3·3. Thus the volume of hydrated cement paste inclusive of gel pores is $8·5 + 3·3 = 11·8$. Now, the volume of dry cement which has hydrated and of mixing water is $5·5 + 19·4 = 24·9$. Hence, the volume of capillary pores is $24·9 − 11·8 = 13·1$.

Thus, the voids are:

capillary pores	= 13·1
gel pores	= 3·3
air	= 2·3
Total void content	= 18·7 per cent.

strength of fully compacted concrete. Perhaps the best statement of the situation is that by Gilkey:[5.74]

"For a given cement and acceptable aggregates, the strength that may be developed by a workable, properly placed mixture of cement, aggre-

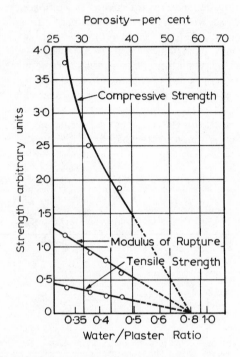

Fig. 5.2. *Strength of plaster as a function of its void content*[5.1]

gate, and water (under the same mixing, curing, and testing conditions) is influenced by the:

(*a*) ratio of cement to mixing water
(*b*) ratio of cement to aggregate
(*c*) grading, surface texture, shape, strength, and stiffness of aggregate particles
(*d*) maximum size of the aggregate."

We can add that factors (*b*) to (*d*) are of lesser importance than factor (*a*) when usual aggregates up to 40 mm (1½ in.) maximum size are employed. They are, nevertheless, present since, as pointed out by Walker and

Bloem,[5.74] "the strength of concrete results from: (1) the strength of the mortar; (2) the bond between the mortar and the coarse aggregate; and (3) the strength of the coarse aggregate particle, i.e., its ability to resist the stresses applied to it."

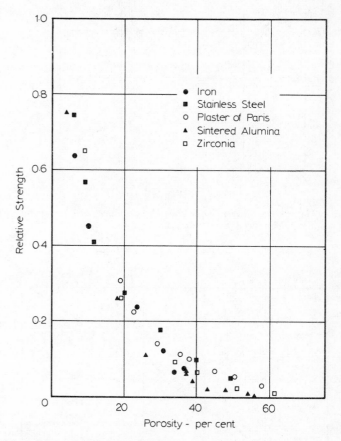

Fig. 5.3. *Influence of porosity on relative strength of various materials*

Fig. 5.4 shows that the graph of strength versus water/cement ratio is approximately in the shape of a hyperbola. This applies to concrete made with any given type of aggregate and at any given age. It is a geometrical property of a hyperbola $y = \dfrac{k}{x}$ that y against $1/x$ plots as a straight line. Thus the relation between the strength and the *cement/water* ratio is

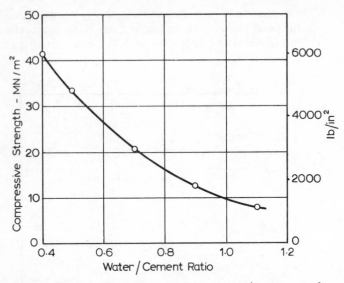

Fig. 5.4. *The relation between 7-day strength and water/cement ratio for concrete made with a rapid-hardening Portland cement*

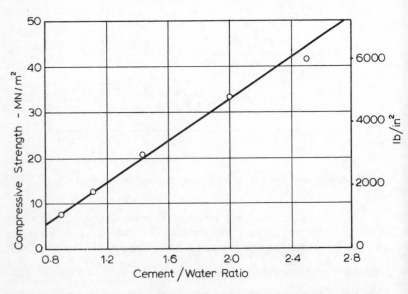

Fig. 5.5. *A plot of strength against cement/water ratio for the data of Fig. 5.4*

approximately linear in the range of cement/water ratios between about 1·2 and 2·5. This linear relationship is clearly more convenient to use than the water/cement ratio curve, particularly when interpolation is desired. Fig. 5.5 shows the data of Fig. 5.4 plotted with the cement/water ratio as abscissa. The values used apply to the given cement only, and in any practical case the actual relation between strength and cement/water ratio has to be determined.

It must be admitted that the relations discussed here are not precise, and other approximations can be made. For instance, it has been suggested that as an approximation the relation between the logarithm of strength and the natural value of the water/cement ratio can be assumed to be linear[5.3] (cf. Abrams' expression). As an illustration, Fig. 5.6 gives

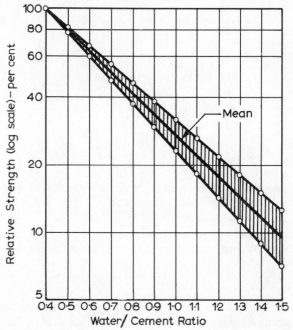

Fig. 5.6. *Relation between logarithm of strength and water/cement ratio*[5.3]

the relative strength of mixes with different water/cement ratios, taking the strength at the water/cement ratio of 0·4 as unity.

The pattern of strength of aluminous cement concrete appears to be different from that of concrete made with Portland cement.[5.4] As can be seen from Fig. 5.7, strength increases with the cement/water ratio at a progressively decreasing rate.

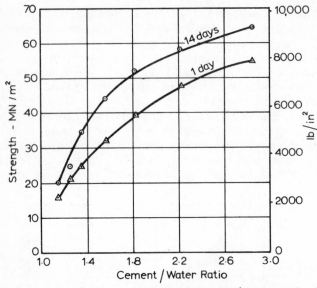

Fig. 5.7. *Relation between compressive strength and cement/water ratio for aluminous cement concrete test cylinders*[5.4]

Gel/Space Ratio

The influence of the water/cement ratio on strength does not truly constitute a *law* as the water/cement ratio rule does not include many qualifications necessary for its validity. In particular, strength at any water/cement ratio depends on the degree of hydration of cement and its chemical and physical properties; the temperature at which hydration takes place; the air content of the concrete; and also the change in the effective water/cement ratio and the formation of fissures due to bleeding.[5.5]

It is more correct, therefore, to relate strength to the concentration of the solid products of hydration of cement in the space available for these products; in this connexion it may be relevant to refer again to Fig. 1.9. Powers[5.6] has determined the relation between the strength development and the gel/space ratio. This ratio is defined as the ratio of the volume of the hydrated cement paste to the sum of the volumes of the hydrated cement and of the capillary pores.

On page 24 it was shown that cement hydrates to occupy more than twice its original volume; in the following calculations the products of hydration of 1 ml of cement will be assumed to occupy 2·06 ml; not all the hydrated material is gel, but as an approximation we can consider it as such.

Let c = weight of cement,

v_c = specific volume of cement,

w_o = volume of mixing water,

and α = the fraction of cement that has hydrated.

Then,[5.7] the volume of gel = $2 \cdot 06 c v_c \alpha$, and the total space available to the gel is $c v_c \alpha + w_o$.

Hence, the gel/space ratio is

$$x = \frac{2 \cdot 06\, v_c \alpha}{v_c \alpha + \dfrac{w_o}{c}} \, .$$

Taking the specific volume of dry cement as $0 \cdot 319$ ml/g, the gel/space ratio becomes—

$$x = \frac{0 \cdot 647 \alpha}{0 \cdot 319 \alpha + \dfrac{w_o}{c}} \, .$$

The compressive strength of concrete tested by Powers[5.7] was found to be $234 x^3$ MN/m² ($34{,}000\ x^3$ lb/in²), and is independent of the age of the concrete or its mix proportions. The actual relation between the compressive strength of mortar and the gel/space ratio is shown in Fig. 5.8: it can be seen that strength is approximately proportional to the cube of the gel/space ratio, and the figure 234 MN/m² (34,000 lb/in²) represents the intrinsic strength of gel for the type of cement and of specimen used.[5.8] Numerical values differ little for the usual range of Portland cements except that a higher C_3A content leads to a lower strength at a given gel/space ratio.[5.5]

If A cm³ of air are present in the set paste, the ratio $\dfrac{w_o}{c}$ in the above expression would be replaced by $\dfrac{w_o + A}{c}$ (*see* Fig. 5.9).

The resulting formula for strength would be similar to that of Feret's but the ratio used here involves a quantity proportional to the volume of hydrated cement instead of the total volume of cement and is thus applicable at any age.

The expression relating strength to the gel/space ratio can be written in a number of ways. It may be convenient to utilize the fact that the volume of non-evaporable water, w_n, is proportional to the volume of the gel; and also that the volume of mixing water, w_o, is related to the

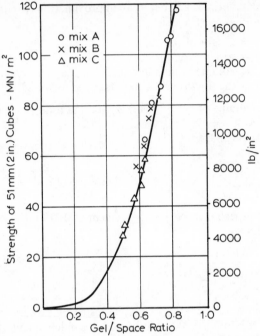

Fig. 5.8. *Relation between the compressive strength of mortar and gel/space ratio*[5.8]

space available for the gel. The strength, S, in pounds per square inch, for S greater than about 2,000 lb/in², when the relation is approximately linear, can then be written in the form[5.6]—

$$S = 34,200 \frac{w_n}{w_o} - 3,600$$

Alternatively, the surface area of gel, V_m, can be used. Then

$$S = 120,000 \frac{V_m}{w_o} - 3,600$$

Fig. 5.10 shows Powers' actual data[5.6] for cements with low C_3A contents.

The above expressions have been found to be valid for many cements but the numerical coefficients may depend on the intrinsic strength of the gel produced by a given cement. In other words, the strength of the paste depends primarily on the physical structure of the gel but the effects of the chemical composition of cement cannot be neglected; however, at later ages these effects become minor only.

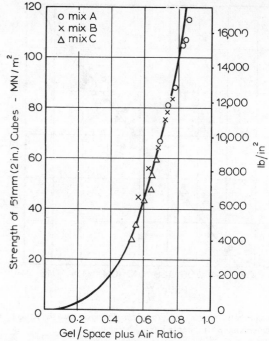

Fig. 5.9. *Relation between the compressive strength of mortar and gel/space ratio, modified to include entrapped air voids*[5.7]

Effective Water in the Mix

The relations discussed so far in this chapter involve the quantity of water in the mix. This needs a more careful definition. We consider as effective that water which occupies space outside the aggregate particles when the gross volume of concrete becomes stabilized, i.e., approximately at the time of setting. Hence the terms *effective* or *nett* water/cement ratio.

Generally, water in concrete consists of that added to the mix and that held by the aggregate at the time when it enters the mixer. A part of the latter water is absorbed within the pore structure of the aggregate while some exists as free water on the surface of the aggregate and is therefore no different from the water added direct into the mixer. Conversely, when the aggregate is not saturated and some of its pores are therefore air-filled, a part of the water added to the mix will be absorbed by the aggregate during the first half hour or so after mixing. Under such circumstances the demarcation between absorbed and free water is a little difficult.

On a site the aggregate is as a rule wet, and the water in excess of that

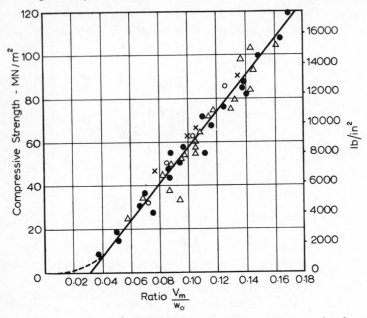

Fig. 5.10. *Relation between the strength of cement paste and the ratio of surface area of gel V_m to the volume of mixing water w_o* [5.6]

required to bring it to a saturated and surface-dry condition is included in the effective water of the mix. For this reason the strength curves of Road Note No. 4[5.9] (*see* Fig. 10.2) are based on the water in excess of that absorbed by the aggregate. On the other hand, McIntosh and Erntroy's data[5.10] refer to the total water added to a dry aggregate. This condition of aggregate is frequently met in the laboratory. Care is therefore necessary in translating laboratory results into mix proportions to be used on a site.

Nature of Strength of Concrete

The paramount influence of voids in concrete on its strength has been repeatedly mentioned, and it should be possible to relate this factor to the actual mechanism of failure. For this purpose, concrete is considered as a brittle material, even though it exhibits a small amount of plastic action, as fracture under static loading takes place at a moderately low total strain; a strain of 0·001 to 0·005 at failure has been suggested as the limit of brittle behaviour.

Strength in Tension

The actual (technical) strength of cement paste or of similar brittle materials such as stone is very much lower than the theoretical strength estimated on the basis of molecular cohesion, and calculated from the surface energy of a solid assumed to be perfectly homogeneous and flawless. The theoretical strength has been estimated to be as high as 10.5 GN/m^2 (1.5×10^6 lb/in^2).

This discrepancy can be explained by the presence of flaws postulated by Griffith.[5.17] These flaws lead to high stress concentrations in the material under load so that a very high stress is reached in very small volumes of the specimen with a consequent microscopic fracture, while the average (nominal) stress in the whole specimen is comparatively low. The flaws vary in size and it is only the few largest ones that cause failure: the strength of a specimen is thus a problem of statistical probability, and the size of the specimen affects the probable nominal stress at which failure is observed.

Cement paste is known to contain numerous discontinuities—pores, fissures and voids—but the exact mechanism through which they affect the strength is not known. The voids themselves need not act as flaws, but the flaws may be cracks in individual crystals associated with the voids[5.14] or caused by shrinkage or bad bond. In unsegregated concrete, voids are distributed in a random manner,[5.15] a condition necessary for the application of Griffith's hypothesis. While we do not know the exact mechanism of rupture of concrete this is probably related to the bond within the cement paste and between the paste and the aggregate.

Griffith's hypothesis postulates microscopic failure at the location of a flaw, and it is usually assumed that the "volume unit" containing the weakest flaw determines the strength of the entire specimen. This statement implies that any crack will spread throughout the section of the specimen subjected to the given stress, or, in other words, an event taking place in an element is identified with the same event taking place in the body as a whole.

This behaviour can be met with only under a uniform stress distribution, with the additional proviso that the "second weakest" flaw is not strong enough to resist a stress of $\dfrac{n}{n-1}$ times the stress at which the weakest flaw failed, where n is the number of elements in the section under load, each element containing one flaw.

While local fracture starts at a point and is governed by the conditions at that point, the knowledge of stress at the most highly stressed point in the body is not sufficient to predict failure. It is necessary to know also the stress distribution in a volume sufficiently extended round this point as the deformational response within the material, particularly near failure, depends on the behaviour and state of the material surrounding the critical point, and the possibility of spreading of failure is strongly affected by this state. This would explain, for instance, why the maximum fibre stresses in flexure specimens at the moment of incipient failure are higher than the strength determined in uniform direct tension: in the latter case the propagation of fracture is not blocked by the surrounding material. Some actual data on the relation between the strength in flexure and in direct tension are given in Fig. 8.10.

We can see then that in a given specimen different stresses will produce fracture at different points, but it is not possible physically to test the strength of an individual element without altering its condition in relation to the rest of the body. If the strength of a specimen is governed by the weakest element in it, the problem becomes that of the proverbial weakest link in a chain. In statistical terms we have to determine the least value (i.e. the strength of the most effective flaw) in a sample of size n, where n is the number of flaws in the specimen. The chain analogy may not be quite correct for in concrete the links may be arranged in parallel as well as in series but computations on the basis of the weakest link assumption yield results of the correct order. It follows that the strength of a brittle material such as concrete cannot be described by an average value only: an indication of the variability of strength must be given, as well as information about the size and shape of the specimens. These factors are discussed in Chapter 8.

Cracking and Failure in Compression

Griffith's hypothesis applies to failure under the action of a tensile force but it can be extended to fracture under bi- and triaxial stress and also under uniaxial compression. Even when two principal stresses are compressive the stress along the edge of the flaw is tensile at some points, so that fracture can take place. Orowan[5.16] calculated the maximum tensile stress at the tip of the flaw of the most dangerous orientation relative to the principal stress axes as a function of the two principal stresses P and Q. The fracture criteria are represented graphically in Fig. 5.11, where K is the tensile strength in direct tension. Fracture occurs under a combination of P and Q such that the point representing the state of stress crosses the curve outwards on to the shaded side.

From Fig. 5.11 it can be seen that fracture may occur when uniaxial compression is applied; this has in fact been observed in tests on concrete compression test specimens.[5.18] The nominal compressive strength in this case is $8K$, i.e. eight times the tensile strength determined in a direct tension test. This figure is in good agreement with the observed values of the ratio of the compressive to tensile strengths of concrete. There are, however, difficulties in reconciling certain aspects of Griffith's hypothesis with the observed direction of cracks in compression specimens. It is possible, though, that failure in such a specimen is governed by the lateral strain induced by Poisson's ratio. The order of values of Poisson's ratio for concrete is such that, for elements sufficiently removed from the platens of the testing machine, the resulting lateral strain can exceed the ultimate tensile strain of concrete. Failure occurs then by splitting at right angles to the direction of the load,* and this has been frequently

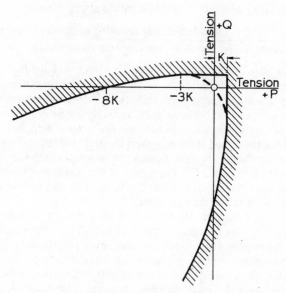

Fig. 5.11. *Orowan's*[5.16] *criteria of fracture under biaxial stress*

observed, especially in specimens whose height is greater than breadth.[5.18] There are strong indications that it is not a limiting stress but a limiting tensile strain that determines the strength of concrete under static loading: this is usually assumed to be between 1×10^{-4} and 2×10^{-4}. It has been found that at the point of initial cracking the strain on the tension face of a beam in flexure and the lateral tensile strain in a cylinder in uniaxial compression are of the same magnitude.[5.21]

The tensile strain in a beam at cracking is

$$\frac{\text{tensile stress at cracking}}{E}$$

---

* This is similar to the situation in the splitting test (see p. 481).

where E is the modulus of elasticity of concrete over the linear range of deformation. Now the lateral strain in a compression specimen when cracking is first observed is

$$\frac{\mu \times \text{compressive stress at cracking}}{E}$$

where μ is the static Poisson's ratio, and E is the same as above. From the observed equality of the two strains it would appear that

$$\mu = \frac{\text{tensile stress at cracking in flexure}}{\text{compressive stress at cracking in a compression specimen}}$$

Poisson's ratio varies generally between about 0·11 for high strength concrete and 0·21 for weak mixes, and it is significant that the ratio of the nominal tensile and compressive strengths for different concretes varies in a similar manner and between approximately the same limits. There is thus a possibility of a connexion between the ratio of nominal strengths and Poisson's ratio, and there are good grounds for suggesting that the mechanism producing the initial cracks in uniaxial compression and in flexure tension is the same.[5.19] The nature of this mechanism has not been established, but cracking is probably due to local breakdowns in adhesion between the cement and the aggregate.[5.20]

The ultimate failure under the action of a uniaxial compression is either a tensile failure of cement crystals or of bond in a direction perpendicular to the applied load, or is a collapse caused by the development of inclined shear planes.[5.20] It is probable that ultimate strain is the criterion of failure, but the level of strain varies with the strength of concrete: the higher the strength the lower the ultimate strain. Some typical, but by no means general,* values are given below—

Nominal compressive strength		Maximum strain at failure
MN/m²	lb/in²	10^{-3}
7	1,000	4·5
14	2,000	4
35	5,000	3
70	10,000	2

Under triaxial compression, failure must take place by crushing: the mechanism is, therefore, quite different from that described above. An increase in lateral compression increases the axial load than can be sustained, as shown for instance in Fig. 5.12. With very high lateral stresses extremely high strengths have been recorded (Fig. 5.13). It may be noted

* The actual values depend on the method of test (cf. Fig. 6.2).

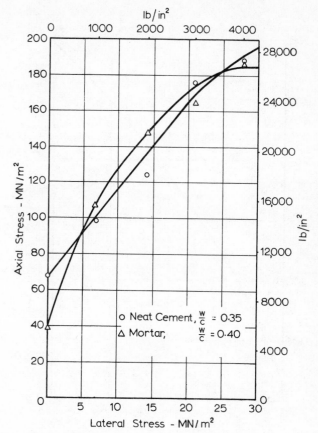

Fig. 5.12. *Influence of lateral stress on the axial stress at failure of neat cement and mortar*[5.26]

that if the development of pore water pressure in concrete is limited by allowing the displaced pore water to escape through the loading platens, then the apparent strength is higher.[5.75]

A lateral tensile stress has a similar influence but, of course, in the opposite direction.[5.11] This behaviour agrees well with the theoretical considerations of page 470.

In practice it is likely that failure of concrete takes place over a range of stress rather than as an instantaneous phenomenon, so that ultimate failure is a function of the type of loading.[5.19] This is of especial interest when repeated loading is applied—a condition frequently met with in practice. Fatigue strength of concrete is considered on page 295.

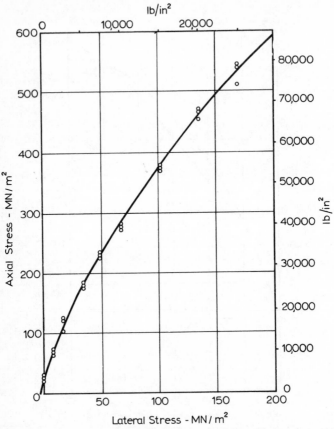

Fig: 5.13. *Influence of high lateral stress on the axial stress at failure of concrete*[5.11]

Microcracking
Investigations in recent years have shown that very fine cracks at the interface between coarse aggregate and cement paste exist in fact even prior to application of the load on concrete.[5.76] These cracks remain stable up to about 30 per cent or more of the ultimate load and then begin to increase in length, width, and number. The overall stress under which they develop is sensitive to the water/cement ratio of the paste. This is the stage of slow crack propagation. At 70 to 90 per cent of the ultimate strength, cracks open through the mortar (cement paste and fine aggregate); they bridge the bond cracks so that a continuous crack pattern is formed.[5.76] This is the fast crack propagation stage and, if the load is sustained, failure may take place with time.

It may be proper at this stage to consider the development of failure of concrete, although general agreement upon it is still remote. One aspect that is not controversial is the influence of the heterogeneity of concrete on its failure behaviour. While in a homogeneous body under a simple state of stress, the stress trajectories are straight or simple curves, this cannot be the case in concrete owing to the presence of aggregate: the properties of the material vary from point to point, and the aggregate–paste bond surfaces may form all the possible angles with the direction of the external force. As a result, the local stress varies substantially above and below the nominal applied stress.

Influence of Coarse Aggregate on Strength
Vertical cracking in a specimen subjected to uniaxial compression starts under a load equal to 50 to 75 per cent of the ultimate load. This has been determined from measurements of the velocity of sound transmitted through the concrete,[5.22] and also using ultrasonic pulse velocity techniques.[5.23] The stress at which the cracks form depends largely on the properties of the coarse aggregate: smooth gravel leads to cracking at lower stresses than rough and angular crushed rock, probably because mechanical bond is influenced by the surface properties and, to a certain degree, by the shape of the coarse aggregate.[5.19]

The properties of aggregate affect thus the cracking, as distinct from the ultimate, load in compression and the flexural strength in the same manner, so that the relation between the two quantities is independent of the type of aggregate used. Fig. 5.14 shows Jones and Kaplan's[5.19] results, each symbol representing a different type of coarse aggregate. On the other hand, the relation between the flexural and compressive *strengths* depends on the type of coarse aggregate used (*see* Fig. 5.15) since (except in high strength concrete) the properties of aggregate, especially its surface texture, affect the ultimate strength in compression very much less than the strength in tension or the cracking load in compression.

The influence of the type of coarse aggregate on the strength of concrete varies in magnitude and depends on the water/cement ratio of the mix. For water/cement ratios below 0·4 the use of crushed aggregate has resulted in strengths up to 38 per cent higher than when gravel is used. With an increase in the water/cement ratio the influence of aggregate falls off, presumably because the strength of the paste itself becomes paramount, and at a water/cement ratio of 0·65 no difference in the strengths of concretes made with crushed rock and gravel has been observed.[5.24]

The influence of aggregate on flexural strength seems to depend also on the moisture condition of the concrete at the time of test.[5.60]

The shape and surface texture of coarse aggregate affect also the

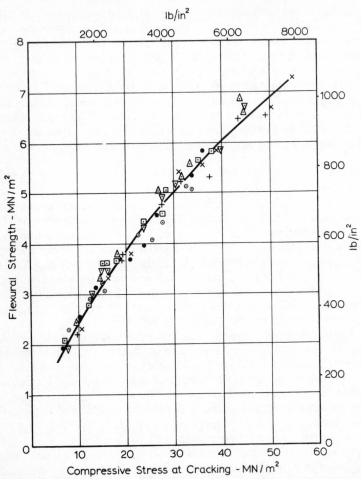

Fig. 5.14. *Relation between flexural strength and compressive stress at cracking for concretes made with different coarse aggregates*[5.19]

(*Crown Copyright*)

impact strength of concrete, the influence being qualitatively the same as on the flexural strength[5.61] (*see* p. 608).

Kaplan[5.25] observed that the flexural strength of concrete is generally lower than the flexural strength of corresponding mortar. Mortar would thus seem to set the upper limit to the flexural strength of concrete and the presence of the coarse aggregate generally reduces this strength. On the other hand, the compressive strength of concrete is higher than that of mortar, which, according to Kaplan, indicates that the mechanical

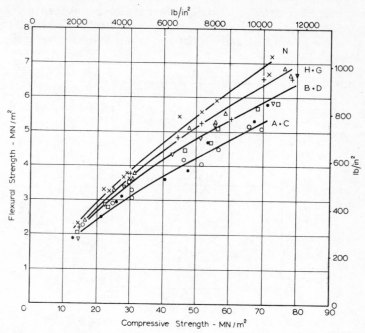

Fig. 5.15. *Relation between flexural and compressive strengths for concretes made with different aggregates*[5.19]

(*Crown Copyright*)

interlocking of the coarse aggregate contributes to the strength of concrete in compression. This behaviour has not, however, been confirmed to apply generally, and the entire problem of strength of mortar and concrete requires further study.

Influence of Richness of the Mix on Strength

The anomalous behaviour of extremely rich mixes was mentioned on page 235, but the aggregate/cement ratio affects the strength of all medium- and high-strength concretes, i.e. those with a strength of about 35 MN/m² (5,000 lb/in²) or more. There is no doubt that the aggregate/cement ratio is only a secondary factor in the strength of concrete but it has been found that for a constant water/cement ratio a leaner mix leads to a higher strength[5.12] (*see* Fig. 5.16).

This behaviour is probably associated with the absorption of water by the aggregate: a larger amount of aggregate absorbs a greater quantity of water, the effective water/cement ratio being thus reduced. It is probable, however, that other factors also play a part: for instance, the total water content per cubic metre of concrete is lower in a leaner mix than in a rich one. As a result, in a leaner mix the voids form a smaller fraction of the total volume of concrete, and it is these voids that have an adverse effect on strength.*

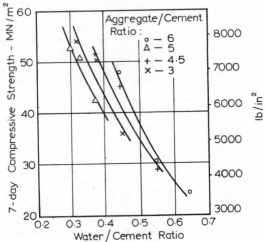

Fig. 5.16. *The influence of the aggregate/cement ratio on strength of concrete*[5.13]

Equation of Strength

Strength of concrete is an inherent property of the material but, as measured in practice, is also a function of the stress system which is acting. Mather[5.77] pointed out that, ideally, it should be possible to express the failure criteria under all possible stress combinations by a single stress parameter, such as strength in uniaxial tension. However, such a solution has not yet been found.

To develop equations for concrete strength, Berg[5.21] considered the stress, σ_{cr}, at the initiation of crack propagation, the maximum nominal stress, σ_{pr}, on a concrete prism, and the cleavage strength, σ_{cl}, of concrete in the direction normal to the applied compressive stress. The cleavage strength is approximately equal to the strength in axial tension. He showed that for the general case of a principal stress system σ_1, σ_2, σ_3 (where $\sigma_1 > \sigma_2 > \sigma_3$ and compression is positive)

$$\frac{\sigma_1}{\sigma_{pr}} = \left[1 + 2Kn_3 + K^2 n_3{}^2 - \frac{n_2{}^2}{c^2} - \frac{n_3{}^2}{c_3} \right]^{\frac{1}{2}}$$

---

* Lean concrete is considered to mean mixes with high aggregate/cement ratios but generally not above about 10, and should be distinguished from "lean-mix concrete base." The latter, used in road construction, may have an aggregate/cement ratio as high as 20 and is suitable for compaction by rolling.

where

$$n_2 = \frac{\sigma_2}{\sigma_{cl}}; \; n_3 = \frac{\sigma_3}{\sigma_{cl}}; \; c = \frac{\sigma_{pr}}{\sigma_{cl}}; \; \text{and } K = \frac{\sigma_{pr} - \sigma_{cr}}{\sigma_{pr}}.$$

This equation can be used for an analytical evaluation of the failure of concrete under combined states of stress with the parameters σ_{pr}, σ_{cr}, and σ_{cl} determined from tests on uniaxial compression and uniaxial tension. Some empirical data for specimens of various types are shown in Fig. 5.17. The criterion ceases to apply, however, when σ_2 and σ_3

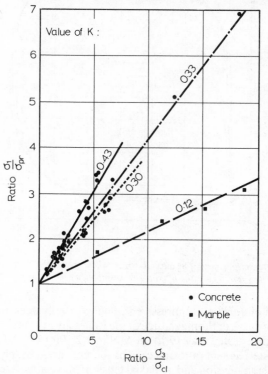

Fig. 5.17. *Berg's equation for strength of concrete*[5.21]

have values such that the transverse cleavage strength cannot be overcome. The behaviour of concrete is then no longer brittle but plastic. Some data on the strength of concrete when $\sigma_2 = \sigma_3$ and $\sigma_1 > \sigma_2$ are shown in Fig. 5.13.

A general biaxial stress interaction curve is shown in Fig. 5.18. A large interaction is observed when there is a considerable frictional restraint at the platens, but when the end restraint of the specimen is effectively eliminated (e.g. by the use of steel brush platens, *see* p. 468), the effect is much smaller. It can be seen from Fig. 5.18 that under a biaxial stress $\sigma_1 = \sigma_2$, strength is only 16 per cent higher than in uniaxial compression; biaxial tensile strength is no different from uniaxial tensile strength.[5.78] These findings were confirmed by Nelissen.[5.102]

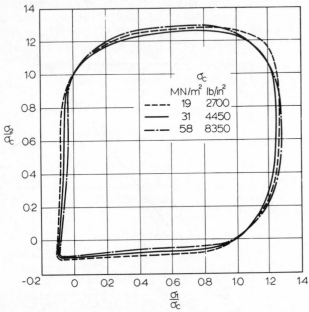

Fig. 5.18. *Interaction curve for biaxial stress when the end restraint is effectively eliminated*[5.78] (σ_1 and σ_2 are the biaxial stresses applied)

The level of uniaxial compressive strength virtually does not affect the shape of the curve or the magnitude of the values given by it;[5.78] the prism strength range tested was 19 to 58 MN/m² (2,700 to 8,350 lb/in²) and both the water/cement ratio and cement content varied widely. However, in compression–tension and in biaxial tension, the relative strength at any particular biaxial stress combination decreases as the level of uniaxial compressive strength increases.[5.78] This accords with the general observation that the ratio of uniaxial tensile strength to uniaxial compressive strength decreases as the compressive strength level rises (*see* p. 261); in these tests, the ratio was 0·11, 0·09, and 0·08 at a uniaxial compressive strength level of 19, 31, and 58 MN/m² (2,700, 4,450, and 8,350 lb/in²) respectively.[5.78]

Recognizing that the strength of concrete cannot be predicted by considering limitations on the compressive, tensile, and shearing stresses independently of each other, Bresler and Pister[5.79] proposed a general equation of strength of concrete. The equation uses criteria independent of chosen cartesian co-ordinates, i.e. so-called invariant terms, specifically

$$I_1 = \sigma_1 + \sigma_2 + \sigma_3$$
$$I_2 = \sigma_1\sigma_2 + \sigma_2\sigma_3 + \sigma_3\sigma_1$$

and

$$I_3 = \sigma_1\sigma_2\sigma_3$$

with the generalized failure criterion $F(I_1, I_2, I_3) = 0$.

If one of the principal stresses is zero, $I_3 = 0$.

Bresler and Pister[5.80] considered, after Nadai,[5.81] failure to be defined by

$$\tau_0 = F(\sigma_0)$$

where σ_0 and τ_0 are the normal and shearing octahedral stresses given by

$$\sigma_0 = \tfrac{1}{3}(\sigma_1 + \sigma_2 + \sigma_3)$$

i.e.

$$\sigma_0 = \frac{I_1}{3}$$

and

$$\tau_0 = \tfrac{1}{3}[(\sigma_1 - \sigma_2)^2 + (\sigma_2 - \sigma_3)^2 + (\sigma_3 - \sigma_1)^2]^{\frac{1}{2}}$$

i.e.

$$\tau_0 = F(I_1, I_2).$$

Octahedral stresses are so named because they occur on the sides of an octahedral element formed by planes whose normals make equal angles with the principal stress axes.

To avoid the influence of the properties of the specimen on the compressive strength of concrete, the octahedral stresses are rendered dimensionless by dividing them by the nominal uniaxial compressive strength of the specimen used, σ_c. An equation of the type

$$-\frac{\tau_0}{\sigma_c} = a\frac{\sigma_0}{\sigma_c} + b$$

is obtained, and this gives the combination of stresses that will just cause failure.[5.82] A typical curve is shown in Fig. 5.19. The effect of I_3 has been found to be significant so that the intermediate stress (e.g. in biaxial compression–tension) cannot be neglected. For this reason, some of the classical failure theories are not considered properly applicable to concrete. While the octahedral stress failure theory has met with considerable success it does not give the same equation for bi- and triaxial stress

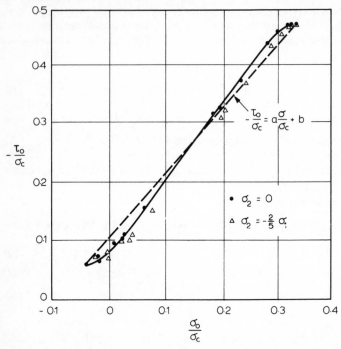

Fig. 5.19. *Relation between octahedral shear stress and octahedral normal stress at failure for 20 MN/m² (3,000 lb/in²) concrete subjected to uniaxial compression and to biaxial compression–tension*[5.82]

so that further development of an equation of strength is still needed. Such information will be of value in the design of structures such as shells, plates, pressure vessels and even in parts of flexural members.

Effect of Age on Strength of Concrete

The relation between the water/cement ratio and the strength of concrete applies to one type of cement and one age only. On the other hand, the strength versus gel/space ratio relationship has a more general application because the amount of gel present in the cement paste at any time is itself a function of age and type of cement. In other words, different cements require a different length of time to produce the same quantity of gel.

The rate of gain of strength of different cements was discussed in Chapter 2, and Figs. 2.1 and 2.2 show typical strength–time curves. The influence of the curing conditions on the development of strength is considered later in this chapter, but here we are concerned with the practical problem of strength of concrete tested at different ages. In the majority of cases the tests are made at the age of 28 days when the strength of concrete is considerably lower than its long-term strength. In the past the gain in strength beyond the age of 28 days was regarded merely as contributing to an increase in the factor of safety of the structure but since 1957 the Codes of Practice for Reinforced and Prestressed Concrete allow the gain in strength to be taken into account in the design of structures that will not be subjected to load until a later age except when no-fines concrete is used; with some lightweight aggregates, verifying tests are advisable. The Code values of permissible stress based on the 28-day compressive strength, are given in Table 5.1, but they do not, of course, apply when accelerators are used.

Table 5.1: *The British Code of Practice CP110: 1972 Factors for Increase in Compressive Strength of Concrete with Age* (*Average values*)

Minimum age of member when full design load is applied months	Age factor
1	1·0
2	1·1
3	1·15
6	1·2
12	1·24

For high strength concrete, slightly lower values should be used.

The rate of gain in the strength of concrete is of interest also in connexion with testing. It is often desirable to check the suitability of a mix long before the results of the 28-day test are available. However, even if the curing conditions are carefully controlled, the prediction of the 28-day strength from that measured at the age of 7 days is difficult, mainly because of the variation in the intrinsic rate of hardening of commercial cements. Furthermore, mixes with a low water/cement ratio gain strength, expressed as a percentage of long-term strength, more rapidly than mixes with higher water/cement ratios[5.83] (Fig. 5.20). This is because in the former case the cement grains are closer to one another and a continuous system of gel is established more rapidly. For this reason, a general extrapolation of, say, the 7-day strength to the 28-day strength is not easy, even when dealing with one cement only.

When no specific data on the materials used are available, the 28-day strength may be assumed to be 1·5 times the 7-day strength and, as an alternative to the specified 28-day strength of test cubes, the Code of Practice CP114 (1969) accepts a 7-day strength equal to not less than $\frac{2}{3}$

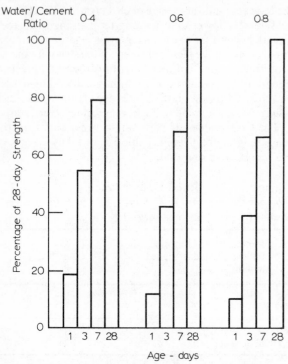

Fig. 5.20. *Relative gain of strength with time of concretes with different water/cement ratios, made with ordinary Portland cement* [5.83]

of the required 28-day strength. Tests have shown that for concretes made with ordinary Portland cement the ratio of the 28-day to 7-day strengths lies generally between 1·3 and 1·7 but the majority of the results fall above 1·5. The extrapolation of 7-day strength according to the Code of Practice is therefore quite reliable. However, in a hot climate the early strength gain is high and the ratio of the 28-day to 7-day strengths tends to be lower than in cooler weather. This is also the case with some light-weight aggregate concretes.

In Germany, the relation between the 28-day strength, σ_{28}, and the 7-day strength, σ_7, is often taken to lie between

$$\sigma_{28} = 1·4\,\sigma_7 + 150$$

and

$$\sigma_{28} = 1·7\,\sigma_7 + 850$$

σ being expressed in pounds per square inch. For strengths expressed in MN/m^2, the two constant terms become respectively 1·0 and 5·9.

Hummel[5.3] recommends the use of an approximately linear relation between the strength and the logarithm of age within the range of 3 days to 2 months. Thus, if the strength is determined at 3 and 7 days, it is possible to estimate the 28-day strength by extrapolation.

Piñeiro[5.84] suggested an expression of the type

$$\sigma_{28} = k_2(\sigma_7)^{k_1}$$

where σ_7 and σ_{28} are strengths at 7 and 28-days respectively, and k_1 and k_2 are coefficients, different for each cement and curing condition used. The value of k_1 ranges from about 0·3 to 0·8, and that of k_2 from 3 to 6.

All the expressions mentioned here apply only to concrete made with ordinary Portland cement. Many of the other cements gain strength at different rates, and when they are used the prediction of strength should be based on experimental results.

Autogenous Healing

Fine cracks in fractured concrete, if allowed to close without tangential displacement, will heal completely under moist conditions. This autogenous healing is probably due to the hydration of the hitherto unhydrated cement, and may also be aided by carbonation. The younger the concrete, i.e. the more unhydrated cement it contains, the higher the re-gain of strength, but healing without a loss of strength has been observed at ages up to three years. The application of pressure across the crack assists in healing.

Relation between Compressive and Tensile Strengths

From the discussion on the strength of compression and tension (both direct and flexure) test specimens it would be expected that the two types of strength are closely related. This is indeed the case but there is no direct proportionality, the ratio of the two strengths depending on the general level of strength of the concrete. In other words, as the compressive strength, σ_c, increases, the tensile strength, σ_t, also increases but at a decreasing rate.

A number of factors affect the relation between the two strengths. The beneficial effect of crushed coarse aggregate on flexural strength was discussed on page 251, but it seems that the properties of fine aggregate also influence the σ_t/σ_c ratio.[5.27] The ratio is also affected by the grading of the aggregate.[5.28] This is probably due to the different magnitude of the wall effect in beams and in compression specimens: their surface/volume ratios are dissimilar so that different quantities of mortar are required for full compaction.

Age is also a factor in the relation between σ_t and σ_c: beyond about 1 month the tensile strength increases more slowly than the compressive strength so that the ratio σ_t/σ_c decreases with time.[5.29] This is in agreement with the general tendency of the ratio to decrease with an increase in σ_c.

The tensile strength of concrete is more sensitive to inadequate curing than the compressive strength,[5.30] possibly because the effects of non-uniform shrinkage of flexure test beams are very serious. Thus air-cured concrete has a lower σ_t/σ_c ratio than concrete cured in water and tested wet.

Air entrainment affects the σ_t/σ_c ratio because the presence of air lowers the compressive strength of concrete more than the tensile strength, particularly in the case of rich and strong mixes.[5.30] The influence of incomplete compaction is similar to that of entrained air[5.31] (Fig. 5.21).

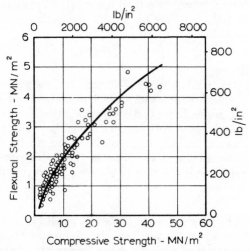

Fig. 5.21. *Relation between compressive and flexural strengths of incompletely compacted concrete*[5.31]

Lightweight concrete conforms to the pattern of the relation between σ_t and σ_c for ordinary concrete. At very low strengths (say, 2 MN/m² (300 lb/in²)) the ratio σ_t/σ_c can be as high as 0·3, but at higher strengths it decreases to values comparable with those for ordinary concrete. In the latter case, the ratio varies generally between 0·16 and 0·07 when the cube crushing strength is taken as σ_c, and the modulus of rupture under third-point loading as σ_t. As shown on page 478, the tensile strength of concrete depends on the type and method of test, so that the means of determining σ_t must be clearly stated.

A number of empirical formulae connecting σ_t and σ_c have been suggested, many of them of the type—

$$\sigma_t = k(\sigma_c)^n$$

where k and n are coefficients. Tests made at the Building Research Station on 100 mm (4 in.) cubes and $100 \times 100 \times 400$ mm ($4 \times 4 \times 16$ in.) beams centrally loaded have yielded values of $n = \frac{1}{2}$ and average $k = 8 \cdot 3$. The latter actually varies between $6 \cdot 2$ for some gravels and $10 \cdot 4$ for crushed rock.[5.32] Other experiments indicate that n may vary between $\frac{1}{2}$ and $\frac{3}{4}$.

Comité Européen du Béton has assumed that the modulus of rupture is related to the compressive strength of cylinders by the expression

$$\sigma_t = 9 \cdot 5 \, (\sigma_c)^{\frac{1}{2}}$$

the strengths being expressed in pounds per square inch. Another formula was suggested at the University of Illinois[5.33]—

$$\sigma_t = \frac{3,000}{4 + \dfrac{12,000}{\sigma_c}}$$

where σ_t is the modulus of rupture and σ_c is determined on standard test cylinders, both expressed in pounds per square inch.

In view of the numerous factors influencing the ratio of the strengths it is not surprising that no simple relation is generally applicable. Data obtained at the laboratories of the Portland Cement Association[5.11] are given in Table 5.2, and Fig. 5.22 shows the results of Walker and Bloem's tests.[5.34]

Table 5.2: *Relation between Compressive and Tensile Strengths of Concrete*[5.11]

Compressive strength of cylinders		Strength ratio		
		Modulus of rupture* to compressive strength	Direct tensile strength to compressive strength	Direct tensile strength to modulus of rupture*
MN/m²	lb/in²			
5	1,000	0·23	0·11	0·48
15	2,000	0·19	0·10	0·53
20	3,000	0·16	0·09	0·57
30	4,000	0·15	0·09	0·59
35	5,000	0·14	0·08	0·59
40	6,000	0·13	0·08	0·60
50	7,000	0·12	0·07	0·61
55	8,000	0·12	0·07	0·62
65	9,000	0·11	0·07	0·63

* Determined under third-point loading.

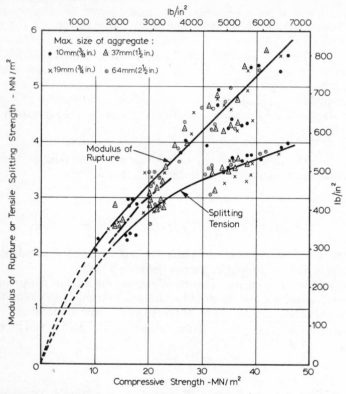

Fig. 5.22. *Relation between compressive and flexural strengths of concrete (both air-
and non-air-entrained)*[5.34]

The modulus of rupture was determined under third-point loading.

Bond between Concrete and Reinforcement

Since concrete is in the vast majority of cases used with steel reinforcement, the strength of bond between the two materials is of considerable
interest. Bond arises primarily from friction and adhesion between concrete and steel, and may also be affected by the shrinkage of concrete
relative to the steel. Bond involves, however, not only the properties of
concrete but also the mechanical properties of steel and its position in the
concrete member. For this reason the subject of bond is largely outside
the scope of this book.

In general terms, bond is related to the quality of the concrete, and
bond strength is approximately proportional to the compressive strength
up to about 20 MN/m² (3,000 lb/in²). For higher strengths of concrete

the increase in bond strength becomes progressively smaller and eventually negligible (*see* Fig. 5.23). This is why most codes of practice restrict the permissible value of bond in high strength concrete. For instance, the British Code of Practice for the Structural Use of Concrete CP110: 1972 gives the values listed in Table 5.3.

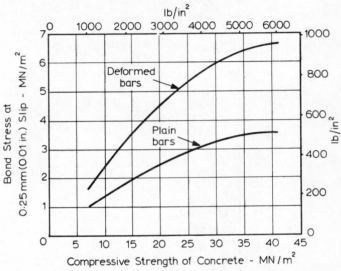

Fig. 5.23. *Influence of the strength of concrete on bond determined by pull-out tests*[5.11]

Galvanizing and other protective treatment of reinforcement generally reduce the bond strength probably because in treated steel the good bond of a rusty surface is absent.

A rise in temperature reduces the bond strength of concrete: at 200 to 300°C (400 to 570°F) there may be a loss of one-half of the bond strength at room temperature.

Curing of Concrete

In order to obtain good concrete the placing of an appropriate mix must be followed by curing in a suitable environment during the early stages of hardening. Curing is the name given to procedures used for promoting the hydration of cement, and consists of a control of temperature and of the moisture movement from and into the concrete. The temperature factor is dealt with on page 276.

More specifically, the object of curing is to keep concrete saturated, or as nearly saturated as possible, until the originally water-filled space in the fresh cement paste has been filled to the desired extent by the

Table 5.3: *Maximum Values of Flexural Bond Stress in Concrete according to the British Code of Practice for the Structural Use of Concrete CP110: 1972*

Type of reinforcement	Bond stress for concrete strength of:							
	20 MN/m² (2,900 lb/in²)		25 MN/m² (3,600 lb/in²)		30 MN/m² (4,350 lb/in²)		40 MN/m² (5,800 lb/in²) and higher	
	MN/m²	lb/in²	MN/m²	lb/in²	MN/m²	lb/in²	MN/m²	lb/in²
Plain bars	1·7	250	2·0	290	2·2	320	2·7	390
Deformed bars	2·1	300	2·5	360	2·8	410	3·4	490

products of hydration of cement. In the case of site concrete, active curing stops nearly always long before the maximum possible hydration has taken place. The order of influence of moist curing on strength can be gauged from Fig. 5.24, obtained for concrete with a water/cement ratio of 0·50.

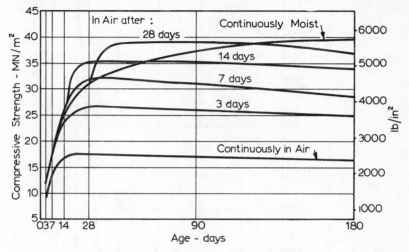

Fig. 5.24. *The influence of moist curing on the strength of concrete with a water/cement ratio of 0·50*[5.11]

The necessity for curing arises from the fact that hydration of cement can take place only in water-filled capillaries. For this reason, a loss of water by evaporation from the capillaries must be prevented. Furthermore, water lost internally by self-desiccation has to be replaced by water from outside, i.e. ingress of water into the concrete must be made possible.

It may be recalled that hydration of a sealed specimen can proceed only if the amount of water present in the paste is at least twice that of the water already combined. Self-desiccation is thus of importance in mixes with water/cement ratios below about 0·5; for higher water/cement ratios the rate of hydration of a sealed specimen equals that of a saturated specimen.[5.35] It should not be forgotten, however, that only half the water present in the paste can be used for chemical combination; this is so even if the total amount of water present is less than the water required for combination.[5.36] This statement is of considerable importance as it was formerly thought that, provided a concrete mix contained water in excess of that required for the chemical reactions with cement, a small loss of water during hardening would not adversely affect the process of hardening and the gain of strength. It is now known that hydration can take place only when the vapour pressure in the capillaries is sufficiently high, about 0·8 of the saturation pressure. Hydration at a maximum rate can proceed only under conditions of saturation. Fig. 5.25 shows the

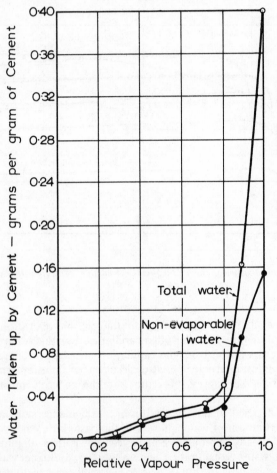

Fig. 5.25. *Water taken up by dry cement exposed for six months to different vapour pressures*[5.36]

degree of hydration after six months' storage at different relative humidities, and it is clear that below a vapour pressure of 0·8 of the saturation pressure the degree of hydration is low, and negligible below 0·3 of the saturation pressure.[5.36]

It must be stressed that for a satisfactory development of strength it is not necessary for all the cement to hydrate, and indeed this is only rarely achieved in practice: as shown earlier, the quality of concrete depends primarily on the gel/space ratio of the paste. If, however, the water-filled space in fresh concrete is greater than the volume that can be filled by the products of hydration, greater hydration will lead to a higher strength and a lower permeability.

Evaporation of water from concrete soon after placing depends on the temperature and relative humidity of the surrounding air and on the velocity of wind which effects a change of air over the surface of the concrete. An indication of the influence of these three factors can be obtained from Figs. 5.26, 5.27 and 5.28, based on Lerch's[5.37] results. The

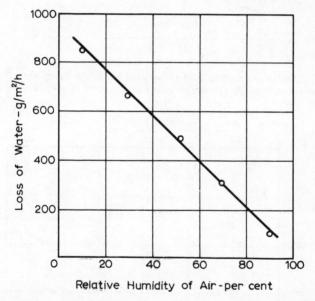

Fig. 5.26. *Influence of relative humidity on the loss of water from concrete in the early stages after placing (air temperature 21°C (70°F); wind velocity 4·5 m/s (10 mph))*

difference between the temperatures of concrete and of air also affects the loss of water, as shown in Fig. 5.29. Thus concrete saturated in day-time would lose water during a cold night, and this would also be the case with concrete cast in cold weather, even in saturated air. The examples quoted are merely typical as the actual loss of water depends on the surface/volume ratio of the specimen.[5.38]

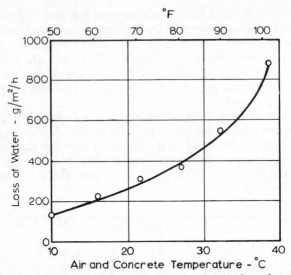

Fig. 5.27. *Influence of temperature of air and concrete on the loss of water from concrete in the early stages after placing (relative humidity of air 70 per cent; wind velocity 4·5 m/s (10 mph))*

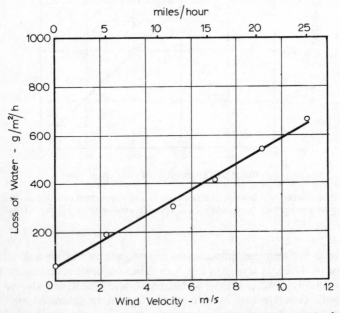

Fig. 5.28. *Influence of wind velocity on the loss of water from concrete in the early stages after placing (relative humidity of air 70 per cent, temperature 21°C (70°F))*

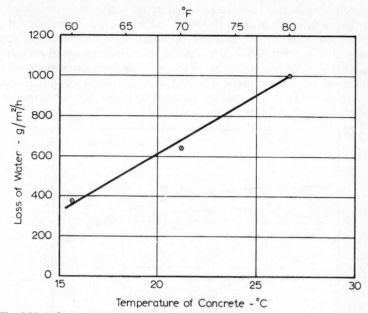

Fig. 5.29. *Influence of temperature of concrete (at an air temperature of 4·5°C (40°F)) on the loss of water from concrete in the early stages after placing (relative humidity of air 100 per cent, wind velocity 4·5 m/s (10 mph))*

Methods of Curing

No more than an outline of the different means of curing will be given here as the actual procedures used vary widely, depending on the conditions on the site and on the size, shape, and position of the member.

In the case of members with a small surface/volume ratio curing may be aided by oiling and wetting the forms before casting; the forms may also be wetted during hardening, and after stripping the concrete should be sprayed and wrapped with polythene sheets or other suitable covering.

Large surfaces of concrete, such as road slabs, present a more serious problem. In order to prevent crazing of the surface on drying out, loss of water must be prevented even prior to setting. As the concrete is at that time mechanically weak it is necessary to suspend a covering above the concrete surface. This protection is required only in dry weather but may also be useful to prevent rain marring the surface of fresh concrete.

Once the concrete has set, wet curing can be provided by keeping the concrete in contact with a source of water. This may be achieved by spraying or flooding (ponding), or by covering the concrete with wet sand or earth, sawdust or straw. Periodically-wetted hessian or cotton mats may be used, or alternatively an absorbent covering with access to water may be placed over the concrete. A continuous supply of water is naturally more efficient than an intermittent one, and Fig. 5.30 compares the strength development of concrete cylinders whose top surface was flooded during the first 24 hours with that of cylinders covered with wet hessian. The difference is greatest at low water/cement ratios where self-desiccation operates rapidly. The influence of curing conditions on strength is lower in the case of air-entrained than non-air-entrained concrete.[5.30]

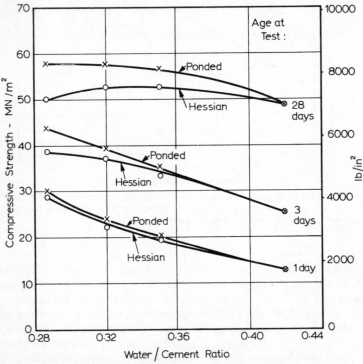

Fig. 5.30. *Influence of curing conditions on strength of test cylinders*[5.39]

Another means of curing is to use an impermeable membrane or waterproof paper. A membrane, provided it is not punctured or damaged, will effectively prevent evaporation of water from the concrete but will not allow ingress of water to replenish that lost by self-desiccation. The membrane is formed by sealing compounds which may be clear, white or black. The opaque compounds have the effect of shading the concrete, and a light colour leads to a lower absorption of the heat from the sun, and hence to a smaller rise in the temperature of the concrete. Details of the curing compounds are outside the scope of this book, A.S.T.M. Method C 156–71 prescribes tests for the efficiency of curing compounds.

Except when used on concrete with a high water/cement ratio, sealing compounds reduce the degree and rate of hydration compared with efficient wet curing. However, wet curing is often applied only intermittently so that in practice sealing may lead to better results.

The period of curing cannot be prescribed simply but it is usual to specify a minimum of seven days for ordinary Portland cement concrete. With slower-hardening cements a longer curing period is desirable. The temperature also affects the length of the required period of curing and the British Code of Practice for the Structural Use of Concrete CP 110: 1972 lays down the normal curing periods for different cements and exposure conditions in terms of maturity of concrete (*see* below).

High-strength concrete should be cured at an early age as partial hydration may make the capillaries discontinuous: on renewal of curing, water would not be able to enter the interior of the concrete and no further hydration would result. However, mixes with a high water/cement ratio always retain a large volume of capillaries so that curing can be effectively resumed at any time. No loss of strength is caused by delaying the curing, as shown for instance by Fig. 5.31 for a 1:3·4 mortar with a

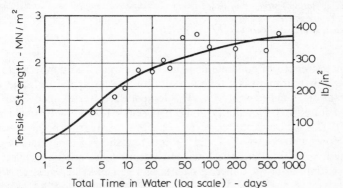

Fig. 5.31. *Relation between total curing time in water and tensile strength of mortar briquettes*[5.41]

The line refers to specimens continuously cured in water; points denote specimens cured 3 days in water, then 28 days in dry air, and remainder of time in water.

water/cement ratio of 0·70 but in practice early drying may lead to shrinkage and cracking, and delaying curing is inadvisable.

Maturity of Concrete

So far we have considered only the time aspect of curing but, as mentioned earlier, the temperature during curing also controls the rate of progress of the reactions of hydration and consequently affects the development of strength of concrete. This influence is shown, for instance, in Fig. 5.32, obtained from tests on specimens cast, sealed and cured at the indicated temperatures. The effect of the temperature at the time of setting is considered on page 276.

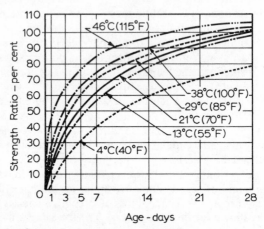

Fig. 5.32. *Ratio of strength of concrete cured at different temperatures to the 28-day strength of concrete cured at 21°C (70°F) (water/cement ratio = 0·50; the specimens were cast, sealed, and cured at the indicated temperature)*[5.11]

Since strength of concrete depends on both age and temperature we can say that strength is a function of Σ(time × temperature), and this summation is called maturity. The temperature is reckoned from an origin found experimentally to be between −12 and −10°C (11 and 14°F). This is because at temperatures below the freezing point of water and down to about −12°C (11°F) concrete shows a small increase in strength with time but the low temperature must not be applied, of course, until after the concrete has set and gained sufficient strength to resist damage due to the action of frost; a "waiting period" of 24 hours is usually required. Below −12°C (11°F) concrete does not appear to gain strength with time.

Maturity is measured in °C-hours (°F-hours) or °C-days (°F-days). Fig. 5.33 shows that strength plotted against the logarithm of maturity gives a straight line. It is, therefore, possible to express strength at any maturity as a percentage of strength of concrete at any other maturity; the latter is often taken as 19,800°Ch (35,600°Fh), being the maturity of concrete cured at 18°C (64°F) for 28 days. The ratio of strengths, expressed as a percentage, can then be written as—

$$A + B \log_{10} (\text{maturity} \times 10^{-3}).$$

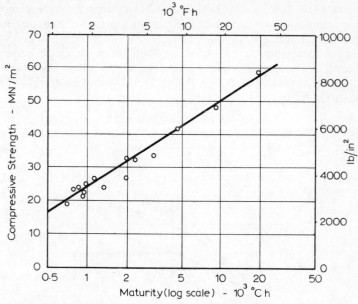

Fig. 5.33. *Relation between logarithm of maturity and strength*[5.42]

The values of the coefficients A and B depend on the strength level of concrete; those suggested by Plowman[5.42] are given in Table 5.4. It can

Table 5.4: *Plowman's[5.42] coefficients for the maturity equation*

Strength after 28 days at 18°C (64°F) (maturity of 19,800°Ch (35,600°Fh))			Coefficient	
			A	
MN/m²	lb/in²	B	for °Ch	for °Fh
<17	<2,500	68	10	−7
17–35	2,500–5,000	61	21	6
35–52	5,000–7,500	54	32	18
52–69	7,500–10,000	46.5	42	30

be seen that the strength–maturity relation depends on the properties of the cement and on the general quality of the concrete, and is valid only within a range of temperatures. This is apparent, for instance, from Fig. 5.34 obtained by Klieger[5.43] who tested ordinary Portland cement concrete with a water/cement ratio of 0·43 and an air content of 4·5 per cent,

cured at 23°C (73°F) from the age of 28 days onwards. A further complication arises from the fact that the effects of a period of exposure to a
higher temperature are not the same when this occurs immediately after
casting or later in the life of the concrete. Specifically, early high temperature leads to a lower strength for a given total maturity than when heating
is delayed for at least a week or is absent. This is of interest in connection
with steam curing.

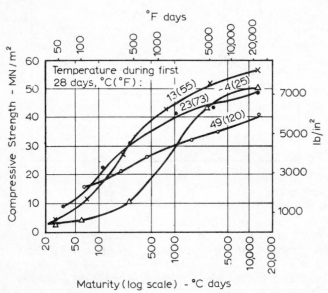

Fig. 5.34. *Influence of the temperature during the first 28 days after casting on the
strength–maturity relation*[5.43]

Nevertheless, the maturity rule applies fairly well when the initial
temperature of concrete is between 16 and 27°C (60 and 80°F) and no
loss of moisture by drying takes place during the period considered.[5.44]

Influence of Temperature on Strength of Concrete

We have seen that a rise in the curing temperature speeds up the chemical
reactions of hydration and thus affects beneficially the early strength of
concrete without any ill-effects on the later strength. However, a higher
temperature during placing and setting, although it increases the very
early strength, may adversely affect the strength from about 7 days
onwards. The explanation is that a rapid initial hydration appears to
form products of a poorer physical structure, probably more porous, so
that a large proportion of the pores will always remain unfilled. It follows

from the gel/space ratio rule that this will lead to a lower strength compared with a less porous, though slowly hydrating, paste in which a high gel/space ratio will eventually be reached. This explanation of the adverse effects of a high early temperature on later strength has been extended by Verbeck and Helmuth[5.85] who suggest that the rapid initial rate of hydration at higher temperatures retards the subsequent hydration and produces a non-uniform distribution of the products of hydration within the paste. The reason for this is that at the high initial rate of hydration there is insufficient time available for the diffusion of the products of hydration away from the cement grain and for a uniform precipitation in the interstitial space (as is the case at lower temperatures). As a result, a high concentration of the products of hydration is built up in the vicinity of the hydrating grains, and this retards the subsequent hydration and adversely affects the long-term strength.

In addition, the non-uniform distribution of the products of hydration *per se* adversely affects the strength because the gel/space ratio in the interstices is lower than would be otherwise the case for an equal degree of hydration: the local weaker areas lower the strength of the paste as a whole.

Fig. 5.35 shows Price's[5.11] data on the effect of the temperature during the first two hours after mixing on the development of strength of concrete with a water/cement ratio of 0·53. The range of temperatures investigated was 4 to 46°C (40 to 115°F), and beyond the age of two hours all specimens were cured at 21°C (70°F). The specimens were sealed so as to prevent movement of moisture.

Tests have also been made on concretes stored in water at different temperatures for a period of 28 days, and thereafter at 23°C (73°F).[5.43] As in Price's tests, a higher temperature was found to result in a higher strength during the first few days after casting, but beyond the age of one to four weeks the situation changed radically (*see* Fig. 5.34). The specimens cured at temperatures between 4 and 23°C (40 and 73°F) up to the age of 28 days all showed a higher strength than those cured at 32 to 49°C (90 to 120°F). Among the latter, retrogression was greater the higher the temperature, but in the lower range of temperatures there appeared to be an optimum temperature that yielded the highest strength. It is interesting to note that even concrete cast at 4°C (40°F) and stored at as low a temperature as − 4°C (25°F) for four weeks and then at 23°C (73°F), is from the age of 3 months onwards stronger than similar concrete stored continuously at 23°C (73°F). Fig. 5.36 shows typical curves for concrete containing 307 kg of ordinary Portland cement per cubic metre (517 lb/yd³) of concrete with 4·5 per cent of entrained air. Similar behaviour has been observed when rapid hardening Portland and modified cement are used. When calcium chloride is added to the mix the adverse effects of a high temperature during setting are attenuated.

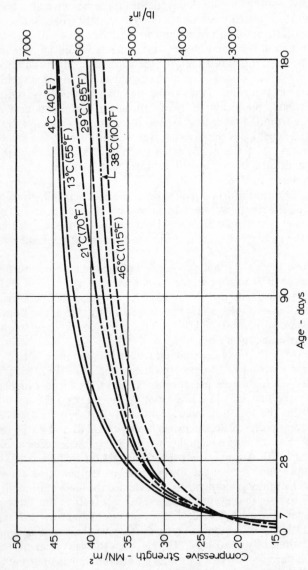

Fig. 5.35. *Effect of temperature during the first two hours after casting on the development of strength (all specimens sealed and after the age of 2 hours cured at 21°C (70°F))*[5.11]

The increase in strength caused by the addition of calcium chloride depends on the temperature of the concrete and is proportionately greater at lower temperatures. For instance, at 13°C (55°F) the addition of 2 per cent of $CaCl_2$ increases the one day strength by about 140 per cent, but the relative increase in the same mix at 49°C (120°F) is only

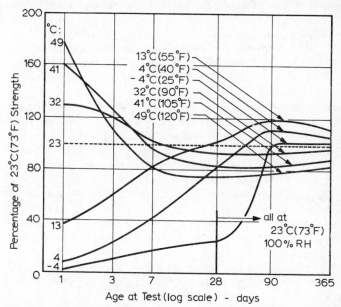

Fig. 5.36. *Effect of temperature during the first 28 days on the strength of concrete (water/cement ratio = 0·41; air content = 4·5 per cent; ordinary Portland cement)*[5.43]

50 per cent.[5.43] This behaviour is, indeed, to be expected, as the rate of hydration at higher temperatures is high even without an accelerator so that little scope is left for the action of $CaCl_2$. Calcium chloride should ordinarily be used only at normal or low temperatures.

Klieger's[5.43] tests indicate that there is an optimum temperature during the early life of concrete that will lead to the highest strength at a desired age. For laboratory-made concrete using ordinary or modified Portland cement the optimum temperature is approximately 13°C (55°F); for rapid hardening Portland cement it is about 4°C (40°F). It must not be forgotten, however, that beyond the initial period of setting and hardening the influence of temperature (within limits) accords with the maturity rule: a higher temperature accelerates the development of strength.

The tests described so far were all made in the laboratory, and it seems that the behaviour on the site in a hot climate may not be the same. Here, there are some additional factors acting: ambient humidity, direct radiation of the sun, wind velocity, and method of curing. It should be remembered also that the quality of concrete depends on *its* temperature and not on that of the surrounding atmosphere, so that the size of the member also enters the picture. Likewise, curing by flooding in windy weather results in a loss of heat due to evaporation so that the temperature of concrete is lower than when a sealing compound is used. Shalon[5.45] found evaporation immediately after casting to be beneficial, possibly because water is drawn out of the concrete while capillaries can still collapse, so that the effective water/cement ratio is decreased. If, however, evaporation is allowed to lead to the drying of the surface, plastic shrinkage and cracking may result.

Tests[5.45] have shown that under hot and dry conditions, such as are encountered in the desert, strength decreases with an increase in temperature down to a critical value at about 30°C (86°F), but between 30°C and 45°C (86°F and 113°F) there may be a slight recovery or no further loss. This behaviour has been observed using concrete without entrained air and stored at a relative humidity of between 20 and 70 per cent. It is possible that the presence or absence of entrained air is responsible, at least in part, for the difference between Klieger's[5.43] and Shalon's[5.45] results. It seems then that we are not yet aware of all the factors involved in the problem, and careful site tests should be made before construction is begun in an unknown climate.

In general terms, however, concrete cast in summer can be expected to have a lower strength than a similar mix cast in winter. Indeed, on many construction sites the strength of test specimens has been found to be lower during hot weather even though from the time of stripping the moulds at the age of 24 hours the specimens were cured in water at 18°C (64°F). Similarly, in tropical countries an apparently lower strength of concrete has been observed[5.46] (Fig. 5.37).

Steam Curing at Atmospheric Pressure

Since an increase in the curing temperature of concrete increases its rate of development of strength, the gain of strength can be speeded up by curing concrete in steam. When steam is at atmospheric pressure, i.e. the temperature is below 100°C (212°F), the process can be regarded as a special case of moist curing. High-pressure steam curing is an entirely different operation and is considered in the next section. Steam curing has been used successfully with different types of Portland cement, but must never be used with aluminous cement because of the adverse effect of hot-wet conditions on the strength of that cement. Concrete with a lower water/cement ratio responds to steam curing much better than a weaker mix.

The primary object of steam curing is to obtain a sufficiently high early strength so that the concrete products may be handled soon after casting: the moulds can be removed, or the prestressing bed vacated, earlier than would be the case with ordinary moist curing, and less curing storage space is required; all these mean an economic advantage. For many applications the long-term strength of concrete is of lesser importance.

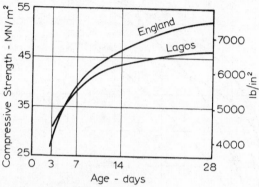

Fig. 5.37. *Comparison of strengths of concrete with a water/cement ratio of 0·40 cast in Lagos, Nigeria and in England*[5.46]

Because of the nature of the operations involved in steam curing, the process is used mainly with precast products. Low pressure steam curing is normally applied in special chambers or in tunnels through which the concrete members are transported on a conveyor belt. Alternatively, portable boxes or plastic covers can be placed over precast members, steam being supplied through flexible connexions.

Owing to the influence of temperature during the early stages of hardening on the later strength, a compromise between the temperatures giving a high early and a high late strength has to be made. Fig. 5.38 shows typical values of strength of concrete made with modified cement and a water/cement ratio of 0·55; steam curing was applied immediately after casting.

A related problem is the rate of increase in temperature at the commencement of steam curing. It has been found that if 49°C (120°F) is reached in a period shorter than about 2 to 3 hours, or 99°C (210°F) in less than 6 to 7 hours from the time of mixing, the gain of strength beyond the first few hours is affected adversely; such a rapid rise in temperature must not be permitted. The values quoted are merely a guide, but excessively rapid heating can cause a loss of strength at later ages of as much as one-third compared with wet curing at room temperature. The adverse effect of the rapid rise is more pronounced the higher the water/cement ratio of the mix, and is also more noticeable with rapid hardening than with ordinary Portland cement. Saul[5.48] found that when the rate of

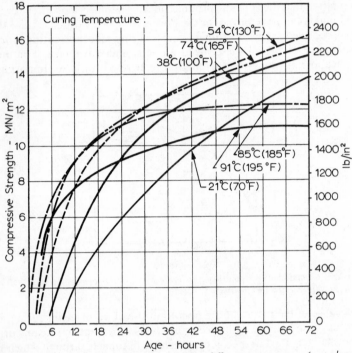

Fig. 5.38. *Strength of concrete cured in steam at different temperatures (water/cement ratio = 0·50; steam curing applied immediately after casting)*[5.47]

rise in the temperature of concrete does not exceed the values mentioned earlier its strength will differ only little from the strength of normally cured concrete, and will fall within zone A of Fig. 5.39. By contrast, the strength of concrete heated too rapidly will lie within zone B of the same Figure.

Since it is the temperature at the time of setting that has the greatest influence on the strength at later ages, a delay in the application of steam curing is advantageous. Some indication of the influence of the delay in heating on strength can be obtained from Fig. 5.40, plotted by Saul[5.49] from the data of Shideler and Chamberlin.[5.50] The concrete used was made with modified cement, and had a water/cement ratio of 0·6. The solid line shows the gain in strength of moist-cured concrete at room temperature plotted against maturity. The dotted lines refer to different curing temperatures between 38 and 85°C (100 and 185°F), and the figure against each point denotes the delay in hours before the higher curing temperature was suddenly applied.

From Fig. 5.40 it can be seen that for each curing temperature there

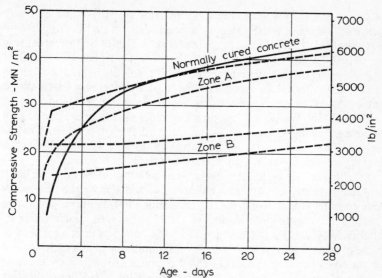

Fig. 5.39. *Gain of strength of steam-cured concrete with time (water/cement ratio =
0·50, rapid-hardening Portland cement)*[5.48]

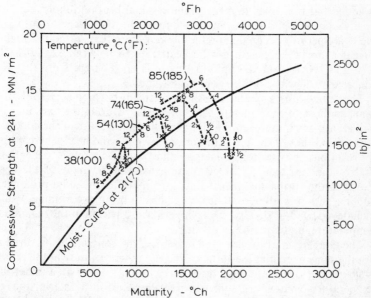

Fig. 5.40. *Effect of delay in steam curing on the early gain of strength with maturity.*[5.49]
Small figures indicate the delay in hours before curing at the temperature indicated.

is a part of the curve showing a normal rate of gain in strength with maturity. In other words, after a sufficient delay, rapid heating has no adverse effect. This delay is approximately 2, 3, 5 and 6 hours respectively for 38, 54, 74 and 85°C (100, 130, 165 and 185°F). If, however, concrete is exposed to the higher temperature with a smaller delay the strength is adversely affected, as shown by the right-hand portion of each dotted curve; this effect is more serious the higher the curing temperature.

Fig. 5.40 shows also that within a few hours of casting the rate of gain in strength is higher than would be expected from the maturity calculations. This confirms the earlier observation that the *age* at which a higher temperature is applied is a factor in the maturity rule.

Practical curing cycles are chosen as a compromise between the early and late strength requirements but are governed also by the time available (e.g. length of work shifts). Economic considerations determine whether the curing cycle should be suited to a given concrete mix or alternatively whether the mix ought to be chosen so as to fit a convenient cycle of steam curing. While details of an optimum curing cycle depend on the type of concrete product treated, a typical satisfactory cycle would consist of the following:[5.51,5.86] a delay period of 3 to 5 hours, heating at the rate of 22 to 33°C per hour (40 to 60°F per hour) up to a maximum temperature of 66 to 82°C (150 to 180°F), then storage at maximum temperature, possibly followed by a period of "soaking" when no heat is added but the concrete takes in the residual heat and moisture, and finally a cooling period (at a moderate rate), the total cycle (exclusive of the delay period) occupying preferably not more than 18 hours. Lightweight aggregate concrete can be heated up to between 82 and 88°C (180 and 190°F), but the optimum cycle is no different from that for concrete made with normal weight aggregate.[5.87]

The temperatures quoted are those of steam but not necessarily the same as those of the concrete being processed. During the first hour or two after placing in the curing chamber the temperature of the concrete lags behind that of the air but later on, owing to the heat generated by the reactions of hydration, the temperature of the concrete is higher than that of the surrounding air. Maximum use can be made of the heat stored in the curing chamber if steam is shut off early and a prolonged cooling period is allowed. Thus an efficient curing programme would include a period of slowly rising temperature, a period at the maximum temperature, and a period of cooling.

The temperature history of the interior of the concrete being cured is not the same as that at the surface. The rise in temperature at the centre is slower but the rate of cooling is lower, too. Thus the area under the temperature–time curve is approximately the same for the interior and for points near the surface of the concrete block, so that all parts of the concrete have the same maturity. This was demonstrated by Ross,[5.52] and

Fig. 5.41 shows the computed temperature–time curves for a long block of concrete subjected to a variation in surface temperature, the effects of the heat of hydration being ignored.

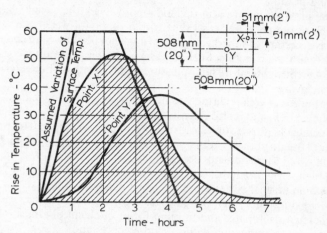

Fig. 5.41. *Ross's*[5.52] *example of two-dimensional flow of heat in concrete with a diffusivity of 0·0037 m²/h (0·04 ft²/h) (by computation)*

In passing, it may be observed that when concrete is cured at high temperatures the heat of hydration of cement is evolved rapidly so that the rise in temperature is increased even in small specimens. On the other hand, with curing at ordinary temperatures the effects of the heat of hydration are significant only in mass structures.

In addition to steam curing, other means of applying higher temperature to the concrete have also been used. For instance, some attempts have been made to cure concrete blocks by combustion gases, with or without humidification; less drying after curing is required than with steam curing but the colour of blocks is affected by the gases. A totally different approach uses electrical methods of heating[5.53] by passing current through the reinforcement (or through the prestressing steel in the long line process) or directly through the concrete; this has been found successful but is generally expensive. The current must be alternating as direct current would lead to hydrolysis of the cement paste. The temperature rise should be gradual, about 10°C (18°F) per hour, up to not more than 80°C (176°F). A period of several hours at this temperature should be followed by slow cooling. Yet another method uses infrared radiation of concrete slabs slowly moving through a tunnel equipped with infra-red generators. With all these methods, it is important to avoid excessive drying of the concrete.

High-pressure Steam Curing

This process is quite different from curing in steam at atmospheric pressure, both in the method of execution and in the nature of the resulting product.

Since pressures above atmospheric are involved, the curing chamber must be of the pressure vessel type with a supply of wet steam; excess water is necessary as superheated steam must not be allowed to come into contact with the concrete. Such a vessel is known as an autoclave, and in the American literature high-pressure steam curing is often referred to as autoclaving.

High-pressure steam curing was first employed in the manufacture of sand-lime brick, and is still extensively used for that purpose. In the field of concrete, high-pressure steam curing is usually applied to precast products (made both of ordinary and lightweight concrete) when any of the following characteristics are desired:

(*a*) high early strength: with high-pressure steam curing the 28-day strength on normal curing can be reached in about 24 hours;

(*b*) high durability: high-pressure steam curing improves the resistance of concrete to sulphates and to other forms of chemical attack, also to freezing and thawing, and reduces efflorescence; and

(*c*) reduced drying shrinkage and moisture movement.

The optimum curing temperature has been found experimentally to be about 177°C (350°F)[5.54] which corresponds to a steam pressure of 0·8 MN/m² (120 lb/in²) gauge.

High-pressure steam curing is most effective when finely ground silica is added to the cement, owing to the chemical reactions between the silica and $Ca(OH)_2$ released on hydration of C_3S (*see* Fig. 5.42). Cements rich in C_3S have a greater capacity for developing high strength when cured at high pressure than those with a high C_2S content, although for short periods of high-pressure steam curing cements with a moderately low C_3S/C_2S ratio give good results.[5.55]

The fineness of the silica should be of the same order as that of the cement, and the two materials must be intimately mixed before they are added to the mixer. The optimum amount of silica depends on the mix proportions but is generally between 0·4 and 0·7 of the weight of cement. This makes the lime/silica ratio of the mixture equal to approximately 1. The high temperature during curing affects also the reactions of hydration of the cement itself. For instance, some of the C_3S may hydrate to C_3SH_x.

High temperature produces a hydrated cement paste of low specific surface, about 7,000 m²/kg. This means that the products of hydration are coarse and largely microcrystalline. Since the specific surface of high-pressure steam-cured paste is only about $\frac{1}{20}$ of that of cement cured at ordinary temperature it appears that no more than 5 per cent of the high-pressure cured paste can be classified as gel.

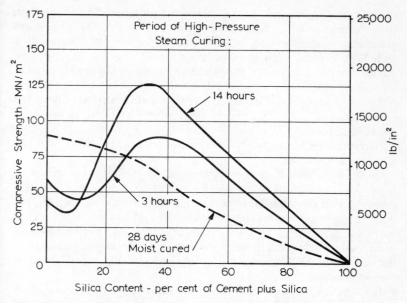

Fig. 5.42. *Influence of pulverized silica content on the strength of high-pressure steam-cured concrete (age at commencement of curing, 24 hours; curing temperature, 177°C (350°F))*[5.54]

Because of the microcrystalline character of the cement paste, high-pressure steam-cured concrete has a considerably reduced shrinkage, about $\frac{1}{6}$ to $\frac{1}{3}$ of that of concrete cured at normal temperatures. When silica is added to the mix, shrinkage is higher, but still only about one-half of the shrinkage of normally cured concrete. By contrast, since low-pressure steam curing does not produce a microcrystalline paste, no reduction in shrinkage is obtained.

The products of hydration of cement subjected to high-pressure steam curing, as well as those of the secondary lime–silica reactions, are stable, and there is no retrogression of strength. At the age of one year the strength of normally cured concrete is approximately the same as that of high-pressure steam-cured concrete of similar mix proportions. The water/cement ratio affects the strength of high-pressure steam-cured concrete in the usual manner, but the actual values of early strength differ, of course, from those for ordinary curing. The coefficient of thermal expansion and the modulus of elasticity of concrete seem unaffected by high-pressure steam curing.[5.54]

High-pressure steam curing improves the resistance of concrete to sulphate attack. This is due to several reasons, the main one being the formation of aluminates more stable in the presence of sulphates than those formed at lower temperatures. For this reason, the relative improvement in resistance to sulphate attack is greater in cements with a high C_3A content than in cements relatively resistant to sulphates. Another important factor is the reduction in lime in the cement paste as a result of the lime–silica reaction. Further improvement in sulphate resistance is due to the increased strength and impermeability of the steam-cured concrete, and also to the existence of hydrates in a well-crystallized form.

High-pressure steam curing reduces efflorescence as there is no lime left to be leached out.

On the debit side, high-pressure steam curing reduces the strength in bond with reinforcement by about one-half compared with ordinary curing so that the application of steam to reinforced concrete members is considered inadvisable. High-pressure steamed concrete tends also to be rather brittle. On the whole, high-pressure steam curing produces good quality, dense and durable concrete. It is whitish in appearance as distinct from the characteristic colour of normally cured Portland cement concrete.

It is essential that the rate of heating during high-pressure curing is not too high as interference with the setting and hardening processes may occur in a manner similar to that discussed in connexion with steam curing at atmospheric pressure. A typical steaming cycle consists of a gradual increase to the maximum temperature of $182°C$ ($360°F$) (which corresponds to a pressure of 0.96 MN/m^2 (14.0 lb/in^2)) over a period of 3 hours. This is followed by 5 to 8 hours at this temperature, and then a release of pressure in about 20 to 30 min.[5.88] A rapid release accelerates the drying of the concrete so that shrinkage *in situ* will be reduced.

The details of the steaming cycle depend on the plant used and also on the size of the concrete members being cured. The length of the period of normal curing preceding placing in the autoclave does not affect the quality of the steam-cured concrete, and the choice of a suitable period is governed by the stiffness of the mix, which must be strong enough to withstand handling. In the case of lightweight concretes the details of the steaming cycle have to be determined experimentally to suit the materials used.

Steam curing should be applied to concretes made with Portland cement only: aluminous and supersulphated cement would be adversely affected by the high temperature.

Within the Portland group, the type of cement affects the strength but not necessarily in the same way as at normal temperatures; no systematic studies have, however, been made. It is known, though, that granulated slag may cause trouble if it has a high sulphur content. High-pressure steam curing accelerates the hardening of concrete containing calcium chloride, but the relative increase in strength is less than when no calcium chloride is used.

Variation in Strength of Cement
Up to now we have not considered the strength of cement as a variable in the strength of concrete. By this we do not mean the differences in the strength-producing properties of cements of different types, but the variation between cements of nominally the same type: they vary fairly widely, and it is this variation that is considered in this section.

The scatter of strength of ordinary and rapid hardening Portland cements is illustrated in Fig. 5.43, which shows also that there is a considerable overlap in the strengths of the two cements. However, these cements nearly always comply with the minimum requirements of B.S. 12: 1958, the strength of some of the cements sometimes reaching values as high as $2\frac{1}{2}$ times the prescribed minimum.

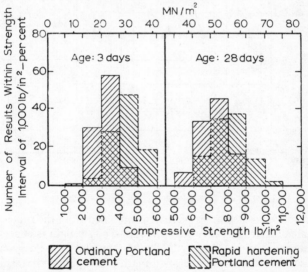

Fig. 5.43. *Histograms for vibrated mortar cubes made with ordinary and rapid hardening Portland cements*[5.56]

Although on a site it is difficult to isolate the influence of cement, there is no doubt that the inherent variation in the strength of cement is reflected in the variation in the strength of concrete. Thus the variability of cement is of considerable practical importance, particularly since a concrete mix has to be designed so as to give a satisfactory strength even when a low-strength batch of cement is encountered: no advantage can be taken of the much higher mean strength. If cement had a lower variability (but the same mean strength) the cement content of the mix could be reduced: there may thus be some advantage in strength grading of cement even if this were accompanied by a higher price of the material.

The variation in strength of cement is due largely to the lack of uniformity in the raw materials used in its manufacture, not only between different sources of supply, but also within a pit or a quarry. Furthermore, differences in details of the processes of manufacture and above all the variation in the ash content of the coal used to fire the kiln contribute to the variation in the properties of commercial cements.

The magnitude of the variation in the strength of cement can be judged from Fig. 5.44, which shows the results of tests on A.S.T.M. standard

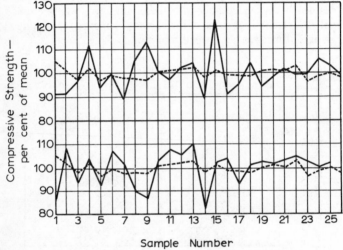

Fig. 5.44. *Variation in strength of cement from the same works*[5.57]

mortar made with samples of cement obtained from the same works in the United States at two-week intervals. The strength is expressed as a percentage of the mean strength of all the samples from the given works, and each curve is an average of strength ratios obtained at 3, 7 and 28 days. The variation in strength due to testing *per se* is indicated by the dotted lines which show the strength ratios for tests made at the same time using a control "stock" cement. The testing error accounts generally for a coefficient of variation of between $2\frac{1}{2}$ and 4 per cent.

The standard deviation of the strength of concrete due to the variation in cement does not increase with age. Since, however, strength increases with age, the coefficient of variation of strength becomes smaller the older the concrete at the time of testing.* This behaviour is not surprising

---

* The statistical terms are defined on page 516.

Table 5.5: *Variation in the Strength of Mortar and Concrete made with Different Batches of Cement from one Works*[5.59]

		Age at test, days					
		1	3	7	28	91	365
B.S.12: 1947 standard mortar	Mean strength, MN/m² (lb/in²)	6·1 (890)	22·0 (3,190)	34·8 (5,050)	50·7 (7,350)	60·4 (8,760)	69·4 (10,030)
	Standard deviation, MN/m² (lb/in²)	3·2 (460)	5·5 (800)	5·0 (720)	3·9 (570)	4·2 (610)	4·4 (640)
	Coefficient of variation, per cent	52	25	14	8	7	6
	Standard deviation between batches, MN/m² (lb/in²)	3·2 (460)	5·7 (830)	5·0 (730)	3·6 (520)	3·6 (530)	4·0 (580)
	Standard deviation within batches, MN/m² (lb/in²)	0·5 (80)	1·0 (150)	1·2 (180)	1·9 (280)	3·4 (500)	3·3 (480)
1:1½:3 concrete with water/cement ratio of 0·5	Mean strength, MN/m² (lb/in²)	15·4 (2,230)	27·2 (3,940)	30·7 (4,450)	39·7 (5,760)	49·8 (7,240)	57·4 (8,330)
	Standard deviation, MN/m² (lb/in²)	4·6 (670)	4·3 (620)	3·9 (560)	4·8 (700)	4·6 (660)	4·6 (660)
	Coefficient of variation, per cent	30	16	13	12	9	8
	Standard deviation between batches, MN/m² (lb/in²)	4·5 (660)	4·3 (620)	3·8 (550)	4·8 (700)	4·5 (700)	4·5 (650)
	Standard deviation within batches, MN/m² (lb/in²)	0·8 (110)	1·1 (160)	1·1 (160)	1·6 (230)	1·7 (250)	1·9 (270)

because a large part of the variation in the strength of cement is due to the differences in fineness and in the C_3S content: the effects of these factors are greatest at early ages and with time cease to be significant. By contrast, the standard deviation within batches increases with an increase in mean strength.[5.58] Table 5.5 gives Wright's[5.59] data on the variability of cement at different ages at test, and Table 5.6 gives approximate values of the standard deviation of strength of site-made concrete.

Table 5.6: *Standard Deviations of Strength of Concrete resulting from Variations in Cement*[5.59]

Cause		Standard deviation			
		Good control		Fair control	
		MN/m²	lb/in²	MN/m²	lb/in²
From variations arising on the site	s_1	3·1	450	4·8	700
From variations in cement from works	s_2	2·7	400	2·7	400
Total, where cement is from one works	s_3	4·1	600	5·5	800
From variation between cement works	s_4	4·8	700	4·8	700
Total, where cement is from several works	s_5	6·2	900	7·2	1,050

Note: The values are related by the equations

$$s_1{}^2 + s_2{}^2 = s_3{}^2 \text{ and } s_3{}^2 + s_4{}^2 = s_5{}^2$$

From the figures quoted it is easy to see that the use of cement from different sources, or even the use of different batches of cement from one works, leads to an appreciable variation in the strength of concrete. This effect may be of considerable importance on a large job: the use of cement all from one batch can result in a decrease in cement content of up to 10 per cent. It must not be forgotten, however, that variation in cement accounts at the most for one-half of the variation in the strength of site cubes. The variation in the strength of site cubes is discussed on page 565.

Finally, it should be stressed that the variation in cement affects most the early strength of concrete, i.e. the strength most often determined by test but not necessarily the strength of greatest practical significance. Furthermore, strength is not the only important characteristic of concrete: from considerations of durability and impermeability, a cement content in excess of that needed for strength may well be required, in which case the variability of cement becomes unimportant.

Quality of Mixing Water

The vital influence of the quantity of water in the mix on the strength of the resulting concrete has been repeatedly mentioned. The quality of the water also plays its rôle: impurities in water may interfere with the setting of the cement, may adversely affect the strength of the concrete or cause staining of its surface, and may also lead to corrosion of the reinforcement. For these reasons, the suitability of water for mixing and curing purposes should be considered. Clear distinction must be made between the effects of mixing water and the attack on hardened concrete by aggressive waters. Some of the latter type of water may be harmless or even beneficial when used in mixing.[5.62]

In many specifications the quality of water is covered by a clause saying that water should be fit for drinking. Such water very rarely contains dissolved solids in excess of 2,000 parts per million (ppm), and as a rule less than 1,000 ppm. For a water/cement ratio of 0·5 the latter content corresponds to a quantity of solids representing 0·05 per cent of the weight of cement, and any effect of the common solids would be small. Thus, while the use of potable water is safe, water not fit for drinking may often also be satisfactorily used in making concrete.

As a rule, water which does not taste saline or brackish is suitable for use, but dark colour or bad smell do not necessarily mean that deleterious substances are present.[5.63] A simple way of determining the suitability of such water is to compare the setting time of cement and the strength of mortar cubes using the water in question with the corresponding results obtained using known "good" water or distilled water; there is no appreciable difference between the behaviour of distilled and ordinary drinking water. A tolerance of about 10 per cent is usually permitted to allow for the chance variations in strength,[5.62] although an appendix to B.S. 3148: 1959 suggests 20 per cent. Such tests are recommended when water for which no service record is available contains dissolved solids in excess of 2,000 ppm or, in the case of alkali carbonate or bicarbonate, in excess of 1,000 ppm. When unusual solids are present a test is also advisable.

Since it is undesirable to introduce large quantities of silt into the concrete, mixing water with a high content of suspended solids should be allowed to stand in a settling basin before use; a turbidity limit of 2,000 ppm has been suggested.[5.47]

Brackish water contains chlorides and sulphates. When chloride does not exceed 500 ppm or SO_3 does not exceed 1,000 ppm the water is harmless, but water with even higher salt contents has been used satisfactorily.[5.64] Sea water has a total salinity of about 3·5 per cent (78 per cent of the dissolved solids being NaCl and 15 per cent $MgCl_2$ and $MgSO_4$), and produces a slightly higher early strength but a lower long-term strength; the loss of strength is usually no more than 15 per cent[5.65] and can therefore often be tolerated. Some tests suggest that sea water slightly accelerates the setting time of cement, others[5.101], a substantial reduction in the initial setting time but not necessarily in the final set. Generally the effects on setting are unimportant if water is acceptable from strength considerations. An appendix to B.S. 3148: 1959 suggests a tolerance of 30 minutes in the initial setting time.

Water containing large quantities of chlorides (e.g. sea water) tends to cause persistent dampness and surface efflorescence. Such water should, therefore, not be used where appearance of the concrete is of importance, or where a plaster finish is to be applied.[5.66]

In the case of reinforced concrete, sea water is believed to increase the *risk* of corrosion of the reinforcement, although there is no experimental evidence that the use of sea water in mixing leads to *attack* on the reinforcing steel.[5.67] The danger is believed to be greater in tropical countries. Corrosion has been observed in structures exposed to humid air when the cover to reinforcement is inadequate or the concrete is not sufficiently dense so that the corrosive action of residual salts in the presence of moisture can take place. On the other hand, when reinforced concrete is permanently in water, either sea or fresh, the use of sea water in mixing seems to have no ill-effects.[5.68] However, in practice it is generally considered inadvisable to use sea water for mixing unless this is unavoidable.[5.66] In prestressed concrete the use of sea water is not permitted[5.69] because the small cross-section of the wire means that the effects of corrosion are relatively more serious.

In this connexion the use of aggregate won from the sea should be considered. Sand dried out in sea water may contain a large amount of salt, but if sand dredged from the sea is washed in sea water and is allowed to drain, and fresh water is used as mixing water, then the salt content represents no more than 1 per cent of the total weight of water[5.70] (*see* p. 135).

Natural waters that are slightly acid are harmless, but water containing humic or other organic acids may adversely affect the hardening of concrete; such water, as well as highly alkaline water, should be tested. The effects of different ions vary, as shown by Steinour.[5.62]

It may be interesting to note that the presence of algae in mixing water results in air entrainment with a consequent loss of strength.[5.71] The use of algae is, however, hardly a practical means of air entrainment.

As far as curing of concrete is concerned, water satisfactory for mixing is also suitable for curing purposes. However, iron or organic matter may cause staining, particularly if water flows slowly over concrete and evaporates rapidly.

Whether or not staining will take place cannot be stated on the basis of a chemical analysis and should be checked by a performance test. U.S. Army Engineers[5.40] recommend a preliminary test in which 300 ml of the water to be used for curing is evaporated from a slight depression, 100 mm (4 in.) in diameter, in the surface of a specimen of neat white cement or plaster of Paris. If the resulting colouring is not considered objectionable a further test is performed. Here, 150 litres (40 U.S. gallons) of water are allowed to flow lengthwise over a 150 by 150 by 750 mm (6 by 6 by 30 in.) concrete beam with a channel-shaped top surface, placed at 15 to 20° to the horizontal; the rate of flow is 4 litres in 3 to 4 hours. Forced circulation of air and heating by electric lamps encourage evaporation and thus deposition of the residue. The test is again evaluated by observation only, and if necessary an actual field test in which a 2 m² (or 20 ft²) slab is cured may be performed.

In some cases discoloration is of no significance, and any water suitable for mixing, or even slightly inferior in quality, is acceptable for curing.

It is essential that curing water be free from substances that attack hardened concrete; these are discussed in Chapter 7.

Fatigue Strength of Concrete

In this chapter, we have considered so far only the strength of concrete under static loading. In many structures, however, repeated loading is applied, and when a material fails under a number of repeated loads, each smaller than the static compressive strength, failure in fatigue is said to take place. Both concrete and steel possess the characteristics of fatigue failure but in this book the behaviour of concrete alone is dealt with.

Let us consider a concrete specimen subjected to alternations of compressive stress between values σ_l ($\geqslant 0$) and σ_h ($> \sigma_l$). The stress–strain curve varies with the number of load repetitions, changing from concave toward the strain axis (with a hysteresis loop on unloading) to a straight line, which shifts at a decreasing rate (i.e., there is some irrecoverable set) and eventually to concave toward the stress axis. The degree of this latter concavity is an indication of how near the concrete is to failure. For practical loadings failure will, however, take place only above a certain limiting value of σ_h, known as fatigue limit or endurance limit. If σ_h is below the fatigue limit the stress–strain curve will indefinitely remain straight, and failure in fatigue will not take place. The changes in the stress–strain curve with the number of applied cycles are illustrated in Fig. 5.45.

If the stress–strain curve for unloading is also drawn, a hysteresis loop in each cycle can be seen. The area of this loop decreases with each successive cycle and then eventually increases prior to fatigue failure.[5.89] There does not seem to be such an increase in specimens which do not fail in fatigue. If we plot the area of each successive hysteresis loop as a percentage of the area of the first loop, the variation with the number of cycles is as shown in Fig. 5.46.

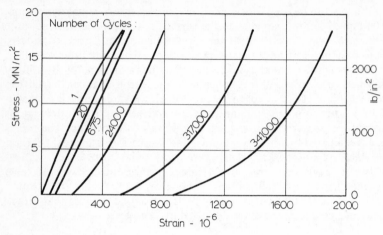

Fig. 5.45. *Stress–strain relation of concrete under cyclic compressive loading*

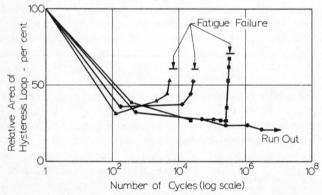

Fig. 5.46. *Variation in the area of the hysteresis loop as a percentage of the first loop with the number of cycles*[5.89]

The interest in the hysteresis loop arises from the fact that its area represents an irreversible energy of deformation, and is manifested by a rise in temperature of the specimen. The irreversible deformation involved may be either in the form of crack formation or irreversible deformation without a loss of continuity, such as viscous flow. Pulse velocity measurements have shown[5.89] that it is the development of cracks that is responsible for the change in behaviour near failure, but it is possible that even earlier there is a slow and stable propagation of one or more cracks,[5.90] although there is some non-elastic deformation without disruption of continuity of the material.

Fatigue cracking is extensive and the observed strain at failure is much larger than in static failure. The non-elastic strain increases with the number of cycles, as shown in Fig. 5.47, and can sometimes be as high as $4,000 \times 10^{-6}$ after 13 million cycles at 200 cycles per minute. Generally, the specimen with a longer fatigue life has a higher non-elastic strain at failure (Fig. 5.48).

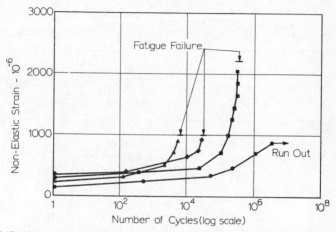

Fig. 5.47. *Variation in non-elastic strain with the number of cycles*[5.89]

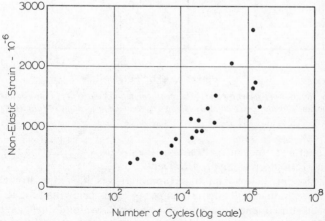

Fig. 5.48. *Relation between non-elastic strain near failure and number of cycles at failure*[5.89]

The elastic strain also increases progressively with cycling. This is shown in Fig. 5.49 by the reduction in the secant modulus of elasticity with increase in the percentage of the "fatigue life" used up. This relation is independent of the level of stress in the fatigue test and is therefore of interest in assessing the remaining fatigue life of a given concrete.

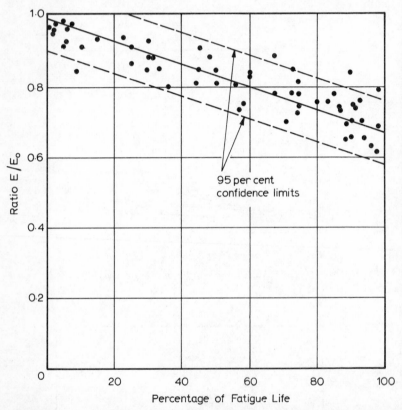

Fig. 5.49. *Relation between the ratio of the secant modulus of elasticity at the given instant (E) to the modulus at the beginning of cycling (E_0) and the percentage of fatigue life used up*[5.89]

The lateral strain is also affected by the progress of cyclic loading, the Poisson's ratio decreasing progressively.

Cyclic loading below the fatigue limit improves the fatigue strength of concrete, i.e. concrete loaded a number of times below its fatigue limit will, when subsequently loaded above the limit, exhibit a higher fatigue strength than concrete which had never been subjected to the initial cycles. The former concrete also exhibits a higher static strength by some

5 to 15 per cent. It is possible that this increase in strength is due to densification of concrete caused by the initial low-stress level cycling, in a manner similar to improvement in strength under moderate sustained loading.[5.91] This property is akin to strain hardening in metals, and is of particular interest since concrete under static loading is a strain-softening rather than strain-hardening material.

Strictly speaking, concrete does not appear to have a fatigue limit, i.e. a fatigue strength at an infinite number of cycles (except when stress reversal takes place). It is usual therefore to refer to fatigue strength at a very large number of cycles, such as 10 million.

The fatigue strength can be represented by means of a modified Goodman diagram (*see* Fig. 5.50). The ordinate from a line at 45° through

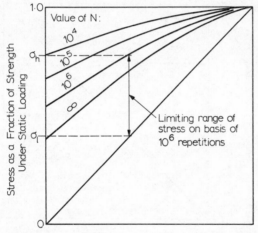

Fig. 5.50. *Modified Goodman diagram for concrete in compression fatigue* (*N is number of repetitions*)

the origin shows the range of stress ($\sigma_h - \sigma_l$) for a given number of cycles; σ_l is generally greater than zero arising from the dead load, while σ_h is due to the dead plus live (transient) load. Thus the range of stress that a given concrete can withstand a specified number of cycles can be read off the diagram. For a given σ_l the number of cycles is very sensitive to the range of stress. For instance, an increase in range from 57·5 to 65 per cent of the ultimate static strength has been found to decrease the number of cycles 40 times.[5.92]

The modified Goodman diagram (*see* Fig. 5.50) shows that for a constant range of stress, the higher the value of the minimum stress the lower the number of cycles that a given concrete can withstand. This is of significance in relation to the dead load. of a concrete member which is to carry a transient load of a certain magnitude.

From the fact that the lines of Fig. 5.50 rise to the right it can also be seen that the fatigue strength of concrete is lower the higher the ratio σ_h/σ_l.

The frequency of the alternating load, at least within the limits of 70 to 2,000 cycles per minute does not affect the resulting fatigue strength;[5.93] higher frequency is of little practical significance. This applies both in compression and in flexure-tension, the similarity between fatigue behaviour in the two types of loading suggesting that the failure mechanism is the same.[5.94] In fact, the fatigue behaviour in flexure parallels closely that in compression (Fig. 5.51). The fatigue strength for 10 million cycles

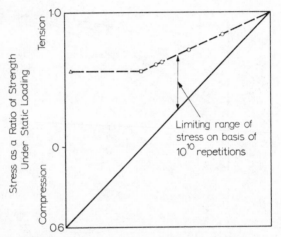

Fig. 5.51. *Modified Goodman diagram for concrete in flexure*[5.90]

is 55 per cent of the static strength. By comparison, in compression fatigue, the fatigue strength is between 60 and 64 per cent after the same number of cycles. Sufficiently accurate test results are not available to state with certainty that these two limits (in compression and in flexure) differ significantly from one another. We should note that semi-random variation in stress limits can adversely affect fatigue strength so that we cannot assume a linear accumulation of fatigue damage in concrete.[5.90]

The important point is that at a given number of cycles, fatigue failure occurs at the same fraction of ultimate strength, and is thus independent of the magnitude of this strength and of age of concrete.[5.93] It can thus be seen that a single parameter is critical in fatigue failure. Murdock[5.93] expressed the view that the deterioration of bond between the cement paste and aggregate is responsible for this failure. Tests have shown that fatigue specimens had fewer broken aggregate particles than specimens which failed in a static test.[5.95] Thus failure at the bond interface is probably dominant in fatigue; in mortar, fatigue failure is believed to take place at the interface of the fine aggregate particles.[5.89]

Air-entrained concrete and lightweight aggregate concrete have the same fatigue behaviour as concrete made with ordinary aggregate.[5.96]

The fatigue strength of concrete is increased by rest periods (this does not apply when there are stress reversals), the increase being proportional to their duration between 1 and 5 minutes; beyond the 5-minute limit there is no further increase in strength. With the rest periods at their maximum effective duration, their frequency determines the beneficial effect.[5.93] The increase in strength caused by rest periods is probably due to relaxation of concrete (primary bonds, which remained intact, restoring the internal structure to its original configuration), as evidenced by a decrease in the total strain; this decrease occurs rapidly after the cessation of cycling.

Murdock[5.93] suggested that fatigue failure occurs at a constant strain, independent of the applied stress level or the number of cycles necessary to produce failure. This fatigue ultimate strain is larger than the static failure strain. Nevertheless, this behaviour of concrete would add further support to the concept of ultimate strain as failure criterion.

In an eccentrically loaded specimen, there are elements stressed to levels lower than the maximum stress, and therefore a greater number of cycles can be sustained than in a uniformly stressed specimen subjected to the same maximum stress. An increase of 17 per cent was observed[5.92] when the eccentricity was 25 mm (1 in.) in a 150 mm (6 in.) face.

Finally, it may be of interest to observe that the behaviour of concrete under fatigue loading is not closely related to behaviour under impact loading, although of course both may occur simultaneously in practice. Failure under fatigue loading occurs at a strength below the static compressive strength while under impact the static strength is not impaired.[5.97]

While this book is not concerned with the fatigue behaviour of reinforced and prestressed concrete, we should note that fatigue cracks in concrete act as stress-raisers, thus magnifying the vulnerability of the steel to fatigue failure[5.98] (if the stress in it is in excess of its critical fatigue stress value).

Impact Strength

Impact strength is of importance primarily in connection with pile driving and with foundations for machines exerting impulsive loading, but accidental impact (e.g. during handling of precast members) is also of interest. Extensive tests on impact strength of concrete were made by Green.[5.99] As principal criteria he considered the ability of a specimen to withstand repeated blows and to absorb energy. In particular, he studied the number of blows the concrete can withstand before reaching the no-rebound condition, this stage indicating a definite state of damage.

Impact tests, when conducted with a relatively small hammer (25 mm (1 in.) diameter face) lead to a greater scatter of results than tests on static compressive strength of the concrete.[5.99] This arises from the fact that in the standard compression test some relief of a highly stressed weak zone is possible owing to creep, while in the impact test no redistribution of stresses is possible during the very short period of deformation. Hence, local weaknesses have a greater influence on the recorded strength of a specimen.

In general, Green found that the higher the static compressive strength of the concrete the lower the energy absorbed per blow before cracking, but the impact strength of concrete increases with its compressive strength (and therefore age) at a progressively increasing rate (Fig. 5.52). The

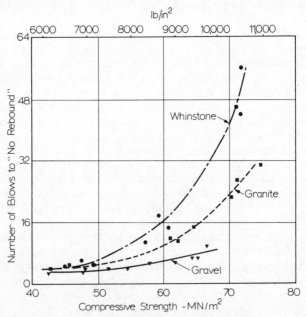

Fig. 5.52. *Relation between compressive strength and number of blows to "no-rebound" for concretes made with different aggregates and Type I cement, stored in water*[5.99]

relation is different for each coarse aggregate and storage condition of the concrete. For the same compressive strength, the impact strength is greater for coarse aggregate of greater angularity and surface roughness. This supports the suggestion[5.100] that impact strength is more closely related to the tensile strength of concrete than to its compressive strength.

Thus concrete made with a gravel coarse aggregate has a low impact strength, failure taking place owing to insufficient bond between mortar and coarse aggregate. On the other hand, when the surface of the aggregate is rough, the concrete is able to develop the full strength of much of the aggregate in the region of failure. Extensive tests on the impact strength of concretes of different properties were made by Hughes and Gregory.[5.103]

The influence of fine aggregate is not well defined but the use of fine sand usually leads to a slightly lower impact strength.

Storage conditions influence the impact strength in a manner different from compressive strength. Specifically, the impact strength of water-stored concrete is lower than when the concrete is dry, although the former concrete can withstand more blows before cracking. Thus, as already stated, the compressive strength without reference to storage conditions, does not give a satisfactory indication of the impact strength.[5.99]

There is evidence that under uniformly applied impact loading (a condition difficult to achieve in practice) the impact strength of concrete is significantly greater than its static compressive strength; a difference of 30 to 80 per cent has been suggested. This increase in strength would explain the greater ability of concrete to absorb strain energy under uniform impact.

REFERENCES

5.1. K. K. SCHILLER, Porosity and strength of brittle solids (with particular reference to gypsum), *Mechanical Properties of Non-metallic Brittle Materials*, pp. 35–45 (London, Butterworth, 1958).

5.2. NATIONAL SAND AND GRAVEL ASSOCIATION, *Joint Tech. Information letter No.* 155 (Washington D.C., 29th April 1959).

5.3. A. HUMMEL, *Das Beton—ABC* (Berlin, W. Ernst, 1959).

5.4. A. M. NEVILLE, Tests on the strength of high-alumina cement concrete, *J. New Zealand Inst. E.,* **14**, No. 3, pp. 73–7 (March 1959).

5.5. T. C. POWERS, The non-evaporable water content of hardened Portland cement paste: its significance for concrete research and its method of determination, *A.S.T.M. Bul. No.* 158, pp. 68–76 (May 1949).

5.6. T. C. POWERS and T. L. BROWNYARD, Studies of the physical properties of hardened Portland cement paste (Nine parts), *J. Amer. Concr. Inst.,* **43** (Oct. 1946 to April 1947).

5.7. T. C. POWERS, The physical structure and engineering properties of concrete, *Portl. Cem. Assoc. Res. Dept. Bul.* 90 (Chicago, July 1958).

5.8. T. C. POWERS, Structure and physical properties of hardened Portland cement paste, *J. Amer. Ceramic Soc.,* **41**, pp. 1–6 (Jan. 1958).

5.9. ROAD RESEARCH, Design of concrete mixes, *D.S.I.R. Road Note No.* 4 (London, H.M.S.O., 1950).

5.10. J. D. MCINTOSH and H. C. ERNTROY, The workability of concrete mixes with ⅜ in. aggregates, *Cement Concr. Assoc. Res. Rep. No.* 2 (London, 1955).

5.11. W. H. PRICE, Factors influencing concrete strength, *J. Amer. Concr. Inst.,* **47**, pp. 417–32 (Feb. 1951).

5.12. H. C. ERNTROY and B. W. SHACKLOCK, Design of high-strength concrete mixes, *Proc. of a Symposium on Mix Design and Quality Control of Concrete*, pp. 55–73 (London, Cement and Concrete Assoc., May 1954).

5.13. B. G. SINGH, Specific surface of aggregates related to compressive and flexural strength of concrete, *J. Amer. Concr. Inst.*, **54**, pp. 897–907 (April 1958).

5.14. A. M. NEVILLE, Some aspects of the strength of concrete, *Civil Engineering* (London), **54**, Part I—Oct. 1959, pp. 1153–6; Part 2—Nov. 1959, pp. 1308–10; Part 3—Dec. 1959, pp. 1435–8.

5.15. A. M. NEVILLE, The influence of the direction of loading on the strength of concrete test cubes, *A.S.T.M. Bul. No.* 239, pp. 63–5 (July 1959).

5.16. E. OROWAN, Fracture and strength of solids, *Reports on Progress in Physics*, **12**, pp. 185–232 (London, Physical Society, 1948–49).

5.17. A. A. GRIFFITH, The phenomena of rupture and flow in solids, *Philosophical Transactions*, Series A, **221**, pp. 163–98 (Royal Society, 1920).

5.18. A. M. NEVILLE, The failure of concrete compression test specimens, *Civil Engineering* (London), **52**, pp. 773–4 (July 1957).

5.19. R. JONES and M. F. KAPLAN, The effects of coarse aggregate on the mode of failure of concrete in compression and flexure, *Mag. Concr. Res.*, **9**, No. 26, pp. 89–94 (August 1957).

5.20. F. M. LEA, Cement research: retrospect and prospect, *Proc. 4th Int. Symp. on the Chemistry of Cement*, Washington D.C. (1960) pp. 5–8.

5.21. O. Y. BERG, Strength and plasticity of concrete, *Doklady Akademii Nauk S.S.S.R.*, **70**, No. 4, pp. 617–20 (1950).

5.22. R. L'HERMITE, Idées actuelles sur la technologie du béton, *Institut technique du Bâtiment et des Travaux Publics* (Paris, 1955).

5.23. R. JONES and E. N. GATFIELD, Testing concrete by an ultrasonic pulse technique, *Road Research Tech. Paper No.* 34 (London, H.M.S.O., 1955).

5.24. W. KUCZYNSKI. Wplyw kruszywa grubego na wytrzymałość betonu (L'influence de l'emploi d'agrégats gros sur la résistance du béton). *Archiwum Inzynierii Ladowej*, **4**, No. 2, pp. 181–209 (1958).

5.25. M. F. KAPLAN, Flexural and compressive strength of concrete as affected by the properties of coarse aggregates, *J. Amer. Concr. Inst.*, **55**, pp. 1193–208 (May 1959).

5.26. U.S. BUREAU OF RECLAMATION, Triaxial strength tests of neat cement and mortar cylinders, *Concrete Laboratory Report No. C–779* (Denver, Colorado, Nov. 1954).

5.27. P. J. F. WRIGHT, Crushing and flexural strengths of concrete made with limestone aggregate, *Road Res. Lab. Note RN/3320/PJFW* (London, H.M.S.O. Oct. 1958).

5.28. L. SHUMAN and J. TUCKER, *J. Res. Nat. Bur. Stand. Paper No. RP1552*, **31**, pp. 107–24 (1943).

5.29. A. G. A. SAUL, A comparison of the compressive, flexural, and tensile strengths of concrete, *Cement Concr. Assoc. Tech. Rep. TRA/333* (London, June 1960).

5.30. B. W. SHACKLOCK and P. W. KEENE, Comparison of the compressive and flexural strengths of concrete with and without entrained air, *Civil Engineering* (London), **54**, pp. 77–80 (Jan. 1959).

5.31. M. F. KAPLAN, Effects of incomplete consolidation on compressive and flexural strength, ultrasonic pulse velocity, and dynamic modulus of elasticity of concrete, *J. Amer. Concr. Inst.*, **56**, pp. 853–67 (March 1960).

5.32. D. C. TEYCHENNÉ, Discussion on: The design of concrete mixes on the basis of flexural strength, *Proc. of a Symposium on Mix Design and Quality Control of Concrete*, p. 153 (London, Cement and Concrete Assoc., 1954).

5.33. M. A. Sozen, E. M. Zwoyer and C. P. Siess, Strength in shear of beams without web reinforcement, *University of Illinois Engineering Experiment Station Bul. No.* 452 (April 1959).

5.34. S. Walker and D. L. Bloem, Effects of aggregate size on properties of concrete, *J. Amer. Concr. Inst.*, **57**, pp. 283–98 (Sept. 1960).

5.35. L. E. Copeland and R. H. Bragg, Self-desiccation in Portland cement pastes, *A.S.T.M. Bul. No.* 204, pp. 34–9 (Feb. 1955).

5.36. T. C. Powers, A discussion of cement hydration in relation to the curing of concrete, *Proc. Highw. Res. Bd.*, **27**, pp. 178–88 (Washington D.C., 1947).

5.37. W. Lerch, Plastic shrinkage, *J. Amer. Concr. Inst.*, **53**, pp. 797–802 (Feb. 1957).

5.38. A. D. Ross, Shape, size, and shrinkage, *Concrete and Constructional Engineering*, pp. 193–9 (London, Aug. 1944).

5.39. P. Klieger, Early high strength concrete for prestressing, *Proc. of World Conference on Prestressed Concrete*, pp. A5–1–A5–14 (San Francisco, July 1957).

5.40. U.S. Army Corps of Engineers, *Handbook for Concrete and Cement* (Vicksburg, Miss., 1954).

5.41. T. Waters, The effect of allowing concrete to dry before it has fully cured, *Mag. Concr. Res.*, **7**, No. 20, pp. 79–82 (July 1955).

5.42. J. M. Plowman, Maturity and the strength of concrete, *Mag. Concr. Res.*, **8**, No. 22, pp. 13–22 (March 1956).

5.43. P. Klieger, Effect of mixing and curing temperature on concrete strength, *J. Amer. Concr. Inst.*, **54**, pp. 1063–81 (June 1958).

5.44. P. Klieger, Discussion on: Maturity and the strength of concrete, *Mag. Concr. Res.*, **8**, No. 24, pp. 175–8 (Nov. 1956).

5.45. R. Shalon and D. Ravina, Studies in concreting in hot countries, *R.I.L.E.M. Int. Symp. on Concrete and Reinforced Concrete in Hot Countries* (Haifa, July 1960).

5.46. T. Ridley, An investigation into the manufacture of high-strength concrete in a tropical climate. *Proc. Inst. C. E.*, **13**, pp. 23–34 (London, May 1959).

5.47. U.S. Bureau of Reclamation, *Concrete Manual*, 6th Ed. (Denver, 1956).

5.48. A. G. A. Saul, Principles underlying the steam curing of concrete at atmospheric pressure, *Cement Concr. Assoc. Tech. Rep. TRA/196* (London, July 1955).

5.49. A. G. A. Saul, Steam curing and its effect upon mix design, *Proc. of a Symposium on Mix Design and Quality Control of Concrete*, pp. 132–42 (London, Cement and Concrete Assoc., 1954).

5.50. J. J. Shideler and W. H. Chamberlain, Early strength of concretes as affected by steam curing temperatures, *J. Amer. Concr. Inst.*, **46**, pp. 273–83 (Dec. 1949).

5.51. C. C. Carlson, Lightweight aggregates for concrete masonry units, *J. Amer. Concr. Inst.*, **53**, pp. 491–508 (Nov. 1956).

5.52. A. D. Ross, Heat flow in the steam curing of concrete, *Proc. Inst. C. E.*, Part 1, **5**, No. 6, pp. 695–702 (London, Nov. 1956).

5.53. J. D. McIntosh, Electrical curing of concrete, *Mag. Concr. Res.*, No. 1, pp. 21–8 (Jan. 1949).

5.54. H. F. Gonnerman, *Annotated bibliography on high-pressure steam curing of concrete and related subjects* (Chicago, National Concrete Masonry Assoc., 1954).

5.55. T. Thorvaldson, Effect of chemical nature of aggregate on strength of steam-cured Portland cement mortars, *J. Amer. Concr. Inst.*, **52**, pp. 771–80 (1956).

5.56. F. M. LEA, Would the strength grading of ordinary Portland cement be a contribution to structural economy? *Proc. Inst. C. E.*, **2**, No. 3, pp. 450–7 (London, Dec. 1953).

5.57. S. WALKER and D. L. BLOEM, Variations in Portland cement, *Proc. A.S.T.M.*, **58**, pp. 1009–32 (1958).

5.58. A. M. NEVILLE, The relation between standard deviation and mean strength of concrete test cubes, *Mag. Concr. Res.*, **11**, No. 32, pp. 57–84 (July 1959).

5.59. P. J. F. WRIGHT, Variations in the strength of Portland cement, *Mag. Concr. Res.*, **10**, No. 30, pp. 123–32 (Nov. 1958).

5.60. S. WALKER and D. L. BLOEM, Studies of flexural strength of concrete, Part 3: Effects of variations in testing procedures, *Proc. A.S.T.M.*, **57**, pp. 1122–39 (1957).

5.61. H. GREEN, Impact testing of concrete, *Mechnical Properties of Non-metallic Brittle Materials*, pp. 300–13 (London, Butterworth, 1958).

5.62. H. H. STEINOUR, Concrete mix water—how impure can it be? *J. Portl. Cem. Assoc. Research and Development Laboratories*, **3**, No. 3, pp. 32–50 (Sept. 1960).

5.63. W. J. McCOY, Water for mixing and curing concrete, *A.S.T.M. Sp. Tech. Publ. No.* 169, pp. 355–60 (1956).

5.64. BUILDING RESEARCH STATION, Analysis of water encountered in construction, *Digest No.* 90 (London, H.M.S.O., July 1956).

5.65. D. A. ABRAMS, Tests of impure waters for mixing concrete, *J. Amer. Concr. Inst.*, **20**, pp. 442–86 (1924).

5.66. F. M. LEA, *The Chemistry of Cement and Concrete* (London, Arnold, 1956).

5.67. J. G. DEMPSEY, Coral and salt water as concrete materials, *J. Amer. Concr. Inst.*, **48**, pp. 157–66 (Oct. 1951).

5.68. R. SHALON and M. RAPHAEL, Influence of sea water on corrosion of reinforcement, *J. Amer. Concr. Inst.*, **55**, pp. 1251–68 (June 1959).

5.69. A.C.I.–A.S.C.E. JOINT COMMITTEE 323, Tentative recommendations for prestressed concrete, *J. Amer. Concr. Inst.*, **54**, pp. 545–78 (Jan. 1958).

5.70. NATIONAL SAND AND GRAVEL ASSOCIATION, *Technical Information Letter No.* 156 (Washington D.C., 10th July 1959).

5.71. B. C. DOELL, Effect of algæ infested water on the strength of concrete, *J. Amer. Concr. Inst.*, **51**, pp. 333–42 (Dec. 1954).

5.72. E. M. KROKOSKY, Strength vs. structure: a study for hydraulic cements, *Materials and Structures*, **3**, No. 17, pp. 313–23 (Paris, Sept.–Oct. 1970).

5.73. E. RYSHKEWICH, Compression strength of porous sintered alumina and zirconia, *J. Amer. Ceramic Soc.*, **36**, pp. 66–8 (Feb. 1953).

5.74. Discussion of paper by H. J. GILKEY: Water/cement ratio versus strength—another look, *J. Amer. Concr. Inst.*, **58**, pp. 1851–78 (Dec. 1961, Part 2).

5.75. D. W. HOBBS, Strength and deformation properties of plain concrete subject to combined stress, Part 1: strength results obtained on one concrete, *Cement Concr. Assoc. Tech. Rep. TRA/42.451* (London, Nov. 1970).

5.76. T. T. C. HSU, F. O. SLATE, G. M. STURMAN and G. WINTER, Microcracking of plain concrete and the shape of the stress–strain curve, *J. Amer. Concr. Inst.*, **60**, pp. 209–24 (Feb. 1963).

5.77. B. MATHER, What do we need to know about the response of plain concrete and its matrix to combined loadings?, *Proc. 1st Conf. on the Behaviour of Structural Concrete Subjected to Combined Loadings; West Virginia Univ.*, 1969, pp. 7–9.

5.78. H. KUPFER, H. K. HILSDORF and H. RÜSCH, Behaviour of concrete under biaxial stresses, *J. Amer. Concr. Inst.*, **66**, pp. 656–66 (Aug. 1969).

5.79. B. BRESLER and K. S. PISTER, Strength of concrete under combined stresses, *J. Amer. Concr. Inst.*, **55**, pp. 321–45 (Sept. 1958).

5.80. B. BRESLER and K. S. PISTER, Failure of plain concrete under combined stresses, *Trans. A.S.C.E.*, **122**, pp. 1049–59 (1957).

5.81. A. NÁDAI, Theory of flow and fracture of solids (New York, McGraw-Hill, 1950).

5.82. D. McHENRY and J. KARNI, Strength of concrete under combined tensile and compressive stresses, *J. Amer. Concr. Inst.*, **54**, pp. 829–40 (April 1958).

5.83. A. MEYER, Über den Einfluss des Wasserzementwertes auf die Frühfestigkeit von Beton, *Betonstein Zeitung* No. 8, pp. 391–4 (1963).

5.84. M. PIÑEIRO, Relacion entre las resistencias a compresion de hormigones a 7 y a 28 dias, *Revista del Idiem*, **2**, No. 1, pp. 33–43 (April 1963).

5.85. G. J. VERBECK and R. H. HELMUTH, Structures and physical properties of cement paste, *Proc. 5th Int. Symp. on the Chemistry of Cement; Tokyo*, 1968; Part III, pp. 1–32.

5.86. A.C.I. COMMITTEE 517, Recommended practice for atmospheric pressure steam curing of concrete, *J. Amer. Concr. Inst.*, **66**, pp. 629–46 (Aug. 1969). **66**, pp. 629–46 (Aug. 1969).

5.87. J. A. HANSON, Optimum steam curing procedures for structural lightweight concrete, *J. Amer. Concr. Inst.*, **62**, pp. 661–72 (June 1965).

5.88. A.C.I. COMMITTEE 516, High pressure steam curing: modern practice, and properties of autoclaved products, *J. Amer. Concr. Inst.*, **62**, pp. 869–908 (Aug. 1965).

5.89. E. W. BENNETT and N. K. RAJU, Cumulative fatigue damage of plain concrete in compression, *Proc. Int. Conf. on Structure, Solid Mechanics and Engineering Design*, Part 2, pp. 1089–1102 *Southampton, April* 1969; (New York, Wiley-Interscience, 1971).

5.90. J. P. LLOYD, J. L. LOTT and C. E. KESLER, Final summary report: fatigue of concrete, *T. & A. M. Report No. 675, Department of Theoretical and Applied Mechanics, University of Illinois*, pp. 33 (Sept. 1967).

5.91. A. M. NEVILLE, Current problems regarding concrete under sustained loading, *Int. Assoc. for Bridge and Structural Engineering, Publications*, **26**, pp. 337–43 (1966).

5.92. F. S. OPLE, JR. and C. L. HULSBOS, Probable fatigue life of plain concrete with stress gradient, *J. Amer. Concr. Inst.*, **63**, pp. 59–81 (Jan. 1966).

5.93. J. W. MURDOCK, The mechanism of fatigue failure in concrete, *Thesis submitted to the University of Illinois for the degree of Ph.D.*, pp. 131 (1960).

5.94. J. A. NEAL and C. E. KESLER, The fatigue of plain concrete, pp. 226–37, *Proc. Int. Conf. on the Structure of Concrete*, (London, Cement and Concrete Assoc., 1968).

5.95. B. M. ASSIMACOPOULOS, R. F. WARNER and C. E. EKBERG, JR., High speed fatigue tests on small specimens of plain concrete, *J. Prestressed Concr. Inst.*, **4**, pp. 53–70 (Sept. 1959).

5.96. W. H. GRAY, J. F. McLAUGHLIN and J. D. ANTRIM, Fatigue properties of lightweight aggregate concrete, *J. Amer. Concr. Inst.*, **58**, pp. 149–62 (Aug. 1961).

5.97. S. C. C. BATE, The strength of concrete members under dynamic loading, *Proc. of a Symp. on the Strength of Concrete Structures*, pp. 487–524 (London, Cement and Concrete Assoc., 1956).

5.98. A. M. OZELL, Discussion of paper by J. P. ROMUALDI and G. B. BATSON: Mechanics of crack arrest in concrete, *J. Eng. Mech. Div., A.S.C.E.*, **89**, No. EM 4, p. 103 (Aug. 1963).

5.99. H. GREEN, Impact strength of concrete, *Proc. Inst. C. E.*, **28**, pp. 383–96 (London, July 1964).

5.100. G. B. WELCH and B. HAISMAN, Fracture toughness measurements of concrete, *Report No. R42, University of New South Wales, Kensington, Australia* (Jan. 1969).

5.101. K. THOMAS and W. E. A. LISK, Effect of sea water from tropical areas on setting times of cements, *Materials and Structures*, **3**, No. 14, pp. 101–105 (1970).

5.102. L. J. M. NELISSEN, Biaxial testing of normal concrete, *Heron*, **18**, No. 1, pp. 1–90 (1972).

5.103. B. P. HUGHES and R. GREGORY, The impact strength of concrete using Green's ballistic pendulum, *Proc. Inst. C. E.*, **41**, pp. 731–750 (London, Dec. 1968).

5.104. M. A. WARD, A. M. NEVILLE and S. P. SINGH, Creep of air-entrained concrete, *Mag. Concr. Res.*, **21**, No. 69, pp. 205–10 (Dec. 1969).

5.105. G. VERBECK, Energetics of the hydration of Portland cement, *Proc. 4th Int. Symp. on the Chemistry of Cement, Washington D.C.* (1960), pp. 453–65.

6. Elasticity, Shrinkage, and Creep

THIS Chapter deals with the different types of deformation of concrete. Like many other structural materials concrete is to a certain degree elastic.*

Under sustained loading, however, strain increases with time, i.e. concrete exhibits creep. In addition, whether subjected to load or not, concrete contracts on drying, undergoing shrinkage.

The magnitudes of shrinkage and creep are of the same order as the elastic strain under the usual range of stresses, so that the various types of strain must be at all times taken into account.

Modulus of Elasticity

Fig. 6.1 shows a typical stress–strain diagram for a concrete specimen loaded and unloaded in compression or tension.†

It can be seen that the term Young's modulus of elasticity can strictly be applied only to the straight part of the stress–strain curve or, when no straight portion is present, to the tangent to the curve at the origin. This is the initial tangent modulus, but it is of little practical importance. It is possible to find a tangent modulus at any point on the stress–strain curve, but this modulus applies only to very small changes in load above or below the load at which the tangent modulus is considered.

The magnitude of the observed strains and the curvature of the stress–strain relation depend, at least in part, on the rate of application of stress. When the load is applied extremely rapidly, say, in less than 0·01 second, recorded strains are greatly reduced, and the curvature of the stress–strain curve becomes extremely small. An increase in loading time from 5 seconds to about 2 minutes can increase the strain by up to 15 per cent, but within the range of 2 to 10 (or even 20) minutes—a time normally required to test a specimen in an ordinary testing machine—the increase in strain is very small.

The increase in strain while the load, or a part of it, is acting is due to creep of concrete, but the dependence of instantaneous strain on the

---

* A material is said to be perfectly elastic if strains appear and disappear immediately on application and removal of stress. This definition does not imply a linear stress–strain relation. Elastic behaviour coupled with a non-linear stress–strain relation is exhibited, for instance, by glass and some rocks.

† A small concave-up part of the curve at the beginning of loading in compression, sometimes encountered, is due to the existence of fine shrinkage cracks.

speed of loading makes the demarcation between elastic and creep strains difficult. For practical purposes an arbitrary distinction is made: the deformation occurring during loading is considered elastic, and the subsequent increase in strain is regarded as creep. The modulus of elasticity satisfying this requirement is the secant modulus of Fig. 6.1.

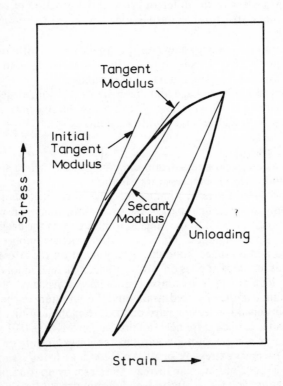

Fig. 6.1. *Typical stress–strain curve for concrete*

There is no standard method of determining the secant modulus; in some laboratories it is measured at stresses ranging from 3 to 14 MN/m^2 (400 to 2,000 lb/in^2), in others at stresses representing 15, 25, 33 or 50 per cent of the ultimate strength.[6.1] Because the secant modulus decreases with an increase in stress, the stress at which the modulus has been determined should always be stated. The secant modulus is a static modulus since it is determined from an experimental stress–strain relation, in contradistinction to the dynamic modulus, considered on page 318.

The determination of the initial tangent modulus is not easy but an approximate value of the modulus can be obtained indirectly: the secant of the stress–strain curve on unloading is often parallel to the initial tangent to the loading curve (Fig. 6.1), but not always. Repeated loading and unloading reduces the subsequent creep, so that the stress–strain curve on third or fourth loading exhibits only small curvature; this method is prescribed in B.S. 1881: 1970. Measuring strain for very small variations in stress also largely eliminates creep, but an accurate determination of strain is usually difficult.

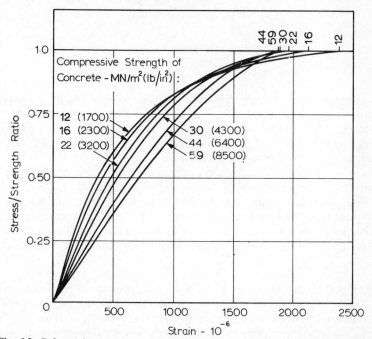

Fig. 6.2. *Relation between stress/strength ratio and strain for concretes of different strengths*[6.2]

Fig. 6.2 shows the strain in concretes of different strengths plotted against the stress/strength ratio. At a load representing the same *proportion* of the ultimate strength (say, one-half) the stronger concrete has a higher deformation; however, for any two concretes the ratio of deformations is considerably smaller than the ratio of strengths, so that the secant modulus of elasticity is greater the stronger the concrete (Table 6.1). This behaviour is in contrast to that of different grades of steel,

Table 6.1: *Modulus of Elasticity of Concretes of Different Strengths Assumed in Code of Practice CP 115: 1969 for the Design of Prestressed Concrete*

Average compressive strength of works cubes		Modulus of elasticity	
MN/m²	lb/in²	GN/m²	10⁶ lb/in²
20	3,000	21	3
27	4,000	28	4
35	5,000	31	4·5
40	6,000	34	5
55	8,000	41	6
70	10,000	45	6·5

possibly because the strength of cement paste is governed by the gel/space ratio which can be expected to affect also the stiffness of the cementitious material. On the other hand, the strength of steel is related to the structure and boundaries of crystals but not to voids, so that the stiffness of the material is unaffected by its strength.

Fig. 6.2 indicates also the maximum strain at failure in compression: this strain is higher the lower the strength of concrete (cf. p. 248).

The modulus of elasticity of concrete increases approximately with the square root of its strength. The British Code of Practice for the Structural Use of Concrete CP110: 1972 relates the tangent modulus of elasticity E_c in GN/m² to the cube strength u in MN/m² by the expression

$$E_c = 4.5\sqrt{u}$$

when the density of concrete is 2,300 kg/m³ or greater. In the corresponding expression of the Comité Européen du Béton, the coefficient is 6·6 instead of 4·5 but the strength is based on cylinders. When the density ρ is between 1,400 and 2,300 kg/m³, the British expression for the modulus of elasticity is

$$E_c = 0.85 \rho^2 \sqrt{u} \times 10^{-6}.$$

It is claimed that the modulus so determined is within ± 7 GN/m² of the tangent modulus which would be given by measurement according to B.S. 1881: Part 5: 1970. The Comité Européen du Béton expression for lightweight aggregate concrete differs primarily in taking the influence of density to the power of 3/2 as in the American approach (*see* below). Thus, the Comité Européen du Béton recommends the value of the modulus in GN/m² as

$$E_c = \frac{1}{1 \cdot 8} \times 10^{-4} \rho^{1.5} \sqrt{f'}_c$$

where f_c' is the cylinder strength in MN/m².

The approach of the American Concrete Institute is similar: the 1971 edition of the ACI Building Code gives E_c in pounds per square inch for normal weight concrete as

$$E_c = 57,000 \sqrt{f_c'}$$

where f_c' is the compressive strength of cylinders in pounds per square inch. When the density of concrete ρ is different from 145 lb/ft³ (assumed to be the value for normal weight concrete) and is expressed in pounds per cubic foot, the modulus of elasticity is given by

$$E_c = 33\rho^{1.5} \sqrt{f_c'}.$$

All these relations are valid only in general terms, and are affected by the condition of the specimen at test: a wet specimen has a higher modulus of elasticity than a dry one,[*][6.3] while strength varies in the opposite sense (Fig. 6.3). The properties of aggregate also influence the modulus of elasticity although generally they do not affect the compressive strength: the higher the modulus of elasticity of the aggregate the higher the modulus of the resulting concrete. The shape of coarse aggregate particles and their surface characteristics may also influence the value of the modulus of elasticity of concrete[6.4] and the curvature of the stress–strain relation.[6.5]

It is interesting to note that the two components of concrete, cement paste and aggregate, when individually subjected to stress exhibit a sensibly linear stress–strain relation (Fig. 6.4). The reason for the curved relation in the composite material—concrete—lies in the presence of interfaces between the cement paste and the aggregate and in the development of microcracking at those interfaces[6.77] (*see* p. 250). Since the cracks develop progressively at interfaces making varying angles with the applied stress, there is a progressive increase in local stress intensity and in the magnitude of strain. Thus strain increases at a faster rate than the applied stress, and so the stress–strain curve continues to bend over, with an apparent pseudo-plastic behaviour.[6.78]

The relation between modulus of elasticity and strength depends also on the mix proportions (since aggregate generally has a higher modulus than the cement paste) and on the age of the specimen: at later ages the modulus increases more rapidly than strength.[6.6] This is shown in Fig. 6.5, which also gives results for concrete made with expanded clay

---

[*] Generally, the opposite is the case when the *dynamic* moduli are compared.

aggregate. The modulus of lightweight aggregate concrete is usually be-tween 40 and 80 per cent of the modulus of ordinary concrete of the same strength. Since the modulus of lightweight aggregate differs little from the modulus of the cement paste, mix proportions do not affect the modulus of elasticity of lightweight aggregate concrete.[6.7]

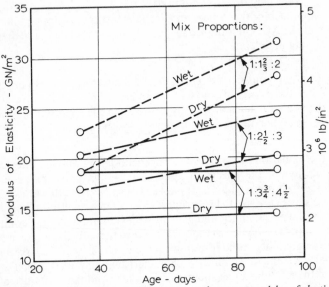

Fig. 6.3. *Influence of moisture condition at test on the secant modulus of elasticity (at 5·5 MN/m² (800 lb/in²)) of concretes at different ages*[6.3]

The relation between the modulus of elasticity and strength is un-affected by temperature of storage up to about 230°C (450°F) or possibly higher (*see* p. 440), since both properties vary with temperature in approximately the same manner.[6.8] However, for the same strength, the modulus is somewhat higher the lower the early curing temperature; thus steam-cured concrete has a lower modulus than moist cured concrete of the same strength, but the difference is under 10 per cent.

So far we have considered the modulus of elasticity in compression, but for any concrete the modulus in tension has sensibly the same, or possibly slightly lower, value. The tensile modulus can also be determined from measurement of deflexion of flexure specimens; where necessary, a correction for shear should be applied.[6.5]

In flexure tests, there is a descending part of the stress–strain curve immediately before failure, i.e. a decrease in stress is accompanied by an increase in strain (Fig. 6.6). This can be observed also in compression tests when the specimen is loaded at a constant rate of strain. In Fig. 6.6, the steeper descending part of the stress–strain curve for concrete made with sintered fly-ash aggregate may be noted; such a shape is characteristic of rather brittle behaviour.

The modulus of elasticity in shear (modulus of rigidity) is not normally determined by direct measurement.

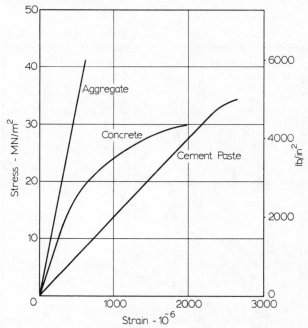

Fig. 6.4. *Stress–strain relations for cement paste, aggregate, and concrete*

It may be useful to consider further the shape of the stress–strain relation such as that in Fig. 6.6. Various attempts have been made to fit an analytical expression to this relation. One that shows considerable promise both in the ascending and the descending part of the curve is the expression suggested by Desayi and Krishnan:[6.79]

$$\sigma = \frac{E\varepsilon}{1 + \left(\dfrac{\varepsilon}{\varepsilon_0}\right)^2}$$

where ε = strain,
σ = stress,
ε_0 = strain at maximum stress,

and E = initial tangent modulus, assumed to be twice the secant modulus at maximum stress σ_{max}, i.e.

$$E = \frac{2\sigma_{max}}{\varepsilon_0}$$

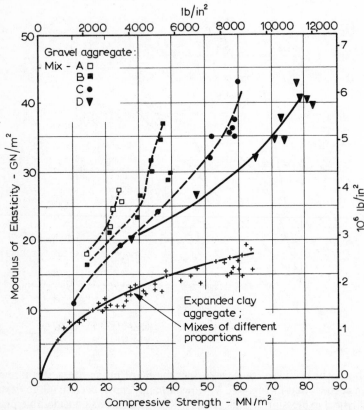

Fig. 6.5. *Static modulus of elasticity of concretes made with gravel aggregate and expanded clay aggregate, and tested at different ages up to one year*[6.6]

Another expression to the stress–strain curve which shows good fit and is easy to apply is one developed by Saenz.[6.80] This is of the form

$$\sigma = \frac{\varepsilon}{A + B\varepsilon + C\varepsilon^2 + D\varepsilon^3}$$

where $A = \dfrac{1}{E}$; $\quad B = \dfrac{R_E + R - 2}{R_E \, \sigma_{max}}$; $\quad C = \dfrac{1 - 2R}{R_E \, \sigma_{max}\varepsilon_0}$;

$$D = \dfrac{R}{R_E \, \sigma_{max}\varepsilon_0{}^2};$$

$$R = \dfrac{R_E(R_\sigma - 1)}{(R_\varepsilon - 1)^2} - \dfrac{1}{R_\varepsilon}; \quad R_E = \dfrac{E}{E_m}; \quad R_\sigma = \dfrac{\sigma_{max}}{\sigma_u}; \quad R_\varepsilon = \dfrac{\varepsilon_{max}}{\varepsilon_0};$$

$$E_m = \dfrac{\sigma_{max}}{\varepsilon_0}$$

and

σ_u = stress at failure.

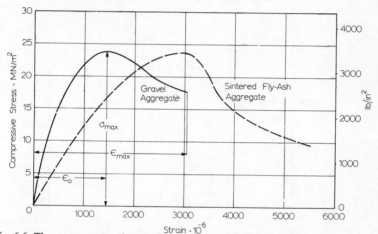

Fig. 6.6. *The stress–strain relation of concrete tested in flexure or in compression at a constant rate of strain*

The expression thus involves three ratios: modular, stress, and strain.

When using either Desayi and Krishnan's or Saenz's expressions, it is important to note that the maximum value of strain, ε_{max}, depends on the rate of loading and the "standard" curve is therefore not a fundamental property of concrete.

It is important to realize that, unfortunately, in many practical cases, certain features of the stress–strain curve are due not to the intrinsic properties of the concrete but to the properties of the testing machine. For instance, as Hognestad, Hanson and McHenry[6.81] pointed out, sudden failure of concrete cylinders occurs when the slope of the descending part of the stress–strain curve of the concrete becomes equal to the slope of the testing machine curve.

Dynamic Modulus of Elasticity

The procedure for determining the dynamic modulus of elasticity is described on page 509. Since during the vibration of the specimen a negligible stress only is applied, the dynamic modulus refers to almost purely elastic effects and is unaffected by creep. For this reason, the dynamic modulus is approximately equal to the initial tangent modulus determined in the static test and is therefore appreciably higher than the secant (static) modulus. The difference between the dynamic and static moduli is also due to the fact that the heterogeneity of concrete affects the two moduli in different ways.[6.1] Fig. 6.7 shows that the ratio of the static to dynamic

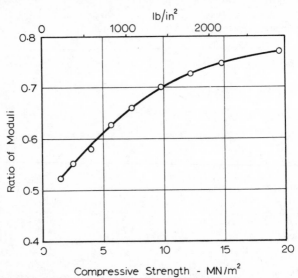

Fig. 6.7. *Ratio of static and dynamic moduli of elasticity of concretes of different strengths*[6.9]

The dynamic modulus determined by a longitudinal vibration of cylinders.

moduli is higher the higher the strength of concrete.[6.9] For a given mix the ratio increases also with age (Fig. 6.8). According to the British Code for Concrete CP 110: 1972, the moduli, expressed in GN/m^2, are related by $E_c = 1 \cdot 25 E_d - 19$ where E_c and E_d are the static and dynamic modulus respectively. The relation does not apply to concretes containing lightweight aggregate or more than 500 kg of cement per cubic metre (850 lb/yd³) of concrete.

Fig. 6.9 shows a typical relation between the dynamic modulus determined by a transverse vibration of cylinders and their compressive strength.[6.10] This relation is unaffected by air entrainment, method of curing, condition at test, or the type of cement used.[6.11]

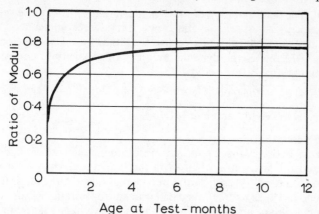

Fig. 6.8. *Ratio of static and dynamic moduli of elasticity of concrete at different ages*[6.1]
The static modulus determined at 7 MN/m² (1,000 lb/in²).

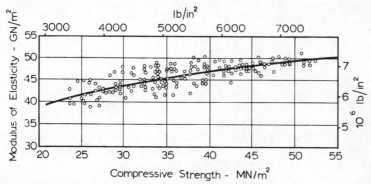

Fig. 6.9. *Relation between the dynamic modulus of elasticity, determined by transverse vibration of cylinders, and their compressive strength*[6.10]

The dynamic modulus can also be determined from the velocity of propagation of a pulse of waves at an ultrasonic frequency (*see* p. 504). The relation between the pulse velocity V and the modulus E is given by—

$$E = \rho V^2 \frac{(1 + \mu)(1 - 2\mu)}{1 - \mu}$$

where ρ = density of concrete, and μ = Poisson's ratio for concrete.

This determination of the dynamic modulus depends on Poisson's ratio, which is generally not known accurately; a change in Poisson's ratio from, say, 0·16 to 0·25 reduces the computed modulus by about 11 per cent. The determination of the dynamic modulus from pulse velocity measurements is not reliable and is not normally recommended.[6.1]

The British Code of Practice for the Structural Use of Concrete CP110: 1972 relates the dynamic modulus of elasticity of concrete E_d to its strength u:

$$E_d = 22 + 2 \cdot 8 \sqrt{u}$$

where E_d is in GN/m^2 and u in MN/m^2.

Poisson's Ratio

This ratio between the lateral strain accompanying an applied axial strain and the latter strain is used in the design and analysis of many types of structures. Poisson's ratio of concrete varies in the range $0 \cdot 11$ to $0 \cdot 21$ (generally $0 \cdot 15$ to $0 \cdot 20$) when determined from strain measurements, both for ordinary and lightweight concrete. A dynamic determination yields higher values, with an average of about $0 \cdot 24$.[6.5]

The latter method requires the measurement of pulse velocity, V, and also the fundamental resonant frequency of longitudinal vibration of a beam of length L (*see* p. 510). Poisson's ratio, μ, can then be calculated from the expression[6.12]—

$$\left(\frac{V}{2nL} \right)^2 = \frac{1 - \mu}{(1 + \mu)(1 - 2\mu)}$$

since $\dfrac{E}{\rho} = (2nL)^2$

where $\rho =$ density of concrete.

Poisson's ratio can also be found from the modulus of elasticity E, as determined in longitudinal or transverse mode of vibration (*see* p. 510), and the modulus of rigidity, G, using the expression:

$$\mu = \frac{E}{2G} - 1.$$

G is normally determined from the resonant frequency of torsional vibration (*see* p. 510). Values of μ obtained by this method are intermediate between those from direct measurements and those from dynamic tests.

No reliable information on the variation in Poisson's ratio with age, strength, or other properties of concrete is available, but it is generally believed that Poisson's ratio is lower in high strength concrete.

Under high loads Poisson's ratio increases rapidly owing to cracking within the specimen, as shown in Fig. 6.10; we are dealing now, however, with an apparent Poisson's ratio, as the specimen is no longer a continuous body. This is reflected in the change in the volumetric strain.

Data on Poisson's ratio under sustained loading are even more scarce. Generally, Poisson's ratio is unaffected, indicating that the longitudinal and lateral deformations due to creep are in the same ratio as the corresponding elastic deformations. This means that the volume of concrete decreases with the progress of creep. Under sustained multiaxial compression, the creep Poisson's ratio is smaller: 0·09 to 0·17.[6.82]

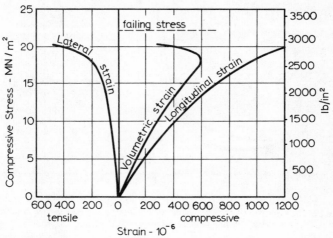

Fig. 6.10. *Strains in a prism tested to failure in compression*[6.71]

Early Volume Changes

When discussing the progress of hydration of cement the resulting changes in volume were mentioned. The chief of these is the reduction in the volume of the system cement plus water. While the cement paste is plastic it undergoes a volumetric contraction whose magnitude is of the order of one per cent of the absolute volume of dry cement.[6.13]

This contraction is known as plastic shrinkage, since it takes place while the concrete is still in the plastic state. Loss of water by evaporation from the surface of the concrete or by suction by dry concrete below aggravates the plastic shrinkage[6.14] (Table 6.2) and can lead to surface cracking, although such cracking is also possible when no evaporation is permitted (*see* p. 214). However, a complete prevention of evaporation immediately on casting eliminates cracking.[6.84] Cracking develops usually over obstructions to uniform settlement, e.g. reinforcement or large aggregate particles, or when a large horizontal area of concrete makes contraction in that direction more difficult than vertically: deep cracks of an irregular pattern are then formed.[6.15]

Early shrinkage is greater the larger the cement content of the mix[6.14] (Fig. 6.11) and the earlier the stiffening of the concrete.[6.16] It has been

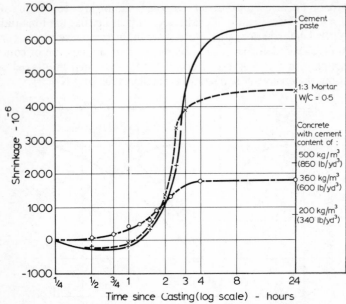

Fig. 6.11. *The influence of cement content of the mix on early shrinkage in air at 20°C (68°F) and 50 per cent relative humidity with wind velocity of 1·0 m/s (2·25 mph)*[6.14]

Table 6.2: *Plastic Shrinkage of Neat Cement Paste Stored in Air at a Relative Humidity of 50 per cent and Temperature of 20°C (68°F)*[6.14]

Wind velocity		Shrinkage 8 hours after placing
m/s	mph	10^{-6}
0	0	1,700
0·6	1·35	6,000
1·0	2·25	7,300
7 to 8	16 to 18	14,000

suggested that a greater bleeding capacity of the concrete decreases the plastic shrinkage,[6.16] but the relation between plastic shrinkage and bleeding has not been confirmed.[6.15]

Volume changes occur also after setting has taken place, and may be in the form of shrinkage or swelling. Continued hydration, when a supply of water is present, leads to expansion (*see* next section) but when no moisture movement to or from the paste is permitted shrinkage occurs. Shrinkage of such a conservative system is known as autogenous volume change or autogenous shrinkage, and it occurs in practice in the interior of a large concrete mass. The magnitude of the movement is between about 40×10^{-6} at the age of one month and 100×10^{-6} after five years (measured as a linear strain).[6.17] The autogenous volume change tends to increase at higher temperatures, with a higher cement content, and possibly also with finer cements.[6.83]

The contraction is thus relatively small and for practical purposes (except in large mass concrete structures) need not be distinguished from shrinkage caused by drying out of concrete. The latter is known as drying shrinkage and normally includes that contraction which is due to autogenous changes.

Swelling

Cement paste or concrete cured continuously in water from the time of casting exhibits a nett increase in volume and an increase in weight. This swelling is due to the adsorption of water by the cement gel: the water molecules act against the cohesive forces and tend to force the gel particles further apart, with a resultant swelling pressure. In addition, the ingress of water decreases the surface tension of the gel, and a further small expansion takes place.[6.18]

Linear expansion of neat cement paste (relative to the dimensions 24 hours after casting) has typical values of[6.14]—

$1,300 \times 10^{-6}$ after 100 days;
$2,000 \times 10^{-6}$ after 1,000 days; and
$2,200 \times 10^{-6}$ after 2,000 days.*

The swelling of concrete is considerably smaller, approximately 100×10^{-6} to 150×10^{-6} for a mix with a cement content of 300 kg/m³ (500 lb/yd³).[6.14] This value is reached 6 to 12 months after casting, and only a very small further swelling takes place.

Swelling is accompanied by an increase in weight of the order of 1 per cent.[6.14] The increase in weight is thus considerably greater than the increase in volume, as water enters to occupy the space created by the decrease in volume on hydration of the system cement plus water.

Drying Shrinkage

Withdrawal of water from concrete stored in unsaturated air causes drying shrinkage. A part of this movement is irreversible and should be distinguished from the reversible moisture movement caused by alternating storage under wet and dry conditions.

---

* Swelling, shrinkage, and creep are expressed as linear strain in metres per metre or inches per inch.

Mechanism of Shrinkage

The change in the volume of drying concrete is not equal to the volume of water removed. The loss of free water, which takes place first, causes little or no shrinkage. As drying continues, adsorbed water is removed and the change in the volume of unrestrained cement paste at that stage is equal approximately to the loss of a water layer one molecule thick from the surface of all gel particles. Since the "thickness" of a water molecule is about 1 per cent of the gel particle size, a linear change in dimensions of cement paste on complete drying would be expected to be of the order of $10,000 \times 10^{-6}$ [6.18]; values up to $4,000 \times 10^{-6}$ have actually been observed.[6.19]

The influence of the grain size on drying is shown by the low shrinkage of the much more coarse-grained natural building stones (even when highly porous) and by the high shrinkage of fine grained shale.[6.18] Also, high-pressure steam cured cement paste, which is microcrystalline and has a low specific surface, shrinks 5 to 10 times,[6.14] and sometimes even 17 times,[6.20] less than a similar paste cured normally.

It is possible also that shrinkage, or a part of it, is related to the removal of zeolitic water. Calcium silicate hydrate has been shown to undergo a change in lattice spacing from 14 to 9 Å on drying[6.21]; hydrated C_4A and calcium sulphoaluminate show similar behaviour.[6.22] It is thus not certain whether the moisture movement associated with shrinkage is inter- or intracrystalline but, since pastes made with both Portland and aluminous cements, and pure ground calcium monoaluminate exhibit essentially similar shrinkage, the fundamental cause of shrinkage must be sought in the physical structure of the gel rather than in its chemical and mineralogical character.[6.22]

The relation between the weight of water lost and shrinkage is shown in Fig. 6.12. For neat pastes, the two quantities are proportional to one another as no capillary water is present and only adsorbed water is removed. However, mixes to which pulverized silica has been added, and which therefore require a higher water/cement ratio, contain capillary cavities even when completely hydrated. Emptying of the capillaries causes a loss of water without shrinkage but, once the capillary water has been lost, the removal of adsorbed water takes place and causes shrinkage in the same manner as in a neat cement paste. Thus the final slope of all the curves of Fig. 6.12 is the same. With concretes which contain some water in aggregate pores and in large (accidental) cavities an even greater variation in the shape of the curves of water loss–shrinkage is found.

Factors Affecting Shrinkage

Typical values of drying shrinkage of mortar and concrete specimens, 127 mm (5 in.) square in cross-section, stored at a temperature of 21°C (70°F) and a relative humidity of 50 per cent for six months are given in

Table 6.3, but these values are no more than a guide since shrinkage is influenced by many factors.

Table 6.3: *Typical Values of Shrinkage of Mortar and Concrete Specimens, 5 in. (127 mm) Square in Cross-section, Stored at a Relative Humidity of 50 per cent and 21°C (70°F)*[6.19]

Aggregate/cement ratio	Shrinkage after six months (10^{-6}) for water/cement ratio of—			
	0·4	0·5	0·6	0·7
3	800	1,200	—	—
4	550	850	1,050	—
5	400	600	750	850
6	300	400	550	650
7	200	300	400	500

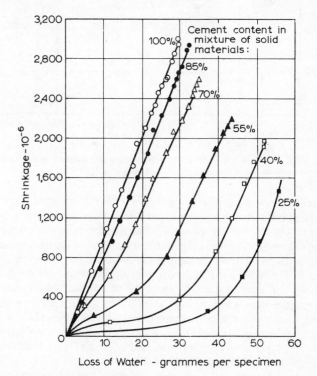

Fig. 6.12. *Relation between shrinkage and loss of water from specimens of cement-pulverized silica pastes cured for 7 days at 21°C (70°F) and then dried*[6.18]

The most important influence is exerted by aggregate, which restrains the amount of shrinkage that can actually be realized. The ratio of shrinkage of concrete, S_c, to shrinkage of neat paste, S_p, depends on the aggregate content in the concrete, a,[6.23] and is

$$S_c = S_p (1 - a)^n$$

The experimental values of n vary between 1·2 and 1·7,[6.14] some variation arising from the relief of stress in the cement paste by creep.[6.35] Fig. 6.13 shows typical results and yields a value of $n = 1·7$.

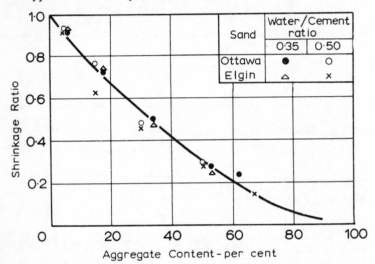

Fig. 6.13. *Influence of the aggregate content in concrete (by volume) on the ratio of the shrinkage of concrete to the shrinkage of neat cement paste*[6.23]

The size and grading of aggregate *per se* do not influence the magnitude of shrinkage but a larger aggregate permits the use of a leaner mix and hence results in a lower shrinkage. If changing the maximum aggregate size from 6·3 mm to 152 mm ($\frac{1}{4}$ in. to 6 in.) means that the aggregate content can rise from 60 to 80 per cent of the total volume of concrete then, as shown in Fig. 6.13, a threefold decrease in shrinkage will result.

Similarly, for a given strength, concrete of low workability contains more aggregate than a mix of high workability made with aggregate of the same size, and as a consequence the former mix exhibits lower shrinkage.[6.18] For instance, increasing the aggregate content of concrete from 71 to 74 per cent (at the same water/cement ratio) will reduce shrinkage by about 20 per cent (Fig. 6.13).

The twin influences of water/cement ratio and aggregate content (Table 6.3 and Fig. 6.13) can be combined in one graph; this is done in

Fig. 6.14 but it must be remembered that the shrinkage values given are no more than typical for drying in a temperate climate.

The elastic properties of aggregate determine the degree of restraint offered; for example, steel aggregate leads to shrinkage one-third less and expanded shale one-third more than ordinary aggregate.[6.6] This influence of aggregate was confirmed by Reichard[6.86] who found a correlation between shrinkage and the modulus of elasticity of concrete, which

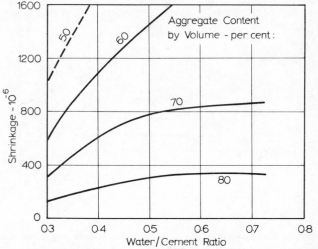

Fig. 6.14. *Influence of water/cement ratio and aggregate content on shrinkage*[6.85]

depends on the compressibility of the aggregate used (Fig. 6.15). The presence of clay in aggregate lowers its restraining effect on shrinkage, and since clay itself is subject to shrinkage, clay coatings on aggregate can increase shrinkage by up to 70 per cent.[6.18]

Even within the range of ordinary aggregates there is a considerable variation in shrinkage (Fig. 6.16). The usual natural aggregate itself is not normally subject to shrinkage, but some Scottish dolerites and some other rocks shrink on drying up to 900×10^{-6}, which is of the same order of magnitude as shrinkage of concrete made with non-shrinking aggregate. Shrinking rocks usually have also high absorption, and this can be treated as a warning sign that the aggregate should be carefully investigated for its shrinkage properties.

Lightweight aggregate usually leads to higher shrinkage, largely because the aggregate, having a lower modulus of elasticity, offers lesser restraint to the potential shrinkage of the cement paste. Those lightweight aggregates that have a large proportion of fine material smaller

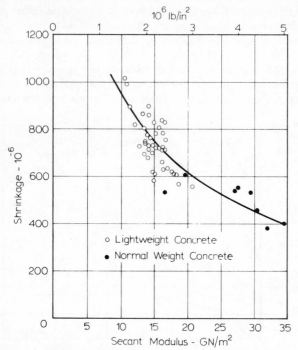

Fig. 6.15. *Relation between drying shrinkage after 2 years and secant modulus of elasticity of concrete (at a stress/strength ratio of 0·4) at 28 days*[6.86]

than 75 μm (No. 200) sieve have still higher shrinkage, as the fines lead to a larger void content.

The water content of concrete affects shrinkage in so far as it reduces the volume of restraining aggregate. Thus in general the water content of a mix would indicate the order of shrinkage to be expected, following the general pattern of Fig. 6.17, but the water content *per se* is not believed to be a primary factor.

The properties of cement have little influence on the shrinkage of concrete, and Swayze[6.26] has shown that a higher shrinkage of neat cement paste does not necessarily mean a higher shrinkage of concrete made with the given cement. Fineness of cement is a factor only in so far as particles coarser than, say, 75 μm (No. 200) sieve, which hydrate comparatively little, have a restraining effect similar to aggregate. Otherwise, contrary to some earlier suggestions, finer cement does not increase shrinkage of concrete[6.26] (cf. Fig. 6.38) although shrinkage of neat cement paste is increased.[6.72] The chemical composition of cement is now believed not to affect shrinkage except that cements deficient in gypsum exhibit a greatly increased shrinkage,[6.27] since the initial framework established in setting determines the subsequent structure of the hydrated paste[6.22] and thus influences also the gel/space ratio, strength, and creep. An optimum gypsum content from the standpoint of retardation of cement is somewhat lower than that leading to least shrinkage.[6.28] The range of gypsum contents which significantly affects shrinkage is narrower than that affecting the setting time.

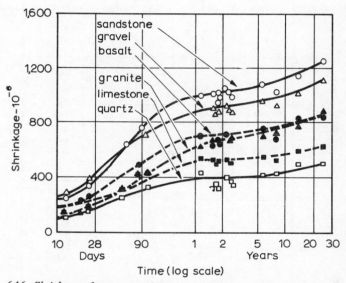

Fig. 6.16. *Shrinkage of concretes of fixed mix proportions but made with different aggregates, and stored in air at 21°C (70°F) and a relative humidity of 50 per cent*[6.24] Time reckoned since end of wet curing at the age of 28 days.

Shrinkage of concrete made with aluminous cement is of the same magnitude, but takes place much more rapidly than when Portland cement is used.[6.19]

Entrainment of air has been found to have no effect on shrinkage.[6.29]

Added calcium chloride increases shrinkage by varying amounts, generally between 10 and 50 per cent,[6.30] probably because a finer gel is produced and possibly because of greater carbonation of the more mature specimens with calcium chloride.[6.87] Other admixtures vary in the amount of increase in shrinkage but the percentage influence on shrinkage is constant for a given admixture, whatever the aggregate used.[6.88] With many plasticizing agents that allow a reduction in the water content of the mix the net result on shrinkage is negligible.[6.89]

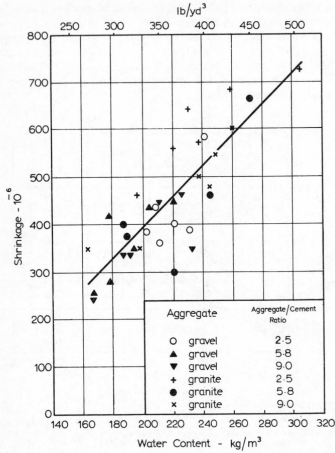

Fig. 6.17. *Relation between the water content of fresh concrete and drying shrinkage*[6.25]

Influence of Curing and Storage Conditions

Shrinkage takes place over long periods: some movement has been observed even after 28 years[6.24] (Fig. 6.18) but a part of the long-term shrinkage may be due to carbonation. The rate of shrinkage decreases rapidly with time—

14 to 34 per cent of the 20-year shrinkage occurs in 2 weeks;
40 to 80 per cent of the 20-year shrinkage occurs in 3 months; and
66 to 85 per cent of the 20-year shrinkage occurs in 1 year.

Prolonged moist curing delays the advent of shrinkage, but the effect of curing on the magnitude of shrinkage is small though rather complex.

As far as neat cement paste is concerned, the greater the quantity of hydrated cement the smaller is the volume of unhydrated cement grains which restrain the shrinkage: thus prolonged curing leads to greater shrinkage[6.18] but the paste becomes stronger with age and is able to attain a larger fraction of its shrinkage tendency without cracking. If,

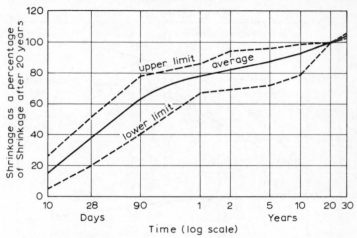

Fig. 6.18. *Range of shrinkage–time curves for different concretes stored at relative humidities of 50 and 70 per cent*[6.24]

however, cracking takes place, e.g. around aggregate particles, the overall shrinkage, measured on a concrete specimen, apparently decreases. Well-cured concrete shrinks more rapidly[6.72] and therefore the relief of shrinkage stresses by creep is smaller; also, the concrete, being stronger, has an inherent low creep capacity. These factors may outweigh the higher tensile strength of well-cured concrete and may lead to cracking. In view of this it is not surprising that contradictory results on the effects of curing on shrinkage have been reported, but in general the length of the curing period is not an important factor in shrinkage. (*See* p. 334.)

The magnitude of shrinkage is largely independent of the rate of drying except that transferring concrete directly from water to a very low humidity can lead to fracture. Rapid drying out does not allow a relief of stress by creep and may lead to more pronounced cracking. However, neither wind nor forced convection have any effect on the rate of drying of hardened concrete (except during very early stages) because the moisture conductivity of concrete is so low that only a very small rate of evaporation is possible: the rate cannot be increased by movement of air.[6.90] This has been confirmed experimentally.[6.91] (*See* p. 269 for evaporation from fresh concrete.)

The relative humidity of the medium surrounding the concrete greatly affects the magnitude of shrinkage, as shown for instance in Fig. 6.19. The same figure illustrates also the greater absolute magnitude of shrinkage compared with swelling in water: swelling is about six times smaller than shrinkage in air of relative humidity of 70 per cent or eight times smaller than shrinkage in air at 50 per cent.

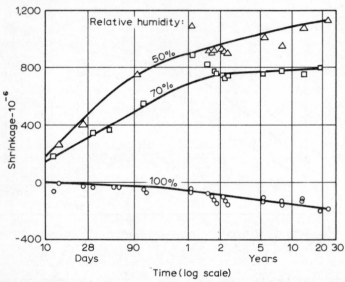

Fig. 6.19. *Relation between shrinkage and time for concretes stored at different relative humidities*[6.24]

Time reckoned since end of wet curing at the age of 28 days:

We see thus that concrete placed in "dry" (unsaturated) air shrinks, but it swells in water or air with a relative humidity of 100 per cent. This would indicate that the vapour pressure within the cement paste is always less than the saturated vapour pressure, and it is logical to expect that there is an intermediate humidity at which the paste would be in hygral equilibrium. In fact, Lorman[6.31] found this humidity to be 94 per cent, but in practice equilibrium is possible only in small and practically unrestrained specimens.

In the shrinkage test prescribed in B.S. 1881 : 1970 the specimens are dried for a specified period under prescribed conditions of temperature and humidity. The shrinkage occurring under these conditions is of the same order as that after a long exposure to air with a relative humidity of approximately 65 per cent,[6.19] and is therefore in excess of the shrinkage met with outdoors in the British Isles. The magnitude of shrinkage can be determined using a measuring frame fitted with a micrometer gauge or a dial gauge reading to 10^{-5} strain, or by means of an extensometer or strain gauges.

B.S. 2028: 1968 prescribes maximum shrinkage of precast blocks as—

500×10^{-6} to 600×10^{-6} for general use concrete blocks;
700×10^{-6} to 900×10^{-6} for load-bearing lightweight concrete blocks; and
800×10^{-6} to 900×10^{-6} for non-load-bearing lightweight concrete.

In each case, the particular limit within the range depends on strength or density. The higher limit for lightweight aggregate concrete is due to its inherently higher shrinkage; in the case of precast concrete this can be reduced by drying during the process of manufacture.[6.19]

Differential Shrinkage

In addition to internal restraints—aggregate and reinforcement—some restraint arises also from non-uniform shrinkage within the concrete member itself. Moisture loss takes place at the surface so that a moisture gradient is established in the concrete specimen, which is thus subject to differential shrinkage. This shrinkage is compensated by strains due to internal stresses, tensile near the surface and compressive in the core. When drying takes place in an unsymmetrical manner, warping can result.

The progress of shrinkage extends gradually from the drying surface into the interior of the concrete but does so only extremely slowly. Desiccation was observed to reach the depth of 75 mm (3 in.) in one month but only 600 mm (2 ft.) after 10 years.[6.14] Ross[6.32] found the difference between shrinkage in a mortar slab at the surface and at a depth of 150 mm (6 in.) to be 470×10^{-6} after 200 days. If the modulus of elasticity of mortar is 21 GN/m^2 (3×10^6 lb/in^2) the differential shrinkage would induce a stress of 10 MN/m^2 (1,400 lb/in^2); since the stress arises gradually it is relieved by creep but even so surface cracking may result. Increasing the volume of aggregate would considerably restrain the shrinkage so that the technical advantage of using concrete rather than neat cement or mortar is clear.

Because drying takes place at the surface of concrete, the magnitude of shrinkage varies considerably with the size and shape of the specimen, being a function of the surface/volume ratio.[6.32] A part of the size effect may also be due to the pronounced carbonation of small specimens. Thus for practical purposes shrinkage cannot be considered as purely an inherent property of concrete without reference to the size of the concrete member.

Many investigations have in fact indicated an influence of the size of the specimen on shrinkage. The observed shrinkage decreases with an increase in the size of the specimen but above some value the size effect is no longer apparent. The shape of the specimen also appears to enter the picture but as a first approximation shrinkage can be expressed as a function of the volume/surface ratio of the specimen. There appears to be a linear relation between this ratio and the logarithm of shrinkage[6.92] (Fig. 6.20). Furthermore, the ratio is linearly related to the logarithm of

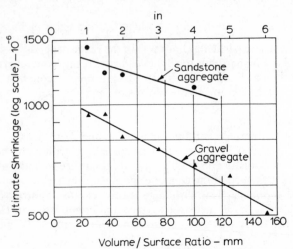

Fig. 6.20. *Relation between ultimate shrinkage and volume/surface ratio* [6.92]

time required for half the shrinkage to be achieved. The latter relation applies to concretes made with different aggregates, so that, while the magnitude of shrinkage is affected by the type of aggregate used, the rate at which the final value of shrinkage is reached is not influenced.[6.92]

The effect of shape is secondary. I-shaped specimens exhibit less shrinkage than cylindrical ones of the same volume/surface ratio, the difference being 14 per cent on the average.[6.92] The difference, which can be explained in terms of variation in the mean distance that the water has to travel to the surface, is thus not significant for design purposes.

Shrinkage-induced Cracking

As mentioned in connection with differential shrinkage, the importance of shrinkage in structures is largely related to cracking. Strictly speaking, we are concerned with the cracking tendency as the advent or absence of cracking depends not only on the potential contraction but also on the extensibility of concrete, its strength and its degree of restraint to the deformation that may lead to cracking.[6.93] Restraint in the form of reinforcing bars or a gradient of stress increases extensibility in that it allows concrete to develop strain well beyond that corresponding to maximum stress.

A high extensibility of concrete is generally desirable because it permits concrete to withstand greater volume changes. The Bureau of Reclamation[6.94] made some thermal cycle tests on concrete at a constant strain and found that tension developed on cooling to the original temperature. The failing (tensile) stress was lower in concretes made with ordinary or modified cement than with low-heat or Portland-pozzolana cement, which could withstand a greater temperature drop before failure.

The schematic pattern of crack development when stress is relieved by creep is shown in Fig. 6.21. Cracking can be avoided only if the stress

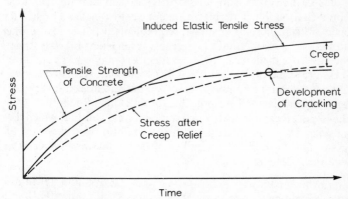

Fig. 6.21. *Schematic pattern of crack development when tensile stress due to restrained shrinkage is relieved by creep*

induced by the free shrinkage strain, reduced by creep, is at all times smaller than the tensile strength of the concrete. Thus time has a two-fold effect: the strength increases, thereby reducing the danger of cracking, but on the other hand the modulus of elasticity also increases so that the stress induced by a given shrinkage becomes greater. Furthermore, the creep relief decreases with age so that the cracking tendency becomes greater. A minor practical point is that if the cracks due to restrained shrinkage form at an early age, and moisture subsequently has access to the crack, many of the cracks will become closed by autogenous healing.

One of the most important factors in cracking is the water/cement ratio of the mix because its increase tends to increase shrinkage and at the same time to reduce the strength of the concrete. An increase in the cement content also increases shrinkage and therefore the cracking tendency but the effect on strength is positive. This applies to drying shrinkage. Carbonation, although it produces shrinkage, reduces subsequent moisture movement, and therefore is advantageous from the standpoint of cracking tendency. On the other hand, the presence of clay in aggregate leads both to higher shrinkage and greater cracking.

The use of admixtures may influence the cracking tendency through an interplay of effects on hardening, shrinkage and creep. Specifically, retarders may allow more shrinkage to be accommodated in the form of plastic shrinkage (*see* p. 321) and also probably increase the extensibility of concrete, and therefore reduce cracking. On the other hand, if concrete has attained rigidity too rapidly it cannot accommodate the would-be plastic shrinkage and, having low strength, cracks.

The temperature at the time of placing determines the dimensions of concrete at the moment when it ceases to deform plastically (i.e. without loss of continuity). A subsequent drop in temperature will produce potential contraction. Thus placing in hot weather means a high cracking tendency. Steep thermal or moisture gradients produce severe internal restraints and thus represent a high cracking tendency. Likewise, restraint by the base of a member or by other members may lead to cracking.

These are some of the factors to be considered. Actual cracking and failure depend on the combination of factors and indeed it is rarely that a single adverse factor is responsible for cracking of concrete.

The importance of cracking and the minimum width at which a crack is considered significant depend on the conditions of exposure of the concrete. On the basis of a number of investigations, Kesler *et al.*[6.95] suggest the following permissible crack widths:

interior members	0·35 mm (0·014 in.)
exterior members under normal exposure conditions	0·25 mm (0·010 in.)
exterior members exposed to particularly aggressive environment	0·15 mm (0·006 in.)

It may be relevant to mention that, although there is a variation between observers, the minimum crack width that can be seen with a naked eye is 0·13 mm (0·005 in.). In the laboratory, cracking is normally detected visually with the aid of a magnifying device. A novel method to detect cracks uses strips of electro-conductive paint applied to the surface of concrete.[6.96] A crack is indicated by the interruption of the circuit owing to a temperature rise followed by burning of the paint. Cracks as small as $0·1 \mu m$ (4×10^{-5} in.) can be detected.

Since under given physical conditions the total crack width per unit length of concrete is fixed and we want the cracks to be as fine as possible, it is desirable to have more cracks. For this reason, the restraint to cracking should be uniform along the length of the member.

We should note that, from energy considerations, it is easier to extend an existing crack than to form a new one. This explains why, under an applied load, each subsequent crack occurs under a higher load than the preceding one. The total number of cracks developed is determined by the size of the concrete member, and the distance between cracks depends on the maximum size of aggregate present.[6.97]

Moisture Movement

If concrete which has been allowed to dry in air with a given relative humidity is subsequently placed in water (or at a higher humidity) it will swell. Not all initial drying shrinkage is, however, recovered, even after prolonged storage in water. For the usual range of concretes the irreversible part of shrinkage represents between 0·3 and 0·6 of the drying shrinkage,[6.14] the lower value being more common.[6.25] The absence of fully reversible behaviour is probably due to the introduction of additional links within the gel during the period of drying, when closer contact between the gel particles is established. If the cement paste has hydrated to a considerable degree before drying it will be less affected by the closer configuration of the gel when dry; in fact, neat cement paste water-cured for six months and then dried was found to have no residual shrinkage on rewetting.[6.33] On the other hand, if drying is accompanied by carbonation the cement paste becomes insensitive to moisture movement so that the residual shrinkage is increased.[6.14]

The influence on moisture movement of curing before drying and of carbonation during drying may explain why there is no simple relation between the magnitude of moisture movement and shrinkage.

Fig. 6.22 shows the moisture movement of cement paste subjected to alternating storage in water and in air at a relative humidity of 50 per cent.[6.33] The magnitude of the moisture movement varies with the range of humidity and composition of concrete (Table 6.4). Lightweight concrete has a higher moisture movement than concrete made with ordinary aggregate.

For a given concrete there is a gradual reduction in the movement during succeeding cycles, probably due to the creation of additional valence bonds.[6.22] If the water storage periods are of sufficient duration, the continued hydration of cement results in some additional swelling so that there is a nett increase in dimensions superimposed on the reversible movement due to drying and wetting.[6.19] (In Fig. 6.22 this would be shown by a slight rise in the upper dotted line.)

An advantage of moisture movement can be taken in the manufacture of precast prestressed concrete members, as suggested by Ross,[6.34] but only when very small sections are used. The concrete is desiccated prior to post-tensioning, but after prestressing is exposed to a humid atmosphere when expansion due to the moisture movement counteracts the contraction caused by creep. A "creepless and shrinkless" concrete is thus produced.

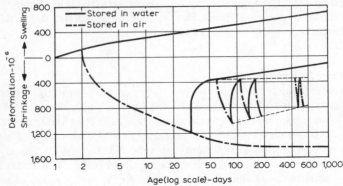

Fig. 6.22. *Moisture movement of a 1:1 cement:pulverized basalt mix stored alter-nately in water and in air at 50 per cent relative humidity; cycle period 28 days*[6.33]

Table 6.4: *Typical Values of Moisture Movement of Mortar and Concrete Dried at 50°C (122°F) and Immersed in Water*[6.19]

Mix proportions by weight	Moisture movement 10^{-6}
neat cement	1,000
1:1 mortar	400
1:2 mortar	300
1:3 mortar	200
1:2:4 concrete	300

Carbonation Shrinkage

In addition to shrinkage upon drying, concrete undergoes shrinkage due to carbonation—a phenomenon only recently recognized—and most of the experimental data on drying shrinkage include the effects of carbonation. Drying shrinkage and carbonation shrinkage are, however, quite distinct in nature.

CO_2 present in the atmosphere reacts, in the presence of moisture, with hydrated cement minerals (the agent being really the carbonic acid). The action of CO_2 takes place even at small concentrations such as are present in rural air, where the partial pressure of CO_2 is about 3×10^{-4} atmospheres; in an unventilated laboratory the pressure may rise to 12×10^{-4} atmospheres. The rate of carbonation increases with an increase in the concentration of CO_2.

$Ca(OH)_2$ carbonates to $CaCO_3$, but other cement compounds are also decomposed, hydrated silica, alumina, and ferric oxide being produced.[6.19] Such a complete decomposition of calcium compounds in hydrated cement is chemically possible even at the low pressure of CO_2 in normal atmosphere[6.36] but carbonation penetrates beyond the exposed surface of concrete only extremely slowly.

The extent of carbonation can be easily determined by treating a freshly broken surface with phenolphthalein*: the free $Ca(OH)_2$ is coloured pink while the carbonated portion is uncoloured; with progress of carbonation of the newly exposed surface the pink colouring gradually disappears.

The rate of carbonation depends also on the moisture content of the concrete and the relative humidity of the ambient medium. The size of the specimen is a factor too, since the moisture released by the reaction of CO_2 with $Ca(OH)_2$ must diffuse out in order to preserve hygral equilibrium between the inside of the specimen and the atmosphere. If diffusion is too slow the vapour pressure within the concrete rises to saturation and the diffusion of CO_2 into the paste is practically stopped.

Carbonation is accompanied by an increase in the weight of the concrete and by shrinkage. Carbonation shrinkage is probably caused by the dissolving of crystals of $Ca(OH)_2$ while under a compressive stress (imposed by the drying shrinkage) and depositing of $CaCO_3$ in spaces free from stress; the compressibility of the cement paste is thus temporarily increased. Carbonation of the hydrates present in the gel does not contribute to shrinkage as the reaction does not involve solution and re-precipitation.[6.75] Fig. 6.23 shows the drying shrinkage of mortar specimens dried in CO_2-free air at different relative humidities, and also the shrinkage after subsequent carbonation. Carbonation increases the shrinkage at intermediate humidities, but not at 100 per cent or 25 per cent. In the latter case there is insufficient water in the pores within the cement paste for CO_2 to form carbonic acid. On the other hand, when the pores are full of water the diffusion of CO_2 into the paste is very slow; it is also possible that the diffusion of calcium ions from the paste leads to precipitation of $CaCO_3$ with a consequent clogging of surface pores.[6.37]

The sequence of drying and carbonation greatly affects the total magnitude of shrinkage. Simultaneous drying and carbonation produces lower total shrinkage than when drying is followed by carbonation (Fig. 6.24) since in the former case a large part of the carbonation occurs at relative humidities above 50 per cent: under such conditions carbonation shrinkage is reduced (Fig. 6.23). Carbonation shrinkage of high-pressure steam cured concrete is very small.

When concrete is subjected to alternating wetting and drying in air containing CO_2, shrinkage due to carbonation (during the drying cycle)

---

* Manganese hydroxide can also be used.

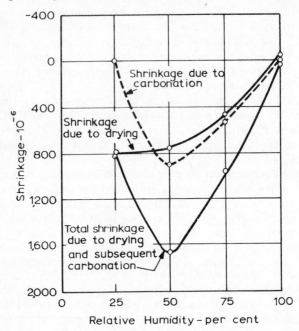

Fig. 6.23. *Drying shrinkage and carbonation shrinkage at different relative humidities*[6.37]

becomes progressively more apparent. The total shrinkage at any stage is greater than if drying took place in CO_2-free air,[6.37] so that carbonation increases the magnitude of irreversible shrinkage and may contribute to crazing of exposed concrete.

However, carbonation of concrete prior to exposure to alternating wetting and drying reduces the moisture movement, sometimes by nearly a half.[6.38] A practical application of this is to pre-carbonate precast products immediately after demoulding. Concrete with a small moisture movement is then obtained, but the humidity conditions during carbonation have to be carefully controlled.

Carbonation of concrete results also in increased strength[6.39] and reduced permeability,[6.37] possibly because water released by carbonation aids the process of hydration and $CaCO_3$ reduces the voids within the cement paste. This applies to concrete made with Portland cement only; with supersulphated cement there is a loss of strength on carbonation but, because this applies to the "skin" of the concrete only, the loss is not structurally significant.[6.98]

There is one other aspect of carbonation which should be mentioned: the protection of steel from corrosion by the alkaline conditions of

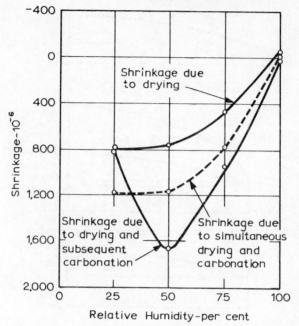

Fig. 6.24. *Influence of the sequence of drying and carbonation on shrinkage*[6.37]

hydrated cement paste is neutralized by carbonation. Thus if the entire concrete cover to steel were carbonated, corrosion of steel would occur if moisture and oxygen could ingress.

Creep of Concrete*

We have seen that the relation between stress and strain for concrete is a function of time: the gradual increase in strain with time is due to creep. Creep can thus be defined as the increase in strain under a sustained stress (Fig. 6.25) and since this increase can be several times as large as the strain on loading, creep is of considerable importance in structural mechanics.

Creep may also be viewed from another standpoint: if the restraints are such that a stressed concrete specimen is subjected to a constant strain, creep will manifest itself as a progressive decrease in stress with time. This form of relaxation is shown in Fig. 6.26.

Creep has in the past been referred to also as flow, plastic flow, plastic yield, plastic deformation, many of these terms arising from imperfect

* For fuller treatment of this topic, see: A. M. NEVILLE, *Creep of Concrete: Plain, Reinforced and Prestressed*, North-Holland Publishing Co. (Amsterdam, 1970).

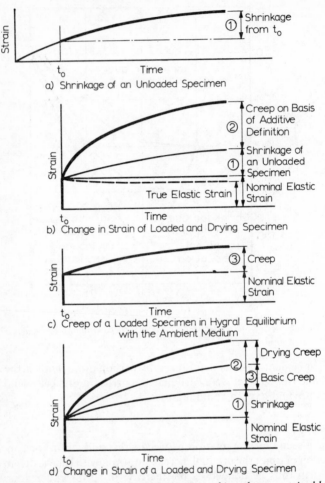

Fig. 6.25. *Time-dependent deformations in concrete subjected to a sustained load*

understanding of the nature of the phenomenon. Nowadays the term creep is universally adopted.

Under normal conditions of loading the instantaneous strain recorded depends on the speed of application of the load and includes thus not only the elastic strain but also some creep. It is difficult to differentiate accurately between the immediate elastic strain and early creep, but this is not of practical importance as it is the total strain induced by the application of load that matters. Since the modulus of elasticity of concrete increases with age the elastic deformation gradually decreases, and strictly speaking creep should be taken as strain in excess of the elastic strain at the time at which creep is being determined (Fig. 6.25). Often, the modulus of elasticity is not determined at different ages and creep is simply taken as an increase in strain above the *initial* elastic strain. This alternative definition, although theoretically less correct, does not introduce a serious error and is often more convenient to use.

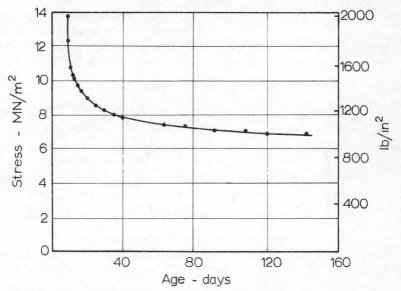

Fig. 6.26. *Relaxation of stress under a constant strain of 360 × 10⁻⁶* (6.41)

So far we have considered the creep of concrete stored under such conditions that no shrinkage or swelling takes place. If a specimen is drying while under load it is usually assumed that creep and shrinkage are additive; creep is thus calculated as the difference between the total time deformation of the loaded specimen and the shrinkage of a similar unloaded specimen stored under the same conditions through the same period (Fig. 6.25). This is a convenient simplification but, as shown on page 345, shrinkage and creep are not independent phenomena to which the principle of superposition can be applied, and in fact the effect of shrinkage on creep is to increase the magnitude of creep. In the case of many actual structures, however, creep and shrinkage occur simultaneously and the treatment of the two together is from the practical standpoint often convenient.

For this reason, and also because the great majority of the available data on creep were obtained on the assumption of the additive properties of creep and shrinkage, the discussion in this chapter will, for the most part, consider creep as a deformation in excess of shrinkage. However, where a more fundamental approach is warranted, distinction will be made between creep of concrete under conditions of no moisture movement to or from the ambient medium (true or basic creep) and the additional creep caused by drying (drying creep).

The terms and definitions involved are illustrated in Fig. 6.25.

If a sustained load is removed, the strain decreases immediately by an amount equal to the elastic strain at the given age, generally lower than the elastic strain on loading. This *instantaneous recovery* is followed by a gradual decrease in strain, called *creep recovery* (Fig. 6.27). The shape

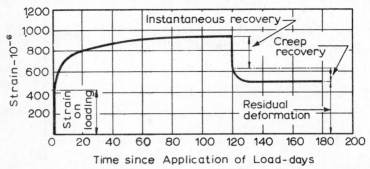

Fig. 6.27. *Creep and recovery of a mortar specimen, stored in air at a relative humidity of 95 per cent, subjected to a stress of 14·8 MN/m² (2,150 lb/in²) and then unloaded*[6.44]

of the creep recovery curve is rather like that of the creep curve, but the recovery approaches its maximum value much more rapidly.[6.44] The reversal of creep is not complete, and creep is not a simply reversible phenomenon.

Factors Influencing Creep

In most investigations creep has been studied empirically in order to determine how it is affected by various properties of concrete. A difficulty in interpreting many of the available data arises from the fact that in proportioning concrete it is not possible to change one factor without altering also at least one other. For instance, the richness and the water/cement ratio of a mix of a given workability vary at the same time. Certain influences are, however, apparent.

One of the most important of these is the relative humidity of the medium surrounding the concrete. For a given concrete, creep is higher the lower the relative humidity. This is illustrated in Fig. 6.28 for specimens cured at a relative humidity of 100 per cent and then loaded and exposed to different humidities. Such treatment results in a greatly varying shrinkage occurring in the different specimens during the early stages after the application of the sustained load. The rates of creep during that period vary correspondingly, but at later ages the rates seem to be close to one another. Thus, drying while under load enhances creep of concrete, i.e. induces the additional drying creep (cf. Fig. 6.25). The influence of relative humidity is much smaller, or absent, in the case of specimens which have reached hygral equilibrium with the surrounding medium prior to the application of the load[6.45] (Fig. 6.29). The creep of

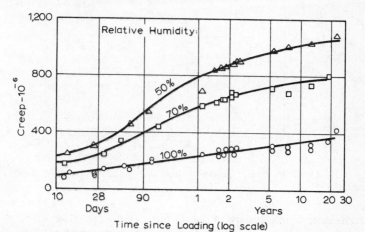

Fig. 6.28. *Creep of concrete cured in fog for 28 days, then loaded and stored at different relative humidities*[6.24]

such concrete is termed basic creep (*see* Fig. 6.25). It follows that creep of concrete loaded at an advanced age is little affected by the ambient relative humidity.[6.46]

L'Hermite[6.47] suggested that the effect of drying on creep can be taken into account using an expression—

$$c = c_i (1 + Qs)$$

where c_i = creep under shrinkage-free conditions,

s = shrinkage at the given relative humidity,

Q = a constant depending on the concrete.

Concrete which exhibits high shrinkage shows generally also a high creep.[6.14] This does not mean that the two phenomena are due to the same cause, but they may both be linked to the same aspect of structure of hydrated cement paste. It should not be forgotten that concrete cured and loaded at a constant relative humidity exhibits creep, and that creep

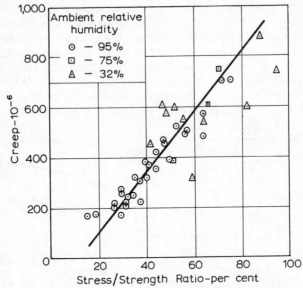

Fig. 6.29. *Creep of mortar specimens cured and stored continuously at different humidities*[6.45]

produces no significant loss of water from the concrete to the surrounding medium;[6.48,6.49] nor is there any gain in weight during creep recovery.*[6.49]

A further indication of the interrelation between shrinkage and creep is given in Fig. 6.30. Specimens which had been loaded for 600 days and then unloaded and allowed to recover their creep, exhibited on subsequent immersion in water swelling proportional to the stress which had been removed over two years previously. The residual deformation after swelling shows a similar proportionality.

Fig. 6.31 shows time deformation of loaded specimens stored alternately in water and in air with a relative humidity of 50 per cent. The ordinates represent the change in deformation from that existing after 600 days under load in air. It can be seen that while in water the loaded specimens show creep relative to the swelling of the unloaded specimen, but in air the change in deformation of all specimens is the same. The increase in creep on immersion of this old concrete in water may be due to the breaking of some of the bonds formed during the period of drying (cf. p. 337). Fig. 6.32 shows the data of Fig. 6.31 plotted as a deformation relative to the deformation of the unloaded specimen. A practical conclusion from these observations is that alternating wetting and drying

---

* Some increase in weight during the period of creep or recovery may be due to carbonation (*see* page 339).

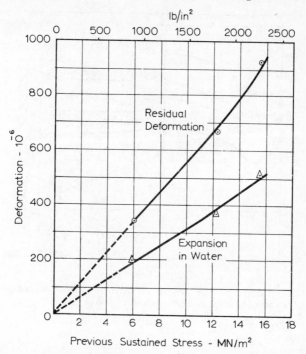

Fig. 6.30. *Relation between original sustained stress and (a) expansion in water, and (b) residual deformation of concrete*[6.50]

increases the magnitude of creep, so that results of laboratory tests may underestimate the creep under normal weather conditions.

As we have seen, creep and shrinkage are not simply additive; it may often be convenient to consider the total time deformation of specimens stored at a *constant* relative humidity. This deformation is proportional to the stress applied, except that for very low stresses or a zero stress the deformation is equal to shrinkage (Fig. 6.33). According to L'Hermite,[6.51] the shrinkage introduces internal stresses which are compensated by stresses induced by the applied load, so that below a certain stress creep is no greater than shrinkage of an unloaded specimen. The realization that shrinkage is not merely a strain but also a three-dimensional system of stress is of considerable importance.

In many tests a direct proportionality between creep and the applied stress has been found to exist[6.50] with a possible exception of specimens loaded at an early age. What is not certain is the upper limit of the relation. (The lower limit is virtually at zero stress as creep is exhibited

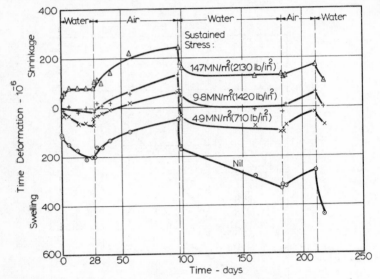

Fig. 6.31. *Time deformation of concrete subjected to different stresses and stored alternately in water and air at a relative humidity of 50 per cent*[6.14]

Strains at origin of time (after 600 days under load in air)—

stress	MN/m²	lb/in²		strain,	10^{-6}
	0	0			280
	4·9	710			1,000
	9·8	1,420			1,800
	14·7	2,130			2,900

by concrete even at low stresses.) In terms of the stress/strength ratio, an upper limit between about 0·3 and 0·75 has been suggested.

It is known that microcracking takes place in a concrete compression specimen at a stress/strength ratio of 0·4 to 0·6, and it is not surprising that, once the cracking has started, the creep behaviour also changes. It is possible that the onset of cracking depends on the degree of heterogeneity of the concrete; for instance, mortars are less grossly heterogeneous than concrete containing a large-size aggregate and exhibit proportionality between creep and stress/strength ratio up to a higher limit, possibly 0·85.[6.52]

It appears safe to conclude that within the range of working stresses the proportionality between creep and stress holds good, and creep expressions and prediction diagrams (considered on p. 364) assume this to be the case.

Above the limit of proportionality, creep increases with an increase in stress at an increasing rate, and there exists a stress/strength ratio above which creep produces time failure. This stress/strength ratio is in the

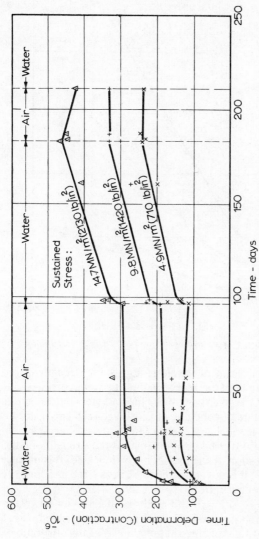

Fig. 6.32. *Time deformation of loaded specimens of Fig. 6.31 plotted relative to the strain of the unloaded specimen*[6.14]

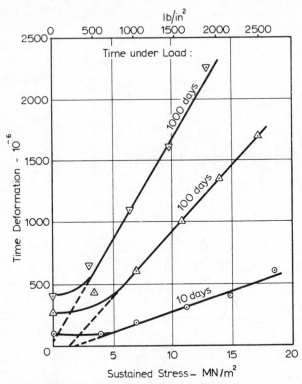

Fig. 6.33. *Time deformation (creep plus shrinkage) of concrete stored in air*[6.43]

region of 0·8 to 0·9 in terms of the short-term static strength. Creep increases the total strain until this reaches a limiting value corresponding to the ultimate strain of the given concrete. This statement implies a limiting strain concept of failure of concrete, at least in the cement paste.

The strength of concrete has a considerable influence on creep: within a wide range, creep is inversely proportional to the strength of concrete at the time of application of the load. This is indicated, for instance, in the data of Table 6.5. It is thus possible to express creep as a linear function of the stress/strength ratio[6.52] (Fig. 6.29).

Since for a *given* mix, strength and modulus of elasticity are related to one another, creep and modulus of elasticity are also related. Fig. 6.34 shows experimental values of creep at any time, t, against the ratio of the modulus of elasticity at the time t to the modulus at the time of application of the load;[6.42] the ages at which the load was applied and at which creep was determined varied widely, but one mix only was used. The modulus at the time of application of the load gives an indication of the

Table 6.5: *Ultimate Specific Creep of Concretes of Different Strengths Loaded at the Age of 7 Days*

Compressive strength of concrete		Ultimate specific creep		Product of specific creep and strength
MN/m²	lb/in²	10^{-6} per MN/m²	10^{-6} per lb/in²	10^{-3}
14	2,000	203	1·40	2·8
28	4,000	116	0·80	3·2
41	6,000	80	0·55	3·3
55	8,000	58	0·40	3·2

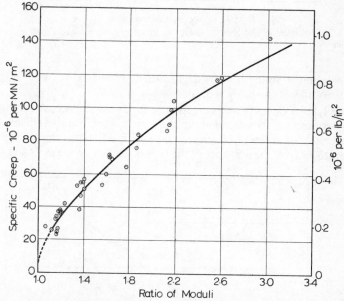

Fig. 6.34. *Relation between creep at any time t and ratio of the modulus of elasticity of concrete at time t to the modulus at the time of application of the load; various concretes, ages at loading, and periods under load*[6.42]

strength at that time, and the increase in the modulus at that time reflects the duration of the load.

From the influence of strength on creep it follows that creep under a given stress is closely related to the water/cement ratio of the mix, but for the same stress/strength ratio, creep is sensibly independent of the water/cement ratio.

The age at which the load is applied greatly affects the magnitude of creep (Fig. 6.35), the influence of age arising probably mainly from an

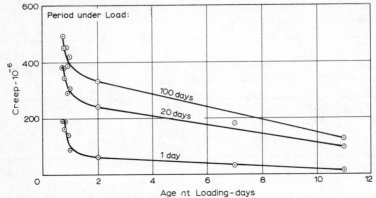

Fig. 6.35. *Influence of the age at loading on creep of concrete made with aluminous cement*[6.54]

increase in the strength of concrete with age.[6.54] For the same reason, maturity affects creep,[6.55] as shown in Fig. 6.36.

The type of cement affects creep in so far as it influences the strength of the concrete at the time of application of the load. For this reason any comparison of creep of concretes made with different cements should take into account the influence of the type of cement on the early strength

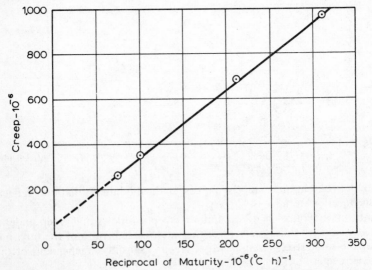

Fig. 6.36. *Influence of maturity of concrete (loaded at the age of 7 days) on creep after 420 days under load*

of concrete. On this basis, both Portland cements of different types and aluminous cement lead to sensibly the same creep,[6.54,6.56] but the rate of gain of strength has some effect as shown on page 354. This applies to creep both in air and in water,[6.54] contrary to some earlier data (Fig. 6.37). Portland blast-furnace cement results in a higher creep than the standard types of Portland cement.[6.57]

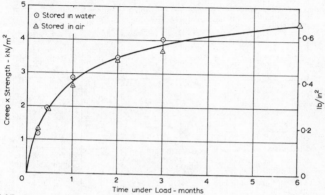

Fig. 6.37. *Creep of aluminous cement concrete stored wet and dry, adjusted for strength of concrete*[6.54]

Fineness of cement affects the strength development at early ages and thus influences creep. It does not seem, however, that fineness *per se* is a factor in creep: many of the contradictory results may be due to the indirect influence of gypsum. The finer the cement the higher its gypsum requirement, so that re-grinding a cement in the laboratory without the addition of gypsum produces an improperly retarded cement, which exhibits high shrinkage and high creep.[6.28]

Recently, studies on the influence of fineness of cement on creep of concrete have been extended to ultra high early strength Portland cement (*see* p. 65) whose fineness is above 700 m²/kg. The creep tests were done at a relative humidity of 55 per cent, all mixes having the same water/cement ratio and the same stress/strength ratio of 0·5. The age at loading was chosen so that all the concretes had the same strength. Figure 6.38 shows that, although the creep with the finest cement was at first greatest, it became least after 1,000 days. This is probably due to the high gain of strength of the finest cement with a resultant rapid drop in the actual stress/strength ratio.[6.99]

The change in strength of concrete while under load is of importance in evaluating the preceding statement that creep is not influenced by the type of cement. For the same stress/strength ratio at the time of application of load, creep is smaller the greater the relative increase in strength beyond the time of application of load.[6.99] Thus creep increases in order for low heat, ordinary, and rapid hardening cements. There is no doubt, however, that for a constant applied stress (not stress/strength ratio) at a fixed (early) age, creep increases in order for rapid hardening, ordinary, and low heat cements. These two statements bring out clearly the need for a full qualification of information about factors in creep.

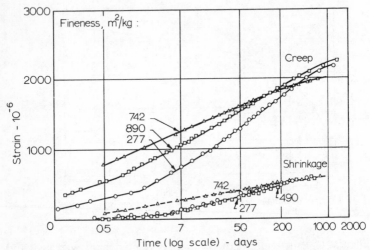

Fig. 6.38. *Creep and shrinkage of concretes made with cements of different fineness. (All concretes had the same mix proportions; in the case of creep specimens, the age was chosen so that the stress–strength ratio was 0·5 for all the concretes)*[6.73]

Water-reducing and set-retarding admixtures have been found to increase creep in many but not all cases[6.103,6.104] (*see* Fig. 6.39). Thus if creep is of importance in a given structure, the influence of any admixture to be used should be carefully checked.

Let us now consider the influence of aggregate on creep.

The usual normal weight aggregate is not liable to creep to an appreciable extent,* so that it is reasonable to assume that the seat of creep is in the cement paste, but the aggregate influences the creep of concrete through a restraining effect similar to that in the case of shrinkage (*see* p. 326) and through some physical properties of the parent rock.

The restraining effect arises from the fact that the cement paste is subject to creep and the aggregate generally is not, so that the effect of the aggregate is to reduce the effective creep of concrete. Creep is thus a

---

* Some aggregates are, however, liable to creep at stresses of no more than several hundred pounds per square inch: McHenry[6.64] demonstrated this for a volcanic agglomerate; the Bureau of Reclamation reported creep of the Glen Canyon sandstone[6.100] and Taiwan greywacke.[6.101]

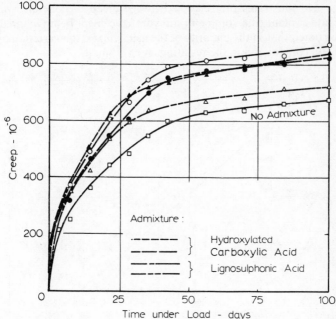

Fig. 6.39. *Influence of water-reducing and set-retarding admixtures on creep of concrete (water/cement ratio = 0·65; age at loading = 28 days; relative humidity of storage = 94 per cent)*[6.103]

function of the volumetric content of cement paste in concrete, but the relation is not linear. It has been shown[6.76] that creep of concrete, c, the volumetric content of aggregate, g, and the volumetric content of unhydrated cement u, are related by:

$$\log \frac{c_p}{c} = \alpha \log \frac{1}{1 - g - u}$$

where c_p is creep of neat cement paste of the same quality as used in concrete, and

$$\alpha = \frac{3(1 - \mu)}{1 + \mu + 2(1 - 2\mu_a)\dfrac{E}{E_a}}$$

Here, μ_a = Poisson's ratio of aggregate, μ = Poisson's ratio of surrounding material (concrete), E_a = modulus of elasticity of aggregate, and E = modulus of elasticity of the surrounding material. This relation applies to concrete made both with normal weight aggregate and lightweight aggregate.[6.102]

Fig. 6.40 illustrates the relation between creep of concrete and its aggregate content (the volume of unhydrated cement being ignored). It may be noted that in the majority of the usual mixes, the variation in the aggregate content and in the resulting creep is small.

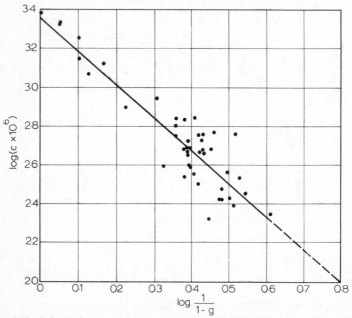

Fig. 6.40. *Relation between creep c after 28 days under load and content of aggregate g for wet-stored specimens loaded at the age of 14 days to a stress/strength ratio of 0·50*[6.76]

The grading, maximum size, and shape of the aggregate have been suggested as factors in creep. It is now believed, however, that their main influence lies in the effect that these properties have directly or indirectly on the aggregate content,[6.76] providing that full consolidation of concrete has been achieved in all cases.

There are, however, certain physical properties of aggregate which influence the creep of concrete. The modulus of elasticity of aggregate is probably the most important factor. The higher the modulus the greater the restraint offered by the aggregate to the potential creep of the cement paste; this is evident from the equation for α on page 355.

Porosity of aggregate has also been found to influence the creep of concrete but, since aggregates with a higher porosity generally have a lower modulus of elasticity, it is possible that porosity is not an independent factor in creep. On the other hand, it can be visualized that the porosity of aggregate, and even more so its absorption, play a direct rôle in the transfer of moisture within concrete; this transfer may be associated with creep.

Because of the great variation in aggregate within any mineralogical and petrological type, it is not possible to make a general statement about the magnitude of creep of concrete made with aggregates of different types. However, the data of Fig. 6.41 are of considerable importance:

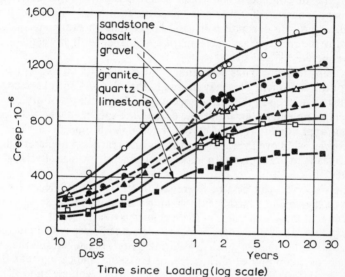

Fig. 6.41. *Creep of concretes of fixed proportions but made with different aggregates, loaded at the age of 28 days, and stored in air at 21°C (70°F) and a relative humidity of 50 per cent*[6.24]

after 20 years' storage at a relative humidity of 50 per cent, concrete made with sandstone aggregate exhibited creep more than twice as great as concrete made with limestone. An even greater difference between the creep strains of concretes made with different aggregates was found by Rüsch *et al.*[6.105] After 18 months under load at a relative humidity of 65 per cent, the maximum creep was five times the minimum value, the aggregates in the increasing order of creep being: basalt; quartz; gravel, marble and granite; and sandstone.

Lightweight aggregate deserves a special mention because of the rather common belief that its use leads to substantially higher creep than normal weight aggregate. Recent work[6.106] indicates that there is no fundamental difference between normal and lightweight aggregates as far as the creep properties are concerned, and the higher creep of concretes made with lightweight aggregates reflects only the lower modulus of elasticity of the aggregate. There is no inherent difference in the behaviour of coated and uncoated aggregates or between those obtained by different manufacturing processes, but this of course does not mean that all the aggregates lead to the same creep.

As a general rule, it can be stated that creep of structural quality lightweight aggregate concrete is about the same as that of concrete made with ordinary aggregate. (It is important in any comparison that the aggregate contents do not differ widely between the lightweight and ordinary concretes.) Furthermore, since the elastic deformation of lightweight aggregate concrete is usually larger than in ordinary concrete, the ratio of creep to elastic deformation is smaller for lightweight aggregate concrete.[6.106]

In many tests creep has been found to decrease with an increase in the size of the specimen. This may be due to the effects of shrinkage and to the fact that creep at the surface occurs under conditions of drying and is therefore greater than within the core of the specimen where the conditions approximate to mass curing. Even if, with time, drying reaches the core, it will have hydrated extensively and reached a higher strength, which leads to lower creep. In mass cured concrete no size effects can be present.

The size effect can best be expressed in terms of the surface/volume ratio of the concrete member: the relation is shown in Fig. 6.42. It can be seen that the actual shape of the specimen is of even lesser importance than in the case of shrinkage. Also, the decrease in creep with an increase in size is smaller than in the case of shrinkage (cf. Fig. 6.20). But the rates of gain of creep and of shrinkage are the same, indicating that both phenomena are the same function of the surface/volume ratio. These data apply to shrinkage and creep at 50 per cent relative humidity.[6.92]

The influence of temperature on creep has become of increased interest in connection with the use of concrete in the construction of prestressed concrete nuclear pressure vessels but the problem is of significance also in other types of structures, e.g. bridges. The rate of creep increases with temperature up to about 70°C (160°F) when, at least for a 1:7 mix with a water/cement ratio of 0·6, it is approximately 3·5 times higher than at 21°C (70°F). Between 70°C (160°F) and 96°C (205°F) the rate drops off to 1·7 times the rate at 21°C (70°F).[6.107] These differences in rate persist at least for 15 months under load. Figure 6.43 illustrates the progress of creep. This behaviour is believed to be due to desorption of water from the surface of the gel so that gradually the gel itself becomes the sole phase subject to molecular diffusion and shear flow; consequently the rate of creep decreases. The behaviour over a wide range of temperatures is shown in Fig. 6.44. Freezing produces a higher initial rate of creep but it quickly drops to zero.[6.109]

Practically all test data on creep have been obtained under a sustained

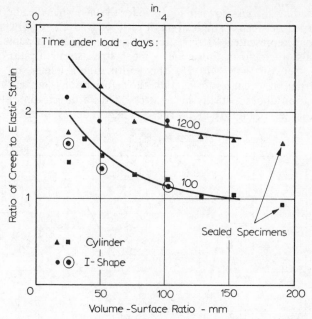

Fig. 6.42. *Relation between ratio of creep to elastic strain and volume/surface ratio*[6.92]

constant stress. Concrete subjected to rapidly alternating loading and unloading also exhibits a progressive increase in deformation (Fig. 6.45). It is important to note that rapid alternating loading (at least 10 cycles per minute) with a maximum stress σ_h produces a higher deformation than a static load of intensity σ_h acting during the same total time as the

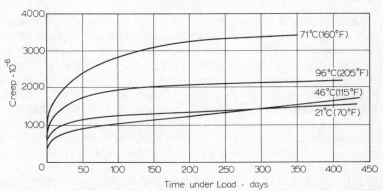

Fig. 6.43. *Relation between creep and time under load for concretes stored at different temperatures (stress/strength ratio of 0·70)*

360   *Properties of Concrete*

cyclic loading. This is shown in Fig. 6.46 for the case when the alternating load varied between a stress/strength ratio of 0·35 and 0·05 while the static load represented a stress/strength ratio of 0·35. The same figure shows also the deformation under a mean stress/strength ratio of 0·35 (varying between 0·45 and 0·25): the deformation is higher still. When the alternations occur less frequently, the deformation is higher than the deformation under a steady mean stress. Thus a direct application of laboratory test results to actual structures may under-estimate creep.[6.110]

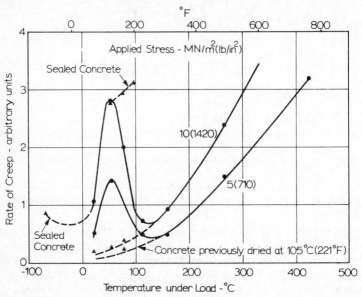

Fig. 6.44. *Influence of temperature on rate of creep*[6.108]

The preceding discussion referred to uniaxial compression but creep also occurs in other loading situations and information about creep behaviour under these conditions is especially helpful in establishing the nature of creep and in some design problems. Unfortunately, experimental data are limited,[6.42] and in many cases quantitative evaluation and comparison with the behaviour in compression are not possible. For this reason no more than broad qualitative statements will be made.

Creep of mass concrete in uniaxial tension is 20 to 30 per cent higher than under a compressive stress of equal magnitude. The difference may be as high as 100 per cent for storage at a relative humidity of 50 per cent. The shape of the creep–time curves in tension and in compression is similar. Drying enhances creep in tension as well as creep in compression.

Creep occurs under torsional loading, and is affected by stress, water/cement ratio, and ambient relative humidity in qualitatively the same

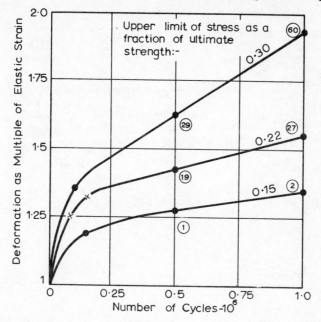

Fig. 6.45. *Increase in strain due to an alternating load applied at the rate of 500 cycles per minute*[6.58]

Ringed figures denote the period under sustained loading (in days) after which the same deformation is reached.

manner as creep in compression. The creep–time curve is also of the same shape.[6.43] The ratio of creep to elastic deformation in torsion was found to be the same as for compressive loading.[6.111]

Under uniaxial compression creep occurs not only in the axial direction but also in the normal directions. This is referred to as lateral creep. The resulting creep Poisson's ratio was considered on page 321. From the fact that there is lateral creep induced by an axial stress, it follows that, under multiaxial stress, in any direction there is creep due to the stress applied in that direction and also creep due to the Poisson's ratio effect of creep strains in the two normal directions. There is evidence[6.22] that the superposition of creep strains due to each stress separately is not valid, so that creep under multiaxial stress cannot be simply predicted from uniaxial creep measurements. Specifically, creep under multiaxial compression is less than under a uniaxial compression of the same magnitude in the given direction (Fig. 6.47). But even under hydrostatic compression there is considerable creep.

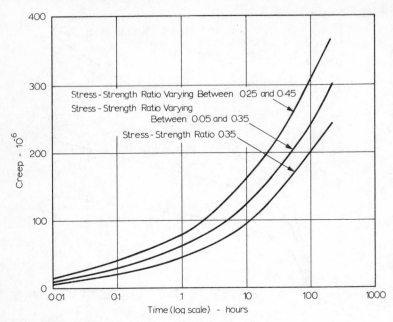

Fig. 6.46. *Creep under alternating and static loading*

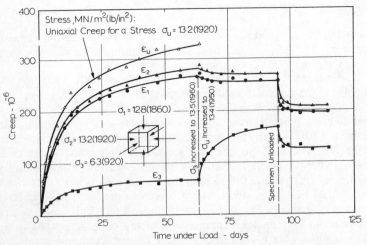

Fig. 6.47. *Typical creep–time curves for concrete under triaxial compression*

Relation between Creep and Time

Creep is usually determined by measuring the change with time in the strain of a specimen subjected to a constant stress and stored under appropriate conditions. A suitable testing device is shown in Fig. 6.48, the spring ensuring that the load is sensibly constant despite the contraction of the specimen.

Under such conditions, creep continues for a very long time, the longest determination to date indicating that a small increase in creep takes place after as long as 30 years[6.24] (Fig. 6.49). The rate of creep decreases, however, at a continuous rate, and it is generally assumed that creep tends to a limiting value after an infinite time under load; this has not, however, been proved.

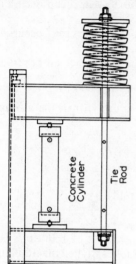

Fig. 6.48. *A loading device for the measurement of creep under a constant stress*[6.40]

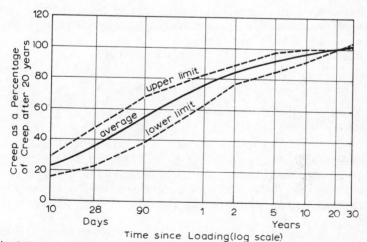

Fig. 6.49. *Range of creep–time curves for different concretes stored at various relative humidities*[6.24]

Fig. 6.49 gives Troxell, Raphael and Davis's[6.24] long-term measurements, and it can be seen that—

18 to 35 per cent (average 26 per cent) of the 20-year creep occurs in 2 weeks;

40 to 70 per cent (average 55 per cent) of the 20-year creep occurs in 3 months; and

64 to 83 per cent (average 76 per cent) of the 20-year creep occurs in 1 year.

If creep after 1 year under load is taken as unity, then the average values of creep at later ages are—

> 1·14 after 2 years;
> 1·20 after 5 years;
> 1·26 after 10 years;
> 1·33 after 20 years; and
> 1·36 after 30 years.

These values show that ultimate creep is in excess of 1·36 times the one-year creep, although for calculations it is often assumed that ultimate creep is equal to $\frac{4}{3}$ of the one-year creep.[6.59] This estimate is reliable within ± 15 per cent for concretes loaded at early ages. Estimating ultimate creep from values at lower ages is more difficult. For instance, creep after 20 years is between 1·5 and 3·0 times the creep after $2\frac{1}{2}$ months under load.[6.24]

Numerous mathematical expressions relating creep and time have been suggested. One of the most convenient is the hyperbolic expression, introduced by Ross[6.53] and Lorman.[6.31]

Ross expresses creep c after time t under load as

$$c = \frac{t}{a + bt}$$

When $t = \infty$, $c = 1/b$, i.e. $1/b$ is the limiting value.

a and b are constants determined from experimental results: by plotting t/c against t a straight line of slope b is obtained, and the intercept on the t/c axis is equal to a. The straight line should be drawn so as to pass through the points at later ages, there being generally some deviation from the straight line during the early period after the application of the load.

The U.S. Bureau of Reclamation, which has made an extensive study of creep of concrete in dams, has found that creep can be represented by an expression of the type

$$c = F(K) \log_e (t + 1)$$

where K is the age at which the load is applied,

   $F(K)$ is a function representing the rate of creep deformation with time,

   and t is time under load, in days.

$F(K)$ is obtained from a plot on semi-logarithmic paper.

Fig. 6.50 shows a comparison of experimental data with the calculated values of total deformation, ξ, i.e. creep c plus elastic deformation $1/E$, for a unit stress of 1 MN/m^2 or 1 lb/in^2.[6.60] Creep under unit stress is referred to as specific creep.

The rate of creep is usually considered in the form[6.61]

$$\frac{dc}{dt} = (c_\infty - c)K$$

where c_∞ is ultimate creep.

Here the rate of creep at any time is proportional to the creep still to appear—a characteristic of visco-elastic flow.

The various empirical expressions, be they hyperbolic, logarithmic, exponential, or power functions, hold only for concretes similar to those for which the formulae have been derived, and a generally valid expression is not available.

Broad estimates can nevertheless be made on the basis of accumulated data, and are indeed necessary for design purposes. An estimate of this type can be made using the data of the Comité Européen du Béton,[6.112] which recommends the use of a creep coefficient ϕ, whereby creep c is expressed as a multiple of the strain on loading, e:

$$c = \phi e$$

The creep coefficient depends on factors allowing for: the composition of the mix k_b, effective thickness of member (cross-section divided by exposed semi-perimeter) k_e, relative humidity of storage k_c, age at loading k_d, and variation with time under load k_t (Fig. 6.51). The estimated creep applies only to the usual range of mixes at stress/strength ratios not exceeding 0.35. Creep is then calculated as:

$$c = e \, k_c \, k_d \, k_b \, k_e \, k_t.$$

If the curing temperature is higher than normal (20°C (68°F)), so that a greater fraction of strength is achieved at a given chronological age, a fictitious age based on the maturity of the concrete should be used.

A modification of this approach recognizes that the relative increase in creep with time depends on the size of the member; accordingly, Fig. 6.52 should be used instead of the relevant part of Fig. 6.51.

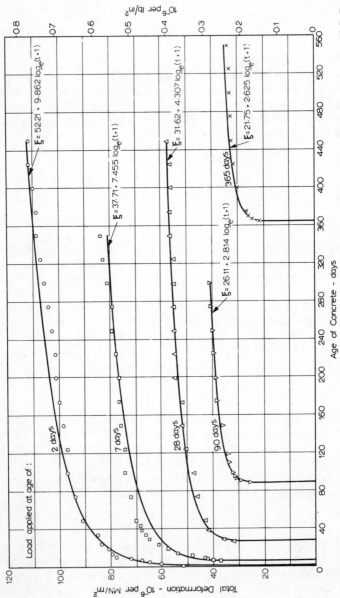

Fig. 6.50. Observed and calculated total deformation (elastic strain plus creep) in Canyon Ferry dam concrete, using U.S. Bureau of Reclamation formula[6.60]

It should be noted that the expression of the Comité Européen du Béton makes no provision for the influence of the type of normal weight aggregate used. An allowance for this is recommended.

Prediction on the basis of the data of Fig. 6.51 or of similar charts[6.62] of Wagner's yields results which are sufficiently accurate for design purposes when the structure is not very sensitive to creep. In other cases, a

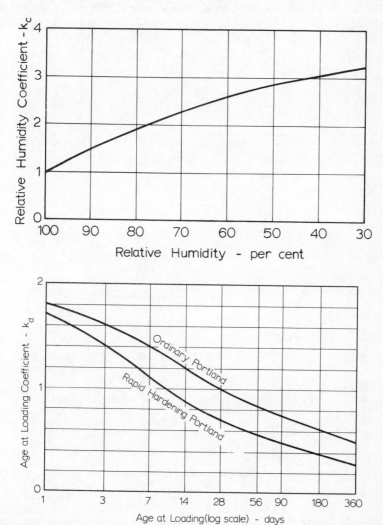

Fig. 6.51. *The Comité Européen du Béton (European Concrete Committee) creep prediction curves*[6.112]

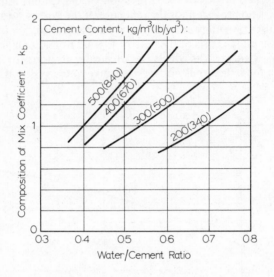

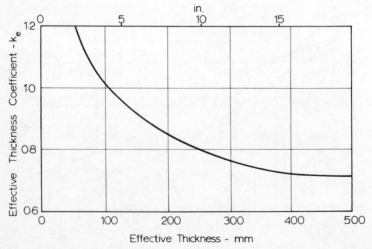

Fig. 6.51. *Continued*

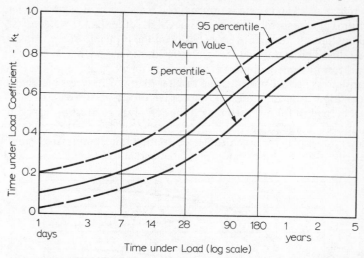

Fig. 6.51. *Continued*

creep test of at least a 60-day duration should be made; extrapolation of creep by an expression such as that of Ross's (*see* p. 364) will generally give satisfactory results.

Nature of Creep

From Fig. 6.27 it is apparent that creep and creep recovery are related phenomena, but their nature is far from clear. Two views are of interest. Dutron[6.63] does not regard recovery as an elastic phenomenon but merely as a manifestation of a slight swelling of the cement paste released

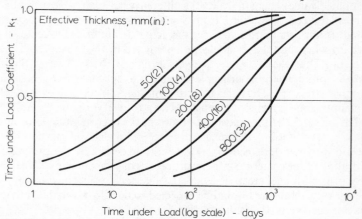

Fig. 6.52. *Modification of the coefficient for time under load, k_t, in Fig. 6.51*

from load as the concrete is returning to the state of hygrometric equilibrium with the surrounding medium. On the other hand, McHenry's[6.64] explanation of creep recovery is based on the principle of superposition of strains. This states that the strains produced in concrete at any time t by a stress increment applied at any time t_o are independent of the effects of any stress applied either earlier or later than t_o. The stress increment is understood to mean either a compressive or a tensile stress, i.e. also a relief of load. It follows then that, if the compressive stress on a specimen is removed at age t_1, the resulting creep recovery will be the same as the creep of a similar specimen subjected to the same compressive stress at the age t_1. Fig. 6.53 illustrates this statement, and it can be seen that the

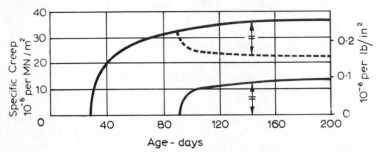

Fig. 6.53. *Example of McHenry's*[6.64] *principle of superposition of strains*

creep recovery is represented by the difference between the actual strain at any time and the strain that would exist at the same time had the specimen continued to be subjected to the original compressive stress.

Fig. 6.54 shows a comparison of actual and computed strains (the computed values being in reality the difference between two experimental curves) for concrete sealed from the surrounding medium, i.e. mass-cured. It appears that in all cases the actual strain after the removal of load is higher than the residual strain predicted by the principle of superposition. Thus the actual creep recovery is less than expected. Similar error is found when the principle is applied to specimens subjected to a variable stress.[6.41] Thus the principle of superposition does not fully explain the phenomena of creep and creep recovery.

The principle of superposition of strains is nevertheless a convenient working assumption. It implies that creep is a delayed elastic phenomenon in which full recovery is generally impeded by the progressive hydration of the cement. Since the properties of old concrete change only very little with age, creep of concrete subjected to sustained loading at the age of several years would be expected to be fully reversible, but this has not been verified experimentally.

The problem of the nature of creep is still controversial[6.66] and cannot

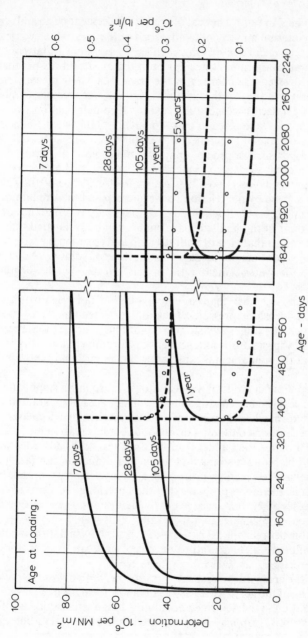

Fig. 6.54. Comparison of measured and computed strains on the basis of McHenry's principle of superposition[6.65]

be discussed here in full.* The seat of creep is the cement paste and creep is related to internal movement of adsorbed or intracrystalline water, i.e. to internal seepage. Tests of Glucklich's[6.74] have shown that concrete from which all evaporable water has been removed exhibits practically no creep. However, the changes in the creep behaviour of concrete at high temperatures suggest that at that stage the water ceases to play a rôle and the gel itself becomes subjected to creep-deformation.

Since creep can take place in mass concrete it follows that seepage of water to the outside of concrete is not essential to the progress of basic creep, although such a process may well take place in drying creep. However, internal seepage of water from the adsorbed layers to voids such as capillary voids is possible. An indirect evidence of the rôle of such voids is given by the relation between creep and the strength of the hydrated cement paste: it would appear that creep is a function of the relative amount of the unfilled space, and it can be speculated that it is the voids in the gel that govern both strength and creep; in the latter case, the voids may be related to seepage. The volume of voids is, of course, a function of the water/cement ratio and is affected by the degree of hydration.

We should remember that capillary voids do not remain full even against full hydrostatic pressure of the ambient medium. Thus internal seepage is possible under any storage conditions. The fact that creep of non-shrinking specimens is independent of the ambient humidity would indicate that the fundamental cause of creep "in air" and "in water" is the same.

The creep–time curve shows a definite decrease in its slope, and the question arises whether this signifies a change, possibly a gradual one, in the mechanism of creep. It is conceivable that the slope decreases with the same mechanism continuing throughout but it is reasonable to imagine that after many years under load the thickness of the adsorbed water layers could be reduced so far that no further reduction can take place under the same stress, and yet creep after as many as thirty years has been recorded. It is, therefore, probable that the slow, long-term part of creep is due to causes other than seepage but the deformation can develop only in the presence of some evaporable water. This would suggest viscous flow or sliding between the gel particles. Such mechanisms are compatible with the influence of temperature on creep, and can explain also the largely irreversible character of long-term creep.

Some of the creep due to seepage is reversible, but only some, because of coarsening of the gel particles associated with formation of new bonds and stabilization in the deformed position. At high stresses, a part of the overall measured creep may be due to growth in microcracks but under

---

* See A. M. Neville, *Creep of Concrete: Plain, Reinforced and Prestressed* (Amsterdam, North-Holland Publishing Co., 1970).

working loads a significant contribution of microcracking to creep is unlikely.

This hypothesis is formulated in phenomenological terms only but, because we are dealing with a phase whose properties are not properly quantitatively described and because it is not certain that the laws of thermodynamics apply to layers of adsorbed water only one or two molecules thick, it is not at present possible to express the phenomena involved by energy equations.

Thus we have to admit that the exact mechanism of creep is still unknown, but it is apparent that creep, and indeed many mechanical properties of cement paste, depend largely on a grosser structure of colloidal dimensions and only indirectly on the chemical constitution.

Rheological Models

Although the nature of creep is still uncertain, its partly reversible character suggests that the deformation may consist of a partly reversible visco-elastic movement (consisting of a purely viscous phase and a purely elastic phase) and possibly also a non-reversible plastic deformation.

An *elastic* deformation is always recoverable on unloading. A *plastic* deformation is never recoverable, can be time-dependent, and there is no proportionality between plastic strain and the applied stress, or between stress and rate of strain. A *viscous* deformation is never recoverable on unloading, is always time-dependent, and there is always proportionality between the rate of viscous strain and the applied stress, and hence between stress and strain at a given time.[6.67] These various types of deformation can be summarized as follows—

Type of deformation	Instantaneous	Time-dependent
reversible	elastic	delayed-elastic
irreversible	plastic set	viscous

Consideration of these types of deformation has led to the conception of rheological models—fictitious devices, combining ideal springs, dashpots, and sometimes friction and non-return valves. For each spring the deformation ε is proportional to the stress applied, f, and for a dashpot the rate of deformation ε_c at time t is given by

$$\frac{d\varepsilon_c}{dt} = f \times \text{(a constant)}$$

Fig. 6.55 illustrates one of the models, suggested by Hansen.[6.67]

The model gives a phenomenological description of the response of the concrete, but says nothing about the actual mechanism of creep.

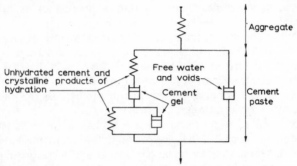

Fig. 6.55. *A rheological model of creep of concrete*[6.67]

Effects of Creep

Creep affects strains and deflexions and often also stress distribution, but the effects vary with the type of structure.[6.68]

Creep of plain concrete does not *per se* affect the strength although under *very* high stresses creep hastens the approach of the limiting strain at which failure takes place; this applies only when the sustained load is above 85 or 90 per cent of the rapidly applied static ultimate load.[6.52] Under a low sustained stress the volume of concrete decreases (since the creep Poisson's ratio is less than 0·5) and this would be expected to increase the strength of the concrete. However, this effect is probably small.

The influence of creep on the ultimate strength of a simply supported reinforced concrete beam subjected to a sustained load is not significant, but the deflexion increases considerably and may in many cases be a critical consideration in design. According to Glanville and Thomas,[6.69] there are two distinct neutral surfaces in a beam subjected to sustained loading: one of zero stress, the other of zero strain. This arises from the fact that an increase in the strain in concrete leads to an increased stress in the steel and a consequent lowering of the neutral axis when an increasing depth of concrete is brought into compression. As a result the elastic strain distribution changes, but the creep strain is not cancelled out, so that at the level of the new stress-neutral-axis a residual tensile strain will remain. At some level above this axis there is a fibre of zero strain at any time although there is a stress acting. This is an interesting example of the influence of the stress history on strain at any time.

In reinforced concrete columns, creep results in a gradual transfer of load from the concrete to the reinforcement. Once the steel yields, any increase in load is taken by the concrete, so that the full strength of both the steel and the concrete is developed before failure takes place—a fact recognized by the design formulae. In statically indeterminate structures creep may relieve stress concentrations induced by shrinkage, temperature changes, or movement of supports. In all concrete structures creep reduces internal stresses due to non-uniform shrinkage so that there is a reduction in cracking. In calculating creep effects in structures it is important to realize that the actual time-dependent deformation is not the "free" creep of concrete (as given in Fig. 6.51) but a value modified by the quantity and position of reinforcement.

On the other hand, in mass concrete, creep in itself may be a cause of cracking when a restrained concrete mass undergoes a cycle of temperature change due to the development of the heat of hydration and subsequent cooling. Creep relieves the compressive stress induced by the rapid rise in temperature so that the remaining compression disappears as soon as some cooling has taken place. On further cooling of concrete, tensile stresses develop and, since the rate of creep is reduced with age, cracking may occur even before the temperature has dropped to the initial (placing) value (Fig. 6.56). For this reason the rise in temperature in the

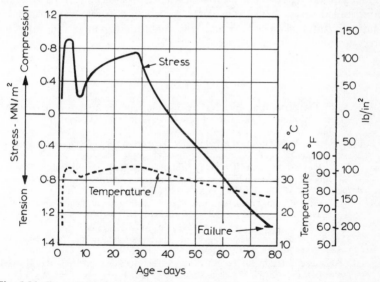

Fig. 6.56. *Stress in concrete subjected to a temperature cycle at a constant length*[6.70]

interior of a large concrete mass must be controlled by the use of low heat cement, a low cement content, precooling of mix ingredients, limiting the height of concrete lifts, and cooling of concrete by circulating refrigerated water through a network of pipes embedded in the concrete mass.

Another instance of the adverse effects of creep is its influence on the stability of the structure due to increase in deformation. Even when creep does not affect the ultimate strength of a structure its effects may be extremely serious as far as the performance of the structure is concerned. This is, for instance, the case in the foundation block of a very large turbo-generator: differential movement of the supports of this indeterminate structure due to creep and the change in the slope of the beams (which may be nearly 50 m (170 ft) long) due to shrinkage would upset the alignment of the shaft of the generator. Likewise, in very tall buildings, differential creep may cause movement and cracking of partitions and also structural effects in beams and slabs.[6.113]

The loss of prestress due to creep is well known and, indeed, accounts for the failure of all early attempts at prestressing. It was only the introduction of high tensile steel, whose elongation is several times the contraction of concrete due to creep and shrinkage, that made prestressing a successful proposition.

The effects of creep may thus be harmful but, on the whole, creep, unlike shrinkage, is beneficial in relieving stress concentrations and has contributed very considerably to the success of concrete as a structural material. Rational design methods allowing for creep in various types of structures have been developed.*

REFERENCES

6.1.   R. E. PHILLEO, Comparison of results of three methods for determining Young's modulus of elasticity of concrete, *J. Amer. Concr. Inst.*, **51**, pp. 461–69 (Jan. 1955).

6.2.   H. RÜSCH, Versuche zur Festigkeit der Biegedruckzone, *Deutscher Ausschuss für Stahlbeton No.* 120.

6.3.   R. E. DAVIS and G. E. TROXELL, Modulus of elasticity and Poisson's ratio for concrete and the influence of age and other factors upon these values, *Proc. A.S.T.M.*, **29**, Part II, pp. 678–710 (1929).

6.4.   D. O. WOOLF, Toughness, hardness, abrasion, strength, and elastic properties, *A.S.T.M. Sp. Tech. Publcn. No.* 169, pp. 314–24 (1956).

6.5.   L. W. TELLER, Elastic properties, *ibid.*, pp. 94–103 (1956).

6.6.   J. J. SHIDELER, Lightweight aggregate concrete for structural use, *J. Amer. Concr. Inst.*, **54**, pp. 299–328 (Oct. 1957).

6.7.   P. KLIEGER, Early high-strength concrete for prestressing, *Proc. World Conference on Prestressed Concrete* (San Francisco, 1957), pp. A5–1–A5–14.

6.8.   J. C. SAEMAN and G. W. WASHA, Variation of mortar and concrete properties with temperature, *J. Amer. Concr. Inst.*, **54**, pp. 385–95 (Nov. 1957).

6.9.   T. TAKABAYASHI, Comparison of dynamic Young's modulus and static Young's modulus for concrete, *R.I.L.E.M. Int. Symp. on Non-destructive Testing of Materials and Structures* (1954), **1**, pp. 34–44.

6.10. M. R. SHARMA and B. L. GUPTA, Sonic modulus as related to strength and static modulus of high strength concrete, *Indian Concrete J.*, **34**, No. 4, pp. 139–41 (April 1960).

---

* A. M. Neville, *Creep of Concrete: Plain, Reinforced, and Prestressed* (Amsterdam, North-Holland Publishing Co., 1970).

6.11. B. W. SHACKLOCK and P. W. KEENE, A comparison of the compressive and flexural strengths of concrete with and without entrained air, *Civil Engineering* (London), pp. 77–80 (Jan. 1959).

6.12. R. JONES, Testing concrete by an ultrasonic pulse technique, *D.S.I.R. Road Research Technical Paper No.* 34 (London, H.M.S.O., 1955).

6.13. M. A. SWAYZE, Early concrete volume changes and their control, *J. Amer. Concr. Inst.*, **38**, pp. 425–40 (April 1942).

6.14. R. L'HERMITE, Volume changes of concrete, *Proc. 4th Int. Symp. on the Chemistry of Cement*, Washington D.C., 1960, pp. 659–94.

6.15. W. LERCH, Plastic shrinkage, *J. Amer. Concr. Inst.*, **53**, pp. 797–802 (Feb. 1957).

6.16. B. W. SHACKLOCK, The early shrinkage characteristics of hand-placed concrete, *Mag. Concr. Res.*, **10**, No. 28, pp. 3–12 (March 1958).

6.17. H. E. DAVIS, Autogenous volume changes of concrete, *Proc. A.S.T.M.*, **40**, pp. 1103–10 (1940).

6.18. T. C. POWERS, Causes and control of volume change, *J. Portl. Cem. Assoc. Research and Development Laboratories*, **1**, No. 1, pp. 29–39 (Jan. 1959).

6.19. F. M. LEA, *The Chemistry of Cement and Concrete* (London, Arnold, 1970).

6.20. J. D. BERNAL, J. W. JEFFERY and H. F. W. TAYLOR, Crystallographic research on the hydration of Portland cement: A first report on investigations in progress, *Mag. Concr. Res.*, No. 11, pp. 49–54 (Oct. 1952).

6.21. J. D. BERNAL, The structures of cement hydration compounds, *Proc. 3rd Int. Symp. on the Chemistry of Cement*, London, 1952, pp. 216–36.

6.22. F. M. LEA, Cement research: Retrospect and Prospect, *Proc. 4th Int. Symp. on the Chemistry of Cement*, Washington D.C., 1960, pp. 5–8.

6.23. G. PICKETT, Effect of aggregate on shrinkage of concrete and hypothesis concerning shrinkage, *J. Amer. Concr. Inst.*, **52**, pp. 581–90 (Jan. 1956).

6.24. G. E. TROXELL, J. M. RAPHAEL and R. E. DAVIS, Long-time creep and shrinkage tests of plain and reinforced concrete, *Proc. A.S.T.M.*, **58**, pp. 1101–1120 (1958).

6.25. B. W. SHACKLOCK and P. W. KEENE, The effect of mix proportions and testing conditions on drying shrinkage and moisture movement of concrete, *Cement Concr. Assoc. Tech. Report TRA/266* (London, June 1957).

6.26. M. A. SWAYZE, Discussion on: Volume changes of concrete, *Proc. 4th Int. Symp. on the Chemistry of Cement*, Washington D.C., 1960, pp. 700–2.

6.27. G. PICKETT, Effect of gypsum content and other factors on shrinkage of concrete prisms, *J. Amer. Concr. Inst.*, **44**, pp. 149–75 (Oct. 1947).

6.28. W. LERCH, The influence of gypsum on the hydration and properties of Portland cement pastes, *Proc. A.S.T.M.*, **46**, pp. 1252–92 (1946).

6.29. P. W. KEENE, The effect of air-entrainment on the shrinkage of concrete stored in laboratory air, *Cement Concr. Assoc. Tech. Report TRA/331* (London, Jan. 1960).

6.30. J. J. SHIDELER, Calcium chloride in concrete, *J. Amer. Concr. Inst.*, **48**, pp. 537–59 (March 1952).

6.31. W. R. LORMAN, The theory of concrete creep, *Proc. A.S.T.M.*, **40**, pp. 1082–102 (1940).

6.32. A. D. ROSS, Shape, size, and shrinkage, *Concrete and Constructional Engineering*, pp. 193–99 (London, Aug. 1944).

6.33. R. L'HERMITE, J. CHEFDEVILLE and J. J. GRIEU, Nouvelle contribution à l'étude du retrait des ciments, *Annales de l'Institut Technique du Bâtiment et de Travaux Publics. No.* 106. Liants Hydrauliques No. 5 (Dec. 1949).

6.34. A. D. ROSS, Shrinkless and creepless concrete, *Civil Engineering*, **46**, No. 545, pp. 853–4 (London, Nov. 1951).

6.35. A. M. NEVILLE, Discussion on Effect of aggregate on shrinkage of concrete and hypothesis concerning shrinkage, *J. Amer. Concr. Inst.*, **52**, Part 2, pp. 1380–81 (Dec. 1956).

6.36. H. H. STEINOUR, Some effects of carbon dioxide on mortars and concrete—discussion, *J. Amer. Concr. Inst.*, **55**, pp. 905–7 (Feb. 1959).

6.37. G. J. VERBECK, Carbonation of hydrated Portland cement, *A.S.T.M. Sp. Tech. Publicn. No.* 205, pp. 17–36 (1958).

6.38. J. J. SHIDELER, Investigation of the moisture-volume stability of concrete masonry units, *Portl. Cem. Assoc. Development Bul. D.*3 (March 1955).

6.39. I. LEBER and F. A. BLAKEY, Some effects of carbon dioxide on mortars and concrete, *J. Amer. Concr. Inst.*, **53**, pp. 295–308 (Sept. 1956).

6.40. A. M. NEVILLE, The measurement of creep of mortar under fully-controlled conditions, *Mag. Concr. Res.*, **9**, No. 25, pp. 9–12 (March 1957).

6.41. A. D. ROSS, Creep of concrete under variable stress, *J. Amer. Concr. Inst.*, **54**, pp. 739–58 (March 1958).

6.42. U.S. BUREAU OF RECLAMATION, A 10-year study of creep properties of concrete, *Concrete Laboratory Report No. SP-38* (Denver, Colorado, 28th July 1953).

6.43. B. LE CAMUS, Recherches expérimentales sur la déformation du béton et du béton armé, *Comptes Rendues des Recherches des Laboratoires du Bâtiment et des Travaux Publics* (Paris, 1945–46).

6.44. A. M. NEVILLE, Creep recovery of mortars made with different cements, *J. Amer. Concr. Inst.*, **56**, pp. 167–74 (Aug. 1959).

6.45. A. M. NEVILLE, Tests on the influence of the properties of cement on the creep of mortar, *R.I.L.E.M. Bul. No.* 4, pp. 5–17 (Oct. 1959).

6.46. M. ROŠ, Vorgespannter Beton, *E.M.P.A. Bericht. No.* 155 (Zurich, 1946).

6.47. R. L'HERMITE, Idées actuelles sur la technologie du béton, *Documentation Technique du Bâtiment et des Travaux Publics* (Paris 1955).

6.48. G. A. MANEY, Concrete under sustained working loads; evidence that shrinkage dominates time yield, *Proc. A.S.T.M.*, **41**, pp. 1021–30 (1941).

6.49. A. M. NEVILLE, Recovery of creep and observations on the mechanism of creep of concrete, *Applied Scientific Research*, Section A, **9**, pp. 71–84 (The Hague, 1960).

6.50. A. M. NEVILLE, The relation between creep of concrete and the stress–strength ratio, *ibid.*, pp. 285–92.

6.51. R. L'HERMITE, What do we know about the plastic deformation and creep of concrete? *R.I.L.E.M. Bul. No.* 1, pp. 21–51 (Paris, March 1959).

6.52. A. M. NEVILLE, Rôle of cement in the creep of mortar, *J. Amer. Concr. Inst.*, **55**, pp. 963–84 (March 1959).

6.53. A. D. ROSS, Concrete creep data, *The Struct. E.*, **15**, pp. 314–26 (London, 1937).

6.54. A. M. NEVILLE and H. W. KENINGTON, Creep of aluminous cement concrete, *Proc. 4th Int. Symp. on the Chemistry of Cement*, Washington D.C., 1960 pp. 703–8.

6.55. A. D. ROSS, A note on the maturity and creep of concrete, *R.I.L.E.M. Bul. No.* 1, pp. 55–57 (Paris, March 1959).

6.56. A. M. NEVILLE, The influence of cement on creep of concrete in mortar, *J. Prestressed Concrete Inst.*, pp. 12–18 (Gainesville, Florida, March 1958).

6.57. A. D. ROSS, The creep of Portland blast-furnace cement concrete, *J. Inst., C.E.*, pp. 43–52 (London, Feb. 1938).

6.58. B. LE CAMUS, Recherches sur le comportment du béton et du béton armé soumis à des efforts repetés, *Comptes Rendues des Recherches des Laboratoires du Bâtiment et des Travaux Publics*, pp. 24–47 (Paris, 1945–46).

6.59. F. G. THOMAS, A conception of the creep of unreinforced concrete and an estimation of the limiting values, *The Struct. E.*, pp. 69–73 (London, 1933).

6.60. U.S. BUREAU OF RECLAMATION, Investigations of creep characteristics of concrete for Canyon Ferry Dam Missouri River Basin Project, *Concrete Laboratory Report No. C-789* (Denver, Colorado, 4th March 1955).

6.61. A. D. ROSS, Creep and shrinkage of plain, reinforced, and prestressed concrete: a general method of calculation, *J. Inst. C.E.*, pp. 38–57 (London, Nov. 1943).

6.62. O. WAGNER, Das Kriechen unbewehrten Betons, *Deutscher Ausschuss für Stahlbeton No. 131* (Berlin, 1958).

6.63. R. DUTRON, Creep in concretes, *R.I.L.E.M. Bul. No. 34*, pp. 11–33 (Paris, 1957).

6.64. D. McHENRY, A new aspect of creep in concrete and its application to design, *Proc. A.S.T.M.*, **40**, pp. 1069–84 (1943).

6.65. U.S. BUREAU OF RECLAMATION, Supplemental Report—5-year creep and strain recovery of concrete for Hungry Horse Dam, *Concrete Laboratory Report No. C-719A* (Denver, Colorado, 6th Jan. 1959).

6.66. A. M. NEVILLE, Theories of creep in concrete, *J. Amer. Concr. Inst.*, **52**, pp. 47–60 (Sept. 1955).

6.67. T. C. HANSEN, Creep of concrete—a discussion of some fundamental problems, *Bul. No. 33* (Swedish Cement and Concrete Research Inst., Sept. 1958).

6.68. A. M. NEVILLE, Non-elastic deformations in concrete structures, *J. New Zealand Inst. E.*, **12**, pp. 114–20 (April 1957).

6.69. W. G. GLANVILLE and F. G. THOMAS, Further investigations on the creep or flow of concrete under load, *D.S.I.R. Building Research Tech. Paper No. 21* (London, H.M.S.O., 1939).

6.70. R. E. DAVIS, H. E. DAVIS and E. H. BROWN, Plastic flow and volume changes of concrete, *Proc. A.S.T.M.*, **37**, Part II, pp. 317–30 (1937).

6.71. E. REINIUS, A theory of the deformation and the failure of concrete, *Betong No. 1* (1955) (London, Cement and Concrete Assoc. Translation No. 63, March 1957).

6.72. A. M. NEVILLE, Shrinkage and creep in concrete, *Structural Concrete*, **1**, No. 2, pp. 49–85 (London, March 1962).

6.73. E. W. BENNETT and D. R. LOAT, Shrinkage and creep of concrete as affected by the fineness of Portland cement, *Mag. Concr. Res.*, **22**, No. 71, pp. 69–78 (June 1970).

6.74. J. GLUCKLICH, Creep mechanism in cement mortar, *J. Amer. Concr. Inst.*, **59**, pp. 923–48 (July 1962).

6.75. T. C. POWERS, A hypothesis on carbonation shrinkage, *Portl. Cem. Assoc. J. of Res. and Development Labs.*, **4**, No. 2, pp. 40–50 (May 1962).

6.76. A. M. NEVILLE, Creep of concrete as a function of its cement paste content, *Mag. Concr. Res.*, **16**, No. 46, pp. 21–30 (March 1964).

6.77. S. P. SHAH and G. WINTER, Inelastic behaviour and fracture of concrete, *Symp. on Causes, Mechanism, and Control of Cracking in Concrete, Amer. Concr. Inst. Sp. Publicn. No. 20*, pp. 5–28 (1968).

6.78. A. M. NEVILLE, Some problems in inelasticity of concrete and its behaviour under sustained loading, *Structural Concrete*, **3**, No. 4, pp. 261–8 (London, July/Aug. 1966).

6.79. P. DESAYI and S. KRISHNAN, Equation for the stress–strain curve of concrete, *J. Amer. Concr. Inst.*, **61**, pp. 345–50 (March 1964).

6.80. L. P. SAENZ, Discussion of Reference 6.79, *J. Amer. Concr. Inst.*, **61**, pp. 1229–35 (Sept. 1964).

6.81. E. HOGNESTAD, N. W. HANSON and D. McHENRY, Concrete stress distribution in ultimate strength design, *J. Amer. Concr. Inst.*, **52**, pp. 455–79 (Dec. 1955).

6.82. K. S. GOPALAKRISHNAN, A. M. NEVILLE and AMIN GHALI, Creep Poisson's ratio of concrete under multiaxial compression, *J. Amer. Concr. Inst.*, **66**, pp. 1008–20 (Dec. 1969).

6.83.   I. E. HOUK, O. E. BORGE and D. L. HOUGHTON, Studies of autogenous volume change in concrete for Dworshak Dam, *J. Amer. Concr. Inst.*, **66**, pp. 560–8 (July 1969).

6.84.   D. RAVINA and R. SHALON, Plastic shrinkage cracking, *J. Amer. Concr. Inst.*, **65**, pp. 282–92 (April 1968).

6.85.   S. T. A. ÖDMAN, Effects of variations in volume, surface area exposed to drying, and composition of concrete on shrinkage, *RILEM/CEMBUREAU Int. Colloquium on the Shrinkage of Hydraulic Concretes*, **1**, 20 pp. (Madrid, 1968).

6.86.   T. W. REICHARD, Creep and drying shrinkage of lightweight and normal weight concretes, *Nat. Bur. Stand. Monograph* **74**, (Washington D.C., March 1964).

6.87.   K. MATHER, High strength, high density concrete, *J. Amer. Concr. Inst.*, **62**, No. 8, pp. 951–62 (Aug. 1965).

6.88.   B. TREMPER and D. L. SPELLMAN, Shrinkage of concrete—comparison of laboratory and field performance, *Highw. Res. Record, No.* 3, pp. 30–61 (1963).

6.89.   M. MAMILLAN, Étude du retrait du béton, *Rev. Matér. Construct. Trav. Publ. No.* 545, pp. 86–9 (Paris, Feb. 1961).

6.90.   S. E. PIHLAJAVAARA, Notes on the drying of concrete, *Reports*, Series 3, No. 79 (The State Institute for Technical Research, Helsinki, 1963).

6.91.   T. C. HANSEN, Effect of wind on creep and drying shrinkage of hardened cement mortar and concrete, *A.S.T.M. Mat. Res. & Stand.*, **6**, pp. 16–19 (Jan. 1966).

6.92.   T. C. HANSEN and A. H. MATTOCK, The influence of size and shape of member on the shrinkage and creep of concrete, *J. Amer. Concr. Inst.*, **63**, pp. 267–90 (Feb. 1966).

6.93.   J. W. KELLY, Cracks in concrete—the causes and cures, *Concrete Construction*, **9**, pp. 89–93 (April 1964).

6.94.   U.S. BUREAU OF RECLAMATION, *Concrete Manual*, 7th Ed. (Denver, Colorado, 1966).

6.95.   E. E. REIS, J. D. MOZER, A. C. BIANCHINI and C. E. KESLER, Causes and control of cracking in concrete reinforced with high-strength steel bars—a review of research, *University of Illinois Engineering Experiment Station Bul. No.* 479 (1965).

6.96.   R. BAREŠ and J. ROSENKRANZ, Nová metoda zjišťování trhlin v materiálech, *Stavebnicky Casopis*, **10**, No. 6, pp. 378–83 (Bratislava, 1962).

6.97.   T. C. HANSEN, Cracking and fracture of concrete and cement paste, Symp. on Causes, Mechanism, and Control of Cracking in Concrete, *Amer. Concr. Inst. Sp. Publicn. No.* 20, pp. 5–28 (1968).

6.98.   W. MANNS and K. WESCHE, Variation in strength of mortars made of different cements due to carbonation, Supplementary Paper III—16, *Proc. 5th Int. Symp. on the Chemistry of Cement*, pp. 385–93 (Tokyo, The Cement Association of Japan, 1969).

6.99.   A. M. NEVILLE, M. M. STAUNTON and G. M. BONN, A study of the relation between creep and the gain of strength of concrete, *Symp. on Structure of Portland Cement Paste and Concrete, Special Report No.* 90, pp. 186–203 (Highw. Res. Bd., Washington D.C., 1966).

6.100. U.S. BUREAU OF RECLAMATION, Creep of Glen Canyon Dam foundation rock cores under sustained load, *Concrete Laboratory Report No. C–948* (Denver, Colorado; 1960).

6.101. U.S. BUREAU OF RECLAMATION, Laboratory tests of rock cores from the foundation of Shikmen Dam—Taiwan, *Concrete Laboratory Report No. C–966* (Denver, Colorado, 1961).

6.102. S. E. RUTLEDGE and A. M. NEVILLE, Influence of cement paste content on creep of lightweight aggregate concrete, *Mag. Concr. Res.*, **18**, No. 55, pp. 69–74 (June 1966).

6.103. B. B. HOPE, A. M. NEVILLE and A. GURUSWAMI, Influence of admixtures on creep of concrete containing normal weight aggregate, *R.I.L.E.M. Int. Symp. on Admixtures for Mortar and Concrete*, pp. 17–32 (Brussels, Sept. 1967).

6.104. E. L. JESSOP, M. A. WARD and A. M. NEVILLE, Influence of water reducing and set retarding admixtures on creep of lightweight aggregate concrete, *R.I.L.E.M. Int. Symp. on Admixtures for Mortar and Concrete*, pp. 35–46 (Brussels, Sept. 1967).

6.105. H. RÜSCH, K. KORDINA and H. HILSDORF, Der Einfluss des mineralogischen Charakters der Zuschläge auf das Kriechen von Beton, *Deutscher Ausschuss für Stahlbeton*, No. 146, pp. 19–133 (Berlin, 1963).

6.106. A. M. NEVILLE, *Creep of Concrete: plain, reinforced, and prestressed* (Amsterdam, North-Holland, 1970).

6.107. K. W. NASSER and A. M. NEVILLE, Creep of concrete at elevated temperatures, *J. Amer. Concr. Inst.*, **62**, pp. 1567–79 (Dec. 1965).

6.108. J. C. MARÉCHAL, Le fluage du béton en fonction de la temperature, *Materials and Structures*, **2**, No. 8, pp. 111–15 (Paris, March–April 1969).

6.109. R. JOHANSEN and C. H. BEST, Creep of concrete with and without ice in the system, *R.I.L.E.M. Bul. No.* 16, pp. 47–57 (Paris, Sept. 1962).

6.110. A. M. NEVILLE, Concrete—a non-elastic material in the laboratory and in structures, Lecture No. 6, *Stanton Walker Lecture Series on the Materials Sciences*, National Sand and Gravel Association and National Ready Mixed Concrete Association (Washington D.C., 1968).

6.111. H. LAMBOTTE, Le fluage du béton en torsion, *R.I.L.E.M. Bul. No.* 17, pp. 3–12 (Paris, Dec. 1962).

6.112. C.E.B.–F.I.P., *International recommendations for the design and construction of concrete structures*, pp. 80 (London, Cement and Concrete Assoc., 1970).

6.113 A. M. NEVILLE, Differential creep and shrinkage of vertical load-carrying elements, *Proc. ASCE-IABSE Intern. Conference on Planning and Design of Tall Buildings*, pp. 765–74 (Lehigh, U.S.A., Aug. 1972).

7. Durability of Concrete

IT is essential that concrete should withstand the conditions for which it has been designed, without deterioration, over a period of years. Such concrete is said to be durable.

The absence of durability may be caused either by the environment to which the concrete is exposed or by internal causes within the concrete itself. The external causes can be physical, chemical, or mechanical: they may be due to weathering, occurrence of extreme temperatures, abrasion, electrolytic action, and attack by natural or industrial liquids and gases. The extent of damage produced by these agents depends largely on the quality of the concrete, although under extreme conditions any unprotected concrete will deteriorate.[7.1]

The internal causes are the alkali-aggregate reaction (p. 139), volume changes due to the differences in thermal properties of aggregate and cement paste, and above all the permeability of the concrete. The last-named largely determines the vulnerability of concrete to external agencies, so that in order to be durable concrete must be relatively impervious.

Deterioration of concrete is rarely due to one isolated cause: concrete can often be satisfactory despite some undesirable features, but with an additional adverse factor damage will occur. For this reason, it is sometimes difficult to assign trouble to a particular factor, but the quality of concrete, in the broad sense of the word, though with a special reference to permeability, nearly always enters the picture.

Permeability of Concrete
Penetration of concrete by materials in solution may adversely affect its durability, for instance when $Ca(OH)_2$ is being leached out or an attack by aggressive liquids takes place. This penetration depends on the permeability of the concrete, and since this determines the relative ease with which concrete can become saturated with water, permeability has an important bearing on the vulnerability of concrete to frost. Furthermore, in the case of reinforced concrete, the ingress of moisture and of air will result in the corrosion of steel. Since this leads to an increase in the volume of the steel, cracking and spalling of the concrete cover may well follow.

Permeability of concrete is also of interest in relation to water-tightness of liquid-retaining and some other structures, and also with reference to the problem of hydrostatic pressure in the interior of dams. Furthermore, ingress of moisture into concrete affects its thermal insulation properties (*see* p. 429).

It may be noted that movement of water through a thickness of concrete can be caused not only by a head of water but also by a humidity differential on the two sides of the concrete or by osmotic effects.

Both the cement paste and the aggregate contain pores. In addition, the concrete as a whole contains voids caused by incomplete compaction or by bleeding. These voids may occupy between a fraction of one per cent and 10 per cent of the volume of the concrete, the latter figure representing a highly honeycombed concrete of very low strength. Such concrete or concrete with leaking joints, will not be further discussed. Since aggregate particles are enveloped by the cement paste, in fully compacted concrete it is the permeability of the paste that has the greatest effect on the permeability of the concrete.

The pores in the cement paste were considered on pages 29–32, and it may be recalled that we distinguished between gel pores and capillary pores. The former constitute about 28 per cent of the paste volume, and the latter between 0 and 40 per cent, depending on the water/cement ratio and the degree of hydration.

The volume of pore space in concrete, as distinct from its permeability, is measured by absorption; the two quantities are not necessarily related. Absorption is usually measured by drying a specimen to a constant weight, immersing it in water and measuring the increase in weight as a percentage of dry weight. Various procedures are used, and widely differing results are obtained as shown in Table 7.1. One reason for this variation in the values of absorption is that, at one extreme, drying at ordinary temperature may be ineffective in removing all the water; on the other hand, drying at high temperatures may remove some of the combined water. Absorption cannot therefore be used as a measure of quality of concrete, but most good concretes have an absorption well below 10 per cent.

The flow of water through concrete is fundamentally similar to flow through any porous body. Since, however, the paste is composed of particles connected over only a small fraction of their total surface, a part of the water is within the field of force of the solid phase, i.e., it is adsorbed. This water has a high viscosity but is, nevertheless, mobile and takes part in the flow.[7.2]

The permeability of concrete is not a simple function of its porosity, but depends also on the size, distribution, and continuity of the pores. Thus, although the cement gel has a porosity of 28 per cent, its permeability is only about 7×10^{-16} m/s.[7.3] This is due to the extremely fine texture of hardened cement paste: the pores and the solid particles are

Table 7.1: *Values of Absorption of Concrete Determined in Various Ways*[7.7]

Drying condition	Immersion condition	Absorption (per cent) for concrete mix					
		A	B	C	D	E	F
100°C (212°F)	Water for 30 minutes	4·7	3·2	8·9	12·3		
100°C (212°F)	Water for 24 hours	7·4	6·9	9·1	12·9		
100°C (212°F)	Water for 48 hours	7·5	7·0	9·2	13·1		
100°C (212°F)	Water for 48 hours plus 5 hours boiling	8·1	7·3	14·1	18·2		
65°C (149°F)	5 hours boiling	6·4	6·4	13·2	17·2		
105°C (221°F) to constant weight	1 hour					3·0	7·4
	24 hours					3·4	7·7
	7 days					3·5	7·8
20°C (68°F) in vacuo over lime for 30 days	1 hour					1·9	5·9
	24 hours					2·2	6·3
	7 days					2·3	6·4

very small and numerous, while in rocks the pores, though fewer in number, are much larger and lead to a higher permeability. For the same reason, water can flow more easily through the capillary pores than through the much smaller gel pores: the cement paste as a whole is 20 to 100 times more permeable than the gel itself.[7.3] It follows that the permeability of cement paste is controlled by the capillary porosity of the paste. The relation between these two quantities is shown in Fig. 7.1. For comparison, Table 7.2 lists the water/cement ratio of pastes having the same permeability as some common rocks.[7.3] It is interesting to see that

Table 7.2: *Comparison between Permeabilities of Rocks and Cement Pastes*[7.3]

Type of rock	Coefficient of permeability m/s	Water/cement ratio of mature paste of the same permeability
Dense trap	$2·47 \times 10^{-14}$	0·38
Quartz diorite	$8·24 \times 10^{-14}$	0·42
Marble	$2·39 \times 10^{-13}$	0·48
Marble	$5·77 \times 10^{-12}$	0·66
Granite	$5·35 \times 10^{-11}$	0·70
Sandstone	$1·23 \times 10^{-10}$	0·71
Granite	$1·56 \times 10^{-10}$	0·71

the permeability of granite is of the same order as that of paste with a water/cement ratio of 0·7, i.e. not of high quality.

The permeability of cement paste varies with the progress of hydration. In a fresh paste, the flow of water is controlled by the size, shape, and concentration of the original cement grains. With the progress of hydration the permeability decreases rapidly because the gross volume of gel

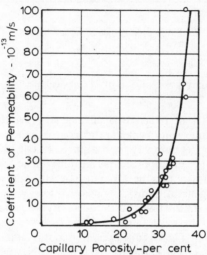

Fig. 7.1. *Relation between permeability and capillary porosity of cement paste*[7.3]

(including the gel pores) is approximately 2·1 times the volume of the unhydrated cement, so that the gel gradually fills some of the original water-filled space. In a mature paste the permeability depends on the size, shape, and concentration of the gel particles and on whether or not the capillaries have become discontinuous.[7.4] Table 7.3 gives values of the coefficient of permeability at different ages for a cement paste with a water/cement ratio of 0·7.[7.5]

For pastes hydrated to the same degree, the permeability is lower the higher the cement content of the paste, i.e. the lower the water/cement ratio. Fig. 7.2 shows values obtained for pastes in which 93 per cent of the cement has hydrated.[7.5] The slope of the line is considerably lower for pastes with water/cement ratios below about 0·6, i.e. pastes in which the capillaries have become segmented (*see* p. 30). From Fig. 7.2 it can be seen that a reduction of water/cement ratio from, say, 0·7 to 0·3 lowers the coefficient of permeability a thousandfold. The same reduction occurs in a paste with a water/cement ratio of 0·7 between the ages of 7 days and one year. In lean mass concrete, age is a more important factor in permeability than cement content.[7.64]

Table 7.3: *Reduction in Permeability of Cement Paste (Water/Cement Ratio = 0·7) with the Progress of Hydration*[7.5]

Age days	Coefficient of permeability, K m/s
fresh	2×10^{-6}
5	4×10^{-10}
6	1×10^{-10}
8	4×10^{-11}
13	5×10^{-12}
24	1×10^{-12}
ultimate	6×10^{-13} (calculated)

The permeability of concrete is affected also by the properties of cement. For the same water/cement ratio, coarse cement tends to produce a paste with a higher porosity than a finer cement.[7.5] The compound composition of the cement affects permeability in so far as it influences the rate of hydration, but the ultimate porosity and permeability are unaffected.[7.5] In general terms, it is possible to say that the higher the strength of the paste the lower its permeability—a state of affairs to be expected, since strength is a function of the relative volume of gel in the space available to it. There is one exception to this statement: drying the cement paste increases its permeability, probably because shrinkage may

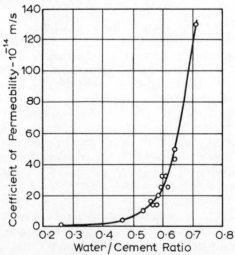

Fig. 7.2. *Relation between permeability and water/cement ratio for mature cement pastes*[7.5] *(93 per cent of cement hydrated)*

rupture some of the gel between the capillaries and thus open new passages to the water.[7.5]

The difference between the permeability of cement paste and of concrete containing the paste of the same water/cement ratio should be appreciated as the permeability of the aggregate itself affects the behaviour of the concrete (*see* Table 7.2). If the aggregate has a very low permeability its presence reduces the effective area over which flow can take place. Furthermore, since the flow path has to circumvent the aggregate particles, the effective path becomes considerably longer so that the effect of aggregate in reducing the permeability may be considerable. Generally, however, the influence of the aggregate content in the mix is small and, since the aggregate particles are enveloped by the cement paste, in fully compacted concrete it is the permeability of the paste that has the greatest effect on the permeability of the concrete.

Typical values of permeability of concrete of the type used in some U.S. dams are given in Table 7.4. Permeability is particularly important

Table 7.4: *Typical Values of Permeability of Concrete used in Dams*[7.64]

Cement content		Water/cement	Permeability
kg/m³	lb/yd³	ratio	10^{-12}m/s
156	263	0·69	8
151	254	0·74	24
138	235	0·75	35
223	376	0·46	28

in this type of structure because it may affect uplift or pore pressure and in some cases also allow leaching by acid waters, but of course the permeability of a dam as a whole is greatly affected by cracks and other water passages. We may note, however, that the permeability of well-designed mass concrete, even as lean as 112 kilogrammes of cement per cubic metre (188 lb/yd³), is so low that equilibrium pore pressure in a dam is not likely to be achieved within its usual life.[7.65]

Permeability of steam-cured concrete is generally lower than that of wet-cured concrete. Typical values of permeability for different heating cycles are shown in Fig. 7.3. Except for concrete subjected to a long temperature cycle, supplemental fog curing may be required to achieve an acceptably low permeability. The magnitude of the values in Fig. 7·3 can be appreciated by reference to an arbitrary acceptable limit of $1·5 \times 10^{-11}$ m/s ($4·8 \times 10^{-11}$ ft/s) used in some Bureau of Reclamation work.

Air entraining would be expected to increase the permeability of concrete. However, since air entraining reduces segregation and bleeding and improves workability, and so permits the use of a lower water/cement ratio, the nett effect of air entraining is not necessarily adverse.

Measurement of Water Permeability

The permeability of concrete can be measured in the laboratory by means of a simple test, but the results are mainly comparative.[7.6] The sides of a test specimen are sealed and water under pressure is applied to the top surface only. Water saturated with air at atmospheric pressure is often used as this is the situation in practical cases. Compressed air is often used to apply the pressure, but care must be taken to prevent air from being absorbed in the water; otherwise, some of the air may be released under the reduced pressure inside the specimen and thus decrease the rate of flow. When a steady regime has been reached (and this may not be achieved until about 10 days after the start of the test), the quantity of water flowing through a given thickness of concrete in a given time is measured, and the permeability is expressed as a coefficient of permeability, K, given by Darcy's equation

$$\frac{dq}{dt}\frac{1}{A} = K\frac{\Delta h}{L}$$

where $\frac{dq}{dt}$ is the rate of flow of water in m^3/s,

A is the cross-sectional area of the sample in m^2,
Δh is the drop in hydraulic head through the sample, measured in m

and L is the thickness of the sample in m.
K is then expressed in metres per second.

The permeability test can be performed on cores in order to study the effects of variations in mix proportions and in mixing, placing, and curing techniques. The test can also give an estimate of the durability of concrete subjected to the corrosive action of percolating water.

Air and Vapour Permeability

Permeability of concrete to air is of interest primarily in structures such as sewage tanks and gas purifiers, the usual requirement being that the concrete should be airtight under a specified internal pressure. More recently, interest in air permeability of concrete arose in connection with pressure vessels in nuclear reactors.

As in the case of water permeability, the rate of flow of air depends on the thickness of the concrete and on the pressure applied but the equilibrium rate of flow is reached within several hours. The magnitude of flow can be gauged from some tests on a 1:2:6 concrete with a water/cement ratio of 0·62. The following values were obtained for different thicknesses of concrete.[7.67]

Thickness, mm (in.)	Rate of flow, $10^{-9}m^3$ per N/m^2 h m^2 (in^3 per lb/in^2 h ft^2)
102 (4)	14·8 (0·58)
178 (7)	9·5 (0·37)
203 (8)	7·7 (0·30)
229 (9)	6·1 (0·24)

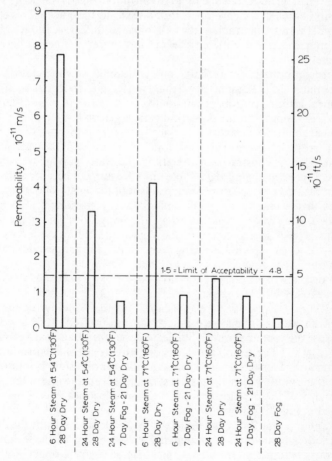

Fig. 7.3. *Permeability of steam-cured concrete*[7.66]

Increasing the cement content of the mix has been found to decrease the air permeability, but probably the true factor is the water/cement ratio. It is also through a reduction in the actual water/cement ratio that air entrainment reduces the air permeability. The addition of fly ash or pozzolana has a beneficial effect.[7.68]

Grading of aggregate seems to be particularly important in reducing air permeability of concrete. With very good grading and a cement content of 300 kilogrammes per cubic metre (500 lb/yd³), concrete has been found to be impermeable to air under a pressure of 0·2 MN/m² (28 lb/in²).[7.69]

As would be expected, prolonged curing reduces the air permeability but even at an advanced age rapid drying out increases the permeability, possibly through the development of fine shrinkage cracks. This effect is considerably greater than in the case of water diffusion. In fact, there is no unique relation between the air and water permeabilities of all concretes.

Little information is available on permeability of concrete to various gases but it has been reported[7.70] that even if a given concrete is relatively permeable to air it may be substantially impermeable to some other gases.

Water vapour transmission of concrete is generally affected in a similar manner to air permeability. Specifically, the transmission is reduced by an increase in the age of concrete and by a decrease in the water/cement ratio. Fig. 7.4 illustrates this relation, the water vapour transmission being expressed in g/m² day. These units are justified because apparently the measured transmission is independent of the length of the flow path.

As in the case of air permeability, the water vapour permeability decreases on prolonged moist curing; after only 1-day curing, the permeability is twice as large as after 3-day curing and 15 times as large as after a 1-year curing period.[7.72] However, unlike air permeability, vapour permeability is unaffected by drying out.

A decrease in water/cement ratio from 0·8 to 0·4 decreases the vapour permeability by 50 to 65 per cent.[7.72]

All these comparisons are applicable only to the vapour permeability at given values of relative humidity on the two sides of the concrete because the vapour permeability decreases as the mean relative humidity increases.[7.73] The reason for this is that an increase in relative humidity decreases the air-filled pore space available for diffusion. It follows then that if the moist side is, for instance, saturated, an increase in the relative humidity of the dry side reduces the vapour permeability. This is illustrated in Fig. 7.4.

Vapour permeability can be determined experimentally by the dry-cup method. This involves a gravimetric measurement of the steady-state vapour transmission through a thickness of concrete under isothermal conditions with a relative humidity of zero on one side and usually

50 per cent on the other.[7.73] Under these conditions virtually all transmission takes place by diffusion, and capillary flow is absent.

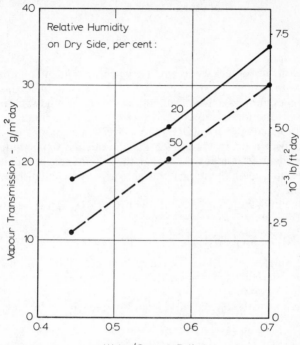

Fig. 7.4. *Relation between vapour transmission and water/cement ratio of concrete at 23°C (73°F)*[7.71]

Chemical Attack of Concrete

Only a small proportion of concrete used in practice is exposed to serious chemical attack. This is fortunate, since the resistance of concrete to attack by chemical agents is generally lower than to other forms of attack.

The more common forms of chemical attack are the leaching out of cement, and the action of sulphates, sea water and natural slightly acidic waters. In general terms, the resistance of concrete varies with the type of cement used; it has been suggested that the resistance increases in the following order[7.1]—

1. Ordinary and rapid hardening Portland cement.
2. Portland blast-furnace cement or low heat Portland cement.
3. Sulphate-resisting Portland cement or pozzolanic cement.
4. Supersulphated cement.
5. Aluminous cement.

It seems, however, that in some cases the density and permeability of the concrete influence its durability to such a degree that they overrule the influence of the type of cement used.

Sulphate Attack

Solid salts do not attack concrete, but when present in solution they can react with hardened cement paste. Some clays contain, for instance, alkali, magnesium and calcium sulphates, and the groundwater in such a clay is in effect a sulphate solution. Attack of cement can thus take place, the sulphate reacting with $Ca(OH)_2$ and with calcium aluminate hydrate.

The products of the reactions, gypsum and calcium sulphoaluminate, have a considerably greater volume than the compounds they replace, so that the reactions with the sulphates lead to expansion and disruption of the concrete.

The reaction of sodium sulphate with $Ca(OH)_2$ can be written as follows—

$$Ca(OH)_2 + Na_2.SO_410H_2O \rightarrow CaSO_4.2H_2O + 2NaOH + 8H_2O$$

In flowing water, $Ca(OH)_2$ can be completely leached out, but if NaOH accumulates, equilibrium is reached, only a part of the SO_3 being deposited as gypsum.

The reaction with calcium aluminate hydrate can be formulated as follows[7.7]—

$$2(3CaO.Al_2O_3.12H_2O) + 3 (Na_2SO_4.10H_2O) \rightarrow$$
$$3CaO.Al_2O_3.3CaSO_4.31H_2O + 2Al(OH)_3 + 6NaOH + 17H_2O$$

Calcium sulphate attacks only calcium aluminate hydrate, forming calcium sulphoaluminate $(3CaO.Al_2O_3.3CaSO_4.31H_2O)$. On the other hand, magnesium sulphate attacks calcium silicate hydrates as well as $Ca(OH)_2$ and calcium aluminate hydrate. The pattern of reaction is

$$3CaO.2SiO_2.aq + MgSO_4.7H_2O \rightarrow$$
$$CaSO_4.2H_2O + Mg(OH)_2 + SiO_2.aq$$

Because of the very low solubility of $Mg(OH)_2$, this reaction proceeds to completion so that under certain conditions the attack by magnesium sulphate is more severe than by other sulphates.

The rate of sulphate attack increases with an increase in the strength of the solution, but beyond a concentration of about 0·5 per cent of $MgSO_4$ or 1 per cent of Na_2SO_4 the rate of increase in the intensity of the attack becomes smaller.[7.7] A saturated solution of $MgSO_4$ leads to serious deterioration of concrete, although with a low water/cement ratio this takes place only after 2 to 3 years.[7.74] The concentration of the sulphates is expressed as the number of parts by weight of SO_3 per million (ppm); 1,000 ppm is considered moderately severe and 2,000 ppm are very severe, especially if $MgSO_4$ is the predominant constituent. The parallel values of soluble sulphate in soil are 0·2 and 0·5 per cent.

In addition to the concentration of the sulphate, the speed with which concrete is attacked depends also on the rate at which the sulphate removed by the reaction with cement can be replenished. Thus, in estimating the danger of sulphate attack the movement of groundwater has to be known. When concrete is exposed to the pressure of sulphate-bearing water on one side, the rate of attack will be highest. Likewise, alternating saturation and drying leads to rapid deterioration. On the other hand, when the concrete is completely buried, without a channel for the groundwater, conditions will be much less severe.

Concrete attacked by sulphate has a characteristic whitish appearance. The damage usually starts at edges and corners and is followed by progressive cracking and spalling which reduce the concrete to a friable or even soft state.

The vulnerability of concrete to sulphate attack can be reduced by the use of cement low in C_3A. Sulphate-resisting cement and the influence of compound composition on resistance to sulphates in general were discussed in Chapter 2. In practice it has been found that a C_3A content of 7 per cent gives a rough division between cements of good and poor performance in sulphate waters.[7.8] There appear, however, to be some other factors, not yet recognized, that influence the resistance of cement to sulphate attack.

Improved resistance to sulphate attack is obtained also by the addition of, or even by partial replacement of the cement by, pozzolanas. They remove free $Ca(OH)_2$ (cf. p. 392) and render the alumina-bearing phases inactive, but sufficient time must be allowed for the pozzolanic activity to be developed before the concrete is exposed to the sulphates. Many pozzolanas have been found very effective in making concrete resistant to sulphate attack, particularly in conjunction with sulphate-resisting cement.[7.102]

The resistance of concrete to sulphate attack depends also on its impermeability. This has been mentioned before, but the importance of this factor cannot be over-emphasized. For instance, the use of lean concrete in haunching or bedding of sewers produces vulnerable parts of a possibly otherwise durable structure. To be dense the concrete must have a low water/cement ratio, i.e. it must be fairly rich. Even concrete made with aluminous cement should not be leaner than 1:8 or 1:9, otherwise a porous structure is obtained and the concrete is easily attacked.[7.1]

Fig. 7.5 shows the influence of a five-year field exposure on the compressive strength of the following mixes—

Mix	Cement content kg/m³	lb/yd³	Average water/cement ratio
A	390	658	0·4
B	307	517	0·5
C	223	376	0·75

Two test basins were used: No. 1 high in sodium sulphate, and No. 2 containing both sodium and magnesium sulphates in varying proportions.

It can be seen that concretes with a cement content of about 390 kg/m³ (650 lb/yd³) are attacked only slowly, even when made with ordinary Portland cement.[7.8] With lower cement contents the attack is more rapid, and the composition of the cement has a greater influence on the durability of the concrete.[7.8]

High-pressure steam curing improves the resistance of concrete to sulphate attack. This applies to concretes made both with sulphate-resisting and ordinary Portland cements, since the improvement is due to the change of C_3AH_6 into a less reactive phase, and also to the removal of $Ca(OH)_2$ by the reaction with silica. On the other hand, the addition of calcium chloride to the mix leads to a reduced resistance to sulphate attack, whatever the type of cement used.

Tests on Sulphate Resistance
The resistance of concrete to sulphate attack can be tested in the laboratory by storing specimens in a solution of sodium or magnesium sulphate, or in a mixture of the two. Alternate wetting and drying accelerates the damage owing to the crystallization of salts in the pores of the concrete. The effects of exposure can be estimated by the loss in strength of the specimen, by changes in its dynamic modulus of elasticity, by its expansion, by its loss of weight, or can even be assessed visually.

Fig. 7.6 shows the change in the dynamic modulus of 1:3 mortar immersed (after 78 days' moist curing) in a 5 per cent solution of different sulphates.[7.9] The results of strength tests on 1:3 mortar bars immersed in a 0·15 molar solution of Na_2SO_4 are shown in Fig. 7.7. The initial increase in strength is probably due to filling of voids by crystals, which increase the density of the concrete. But, as the tendency of the crystals to grow and expand is impeded, the resulting internal stresses destroy the concrete.

Tests on 1:6 cement/Ottawa sand mortar specimens stored in a similar solution have shown that an expansion of about 0·10 per cent after four weeks gives a rough dividing line between good and unsatisfactory performance in the field.[7.8] The test is not, however, sufficiently accurate to give a prediction of behaviour in borderline cases, and no

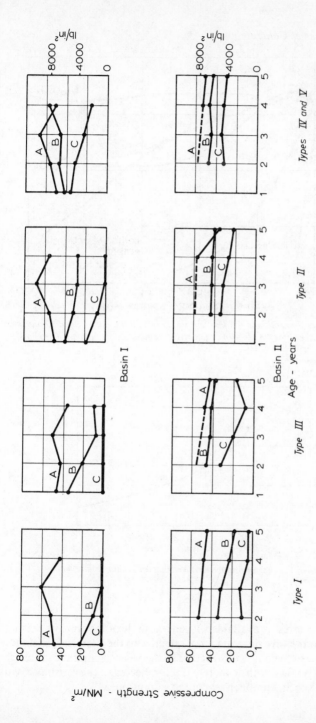

Fig. 7.5. *Effects of field exposure on strength of concretes made with different cements*[7.8]

Mixes and conditions described on page 394.

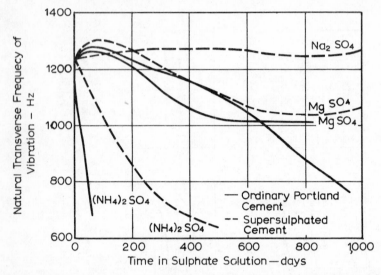

Fig. 7.6. *Effect of immersion in a 5 per cent sulphate solution on the dynamic modulus of elasticity of 1:3 mortars made with ordinary Portland and supersulphated cements*[7.9]

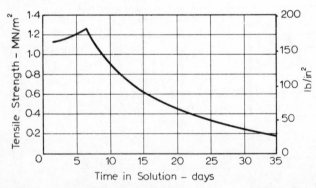

Fig. 7.7. *Strength of 1:3 mortar briquettes stored in a 0·15 molar solution of Na_2SO_4*[7.10]

fully satisfactory test is available. A general fault of laboratory tests on mortars is that they are slow and more sensitive to the chemical resistance of cement than to the physical structure of the paste[7.7] since the influence of the general quality of the concrete is not reflected in the resistance of a small mortar specimen under severe test conditions.

Sea-Water Attack

Sea water contains sulphates and attacks concrete in a manner similar to that described on page 392. In addition to the chemical action, crystallization of the salts in the pores of the concrete may result in its disruption owing to the pressure exerted by the salt crystals. Because crystallization takes place at the point of evaporation of water this form of attack occurs in the concrete above the water level.[7.7] Since, however, the salt solution rises in the concrete by capillary action, the attack takes place only when water can penetrate into the concrete so that impermeability of the concrete is once again its most important attribute.

Concrete between the tide marks, subjected to alternating wetting and drying, is severely attacked, while permanently immersed concrete is attacked least. The actual progress of attack by sea water varies, and is slowed down by the blocking of the pores in the concrete through deposition of magnesium hydroxide. In tropical climates the attack is more rapid.

In some cases the action of sea water on concrete is accompanied by the destructive agencies of frost, wave impact and abrasion, and all these tend to aggravate the damage of the concrete.

Although the action of sulphates in sea water is similar to that of sulphate-bearing groundwater, the attack in the former case is not accompanied by expansion of the concrete, encountered in laboratory sulphate-immersion tests. This absence of expansion is mainly due to the presence in the sea water of a large quantity of chlorides which inhibit the expansion: gypsum and calcium sulphoaluminate are more soluble in a chloride solution than in water, and are thus leached out by the sea water, while in the laboratory test they are left *in situ* and consequently cause expansion.[7.7] This behaviour offers another example of the difficulty of relating the results of laboratory tests to behaviour under real conditions of exposure.

In the case of reinforced concrete, the absorption of salt establishes anodic and cathodic areas; the resulting electrolytic action leads to an accumulation of the corrosion products on the steel with a consequent rupture of the surrounding concrete, so that the effects of sea water are more severe on reinforced concrete than on plain concrete It is therefore essential to provide a sufficient cover to reinforcement (50 mm (2 in.), or preferably 75 mm (3 in.)) and to use dense, impermeable concrete. A cement content of 350 kg/m^3 (600 lb/yd^3) above the low-water mark[7.11] and 300 kg/m^3 (500 lb/yd^3) below it,[7.7] or a water/cement ratio of not more than 0·40 to 0·45[7.12] are recommended. The water/cement ratio is believed to be the vital factor, the cement content being of importance primarily in so far as a high content makes it possible to achieve full compaction of mixes with low water/cement ratios.[7.13] A well-compacted concrete and good workmanship, especially in the construction of joints, are of vital importance. The type of cement used comes second, aluminous, sulphate-resisting, Portland blast-furnace, and Portland-pozzolana cements giving good results.

Acid Attack

In damp conditions SO_2, CO_2, and other acid fumes present in the atmosphere attack concrete by dissolving and removing part of the set cement, a soft and mushy mass being ultimately left behind. This form of attack occurs in chimneys and steam railway tunnels. Acid attack is encountered also under industrial conditions. The action of various acids was considered in detail by Lea,[7.7] and it should be remembered that no Portland cement is acid resistant.

Concrete is also attacked by water containing free CO_2, such as moorland water. Flowing pure water, formed by melting ice or by condensation, and containing little CO_2, also dissolves $Ca(OH)_2$, thus causing surface erosion. This type of attack may be of importance in conduits in mountain regions, not only from the standpoint of durability but also because the leaching out of cement leaves behind protruding aggregate and thus increases the roughness of the pipe. For this reason the use of calcareous rather than siliceous aggregate is advantageous.[7.14]

Although domestic sewage by itself is alkaline and does not attack concrete, severe damage of sewers has been observed in many cases, especially at fairly high temperatures[7.14] when sulphur compounds become reduced by anaerobic bacteria to H_2S. This is not a destructive agent in itself, but is dissolved in moisture films on the exposed surface of the concrete and undergoes oxidation by aerobic bacteria, finally producing sulphuric acid. The attack occurs, therefore, above the level of flow of the sewage. The cement is gradually dissolved, and progressive deterioration of concrete takes place.

Various physical and chemical tests on the resistance of concrete to acids have been developed,[7.7] but there are no standard procedures. It is essential that tests are performed under realistic conditions as when a concentrated acid is used all cements dissolve and no assessment of their relative quality is possible. For this reason care is required in interpreting the results of accelerated tests.

The resistance of concrete to chemical attack is increased by allowing it to dry out before exposure. A film of calcium carbonate (produced by the action of CO_2 on lime) then forms, blocking the pores and reducing the permeability of the surface layer. It follows that precast concrete is generally less vulnerable to attack than concrete cast *in situ*. An illustration of the effects of curing is shown in Fig. 7.8.

$Ca(OH)_2$ can also be fixed by treatment with diluted water glass (sodium silicate). Calcium silicates are then formed, filling the pores, and the resistance of the concrete to acid is also slightly increased, probably due to the formation of colloidal silicofluoric gel. Artificial surface treatment with coal-tar pitch, rubber or bituminous paints, epoxy resins, magnesium silicofluoride and other agents has also been used successfully. The degree of protection of the different treatments varies, but in all cases it is essential that the protective coat produced by the treatment

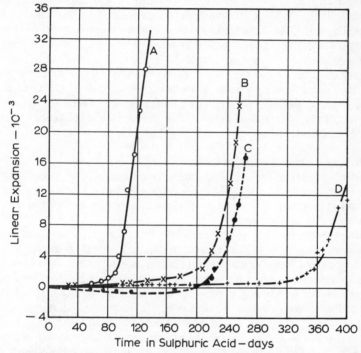

Fig. 7.8. *Influence of curing conditions on the expansion of mortar immersed in a 1 per cent sulphuric acid*[7.9]

A, water at 21°C (70°F) for 78 days; *B*, steam at 95°C (203°F) for 4 hours, then laboratory air for 78 days; *C*, air at 21°C (70°F) and a relative humidity of 80 per cent for 78 days; *D*, steam at 1 MN/m² (150 lb/in²) for 4 hours, then laboratory air for 78 days.

remains undamaged by mechanical agencies, so that access for inspection and renewal of the coating is generally necessary. Good protection of concrete from acid attack is obtained by subjecting concrete in a vacuum to the action of silicon tetrafluoride gas.[7.15] This gas reacts thus with lime—

$$2Ca(OH)_2 + SiF_4 \rightarrow 2CaF_2 + Si(OH)_4 .$$

The treatment can be applied to precast concrete only, which is then known as *Ocrat-concrete*.

Efflorescence

Leaching of lime compounds, mentioned earlier, may under some circumstances lead to the formation of salt deposits on the surface of the concrete, known as efflorescence. This is found, for instance, when water percolates through poorly compacted concrete or through cracks or along badly made joints, and when evaporation can take place at the surface of the concrete. Calcium carbonate formed by the reaction of $Ca(OH)_2$ with CO_2 is left behind in the form of a white deposit. Calcium sulphate deposits are encountered as well.

Efflorescence can also be caused by the use of unwashed seashore aggregate. The salt coating on the surface of the aggregate particles may in due course lead to a white deposit on the surface of the concrete. Gypsum and alkalis in the aggregate have a similar effect.

Apart from the leaching aspect, efflorescence is of importance only in so far as it mars the appearance of concrete.

Effects of Frost on Fresh Concrete

Before discussing the adverse effects of freezing and thawing on hardened concrete—one of the main problems in durability—we should consider the action of frost on fresh concrete and the associated problem of concreting in cold weather.

If concrete which has not yet set is allowed to freeze, the action of frost is somewhat similar to that in a saturated soil subject to heaving: the mixing water freezes with a consequent increase in the overall volume of the concrete. Furthermore, since no water is available for chemical reactions, the setting and hardening of concrete are delayed. It follows from the latter observation that, if concrete freezes immediately after it has been placed, setting will not have taken place, and thus there is no paste that can be disrupted by the formation of ice. While the low temperature continues, the process of setting will remain suspended. When at a later date thawing takes place the concrete should be revibrated, and it will then set and harden without loss of strength. However, because of the expansion of the mixing water on freezing, a lack of revibration would allow the concrete to set with a large volume of pores present, and consequently the strength of the concrete would be very low. But revibration on thawing would produce a satisfactory concrete.

If freezing takes place after the concrete has set but before it has developed appreciable strength, the expansion associated with the formation of ice causes disruption and an irreparable loss of strength. If, however, the concrete has acquired sufficient strength it can resist the freezing temperature without damage, not only by virtue of the higher resistance to the pressure of the ice but also because a large part of the mixing water will have become combined with the cement or located in the gel pores, and would thus not be able to freeze. It is difficult, however, to establish when this situation has been reached, as setting (Fig. 7.9) and hardening of cement depend on the temperature during the period preceding the actual advent of freezing. Generally, the more advanced the hydration of concrete and the higher its strength the less vulnerable it is to frost.

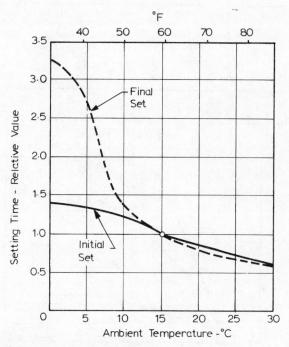

Fig. 7.9. *Setting time of ordinary Portland cement at different temperatures as a proportion of setting time at 15°C (59°F)*[7.16]

Values of minimum strength ranging from 5 MN/m² (700 lb/in²) up to as much as 14 MN/m² (2,000 lb/in²) have been suggested,[7.16] but no reliable data are available on the strength at which concrete can successfully resist temperatures below freezing point. An alternative approach is to consider the minimum age of concrete stored at a given temperature when exposure to frost will not cause damage; typical values (averaged from various sources[7.104,7.105]) are given in Table 7.5. Fig. 7.10 shows the influence of the age at which freezing starts on the expansion of concrete: the considerable decrease in the magnitude of expansion of concrete allowed to harden for about 24 hours is noticeable, and protecting concrete from frost during this period is clearly highly advisable.

Table 7.5: *Age of Concrete at which Exposure to Frost does not Cause Damage*

Type of cement	Water/cement ratio	Age (hours) at exposure when preceding curing temperature was			
		5°C (41°F)	10°C (50°F)	15°C (59°F)	20°C (68°F)
Ordinary	0·4	35	25	15	12
Portland	0·5	50	35	25	17
	0·6	70	45	35	25
Rapid	0·4	20	15	10	7
hardening	0·5	30	20	15	10
Portland	0·6	40	30	20	15

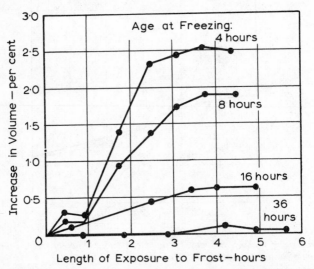

Fig. 7.10. *Increase in volume of concrete during prolonged freezing as a function of age when freezing starts*[7.17]

The resistance to alternating freezing and thawing also depends on the age of the concrete when the first cycle is applied (Fig. 7.11) but this type of exposure is more severe than prolonged freezing without periods of thaw, and several cycles can cause damage even to concrete cured at 20°C (68°F) for 24 hours.[7.17] It may be noted that there is no direct relation between the frost resistance of young concrete and the durability of mature concrete subjected to numerous cycles of freezing and thawing.[7.18]

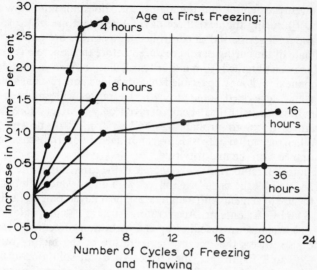

Fig. 7.11. *Increase in volume of concrete subjected to freezing and thawing as a function of age at which first freezing starts*[7.17]

Concreting in Cold Weather

To avoid the ill effects of frost in fresh concrete, certain precautions can be taken. The temperature at the time of casting can be raised by heating the ingredients of the mix. Water can be heated easily, but it is inadvisable to exceed a temperature of 60 to 80°C (140 to 180°F) as flash set of the cement may result; the likelihood of this happening depends on the difference between the temperatures of water and cement. It is also important to prevent the cement from coming into contact with the hot water, and for this reason the order of feeding the mix ingredients into the mixer must be suitably arranged.

If heating the water does not sufficiently raise the temperature of the concrete the aggregate may also be heated. This is done preferably by passing steam through coils rather than by the use of live steam since the latter method leads to a variable moisture content of the aggregate. Heating the aggregate above 52°C (125°F) is inadvisable.[7.19]

The temperature of the mix ingredients must be controlled, and the temperature of the resulting concrete should be calculated in advance in order to make sure that setting does not occur at too high a temperature. This would adversely affect the development of strength of the concrete (*see* p. 276). In addition, a high temperature of fresh concrete lowers its workability and may lead to high thermal contraction.

It is thus desirable for the concrete to set at, say, 7 to 21°C (45 to

70°F), but it is essential that the temperature does not fall below about 10°C (50°F) during the next three days. A temperature of up to 21°C (70°F) and a longer period of controlled temperature are preferable.[7.19] An estimate of the curing period required before the concrete is exposed to frost may be obtained from maturity calculations (*see also* Table 7.5). In this connexion, it may be recalled that, provided concrete has set above freezing and developed some strength, hydration will continue in all unfrozen cavities (with heat being evolved only at a very low rate) at temperatures down to probably − 4°C (25°F),[7.20] or possibly even lower. When the period of frost is over, normal gain of strength continues in accordance with the maturity rule.

The control of temperature of concrete in the early stages after placing can be assisted by the use of rich mixes with a low water/cement ratio, and by the use of cement with a high rate of heat development, i.e. having high C_3S and C_3A contents. Accelerators, such as calcium chloride, are clearly of assistance in speeding up the hydration of cement. Calcium chloride also lowers the freezing temperature of the mixing water by 1 to 2°C (2 to 3°F). In fact, the "water" is a salt solution, so that its freezing point is up to 3°C (5°F) below the freezing point of pure water.

There are numerous precautions that should be taken in practice. For instance, concrete should not be allowed to cool unduly while being transported from the mixer to the forms, and should not be placed against a frozen surface.

Control of temperature after placing is obtained by insulating the concrete from the atmosphere, and if necessary by constructing enclosures around the concrete and providing a source of heat within the enclosure. The form of heating should be such that the concrete does not dry out rapidly, that no part of it is heated excessively, and that no high concentration of CO_2 in the atmosphere results. For these reasons, exhaust steam is probably the best source of heat. Jacket-like steel forms with circulating hot water have been used very successfully. Concrete can also be heated by electric current with electrodes within the concrete or using the reinforcing steel as electrodes. The thinner the section the more easily it freezes and therefore the greater the care required in protecting it from frost. The effects of frost depend also on the rate and amplitude of changes in temperature, and the severity of frost action is aggravated when a sudden lowering of temperature is accompanied by wind. On the other hand, snow acts as an insulator and thus provides natural protection.

A different aid in concreting in freezing weather has recently been developed in U.S.S.R.: potash (K_2CO_3) is added to the mix in order to depress the freezing point of the mixing water and thus enable the concrete to develop some strength.[7.75] Since, however, potash accelerates setting, a retarder also has to be used. It seems that potash does not adversely affect bond with steel or encourage corrosion so that the use of potash in winter concreting may become more widespread in future.

It may be noted that the use of lightweight aggregate is advantageous in concreting in cold weather because lightweight aggregate concrete has a lower thermal conductivity than normal concrete (*see* p. 528) and therefore acts as a self-insulator; such concrete also has a lower specific heat so that a given heat of hydration of cement more effectively keeps the lightweight aggregate concrete from freezing than is the case with normal weight aggregate.

Frost Attack of Hardened Concrete

We shall now consider mature concrete subjected to alternating freezing and thawing—a temperature cycle frequently met with in nature.

Action of Frost

As the temperature of saturated hardened concrete is lowered, the water held in the capillary pores in the cement paste freezes in a manner similar to the freezing in the capillaries in rock, and expansion of the concrete takes place. On re-freezing, further expansion takes place so that repeated cycles of freezing and thawing have a cumulative effect. The larger pores in concrete, arising from incomplete compaction, are usually air-filled and therefore not appreciably subject to the action of frost.[7.21]

Freezing is a gradual process, partly because of the rate of heat transfer through concrete, partly because of a progressive increase in the concentration of dissolved alkalis in the still unfrozen water, and partly because the freezing point varies with the size of the cavity. Since the surface tension of the bodies of ice in the capillaries puts them under pressure that is higher the smaller the body, freezing starts in the largest cavities and gradually extends to smaller ones. Gel pores are too small to permit the formation of nuclei of ice above $- 78°C$, so that in practice no ice is formed in them.[7.21] However, with a fall in temperature, because of the difference in entropy of gel water and ice, the gel water acquires an energy potential enabling it to move into the capillary cavities containing ice. The diffusion of gel water which takes place leads to a growth of the ice body and to expansion.[7.21]

We have thus two sources of dilating pressure. First, freezing of water results in an increase in volume of approximately 9 per cent, so that the excess water in the cavity is expelled. The rate of freezing will determine the velocity with which water displaced by the advancing ice front must flow out, and the hydraulic pressure developed will depend on the resistance to flow, i.e. on the length of path and the permeability of the paste between the freezing cavity and a void that can accommodate the excess water.[7.22]

The second dilating force in concrete is caused by diffusion of water leading to a growth of a relatively small number of bodies of ice. On the basis of numerous investigations the latter mechanism is believed to be particularly important in causing frost damage of concrete.[7.23] This diffusion is caused by osmotic pressure brought about by local increases in solute concentration due to the separation of frozen (pure) water from the solution. For instance, a slab freezing from the top will be seriously damaged if water has access from the bottom and can travel through the thickness of the slab due to osmotic pressure. The total moisture content of the concrete will then become greater than before freezing, and in some cases damage by segregation of ice crystals into layers has actually been observed.[7.24]

Osmotic pressure arises also in another connexion. When salts are used for de-icing roads, some of these salts become absorbed by the upper part of the concrete. This produces a high osmotic pressure with a consequent movement of water toward the coldest zone where freezing takes place. Experimental data on these processes are, however, still lacking.

When the dilating pressure in the concrete exceeds its tensile strength, damage occurs. The extent of the damage varies from surface scaling to complete disintegration as layers of ice are formed, starting at the exposed surface of the concrete and progressing through its depth. Under the conditions prevailing in England, road kerbs (which remain wet for long periods) are more vulnerable to frost than any other concrete. The second most severe conditions are those in a road slab, particularly when salt is used for de-icing. In countries with a colder climate the damage due to frost is more general and, unless suitable precautions are taken in the manufacture of concrete, more serious.

While the resistance of concrete to frost depends on its various properties (e.g. strength of the paste, extensibility, and creep), the main factors are the degree of saturation and the pore structure of the cement paste.

The general influence of saturation of concrete is shown in Fig. 7.12: below some critical value of saturation, concrete is highly resistant to frost,[7.16] and dry concrete is totally unaffected (Table 7.6). It may be noted that even in a water-cured specimen not all residual space is water-filled, and indeed this is why such a specimen does not fail on first freezing.[7.26] A large proportion of concrete in the field dries partially at least at some time in its life, and on rewetting such concrete will not re-absorb as much water as it has lost.[7.27] It is wise, therefore, to allow concrete to dry out before exposure to winter conditions, and failure to do so will increase the severity of frost damage.

What is the critical value of saturation? A closed container with more than 91·7 per cent of its volume occupied by water will on freezing become filled with ice, and will become subjected to bursting pressure. Thus 91·7 per cent can be considered to be the critical saturation in a closed

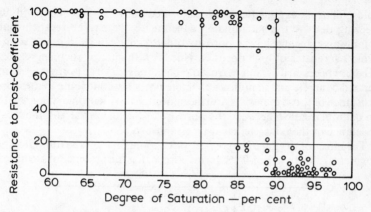

Fig. 7.12. *Influence of saturation of concrete on its resistance to frost*[7.16]

Table 7.6: *Effect of Wet and Dry Freezing on the Compressive Strength of Concrete*[7.25]

Number of cycles of freezing and thawing	Compressive strength as a percentage of the 7-day strength		
	Frozen wet	Frozen dry	Cured at normal temperature
0	100	100	100
10	141	165	189
20	137	189	240
30	119	201	263
40	99	211	304
50	63	220	332
60	0	228	354

All specimens cured at normal temperatures for 7 days prior to freezing.

vessel. This is not, however, the case in a porous body, where the critical saturation depends on the size of the body, on its homogeneity, and on the rate of freezing. Space available for expelled water must be close enough to the cavity in which ice is being formed, and this is the basis of air entrainment: if the paste is subdivided into sufficiently thin layers by air bubbles the paste has no critical saturation (*see* p. 418). Similarly, an aggregate particle by itself will have no critical size if it has a very low porosity, or if its capillary system is interrupted by a sufficient number of macropores. However, an aggregate particle in concrete can be considered as a closed container, since the low permeability of the surrounding cement paste will not allow water to move sufficiently rapidly into air voids. Thus an aggregate particle saturated above 91·7 per cent will on freezing destroy the surrounding concrete.[7.21] It may be recalled that common aggregates have a porosity of 0 to 5 per cent, and it is usual to avoid aggregates of high porosity. However, the use of such aggregates need not necessarily result in frost damage. Indeed, large pores present in aerated concrete and in no-fines concrete probably contribute to the frost resistance of those materials. Furthermore, even with ordinary aggregate no definite relation between the porosity of the aggregate and the frost resistance of the concrete has been established.

The effect of drying of aggregate, prior to mixing, on the durability of concrete is shown in Fig. 7.13. It can be seen that the presence of saturated

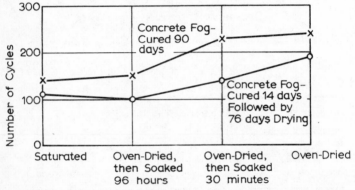

Fig. 7.13. *Relation between the condition of aggregate before mixing and the number of cycles of freezing and thawing to produce a 25 per cent loss in the weight of the specimen*[7.28]

aggregate, particularly of large size, can result in the destruction of concrete, whether or not the concrete is air entrained. On the other hand, if the aggregate is not saturated at the time of mixing or if it is allowed to dry partially after placing and the capillaries in the paste are discontinuous, re-saturation is not easily achieved except during a prolonged period of cold weather.[7.4] On rewetting of concrete it is the cement paste that tends to be more nearly saturated than the aggregate, as water can reach the aggregate only through the paste, and also because the finer-textured paste has a greater capillary attraction. As a result, the cement paste is more vulnerable but the paste can be protected by air entrainment.

Frost-Resistant Concrete
The use of air entrainment is discussed in detail on page 417, but at this stage another means of preventing frost damage is considered: the use of mixes with water/cement ratios sufficiently low for the paste to have

only small capillaries and only little freezable water. It is essential though, that substantial hydration takes place before exposure to frost. Such concrete has a low permeability and does not imbibe water in wet weather.

Fig. 7.14 shows the general effect of the absorption of concrete on its

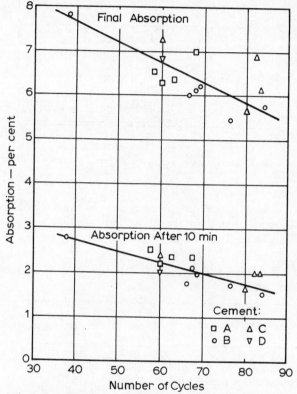

Fig. 7.14. *Relation between absorption of concrete and the number of cycles of freezing and thawing required to cause a 2 per cent reduction in the weight of the specimen*[7.13]

resistance to freezing and thawing, and Fig. 7.15 illustrates the influence of the water/cement ratio on the frost resistance of concrete moist-cured for 28 days. This influence may exist both in air-entrained and non-air-entrained concrete as the water/cement ratio has a definite influence on the size and spacing of air voids in the cement paste.[7.29] The influence of the water/cement ratio on the frost resistance of air-entrained concrete is apparent also from Fig. 7.16, based on tests on concrete moist-cured for 14 days and then stored in air of 50 per cent relative humidity for 76 days prior to exposure to freezing and thawing.[7.29]

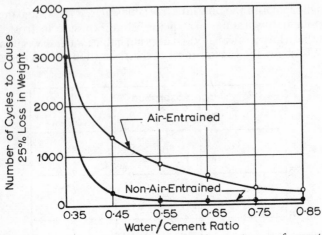

Fig. 7.15. *Influence of water/cement ratio on the frost resistance of concrete moist-cured for 28 days*[7.29]

The importance of the water/cement ratio in the frost resistance of concrete is recognized in practical recommendations, e.g., those of Table 7.7 and Table 10.5. The values refer to concrete cured at a normal

Table 7.7: *Maximum Water/Cement Ratio for Frost Resistance under various conditions*[7.30]

Type of structure and degree of exposure	Maximum water/cement ratio by weight
Concrete exposed to air but not liable to be saturated (provided the cover to reinforcement is not less than 40 to 50 mm (1½ to 2 in.))	0·70
Structural concrete liable to be constantly wet and exposed to freezing; Road slabs exposed to the climate of Great Britain	0·60
Road kerbs	0·55
Road slabs liable to repeated clearance of surface ice by means of salt	0·50 but the use of entrained air is preferable

temperature for at least seven days prior to exposure to frost, and are not applicable when other destructive agencies accompany the action of frost.

Fig. 7.16 shows also results for non-air-entrained concrete. From a comparison of these results with the values of Fig. 7.15 it appears that drying before exposure slightly raises the resistance of concretes with high water/cement ratios, but in the case of mixes with a water/cement ratio below about 0·45 the effect of inadequate hydration is predominant: the concrete with the shorter period of wet curing has a lower resistance to frost.

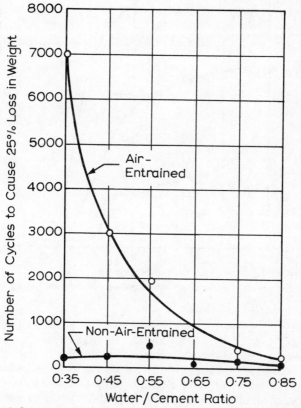

Fig. 7.16. *Influence of water/cement ratio on the frost resistance of concrete moist-cured for 14 days and then stored for 76 days at a relative humidity of 50 per cent*[7.29]

The effect of greater hydration is to reduce the amount of freezable water in the paste; this is illustrated in Fig. 7.17 for concrete with a water/cement ratio of 0·41. This figure shows also that the freezing temperature decreases with age because of an increase in the concentration of alkalis in the still remaining freezable water. In all cases a small amount of water freezes at 0°C (32°F), but this is probably free surface water on the specimen. The temperatures at which freezing of capillary

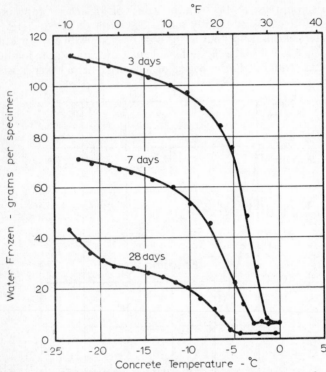

Fig. 7.17. *Effect of age of concrete on amount of water frozen, as a function of temperature*[7.31]

water starts were found to be, approximately − 1°C (30°F) at 3 days, − 3°C (27°F) at 7 days, and − 5°C (23°F) at 28 days.[7.31]

The chemical composition of cement and its fineness have no effect on the frost resistance of concrete[7.32] except at very early ages when these characteristics would affect the degree of hydration, and thus influence both the strength of the paste and the quantity of freezable water in it. The lack of influence of cement was illustrated by field tests of the Portland Cement Association of the United States. Table 7.8 shows the random character of the order of resistance of different types of cement used in different mixes. A possible interpretation, that the relative resistance to frost of cements of different types varies with the water/cement ratio of the mix, is not supported by our knowledge of the phenomena involved.

To reduce the danger of frost attack good compaction of concrete is essential, and for this reason aggregates and techniques that lead to segregation and honeycombing must be avoided. The use of aggregate

Table 7.8: *Frost Resistance of 200 mm (8 in.)-Slump Concrete made with Cements of Different Types*

Cement content		Type of cement in order of performance from best to worst				
kg/m³	lb/yd³					
251	423	V	IV	I	II	III
334	564	III	V	IV	II	I
251	423	IV	II	I	III	V

with a large maximum size or a large proportion of flat particles is inadvisable as pockets of water may collect on the underside of the coarse aggregate.

Whether or not a given concrete is vulnerable to frost, be it due to the paste or to the aggregate, can be determined by cooling the specimen through the freezing range and measuring the change in volume: frost-resistant concrete will contract when water is transferred by osmosis from the paste to the air bubbles but vulnerable concrete will dilate, as shown in Fig. 7.18. Vuorinen[7.103] has shown this one-cycle test to be very useful.

Tests of Frost Resistance of Concrete
Of the four A.S.T.M. tentative methods of test which formerly existed, two have become accepted. In both of these, rapid freezing is applied but in one both freezing and thawing take place in water, while in the other freezing takes place in air and thawing in water. Both methods are covered by A.S.T.M. Standard C 666–71. Freezing saturated concrete in water is several times more severe than in air,[7.49] and the degree of saturation of the specimen at the beginning of the test also affects the rate of deterioration.

The frost damage can be assessed in several ways. The most common method is to measure the change in the dynamic modulus of elasticity of the specimen, the reduction in the modulus after a number of cycles of freezing and thawing expressing the deterioration of the concrete. This method indicates frost damage before it has become apparent either visually or by other methods, although there are some doubts about this interpretation of the decrease in the modulus after the first few cycles of freezing and thawing.[7.48]

With the A.S.T.M. methods it is usual to continue freezing and thawing for 300 cycles or until the dynamic modulus of elasticity is reduced to 60 per cent of its original value, whichever occurs first. The durability can then be assessed by a

$$\text{durability factor} = \frac{\begin{array}{c}\text{number of cycles} \\ \text{at end of test}\end{array} \times \begin{array}{c}\text{percentage of} \\ \text{original modulus}\end{array}}{300}.$$

There are no established criteria for acceptance or rejection of concrete in terms of the durability factor; its value is thus primarily in a comparison of different concretes, preferably when only one variable (e.g. aggregate) is changed. However, some guidance in interpretation can be obtained from the following: a factor smaller than 40 means that the concrete is

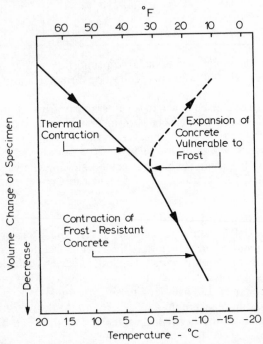

Fig. 7.18. *Change in volume of frost-resistant and vulnerable concretes on cooling*[7.21]

probably unsatisfactory with respect to frost resistance; 40 to 60 is the range for concretes with doubtful performance; and above 60, the concrete is probably satisfactory.

The effects of frost can also be assessed from measurements of loss of compressive or flexural strength or from observations on the change in length[7.48] or weight of the specimen. The last method is applicable when frost damage takes place mainly at the surface of the specimen, but is not reliable in cases of internal failure; the results depend also on the size of the specimen. It may be noted that if failure is primarily due to unsound aggregate it is more rapid and more severe than when the cement paste is disrupted first.

We can see that a number of tests and means of assessing the results is available, and it is not surprising that the interpretation of test data is difficult. If the tests are to yield information indicative of the behaviour of concrete in practice, the test conditions must not be fundamentally different from the field conditions. One difficulty lies in the fact that a test must be accelerated in comparison with the conditions of outdoor freezing, and it is not known at what stage acceleration affects the significance of the test results. There is no doubt, however, that some accelerated freezing and thawing tests result in the destruction of concrete that in practice would be satisfactorily durable.[7.50] One difference between the conditions in the laboratory and actual exposure lies in the fact that in the latter case there is seasonal drying during the summer months, but with permanent saturation imposed in some of the laboratory tests all the air bubbles can eventually become saturated with a consequent failure of the concrete. Also, in the A.S.T.M. tests, cooling takes place at up to 14°C per hour (26°F/h) while in practice 3°C per hour (5°F/h) is rarely exceeded. This severity of conditions is of course unrealistic, but the ability of a concrete to withstand a considerable number of laboratory freezing and thawing cycles (say, 150) is a probable indication of its high degree of durability under service conditions. The A.S.T.M. methods show, however, a high scatter in the middle range of durability. While the numbers of cycles of freezing and thawing in a test and in actual concrete are not related, it may be interesting to note that in most of the U.S. there are more than 50 cycles per annum.

Effects of De-icing Salts

In the case of road slabs frost not only affects the durability of concrete directly but leads also to the use of de-icing salts, which exert an adverse effect on concrete, probably by increasing the severity of the freezing and thawing cycles.

The salts normally used are $NaCl$ and $CaCl_2$, and their repeated application with intervening periods of freezing or drying[7.33] results in surface scaling of the concrete. Sometimes urea is used; it is less deleterious but also less effective in removing ice. Ammonium salts, even in small concentration, are very harmful and should not be used. The salts produce osmotic pressure and cause movement of water toward the top layer of the slab where freezing takes place.[7.21] Since greatest damage occurs when concrete is exposed to relatively low concentrations of salts (2 to 4 per cent solution) (Fig. 7.19) Verbeck and Klieger[7.34] believe the attack to be primarily physical and not chemical in nature. However, the exact mechanism through which de-icers cause scaling has not been determined.

Air entrainment makes concrete very much more resistant to surface scaling, but the reason for this beneficial influence of entrained air is not clear. Use of rich mixes is also advantageous.

Numerous tests on salt scaling have shown that the extent of damage is sensitive to the procedure adopted. For instance, air drying of the concrete after wet curing but prior to exposure cycles increases the resistance to surface scaling.[7.34] The drying out must, however, be preceded by moist-curing of sufficient duration for the cement paste to hydrate extensively. In the case of air-entrained concrete, the period of curing

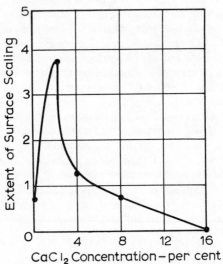

Fig. 7.19. *Effect of concentration of CaCl₂ on scaling of non-air-entrained concrete after 50 cycles of freezing and thawing* (*without removal of the solution*)[7.34]
The extent of surface scaling is rated from 0 = no scaling *to* 5 = bad scaling.

required for high resistance to salt scaling is about the same as that necessary for an adequate strength of the concrete to resist applied loads.[7.35]

The most severe damage occurs when concrete is subjected to alternating freezing and thawing with the de-icer solution remaining on top of the specimen rather than being replaced with fresh water prior to each re-freezing.[7.34] On the other hand, if the liquid is removed from the surface of the concrete prior to re-freezing no scaling takes place, even with non-air-entrained concrete.[7.34]

Air-entrained Concrete
In the British Isles concrete is seldom damaged by frost to any appreciable extent, not only because severe or prolonged frost is rare but also because as a rule very much drier mixes are used than, for instance, in the United States, and such mixes, having a lower water/cement ratio, are *per se* less vulnerable to frost. Recommended values of maximum water/cement ratio for various conditions of exposure are given in

Table 7.7. Otherwise, in the United Kingdom mixes are not usually specifically designed for frost resistance, and air entrainment is only now coming into use. Air-entrained concrete offers, however, some advantages not only from the standpoint of durability but also of workability: a wider use of air-entrained concrete in the future is highly probable.

Air Entrainment
Entrained air in concrete is defined as air intentionally incorporated by means of a suitable agent. This air should be clearly distinguished from accidentally entrapped air: the two kinds differ in the magnitude of the air bubbles, those of entrained air being of the order of 0·05 mm (0·002 in.), while accidental air usually forms very much larger bubbles, some as large as the familiar pockmarks on the surface of the concrete.

Entrained air produces discrete cavities in the cement paste so that no channels for the passage of water are formed and the permeability of the concrete is not increased. The voids never become filled with the products of hydration of cement as gel can form only in water.

The improved resistance of air-entrained concrete to frost attack was discovered accidentally when cement ground with beef tallow, added as a grinding aid, was observed to make more durable concrete than when no grinding aid was used. The main types of air-entraining agents are—

(*a*) animal and vegetable fats and oils and their fatty acids (beef tallow being an example of this group);
(*b*) natural wood resins, which react with lime in the cement to form a soluble resinate. The resin may be pre-neutralized with NaOH so that a water-soluble soap of a resin acid is obtained (e.g. Vinsol resin); and
(*c*) wetting agents such as alkali salts of sulphated and sulphonated organic compounds (e.g. Darex).[7.36]

Numerous proprietary brands of air-entraining agents are available commercially but the performance of the unknown ones should be checked by trial mixes. A.S.T.M. Standard C260–69 lays down the performance requirements of air-entraining admixtures. The essential requirements of an air-entraining agent are that it rapidly produces a system of finely divided and stable foam, the individual bubbles of which resist coalescence. The foam must have no harmful chemical effect on the cement.

The air-entraining agent can be dispensed either as an admixture, i.e. a material added to the mix when the ingredients are fed into the mixer, or else as an addition to the cement, in which case the agent is interground with the cement in fixed proportions. The latter method permits less flexibility in altering the air content of different mixes, particularly when widely varying aggregate/cement ratios or aggregates of different types are used. On the other hand, the use of admixtures complicates somewhat the batching operation. The entraining agent represents between 0·005 and 0·05 per cent of the weight of the cement, but to facilitate the dispensing operation a solution of the agent in water is usually made up so that the actual quantities handled are greater. Nevertheless, careful control at the batching stage is required since the advantages of air-entrained concrete cannot be relied upon unless the quantity of entrained air is within specified limits.

Air Content

For each mix there is a minimum volume of voids required for protection from frost (Fig. 7.20). Klieger[7.37] found this volume to correspond to

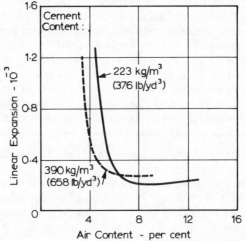

Fig. 7.20. *Influence of air content of concrete on expansion after 300 cycles of freezing and thawing*[7.37]

9 per cent of the volume of *mortar*, and it is of course essential that the air be distributed throughout the cement paste. The actual controlling factor is the spacing of the bubbles, i.e. the thickness of the cement paste between adjacent air voids. A spacing of 0·25 mm (0·01 in.) between the voids is believed to be required for full protection from frost damage[7.38] (Fig. 7.21).

Since the total volume of voids in a given volume of concrete affects the strength of concrete (cf. p. 182) it follows that the air bubbles should be as small as possible. Their size depends to a large degree on the foaming process used. In fact, the voids are not all of one size, and it is convenient to express their size in terms of specific surface (square millimetres per cubic millimetre or square inches per cubic inch).

We must not forget that accidental air is present in any concrete, whether air entrained or not, and as the two kinds of voids cannot be distinguished, the specific surface represents an average value for all voids in a given paste. For air-entrained concrete of satisfactory quality the specific surface of voids is in the range of approximately 16 to 24 mm⁻¹ (400 to 600 in⁻¹), but sometimes it is as high as 32 mm⁻¹ (800 in⁻¹). By

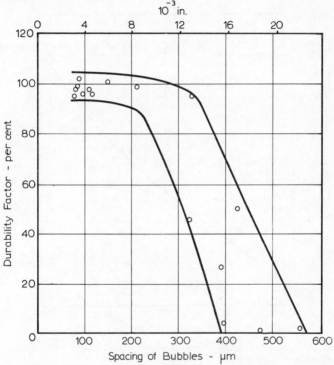

Fig. 7.21. *Relation between durability and spacing of bubbles of entrained air*[7.39]

contrast, the specific surface of accidental air is less than 12 mm⁻¹ (300 in⁻¹).[7.38] The size of the entrained bubbles ranges usually from 0·05 to 1·27 mm (0·002 to 0·05 in.).

For a given air content of the concrete the spacing of air bubbles depends on the water/cement ratio of the mix, as shown in Fig. 7.22. Typical values of the amount of air required for a 0·25 mm (0·01 in.) spacing for different mixes are given in Table 7.9, based on Powers' results.[7.38] Although the air is present only in the cement paste it is usual to specify the air content as a percentage of the volume of the *concrete**

---

* Appropriately higher values may be required in grout in prestressed concrete ducts; the voids induced by aluminium powder (see page 542) are insufficient for frost protection.

(*see* Table 7.10), and a physical determination of air content gives this value and not the void ratio in the cement paste alone.

Table 7.9 indicates that for a particular value of the specific surface of the voids richer mixes require a greater volume of entrained air than lean ones. However, the richer the mix the greater the specific surface of the

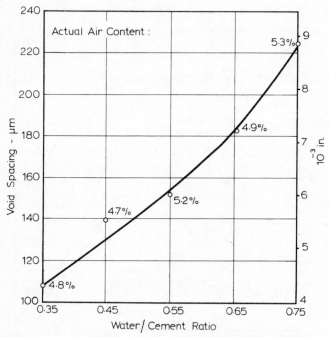

Fig. 7.22. *Influence of the water/cement ratio on the void spacing in concrete with an average air content of 5 per cent*[7.29]

voids for a given air content. For instance, actual mixes with a maximum aggregate size of 19·05 mm ($\frac{3}{4}$ in.) (some of them referred to in Fig. 7.20) have been found to yield the following values of specific surface of voids—

Cement content		Optimum air content per cent	Specific surface of voids	
kg/m³	lb/yd³		mm⁻¹	in⁻¹
223	376	6·5	13	330
307	517	6·0	17	420
391	658	6·0	23	580

Table 7.9: *Air Content required for a Void Spacing of 0·25 mm (0·01 in.)*[7.38]

Approximate cement content of concrete		Water/cement ratio	Air requirement as a percentage of volume of concrete for specific surface of voids, mm^{-1} (in^{-1}), of—				
kg/m³	lb/yd³		14 (350)	18 (450)	20 (500)	24 (600)	31 (800)
445	750		8·5	6·4	5·0	3·4	1·8
390	660	0·35	7·5	5·6	4·4	3·0	1·6
330	560		6·4	4·8	3·8	2·5	1·3
445	750		10·2	7·6	6·0	4·0	2·1
390	660	0·49	8·9	6·7	5·3	3·5	1·9
330	560		7·6	5·7	4·5	3·0	1·6
280	470		6·4	4·8	3·8	2·5	1·3
445	750		12·4	9·4	7·4	5·0	2·6
390	660		10·9	8·2	6·4	4·3	2·3
330	560	0·66	9·3	7·0	5·5	3·7	1·9
280	470		7·8	5·8	4·6	3·1	1·6
225	380		6·2	4·7	3·7	2·5	1·3

Table 7.10: *Recommended Air Content of Concretes Containing Aggregates of Different Maximum Size*[7.12]

Maximum size of aggregate		Average air content of concrete per cent
mm	in.	
10	$\frac{3}{8}$	8
12·5	$\frac{1}{2}$	7
20	$\frac{3}{4}$	6
25	1	5
40	$1\frac{1}{2}$	$4\frac{1}{2}$
50	2	4
70	3	$3\frac{1}{2}$
150	6	3

The British values are slightly lower.

The volume of air entrained in a given concrete is independent of the volume of accidental air and depends primarily on the amount of air-entraining agent added. The larger the quantity of the agent the more air is entrained, but there is a maximum amount of any agent beyond which there is no increase in the volume of voids.

There are other factors influencing the amount of air actually entrained when a given amount of air-entraining agent is added. A more workable mix holds more air than a drier one. An increase in the fineness of cement decreases the effectiveness of air entraining,[7.40] but the exact rôle of the various properties of cement has not yet been determined. The aggregate grading also affects the volume of voids. This volume is decreased by an excess of very fine sand particles, but the material in the 300 to 600 μm (No. 52 to 25 sieve) size range increases the amount of entrained air,[7.41] and so does the use of angular rather than rounded aggregate. Generally, it seems that the solid constituents of concrete affect the amount of entrained air as a result of competition for water: each air bubble requires a boundary film of water, and the larger the surface of solids to be wetted the less water is available for the void boundaries.

The actual mixing operation also affects the resultant air content. If the mixing time is too short the air entraining agent is not sufficiently dispersed, but over-mixing gradually expels some air so that there is an optimum value of mixing time. In practice, the mixing time is fixed from other considerations, usually at a value shorter than the minimum necessary for the agent to become fully dispersed, and the amount of the air-entraining agent must be adjusted accordingly. Higher temperature leads to a greater loss of air so that less of it remains actually entrained (although this effect is slightly offset by the higher volume occupied by a given weight of air at a higher temperature[7.41]). An increase in temperature from 10 to 32°C (50 to 90°F) would approximately halve the nett amount of air actually entrained.

Vibrating the concrete expels some of the entrained air, though mainly the larger bubbles which are of lesser importance.[7.42] However, prolonged vibration results in a considerable loss of air so that after 3 minutes only half the original amount may be left, and as little as 20 per cent may remain after 9 minutes.[7.43]

The adequacy of air entrainment in a given concrete can be estimated by a spacing factor, prescribed by the A.S.T.M. Standard C457–71. The spacing factor is a useful index of the maximum distance of any point in the cement paste from the periphery of a nearby air void. The calculation of the factor is based on the assumption that all air voids are equal-sized spheres arranged in a simple cubic lattice. The calculation is laid down by A.S.T.M. Standard C 457–71, and requires the knowledge of: the air content of the concrete, determined by a linear traverse microscope; the average number of air void sections per inch or the average chord intercept of the voids; and the cement paste content by volume. The factor is expressed in inches although clearly a metric determination is possible; usually a value of not more than 0·008 in. (0·2 mm) is a maximum value required for satisfactory frost protection.

Effects of Air Entrainment

As mentioned before, the original purpose of air entrainment was to make frost-resistant concrete. In the initial stages of freezing, the voids relieve the hydraulic pressure being developed in the channels in the paste, and during the progress of freezing the voids prevent or limit the growth of microscopic bodies of ice in the cement paste. Each void protects only a thin "shell" surrounding it, and with too large a spacing of voids some expansion of the paste would take place. When, however, the protected shells overlap, no part of the paste is vulnerable to frost and, in fact, due to the accommodation of excess water in the voids, a freezing cement paste contracts as the temperature drops (*see* Fig. 7.18), as would any solid body on cooling. During thawing the water returns from the voids to the cement paste so that protection by air entrainment continues permanently for repeated freezing and thawing.[7.44] Air entraining increases also the resistance of concrete to the destructive action of de-icing agents.

There are some further effects of air entrainment on the properties of concrete, some beneficial, others not. One of the most important is the influence of voids on the strength of concrete at all ages. It will be remembered that the strength of concrete is a direct function of its density ratio, and voids caused by entrained air will affect the strength in the same way as voids of any other origin. Fig. 7.23 shows that when entrained air is added to a mix without any other change in the mix proportions being made, the decrease in the strength of concrete is proportional to the volume of air present. The range considered is up to 8 per cent of air and this is why the curved part of the strength–void ratio relation is not apparent (cf. Fig. 4.1). That the origin of the air is not significant is apparent from the dotted curve in Fig. 7.23 which shows the strength–void ratio relation for the case when the voids are due to inadequate compaction and not to entrainment. The range of tests covered mixes with water/cement ratios between 0·45 and 0·72, and this shows that the loss of strength expressed as a fraction of the strength of air-free concrete is independent of the mix proportions. The average loss of compressive strength is 5·5 per cent for each per cent of air present.[7.45] The effect on the modulus of rupture is much smaller.

It should be noted that strength is affected by the total volume of all the voids present: entrapped air, entrained air, capillary pores, and gel pores. When entrained air is present in a mix the total volume of capillary pores is smaller because a part of the gross volume of cement paste consists of entrained air. This is not a negligible factor because the volume of entrained air represents a significant proportion of the gross volume of the paste. For instance, in a 1 : 3·4 : 4·2 mix with a water/cement ratio of 0·80, the capillary pores at the age of 7 days were found to occupy 13·1 per cent of the volume of concrete. With entrained air in a mix of the same workability (1:3·0:4·2 with a water/cement ratio of 0·68), the capillary pores occupied 10·7 per cent, but the volume of air (entrained and entrapped) was 6·8 per cent (compared with 2·3 per cent in the former mix).[7.106]

This is one reason why air entrainment does not cause as large a loss of strength as might be expected. But a more important reason is that the

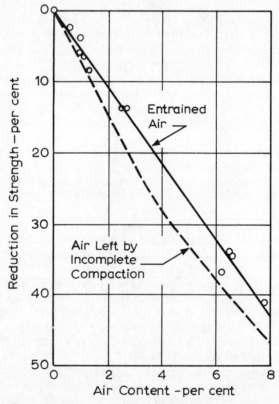

Fig. 7.23. *Effect of entrained and accidental air on the strength of concrete*[7.45]

entrainment of air has a considerable beneficial effect on the workability of the concrete. As a result, in order to keep the workability constant, the addition of entrained air can be accompanied by a reduction in the water/cement ratio, compared with a similar mix without entrained air. For lean mixes, say, with an aggregate/cement ratio of 8 or more, and particularly when angular aggregate is used, the improvement in workability due to air entrainment is such that the possible decrease in the water/cement ratio compensates fully for the loss of strength due to the presence of the voids. In the case of mass concrete, where the development of heat of hydration of cement, and not strength, is of primary importance, air entrainment permits the use of mixes with low cement contents and therefore a low temperature rise. In richer mixes the effect of air entrainment on workability is smaller, so that the water/cement ratio can be lowered only a little, and there is a nett loss in strength. In general terms, entrainment of 5 per cent of air increases the compacting factor of concrete by about 0·03 to 0·07, and the slump by 15 to 50 mm ($\frac{1}{2}$ to 2 in.) (Fig. 7.24) but actual values vary with the properties of the mix. Air entrainment is also effective in improving the workability of the rather harsh mixes made with lightweight aggregate.

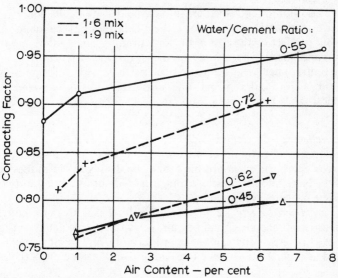

Fig. 7.24. *Effect of entrained air on the compacting factor of concrete (for constant mix proportions)*[7.45]

The reason for the improvement of workability by the entrained air is probably that the air bubbles, kept spherical by the surface tension, act as a fine aggregate of very low surface friction and considerable elasticity. Entraining air in the mix makes it actually behave like an over-sanded mix, and for this reason the addition of entrained air should be accompanied by a reduction in the sand content. The latter change allows a further reduction in the water content of the mix, i.e. a further compensation of the loss of strength due to the presence of voids is possible.

It is interesting to note that air entrainment affects the consistence or "mobility" of the mix in a qualitative manner: the mix can be said to be more "plastic," so that for the *same* workability, as measured, say, by the compacting factor, the mix containing entrained air is easier to place and compact than an air-free mix.

The presence of entrained air is also beneficial in reducing bleeding: the air bubbles appear to keep the solid particles in suspension so that sedimentation is reduced and water is not expelled. For this reason permeability and the formation of laitance are also reduced, and this results in an improved frost resistance of the top layer of a slab or a lift.

It is claimed sometimes that air entrainment reduces segregation. This is true as far as segregation during handling and transporting is concerned, as the mix is more cohesive but segregation due to over-vibration is still possible, particularly as under those conditions the air bubbles are rapidly expelled.

The addition of entrained air lowers the density of the concrete and makes cement and aggregate go further. This offers an economic advantage but is offset by the cost of the air-entraining agent (although this is cheap) and possibly by the increased cost of supervision.

Air entrainment has been used successfully with sulphate-resisting and other Portland cements, and also when $CaCl_2$ is added as an accelerator.[7.46]

Measurement of Air Content

There are three methods of measuring the *total* air content of fresh concrete. Since the entrained air cannot be distinguished in these tests from the large bubbles of accidental air it is important that the concrete tested be properly compacted.

The gravimetric method is the oldest one. It relies simply on comparing the density of concrete containing air, ρ_a, with the calculated density of air-free concrete of the same mix proportions, ρ. The air content, expressed as a percentage of the total volume of the concrete is then

$$1 - \frac{\rho_a}{\rho}.$$

This method is covered by A.S.T.M. Standard C 138–71T and can be used when the specific gravity of the aggregate and the mix proportions are constant. An error of 1 per cent in the calculated air content is not uncommon; this order of error would be expected from the simple experience of determining the density of nominally similar test specimens of non-air-entrained concrete.

In *the volumetric method* the difference in the volumes of a sample of concrete before and after the air has been expelled is determined. The air is removed by agitation, rolling, and stirring, the operation being performed in a pycnometer or other suitable vessel. The details of the test are prescribed by A.S.T.M. Standard C 173–71. The main difficulty lies in the fact that the weight of water replacing the air is small compared with the total weight of the concrete.

The most popular method and one best suited for site use is the *pressure method*. It is based on the relation between the volume of air and the applied pressure (at a constant temperature) given by Boyle's law. The

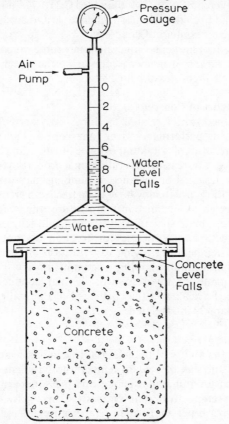

Fig. 7.25. *Pressure-type air meter*

mix proportions or the properties of the materials need not be known, and when commercial air meters* are used, no calculations are required as direct graduations in percentage of air are provided.

A typical air meter is shown in Fig. 7.25. The procedure consists essentially of observing the decrease in the volume of a sample of concrete when subjected to a known pressure. The pressure is applied by a

---

* The meter is not suitable for use with porous aggregates.

small pump, such as a bicycle pump, and measured by a pressure gauge. Due to the increase in pressure over atmospheric, the volume of air in the concrete decreases and this causes a fall in the level of the water above the concrete. By arranging the level of the water to vary within a calibrated tube the air content can be read direct by an unskilled operator.

The test is covered by A.S.T.M. Standard C 231–72T and by B.S. 1881: Part 2: 1970, and provides the most dependable and accurate method of determining the air content of concrete.

The air content of hardened concrete is measured on polished sections of concrete by means of a microscope using the linear traverse technique,[7.47] or by a high-pressure air meter.

Thermal Properties of Concrete

Although not necessarily related to durability, thermal properties of concrete affect its performance over long periods under varying conditions and are also of vital importance in planning mass concrete construction. For this reason data on thermal properties of concrete may be required in design. For instance, a building may require a specific degree of insulation; a slab may have to be free from cracking and warping when subjected to a change in temperature; the stresses induced in statically indeterminate structures by a change in temperature must be calculated; and in mass concrete construction the potential rise and distribution of temperature owing to hydration of cement have to be estimated so that a suitable cooling system may be designed.

The basic quantities involved are: thermal conductivity, thermal diffusivity, specific heat, and the coefficient of thermal expansion. The first three are largely interrelated.

Thermal Conductivity

This measures the ability of the material to conduct heat and is defined as the ratio of the flux of heat to temperature gradient. Thermal conductivity is measured in Joules per second per square metre of area of body when the temperature difference is 1°C per metre of thickness of the body (Btu per h per sq. ft when temperature difference is 1°F per ft of thickness).

The conductivity of ordinary concrete depends on its composition, and when the concrete is saturated the conductivity ranges generally between about 1·4 and 3·6 Jm/m²s°C (0·8 and 2·1 Btu/ft²h°F/ft).[7.51] Density does not appreciably affect the conductivity of ordinary concrete but, due to the low conductivity of air, the thermal conductivity of light-weight concrete varies with its density[7.52] (*see* Fig. 9.1). Typical values of conductivity are listed in Table 7.11. It can be seen that the mineralogical character of the aggregate greatly affects the conductivity of the concrete made with it. In general terms, basalt and trachyte have a low

Table 7.11: *Typical values of Thermal Conductivity of Concrete*[7.51]

Type of aggregate	Density of concrete		Conductivity	
	kg/m³	lb/ft³	Jm/m²s°C	Btu ft/ft²h°F
Barytes	3,640	227	1·38	0·8
Igneous	2,540	159	1·44	0·83
Dolomite	2,560	160	3·68	2·13
Lightweight concrete (oven-dried)	480–1,760	30–110	0·14–0·60	0·08–0·35

conductivity, dolomite and limestone are in the middle range, and quartz exhibits the highest conductivity, which depends also on the direction of heat flow relative to the orientation of the crystals.

The degree of saturation of concrete is a major factor, since the conductivity of air is lower than that of water. For instance, in the case of lightweight concrete an increase in moisture content of 10 per cent increases conductivity by about one-half. On the other hand, the conductivity of water is less than half that of the cement paste, so that the lower the water content of the *mix* the higher the conductivity of the hardened concrete.

A frequent practical difficulty is to know the actual moisture content of the concrete. Loudon and Stacy[7.76] suggest as typical the following values of moisture content in per cent by volume:

	For concrete exposed to weather	For concrete protected from weather
Normal weight aggregate concrete	5	2·5
Lightweight aggregate or aerated concrete	8	5

On that basis, they suggested the use of the values of conductivity given in Table 7.12.

Conductivity is little affected by temperature within the usual ambient range, the general effect of an increase in temperature being to decrease slightly the conductivity of ordinary concrete, but the reverse is the case with lightweight concrete.[7.51] However, at high temperatures, there is a decrease in conductivity until at 800°C (1,470°F) it is about one-half of the value at 20°C (68°F).[7.77]

Thermal conductivity is usually calculated from the diffusivity, the latter being easier to measure, but a direct determination of conductivity

Table 7.12: *Values of Conductivity recommended by Loudon and Stacey*[7.76]

Conductivity, Jm/m²s°C (Btu ft/ft²h°F)

Unit weight		For concrete protected from weather				For concrete exposed to weather			
kg/m³	lb/ft³	Aerated concrete	Lightweight concrete with foamed slag	Lightweight concrete with expanded clay or sintered fly ash	Normal weight aggregate concrete	Aerated concrete	Lightweight concrete with foamed slag	Lightweight concrete with expanded clay or sintered fly ash	Normal weight aggregate concrete
320	(20)	0·109 (0·063)	0·087 (0·050)	0·130 (0·075)		0·123 (0·071)	0·100 (0·058)	0·145 (0·084)	
480	(30)	0·145 (0·084)	0·116 (0·067)	0·173 (0·100)		0·166 (0·096)	0·130 (0·075)	0·187 (0·108)	
640	(40)	0·203 (0·117)	0·159 (0·092)	0·230 (0·133)		0·223 (0·129)	0·173 (0·100)	0·260 (0·150)	
800	(50)	0·260 (0·150)	0·203 (0·117)	0·303 (0·175)		0·273 (0·158)	0·230 (0·133)	0·332 (0·192)	
960	(60)	0·315 (0·182)	0·260 (0·150)	0·376 (0·217)		0·360 (0·208)	0·289 (0·167)	0·433 (0·250)	
1,120	(70)	0·389 (0·225)	0·315 (0·182)	0·462 (0·267)		0·433 (0·250)	0·360 (0·208)	0·519 (0·300)	
1,280	(80)	0·476 (0·275)	0·389 (0·225)	0·562 (0·325)		0·533 (0·308)	0·433 (0·250)	0·635 (0·367)	
1,440	(90)		0·462 (0·267)	0·678 (0·392)					
1,600	(100)		0·549 (0·317)	0·794 (0·459)	0·706 (0·408)				0·808 (0·467)
1,760	(110)		0·649 (0·375)	0·952 (0·550)	0·838 (0·484)				0·952 (0·550)
1,920	(120)				1·056 (0·610)				1·194 (0·690)
2,080	(130)				1·315 (0·760)				1·488 (0·860)
2,240	(140)				1·696 (0·980)				1·904 (1·100)
2,400	(150)				2·267 (1·310)				2·561 (1·480)

is of course possible. However, the method of test may affect the value obtained. For instance, the steady-state methods (hot plate and hot box) yield the same thermal conductivity for dry concrete, but give too low a value for moist concrete because the temperature gradient causes migration of moisture. For this reason, it is preferable to determine the conductivity of moist concrete by transient methods; the hot wire test has been found successful.[7.78]

Thermal Diffusivity

Diffusivity represents the rate at which temperature changes within a mass can take place, and is thus an index of the facility with which concrete can undergo temperature changes. Diffusivity, δ, is simply related to the conductivity K by the equation

$$\delta = \frac{K}{c\rho}$$

where c is the specific heat,
and ρ is the density of concrete.

From this expression it can be seen that conductivity and diffusivity vary in step. The range of typical values of diffusivity of ordinary concrete is between 0·002 and 0·006 m²/h (0·02 and 0·06 ft²/h), depending on the type of aggregate used. The following rock types are listed in order of increasing diffusivity: basalt, rhyolite, granite, limestone, dolerite, and quartzite.[7.53]

The measurement of diffusivity consists essentially of determining the relation between time and the temperature differential between the interior and the surface of a concrete specimen initially at a constant temperature when a change in temperature is introduced at the surface. Details of procedure and calculation are given in U.S. Corps of Engineers Standard CRD–C36–48.[7.61] Because of the influence of moisture in the concrete on its thermal properties, diffusivity should be measured on specimens with a moisture content which will exist in the actual structure.

Specific Heat

Specific heat, which represents the heat capacity of concrete, is little affected by the mineralogical character of the aggregate, but is considerably increased by an increase in the moisture content of the concrete. Specific heat increases with an increase in temperature. The common range of values for ordinary concrete is between 840 and 1,170 J/kg/°C (0·20 and 0·28 Btu/lb/°F). The specific heat of concrete is determined by elementary methods of physics.

Coefficient of Thermal Expansion

Like most engineering materials, concrete has a positive coefficient of thermal expansion, but its value depends both on the composition of the mix and on its hygral state at the time of the temperature change.

The influence of the mix proportions arises from the fact that the two main constituents of the concrete, cement paste and aggregate, have dissimilar thermal coefficients, and the coefficient for concrete is a resultant of the two values. The coefficient of thermal expansion of cement paste varies between about 11×10^{-6} and 20×10^{-6} per °C (6×10^{-6} and 11×10^{-6} per °F),[7.53] and is higher than the coefficient of aggregate. In general terms, the coefficient of concrete is a function of the

Table 7.13: *Influence of Aggregate Content on the Coefficient of Thermal Expansion*[7.60]

Cement/sand ratio	Linear coefficient of thermal expansion at the age of 2 years	
	10^{-6} per °C	10^{-6} per °F
neat cement	18·5	10·3
1:1	13·5	7·5
1:3	11·2	6·2
1:6	10·1	5·6

quantity of aggregate in the mix (Table 7.13) and of the coefficient of the aggregate by itself.[7.54] The influence of the latter factor is apparent from Fig. 7.26, and Table 7.11 gives the values of the coefficient of thermal expansion of 1:6 concretes made with different aggregates.[7.55] The significance of the difference between the coefficients of the aggregate and the cement paste was discussed on page 144.

The influence of the moisture condition applies to the paste component and is due to the fact that the thermal coefficient is made up of two movements: the true kinetic thermal coefficient and swelling pressure. The latter arises from a decrease in the capillary tension of water held by the paste with an increase in temperature.[7.56] No swelling is possible, however, when the specimen is dry, i.e. when it contains no water, or when it is saturated. It follows that at these two extremes the coefficient of thermal expansion is lower than when the paste is partially saturated. When the paste is self-desiccated the coefficient is higher because there is not enough water for free exchange of moisture to occur between capillary and gel pores after the temperature change.

When saturated paste is heated, the moisture diffusion from gel to capillary pores at constant gel water content is partially offset by contraction as gel loses water so that the apparent coefficient is smaller.[7.79]

Table 7.14: *Coefficient of Thermal Expansion of 1:6 Concretes made with Different Aggregates*[7.55]

	Linear coefficient of thermal expansion					
Type of aggregate	Air-cured concrete		Water-cured concrete		Air-cured and wetted concrete	
	10^{-6} per °C	10^{-6} per °F	10^{-6} per °C	10^{-6} per °F	10^{-6} per °C	10^{-6} per °F
Gravel	13·1	7·3	12·2	6·8	11·7	6·5
Granite	9·5	5·3	8·6	4·8	7·7	4·3
Quartzite	12·8	7·1	12·2	6·8	11·7	6·5
Dolerite	9·5	5·3	8·5	4·7	7·9	4·4
Sandstone	11·7	6·5	10·1	5·6	8·6	4·8
Limestone	7·4	4·1	6·1	3·4	5·9	3·3
Portland stone	7·4	4·1	6·1	3·4	6·5	3·6
Blast-furnace slag	10·6	5·9	9·2	5·1	8·8	4·9
Foamed slag	12·1	6·7	9·2	5·1	8·5	4·7

(*Crown Copyright*)

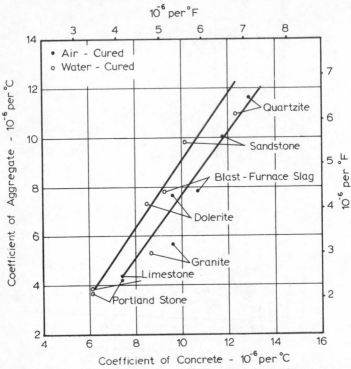

Fig. 7.26. *Influence of the linear coefficient of thermal expansion of aggregate on the coefficient of thermal expansion of a 1:6 concrete*[7.55]

(*Crown Copyright*)

Conversely, on cooling the contraction due to moisture diffusion from capillary to gel pores at constant gel water content is partially offset by the expansion which occurs when the gel absorbs water.[7.79]

Actual values are shown in Fig. 7.27, and it can be seen that for young pastes the coefficient is a maximum at a relative humidity of about 70 per cent. The relative humidity at which the coefficient is a maximum decreases with age, down to about 50 per cent for very old pastes[7.53] (Fig. 7.28). Likewise, the coefficient itself decreases with age due to a reduction in the potential swelling pressure owing to an increase in the amount of crystalline material in the hardened paste. No such variation in the coefficient of thermal expansion is found in high-pressure steam cured pastes since they contain no gel (Fig. 7.27).

Figs. 7.27 and 7.28 refer to neat cement pastes but the effects are apparent also in concrete; here, though, the variation in the coefficient is smaller as only the paste component is affected by the relative humidity and ageing. Table 7.14 gives values of the coefficient for 1:6 concretes:

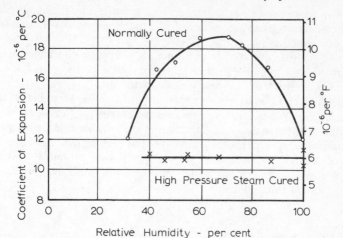

Fig. 7.27. *Relation between ambient relative humidity and the linear coefficient of thermal expansion of neat cement paste cured normally and high-pressure steam cured*[7.53]

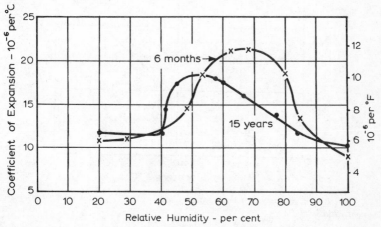

Fig. 7.28. *The linear coefficient of thermal expansion of neat cement paste at different ages*[7.53]

cured in air at 64 per cent relative humidity, saturated (water-cured), and wetted after air-curing.

Only the values determined on saturated or desiccated specimens can be considered to represent the "true" coefficient of thermal expansion but it is the values at intermediate humidities that are applicable to many concretes under practical conditions. When the increase in temperature from winter to summer is associated with some drying, shrinkage enters the picture and the nett expansion is lower than when no loss of water from the concrete takes place.[7.1]

The chemical composition and fineness of cement affect the thermal expansion only in so far as they influence the properties of gel at early ages. The presence of air voids is not a factor.

The data considered so far apply only to temperatures below, say, 65°C (150°F). Considerably higher temperatures can, however, be encountered in the vicinity of jet exhausts on airfields and in industrial applications. Fig. 7.29 shows that above about 320°C (600°F) the coefficient of thermal expansion of concrete increases, probably owing to dehydration of the cement paste; values of the coefficient of thermal expansion are listed in Table 7.15.

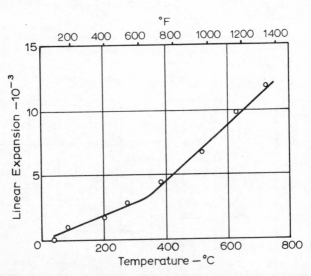

Fig. 7.29. *Linear expansion of concrete on heating*[7.57] *(calcareous aggregate; water/ cement ratio = 0·4)*

Laboratory tests have shown that concretes with a higher coefficient of thermal expansion are less resistant to temperature changes than concretes with a lower coefficient.[7.54] Fig. 7.30 shows the results of tests on concrete heated and cooled repeatedly between 4 and 60°C (40 and 140°F) at the rate of 2·2°C (4°F) per minute. However, the data are not sufficient for the coefficient of thermal expansion to be considered as a quantitative measure of durability of concrete subjected to frequent or rapid changes in temperature (cf. p. 144).

Table 7.15: *Coefficient of Thermal Expansion of Concrete at High Temperatures*[7.57]

Curing condition	Water/cement ratio	Cement content		Aggregate	Linear coefficient of thermal expansion, at the age of:							
					28 days				90 days			
					below 260°C (500°F)		above 430°C (800°F)		below 260°C (500°F)		above 430°C (800°F)	
		kg/m³	lb/yd³		10⁻⁶ per °C	10⁻⁶ per °F	10⁻⁶ per °C	10⁻⁶ per °F	10⁻⁶ per °C	10⁻⁶ per °F	10⁻⁶ per °C	10⁻⁶ per °F
Moist	0·4	435	735	calcareous gravel	7·6	4·2	20·3	11·3	6·5	3·6	11·2	6·2
	0·6	310	520		12·8	7·1	20·5	11·4	8·4	4·7	22·5	12·5
	0·8	245	415		11·0	6·1	21·1	11·7	16·7	9·3	32·8	18·2
Air, 50 per cent relative humidity	0·4	435	735	calcareous gravel	7·7	4·3	18·9	10·5	12·2	6·8	20·7	11·5
	0·6	310	520		7·7	4·3	21·1	11·7	8·8	4·9	20·2	11·2
	0·8	245	415		9·6	5·3	20·7	11·5	11·7	6·5	21·6	12·0
Moist air	0·68	355	600	expanded shale	6·1	3·4	7·5	4·2	—	—	—	—
	0·68	355	600		4·7	2·6	9·7	5·4	5·0	2·8	8·8	4·9

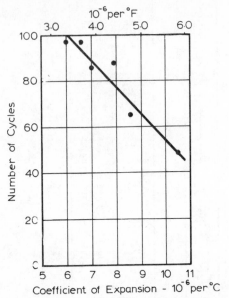

Fig. 7.30. *Relation between the linear coefficient of thermal expansion of concrete and the number of cycles of heating and cooling required to produce a 75 per cent reduction in the modulus of rupture*[7.54]

Nevertheless, rapid changes in temperature, generally faster than encountered under normal conditions, may lead to deterioration of concrete: Fig. 7.31 shows the effects of quenching after heating to the indicated temperature.[7.58]

Resistance of Concrete to Fire

The resistance of concrete to fire is a topic outside the scope of this book as fire endurance applies really to a building element rather than to a building material. We can say, however, that, in general, concrete has good properties with respect to fire resistance; that is, the period of time under fire during which concrete continues to perform satisfactorily is relatively high and no toxic fumes are emitted. The relevant criteria of performance are: load-carrying capacity, resistance to flame penetration, and resistance to heat transfer when concrete is used as a protective material for steel.

Considering purely the behaviour of concrete as a material, we should note that fire introduces high temperature gradients, and as a result the hot surface layers tend to separate and spall from the cooler interior of the body. The formation of cracks is encouraged at joints, in poorly compacted parts of the concrete, or in the planes of reinforcing bars;

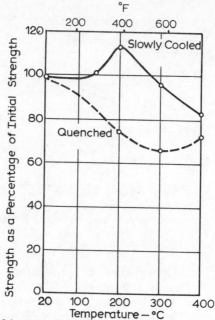

Fig. 7.31. *Effect of the rate of cooling on the strength of concrete made with a sandstone aggregate and previously heated to different temperatures*[7.58]

once the reinforcement has become exposed, it conducts heat and accelerates the action of heat.

The effect of increase in temperature on the strength of concrete is small and somewhat irregular below 250°C (Fig. 7.32) but above about 300°C a definite loss of strength takes place, as shown in Fig. 7.33. If the high temperature is of short duration (e.g. one hour), a slow recovery of strength may take place.[7.77] At low temperatures the strength of concrete is higher than at room temperature. For instance, at −60 to −157°C (−75 to −250°F) and at −80 to −196°C (−112 to −321°F) the strength of moist concrete is two to three times higher than at room temperature,[7.80] but dry concrete is only 20 per cent stronger.[7.63] The loss in strength at higher temperatures is greater in saturated than in dry concrete, and it is probably the moisture content at the time of compression testing that is responsible for the difference.[7.81] The strength of mass-cured concrete beyond the age of 14 days seems to be unaffected by temperature within the range 21 to 96°C (70 to 205°F).[7.82] This behaviour is probably due to an absence of a change in moisture content and an absence of shrinkage. The influence of moisture content on strength is apparent also in fire tests on concrete, where excessive moisture

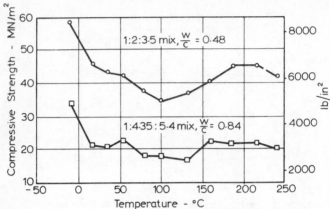

Fig. 7.32. *Compressive strength of concrete after heating to different temperatures*[7.59] (*calcareous aggregate*)

at the time of fire is the primary cause of spalling; when concrete is in hygral equilibrium with air, spalling does not take place.

It is interesting to note that dolomitic gravel leads to a very good fire resistance of concrete. The reason for this is that the calcination of the carbonate aggregate is endothermic.[7.83] Also, the calcined material has a lower density and therefore provides a measure of surface insulation. This effect is significant in thick members.

Leaner mixes appear to suffer a relatively lower loss of strength than richer ones.[7.62] Flexural strength is more affected than compressive strength.[7.58] The loss of strength is considerably lower when the aggregate does not contain silica, e.g. with limestone, basic igneous rocks, and particularly crushed brick and blast-furnace slag. Low conductivity of concrete improves its fire resistance so that, for instance, lightweight concrete stands up better to fire than ordinary concrete (*see* p. 543).

Concretes made with siliceous or limestone aggregate show a change in colour with temperature (Fig. 7.33). This change is permanent, so that the maximum temperature during a fire can be estimated *a posteriori*. Thus the residual strength can be approximately judged: generally, concrete whose colour has changed beyond pink is suspect, and concrete past the grey stage is probably friable and porous.[7.1]

The pattern of influence of temperature on the modulus of elasticity is shown in Fig. 7.34. For mass-cured concrete there is no difference in modulus in the range 21 to 96°C (70 to 205°F).[7.82] However, when water can be expelled from concrete, there is a progressive decrease in the modulus of elasticity between about 50 and 400°C (120 and 750°F) (*see* Fig. 7.34);[7.84] relaxation of bonds may be a factor in this. The extent of

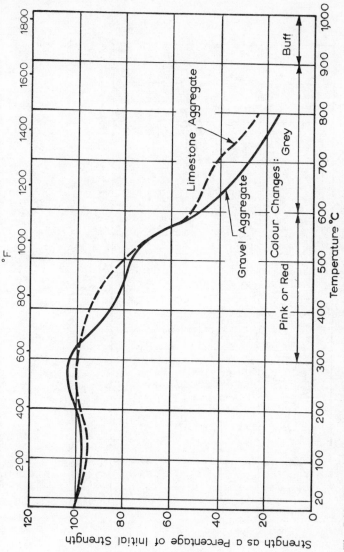

Fig. 7.33. *Compressive strength of concrete after heating to high temperatures*[7.58]

Note: Above about 1,200°C concrete turns yellow.

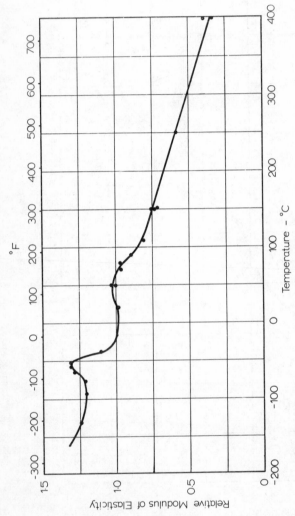

Fig. 7.34. *Influence of temperature on modulus of elasticity of concrete*[7.84]

the decrease in the modulus depends on the aggregate used, but a generalization on this subject is difficult. In broad terms, the variation of strength and of modulus with temperature is of the same form.

Much of the information on the behaviour of concrete at high temperatures has been obtained from investigations on nuclear reactors and guided missiles launching sites and is therefore not accessible.

Resistance to Abrasion and Cavitation

Under many circumstances, concrete surfaces are subjected to wear. This may be due to attrition by sliding, scraping or percussion.[7.85] In the case of hydraulic structures, the action of the abrasive materials carried by water generally leads to erosion. Another cause of damage to concrete in flowing water is cavitation.

Resistance of concrete to abrasion is difficult to assess as the damaging action varies depending on the exact cause of wear, and no one test procedure is satisfactory in evaluating all the conditions: rubbing test, including rolling balls, dressing wheel, or sandblast may each be appropriate in different cases. The A.S.T.M. Standard C418–68 prescribes the procedure for one type of wear but it is not obvious how this should be used as a criterion of wear resistance under different conditions. For this reason, a full treatment of the topic of abrasion resistance and the related tests are given here.

Erosion Resistance

Erosion of concrete is an important type of wear which may occur in concrete in contact with flowing water. It is convenient to distinguish between erosion due to solid particles carried by the water and that due to pitting resulting from cavities forming and collapsing in water flowing at high velocities. The latter is considered on page 444.

The rate of erosion depends on the quantity, shape, size, and hardness of the particles being transported, on the velocity of their movement and on the presence of eddies, and also on the quality of concrete.[7.86] As in the case of abrasion in general, this quality appears to be best measured by the compressive strength of concrete but the mix composition is also relevant. In particular, concrete with large aggregate erodes less than mortar of equal strength, and hard aggregate improves the abrasion resistance. In general, at a constant slump, the abrasion resistance increases with a decrease in the cement content.[7.87] At a constant cement content, the resistance improves with a decrease in slump:[7.87] this is probably in agreement with the general influence of compressive strength. The suggestion that ordinary Portland cement is preferable to rapid hardening cement[7.87] is probably not generally valid.

In all cases, of course, it is only the quality of the concrete at and near the surface that is relevant, but even the best concrete will rarely withstand severe.abrasion over prolonged periods. Vacuum processing may be advantageous (*see* p. 225).

Cavitation Resistance
While good quality concrete can withstand steady, tangential, high velocity flow of water, severe damage rapidly occurs in the presence of cavitation. By this is meant the formation of vapour bubbles when the local absolute pressure drops to the value of the ambient vapour pressure of water at the ambient temperature. The cavities may be large, single voids, which later break up, or clouds of small bubbles.[7.88] They flow downstream and, on entering an area of higher pressure, collapse with great impact. Because the collapse of the cavities means entry of high-velocity water into the previously air-occupied space, extremely high pressure is generated during very short time intervals, and it is the repeated collapse over a given part of the concrete surface that causes pitting. Greatest damage is caused by clouds of minute cavities found in eddies. They usually coalesce momentarily into a large amorphous cavity which collapses extremely rapidly.[7.89] Many of the cavities pulsate at a high frequency and this seems to aggravate damage over an extended area.[7.90]

Cavitation damage occurs in open channels generally only at velocities in excess of 12 m/s (40 ft/s) but in closed conduits even at 8 m/s (25 ft/s).[7.86] The necessary drop in pressure may be caused by flow over boundary irregularities or by divergence of flow from the concrete surface. Although the advent of cavitation depends primarily on pressure changes (and consequently also on velocity changes), it is especially likely to occur in the presence of small quantities of undissolved air in the water. These bubbles of air behave as nuclei at which the change of phase from liquid to vapour can more readily occur. Dust particles have a similar effect, possibly because they "house" the undissolved air. On the other hand, free air in large quantities (up to 2 per cent by volume), while promoting cavitation, may cushion the collapse of the cavities and hence reduce the cavitation damage.[7.91] Deliberate air entrainment in water may therefore be advantageous.

The surface of concrete affected by cavitation is irregular, jagged and pitted, in contrast to the smoothly worn surface of concrete eroded by water-borne solids. The damage does not progress steadily: usually, after an initial period of small damage, rapid deterioration occurs, followed by damage at a slower rate.[7.91]

Best resistance to cavitation damage is obtained by the use of high strength concrete, possibly formed by an absorptive lining (which reduces the local water/cement ratio). The maximum size of aggregate near the surface should not exceed 20 mm ($\frac{3}{4}$ in.)[7.91] because cavitation tends to remove large particles. Hardness of aggregate is not important (unlike the case of erosion resistance) but good bond between aggregate and mortar is vital.

Concrete with polymers shows good resistance, depending chiefly on the type of the filler, cohesion of binder and adhesion to the filler.[7.92] Resilient coatings may also be used so as to reduce the stress to a level which can be tolerated by the underlying concrete. Some neoprene and polyurethane coatings have been found satisfactory[7.88] but they must adhere strongly to the concrete base.

While the use of suitable concrete may reduce cavitation damage, not even the best concrete can withstand cavitation forces for an indefinite time. The solution of the cavitation damage problem lies therefore primarily in reducing cavitation. This can be achieved by the provision of smooth and well-aligned surfaces free from irregularities such as depressions, projections, joints and misalignments, and by the absence of abrupt changes in slope or curvature that tend to pull the flow away from the surface. If possible, local increase in velocity of water should be avoided as damage is proportional to the sixth or seventh power of velocity.[7.91]

Tests for Abrasion Resistance

The resistance of concrete to abrasion can be determined by several methods, each trying to simulate a mode of abrasion found in practice. In all tests the loss of weight of the specimen is used as a measure of abrasion.

In the *steel ball abrasion test* a load is applied to a rotating head which is separated from the specimen by steel balls. The test is performed in circulating water in order to remove the eroded material.

The *dressing wheel test* uses a drill press modified to apply a load to 32 rotating dressing wheels which are in contact with the specimen. The driving head is rotated 5,000 times at 190 rev/min, silicon carbide being fed as an abrasive material.

The dressing wheel and the steel ball tests serve to estimate the resistance of concrete to wheeled and heavy foot traffic. By contrast, the proneness to erosion by solids in flowing water is measured by means of a *shot-blast test*. Here, 2,000 pieces of broken steel shot (of 850 μm (No. 20) size) are ejected under air pressure of $0 \cdot 62$ MN/m^2 (90 lb/in^2) from a $6 \cdot 3$ mm ($\frac{1}{4}$ in.) nozzle against a concrete specimen 102 mm (4 in.) away.

The simulation of the real conditions of wear is not easy, and indeed the main difficulty in abrasion testing is to make sure that the results of a test represent the comparative resistance of concrete to a given type of wear.

Fig. 7.35 shows the results of the three tests on different concretes. Because of the arbitrary conditions of test, the values obtained are not comparable quantitatively but in all cases the resistance to abrasion was found to be proportional to the compressive strength of concrete.[7.93] The steel ball test appears more consistent and more sensitive than the other two. B.S. 368: 1956 prescribes testing concrete flags by free-falling steel balls in a rotating container.

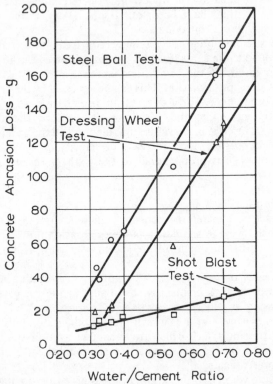

Fig. 7.35. *Influence of the water/cement ratio of the mix on the abrasion loss of concrete for different tests*[7.93]

The influence of the physical properties of the aggregate varies with the type of test: in the steel ball and dressing wheel tests softer aggregate results in greater wear, while in the shot-blast test harder aggregate tends to splinter and causes a greater loss. Grading of aggregate has also some influence, an over-sanded mix leading to a greater loss on abrasion. Special tests may be necessary if metallic or other hardeners are used in the surface layer of concrete.

The best available guide for selection of abrasion resistant concrete is that the compressive strength of concrete is the paramount factor in its resistance to abrasion; this resistance can be increased by the use of fairly lean mixes. Lightweight concrete is obviously unsuitable when surface wear may occur. Concrete which bleeds only little has a stronger surface layer and is therefore more resistant to abrasion. For high resistance proper moist curing is essential: some types of membrane curing are detrimental while absorptive forms improve the resistance of the surface to abrasion.[7.94]

Electrical Properties of Concrete

Although, strictly speaking, not related to durability, there are two properties of concrete that may be of interest with respect to the behaviour of concrete in service; these will now be considered.

Electrical properties are of concern in some specific applications such as railway sleepers (ties) (where inadequate resistivity affects the signalling system) or in structures in which concrete is used for protection from stray currents.

In the vicinity of underground pipelines, concrete may be subjected to impressed electrical activity but, under the usual operating conditions, concrete offers a high resistance to the passage of electric current to or from embedded steel. This is largely due to the electro-chemical effect which concrete has on steel in contact with it, arising from the alkalinity of the electrolyte within the concrete. Such a protection applies within the potential range of about $+0.6$ to -1.0 V (with respect to a copper sulphate electrode), the current being primarily controlled by polarization effects and not by the ohmic resistance of concrete.[7.95]

Moist concrete behaves essentially as an electrolyte with the resistivity of the order of 10^4 ohm-cm; this is within the range of semi-conductors. On the other hand, oven-dry concrete has a resistivity of about 10^{11} ohm–cm, which means that such concrete is a reasonably good insulator.[7.96]

This large increase in resistivity of concrete on removal of water is interpreted to mean that electric current is conducted through moist concrete essentially by electrolytic means, that is by ions in the evaporable water. It can therefore be expected that any increase in water and ions present decreases the resistivity of cement paste, and indeed resistivity decreases sharply with an increase in the water/cement ratio (*see* Table 7.16) or with the addition of calcium chloride. For instance, mortar containing calcium chloride was found to have a resistivity some 15 times lower than similar mortar without calcium chloride.[7.97]

The influence on resistivity of salinity of the mixing water (due to the addition of chlorides or the use of sea water) is greatest in concretes with high water/cement ratios and is quite small in high strength concrete.[7.98] The influence on resistivity of the quantity of electrolyte present is shown also in the increase in resistivity with age, at least during the first few months (Fig. 7.36).

The relation between resistivity of concrete and the volume fraction

Table 7.16: *Influence of Water/Cement Ratio and Length of Moist Curing on Resistivity of Cement Paste*[7.96]

Cement type	Equivalent Na$_2$O content, per cent	Water/cement ratio	Resistivity (at 1,000 Hz, 4V), ohm-cm at the age of		
			7 days	28 days	90 days
Ordinary Portland	0·19	0·4	1,030	1,170	1,570
		0·5	790	880	1,090
		0·6	530	700	760
Ordinary Portland	1·01	0·4	1,230	1,360	1,660
		0·5	820	950	1,200
		0·6	570	730	790

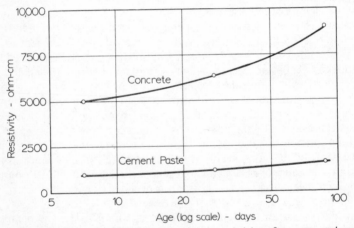

Fig. 7.36. *Influence of prolonged moist curing on resistivity of concrete and cement paste with a water/cement ratio of 0·41, made with ordinary Portland cement (with a very low alkali content) under a current of 1,000 Hz, 4V*[7.96]

occupied by water can be derived from the laws of conductivity of heterogeneous conductors. However, for the range of the usual concrete mixes, the water content varies comparatively little for a given grading and workability, and the resistivity becomes more dependent on the cement used[7.99] because the chemical composition of the cement controls the quantity of ions present in the evaporable water. Some idea of the influence of cement on resistivity can be obtained from Table 7.17, from which it can be seen that the resistivity of concrete made with aluminous cement is 10 to 15 times higher than when Portland cement in the same proportions is used[7.99] (*see* Fig. 7.37).

Table 7.17: Typical Electrical Properties of Concrete under an Alternating Current[7.100]

Type of specimen	Mix proportions	Water/cement ratio	Type of cement	Period of air drying, days	Resistance, 10³ ohm d.c.	50 Hz	500 Hz	5,000 Hz	25,000 Hz	Capaciative reactance, 10³ ohm 50 Hz	500 Hz	5,000 Hz	25,000 Hz	Capacitance, microfarad 50 Hz	500 Hz	5,000 Hz	25,000 Hz
4 in. cube; external plate electrodes*	1:2:4	0·49	Ordinary Portland	7	1·0	0·9	0·9	0·9	0·9	159	159	64	32	0·020	0·0020	0·0005	0·0002
				42		3·1	3·1	3·0	3·0	637	455	106	64	0·005	0·0007	0·0003	0·0001
				113	9·0	8·2	8·0	7·6	7·3	1,061	398	106	64	0·003	0·0008	0·0003	0·0001
			Rapid-hardening Portland	39	2·7	2·7	2·7	2·7	2·7	796	398	106	64	0·004	0·0008	0·0003	0·0001
			Aluminous	5	18·9	18·9	17·3	15·2	13·9	398	228	159	106	0·008	0·0014	0·00020	0·00006
				6	21·6	21·6	19·7	17·1	15·9	423	245	177	106	0·007	0·0013	0·00018	0·00005
				18	39·0	39·0	35·1	30·4	27·5	664	398	245	127	0·005	0·0008	0·00013	0·00005
				40	65·2	65·2	57·7	49·4	44·1	910	569	398	159	0·003	0·0006	0·00008	0·00004
			Ordinary Portland	126	2·0	2·0	1·9	1·9	1·9	118	228	106	127	0·027	0·0014	0·0003	0·00005
6 in. cube; embedded plate electrodes**	1:2:4	0·49	Rapid-hardening	123	1·6	1·6	1·6	1·5	1·5	118	212	80	32	0·027	0·0015	0·0004	0·00020
			Aluminous	138	41·2	41·2	36·0	31·0	28·0	531	398	228	106	0·006	0·0008	0·00014	0·00006
				182	52·6	52·6	46·0	39·4	35·3	692	424	289	127	0·005	0·0007	0·00011	0·00005
			Ordinary Portland	9	0·2	0·2	0·1	0·1	0·1	9	10	6	3	0·350	0·0300	0·0066	0·0020
4 in. × 1 in. prism; external plate electrodes†	neat cement paste	0·23	Rapid-hardening	9	0·1	0·1	0·1	0·1	0·1	6	6	6	2	0·500	0·0540	0·0050	0·0026
			Aluminous	13	6·0	5·5	4·8	4·1	3·7	80	41	26	21	0·040	0·0077	0·0012	0·0003

* Resistivity = 10 × measured resistance
** Resistivity = 30 × measured resistance
† Resistivity = 40 × measured resistance

Although the majority of admixtures do not reduce the resistivity of concrete significantly and reliably,[7.96] some admixtures are effective. For instance, the addition to concrete of finely divided bituminous material, with subsequent heat treatment at 138°C (280°F), increases the resistivity, especially under wet conditions.[7.101] Conversely, in cases where static

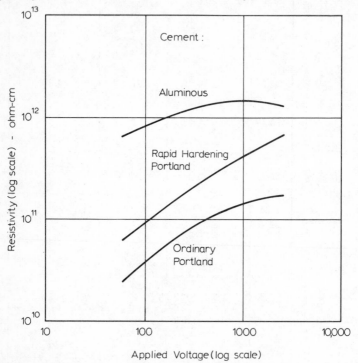

Fig. 7.37. *Relation between resistivity and applied voltage for a 1:2:4 concrete with a water/cement ratio of 0·49, oven-dried and cooled in a desiccator*[7.100]

electricity is undesirable and a decrease in the insulation resistance of concrete is required, satisfactory results can be achieved by the addition of acetylene carbon black (2 to 3 per cent by weight of cement).[7.101]

The resistivity of cement pastes with a pozzolana additive tested in U.S.S.R.[7.99] is shown in Fig. 7.38. The specimens were sealed and were stored at room temperature.

The resistivity of concrete increases with an increase in voltage.[7.100] Fig. 7.37 illustrates this relation for oven-dried specimens not allowed to absorb moisture during the test.

The majority of values quoted in this section are given for alternating current. The resistivity to direct current may be different since it has a polarizing effect, but at 50 Hz there is no significant difference between resistivity to a.c. and d.c.[7.100] In general, for concrete matured in air, the d.c. resistance is approximately equal to the a.c. impedance.[7.100] Hammond and Robson[7.100] interpreted this to mean that the capacitative reactance of concrete is so much larger than its resistance that it is only

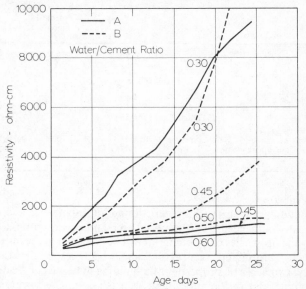

Fig. 7.38. *Relation between resistivity and age for sealed specimens of concrete with a pozzolana additive (A, 14 per cent; B, 7 per cent) stored at room temperature*[7.99]

the latter that contributes significantly to impedance; as a consequence, the power factor is nearly unity. Typical data for alternating current are given in Table 7.17.

The capacitance of concrete decreases with age and with an increase in frequency.[7.100] Neat cement paste with a water/cement ratio of 0·23 has a much higher capacitance than concrete with a water/cement ratio of 0·49 at the same age.[7.100]

Data on the dielectric strength of concrete are given in Table 7.18. It can be seen that the dielectric strength of concrete made with aluminous cement is slightly greater than when Portland cement is used. The table shows also that, despite the much higher moisture content (and therefore lower resistivity) of air-stored concrete compared with oven-dried concrete, the dielectric strength was approximately the same for the two storage conditions, and seems thus to be unaffected by moisture content.

Table 7.18: *Dielectric Strength of Concrete (1:2:4 Mix with Water/ Cement Ratio of 0·49)*[7.100]

| Condition of concrete | Current | Breakdown | Dielectric strength, 10^6V per m | | |
			Ordinary Portland cement	Rapid-hardening Portland cement	Aluminous cement
Stored in air	Positive impulses 1/44 μsec		1·44	1·46	1·84
Dried at 104°C (220°F) air cooled	d.c. negative	First	1·59	1·33	1·77
		Second	1·18	1·06	1·24
		Third	1·25	0·79	1·28
	a.c. (50 Hz) peak values	First	1·43	1·19	1·58
		Second	1·03	1·00	1·21
		Third	1·00	0·97	0·95

Acoustic Properties

In many buildings, acoustic characteristics are of importance and these may be greatly influenced by the material used and by structural details. Here, only the properties of the material will be considered, the influence of the structural form and construction details being a specialized topic.

Basically, two acoustic properties of a building material can be distinguished: sound absorption and sound transmission. The former is of interest when the source of sound and the listener are in the same room. Energy of sound waves, when they hit a wall, is partly absorbed and partly reflected, and we can define a sound absorption coefficient as a measure of the proportion of the sound energy striking a surface which is absorbed by that surface. The coefficient is usually given for a particular frequency. Sometimes, the term noise reduction coefficient is used to denote the average of sound absorption coefficients at 250, 500, 1,000 and 2,000 Hz in octave steps. A typical value for normal-weight-aggregate concrete of medium texture, unpainted is 0·27. The corresponding value for concrete made with expanded shale is 0·45. The difference is related to the variation in porosity and texture, structure which makes air flow possible greatly increasing sound absorption through conversion of sound energy into heat by friction. Thus, foamed concrete, which has discrete air bubbles, would exhibit lower sound absorption than concrete made with porous lightweight aggregate.

Sound transmission is of interest when the listener is in a room adjacent to that in which the source of sound is located. We define the sound transmission loss (or airborne sound insulation) as the difference, measured in decibels, between the incident sound energy and the transmitted sound energy (which radiates into an adjoining room). What constitutes a satisfactory transmission loss depends on the use of the given space.

The primary factor in transmission loss is the unit weight of the partition per square metre of area. This relation, in rough terms, is independent of the type of material used, provided no continuous pores are present, and is sometimes referred to as the "mass law." Fig. 7.39 illustrates the relation for the case when the partition edges are "firmly fixed," i.e.

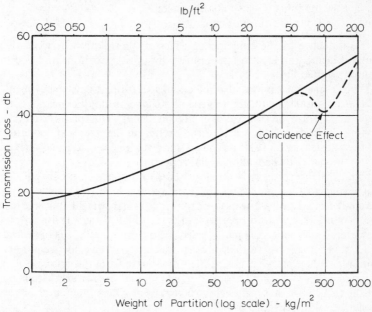

Fig. 7.39. *Relation between transmission loss and unit weight of partition*[7.76]

the flanking walls are of similar material. The sound transmission around the "sound obstacle" has of course to be considered but, as far as the partition itself is concerned, there are some factors additional to the weight: airtightness, bending stiffness, and the presence of cavities.

The stiffness of the partition is relevant because, if the wavelength of the forced bending wave imposed on the wall is equal to the wavelength of free bending waves in a wall, a condition of total sound transmission through the wall arises. This coincidence of wavelengths can occur only above a critical value of frequency at which the velocity of free bending waves in the wall is the same as that of air waves parallel to the wall. Above that frequency, a combination of air-wave incidence and frequency is possible at which there can occur the coincidence of air wave at interface and of the structure bending wave. The effect is usually limited to thin walls.[7.76] The critical frequency is given by

$$q_c = \frac{v^2}{2\pi h} \left[\frac{12\rho(1 - \mu^2)}{E} \right]^{\frac{1}{2}}$$

where v = velocity of sound in air,
$\quad h$ = thickness of the partition,
$\quad \rho$ = density of the concrete,
$\quad E$ = modulus of elasticity of concrete,
and $\quad \mu$ = Poisson's ratio of concrete.

The influence of the coincidence effect on the relation between the sound transmission loss and the unit weight of the partition can be seen from the dotted line in Fig. 7.39.

The presence of cavities also affects this relation, a cavity increasing the transmission loss, so that the use of the given total thickness of concrete in the form of two leaves is advantageous. The quantitative behaviour depends on the width of the cavity, on the degree of isolation between the leaves, and also on the presence or absence of a sealed surface facing the cavity if the wall material is porous.

From the foregoing it is apparent that to a considerable extent the requirements of a high sound absorption and a high transmission loss are conflicting. For instance, the porous type of lightweight concrete has good sound-absorbing properties but a very high sound transmission. However, if one concrete face is sealed the transmission loss is increased and can become equal to that of other materials of the same weight per unit area. It is preferable to seal the side remote from the source of the sound as otherwise sound absorption is impaired. However, there is no reason to believe that lightweight concrete provides inherently better insulation with respect to sound transmission.

REFERENCES

7.1.   F. M. LEA and N. DAVEY, The deterioration of concrete in structures, *J. Inst. C. E.*, No. 7, pp. 248–95 (London, May 1949).

7.2.   T. C. POWERS, H. M. MANN and L. E. COPELAND, The flow of water in hardened Portland cement paste, *Highw. Res. Bd. Sp. Rep. No.* 40, pp. 308–23 (Washington D.C., July 1959).

7.3.   T. C. POWERS, Structure and physical properties of hardened Portland cement paste, *J. Amer. Ceramic Soc.*, **41**, pp. 1–6 (Jan. 1958).

7.4.   T. C. POWERS, L. E. COPELAND and H. M. MANN, Capillary continuity or discontinuity in cement pastes, *J. Portl. Cem. Assoc. Research and Development Laboratories*, **1**, No. 2, pp. 38–48 (May 1959).

7.5.   T. C. POWERS, L. E. COPELAND, J. C. HAYES and H. M. MANN, Permeability of Portland cement paste, *J. Amer. Concr. Inst.*, **51**, pp. 285–98 (Nov. 1954).

7.6.   G. J. VERBECK, Pore structure, *A.S.T.M. Sp. Tech. Publicn. No.* 169, pp. 136–142 (1956).

7.7.   F. M. LEA, *The Chemistry of Cement and Concrete* (London, Arnold, 1970).

7.8.   F. R. MCMILLAN, T. E. STANTON, I. L. TYLER and W. C. HANSEN, Long-time study of cement performance in concrete (Chapter 5: Concrete exposed to sulfate soils), *Amer. Concr. Inst. Sp. Publicn.* (1949).

7.9.   J. H. P. VAN AARDT, The resistance of concrete and mortar to chemical attack—progress report on concrete corrosion studies, *Bul. No.* 13, pp. 44–60, National Building Research Institute (South African Council for Scientific and Industrial Research) (March 1955).

7.10. R. H. BOGUE, *The Chemistry of Portland Cement* (New York, Reinhold 1955).

7.11. I. L. TYLER, Long-time study of cement performance in concrete (Chapter 12: Concrete exposed to sea water and fresh water), *J. Amer. Concr. Inst.*, **56**, pp. 825–36 (March 1960).

7.12   A.C.I. COMMITTEE 211, Recommended practice for selecting proportions for normal weight concrete (A.C.I. 211. 1–70), *J. Amer. Concr. Inst.*, **66**, pp. 612–628 (Aug. 1969), *J. Amer. Concr. Inst.*, **67**, pp. 192–194 (Feb. 1970), *J. Amer. Concr. Inst.*, **67**, p. 953 (Dec. 1970).

7.13. P. W. KEENE, Some tests on the durability of concrete mixes of similar compressive strength, *Cement Concr. Assoc. Tech. Rep. TRA/330* (London, Jan. 1960).

7.14. J. H. P. VAN AARDT, Chemical and physical aspects of weathering and corrosion of cement products with special reference to the influence of warm climate, *R.I.L.E.M. Symposium on Concrete and Reinforced Concrete in Hot Countries* (Haifa, 1960).

7.15. L. H. TUTHILL, Resistance to chemical attack, *A.S.T.M. Sp. Tech. Publicn. No.* 169, pp. 188–200, 1956.

7.16. CENTRE D'INFORMATION DE L'INDUSTRIE CIMENTIERE BELGE, Le béton et le gel, *Bul. No.* 61 (Sept. 1957), *No.* 62 (Oct.), *No.* 63 (Nov.), *No.* 64 (Dec.).

7.17. G. MOLLER, Tests of resistance of concrete to early frost action, *R.I.L.E.M. Symposium on Winter Concreting* (Copenhagen, 1956).

7.18. E. G. SWENSON, Winter concreting trends in Europe, *J. Amer. Concr. Inst.*, **54**, pp. 369–84 (Nov. 1957).

7.19. NATIONAL READY MIXED CONCRETE ASSOCIATION, Cold weather ready mixed concrete, *Publicn. No.* 34 (Washington D.C., Sept. 1960).

7.20. T. C. POWERS, Discussion on: Maturity and the strength of concrete by J. M. Plowman, *Mag. Concr. Res.*, **8**, No. 24, pp. 178–79 (Nov. 1956).

7.21. T. C. POWERS, Resistance to weathering—freezing and thawing, *A.S.T.M. Sp. Tech. Publicn. No.* 169, pp. 182–87 (1956).

7.22. T. C. POWERS, What resulted from basic research studies, *Influence of cement characteristics on the frost resistance of concrete*, pp. 28–43 (Chicago, Portland Cement Assoc., Nov. 1951).

7.23. R. A. HELMUTH, Capillary size restrictions on ice formation in hardened Portland cement pastes, *Proc. 4th Int. Symp. on the Chemistry of Cement*, Washington D.C., 1960, pp. 855–69.

7.24. A. R. COLLINS, Discussion on: A working hypothesis for further studies of frost resistance of concrete by T. C. Powers, *J. Amer. Concr. Inst.*, **41**, (Supplement) pp. 272–12—272–14 (Nov. 1945).

7.25. A. R. COLLINS, The destruction of concrete by frost, *J. Inst. C. E.*, **23**, No. 1, pp. 29–41 (London, Nov. 1944).

7.26. T. C. POWERS, Some observations on using theoretical research, *J. Amer Concr. Inst.*, **43**, pp. 1089–94 (June 1947).

7.27. G. J. VERBECK, What was learned in the laboratory, *Influence of cement characteristics on the frost resistance of concrete*, pp. 14–27 (Chicago, Portland Cement Assoc., Nov. 1951).

7.28. U.S. BUREAU OF RECLAMATION, Relationship of moisture content of aggregate to durability of the concrete, *Materials Laboratories Report No. C–513* (Denver, Colorado, 1950).

7.29. U.S. BUREAU OF RECLAMATION, Investigation into the effect of water/cement ratio on the freezing–thawing resistance of non-air and air-entrained concrete, *Concrete Laboratory Report No. C–810* (Denver, Colorado, 1955).

7.30. A. R. COLLINS, Mix design for frost resistance, *Proc. of a Symposium on Mix Design and Quality Control of Concrete*, pp. 92–6 (London, Cement and Concrete Assoc., 1954).

7.31. G. J. VERBECK and P. KLIEGER, Calorimeter-strain apparatus for study of freezing and thawing concrete, *Highw. Res. Bd. Bul. No.* 176, pp. 9–12 (Washington D.C., 1958).

7.32. I. L. TYLER, What was found in the field, *Influence of cement characteristics on the frost resistance of concrete*, pp. 6–13 (Chicago, Portland Cement Assoc., Nov. 1951).

7.33. H. F. GONNERMAN, A. G. TIMMS and A. G. TAYLOR, Effects of calcium and sodium chlorides on concrete when used for ice removal, *J. Amer. Concr. Inst.*, **33**, pp. 107–22 (Nov.–Dec. 1936).

7.34. G. J. VERBECK and P. KLIEGER, Studies of "salt" scaling of concrete, *Highw. Res. Bd. Bul. No.* 150, pp. 1–13 (Washington D.C., 1957).

7.35. P. KLIEGER, Curing requirements for scale resistance of concrete, *ibid.*, pp. 18–31.

7.36. W. LERCH, The use of air-entraining admixtures in concrete in large dams in the United States, *6th Congress on Large Dams*, vol. III, pp. 913–24 (New York, 1958).

7.37. P. KLIEGER, Further studies on the effect of entrained air on strength and durability of concrete with various sizes of aggregates, *Highw. Res. Bd. Bul. No.* 128, pp. 1–19 (Washington D.C., 1956).

7.38. T. C. POWERS, Void spacing as a basis for producing air-entrained concrete, *J. Amer. Concr. Inst.*, **50**, pp. 741–60 (May 1954), and Discussion, pp. 760–6—760–15 (Dec. 1954).

7.39. U.S. BUREAU OF RECLAMATION, The air-void systems of Highway Research Board co-operative concretes, *Concrete Laboratory Report No. C–824* (Denver, Colorado, April 1956).

7.40. E. W. SCRIPTURE, S. W. BENEDICT and F. J. LITWINOWICZ, Effect of temperature and surface area of the cement on air entrainment, *J. Amer. Concr. Inst.*, **48**, pp. 205–12 (Nov. 1951).

7.41. S. WALKER and D. L. BLOEM, Studies of concrete containing entrained air, *J. Amer. Concr. Inst.*, **42**, pp. 629–39 (June 1946).

7.42. P. KLIEGER, Early high strength concrete for prestressing, *Proc. of World Conference on Prestressed Concrete*, pp. A5–1—A5–14 (San Francisco, July 1957).

7.43. L. H. TUTHILL, Entrained air loss in handling, placing, vibrating, *J. Amer. Concr. Inst.*, **44**, p. 504 (Feb. 1948).

7.44. T. C. POWERS and R. A. HELMUTH, Theory of volume changes in hardened Portland cement paste during freezing, *Proc. Highw. Res. Bd.*, **32**, pp. 285–97 (Washington D.C., 1953).

7.45. P. J. F. WRIGHT, Entrained air in concrete, *Proc. Inst. C. E.*, Part I, **2**, No. 3, pp. 337–58 (London, May 1953).

7.46. U.S. BUREAU OF RECLAMATION, Effect of air-entraining agents with sulphate-resistant cement on the durability and other properties of concrete, *Materials Laboratories Report No. C*–362 (Denver, Colorado, Oct. 1947).

7.47. L. S. BROWN and C. U. PIERSON, Linear traverse technique for measurement of air in hardened concrete, *J. Amer. Concr. Inst.*, **47**, pp. 117–23 (Oct. 1950).

7.48. T. C. POWERS, Basic considerations pertaining to freezing and thawing tests, *Proc. A.S.T.M.*, **55**, pp. 1132–54 (1955).

7.49. HIGHWAY RESEARCH BOARD, Report on co-operative freezing and thawing tests of concrete, *Special Report No.* 47 (Washington D.C., 1959).

7.50. H. WOODS, Observations on the resistance of concrete to freezing and thawing, *J. Amer. Concr. Inst.*, **51**, pp. 345–49 (Dec. 1954).

7.51. L. J. MITCHELL, Thermal properties, *A.S.T.M. Sp. Tech. Publcn. No.* 169, pp. 129–35 (1956).

7.52. N. DAVEY, Concrete mixes for various building purposes, *Proc. of a Symposium on Mix Design and Quality Control of Concrete*, pp. 28–41 (London, Cement and Concrete Assn., 1954).

7.53. S. L. MEYERS, How temperature and moisture changes may affect the durability of concrete, *Rock Products*, pp. 153–57 (Chicago, Aug. 1951).

7.54. S. WALKER, D. L. BLOEM and W. G. MULLEN, Effects of temperature changes on concrete as influenced by aggregates, *J. Amer. Concr. Inst.*, **48**, pp. 661–79 (April 1952).

7.55. D. G. R. BONNELL and F. C. HARPER, The thermal expansion of concrete, *National Building Studies, Technical Paper No.* 7 (London, H.M.S.O., 1951).

7.56. T. C. POWERS and T. L. BROWNYARD, Studies of the physical properties of hardened Portland cement paste (Nine parts), *J. Amer. Concr. Inst.*, **43** (Oct. 1946 to April 1947).

7.57. R. PHILLEO, Some physical properties of concrete at high temperatures, *J. Amer. Concr. Inst.*, **54**, pp. 857–64 (April 1958).

7.58. N. G. ZOLDNERS, Effect of high temperatures on concretes incorporating different aggregates, *Mines Branch Research Report R.64* (Department of Mines and Technical Surveys, Ottawa, May 1960).

7.59. J. C. SAEMANN and G. W. WASHA, Variation of mortar and concrete properties with temperature, *J. Amer. Concr. Inst.*, **54**, pp. 385–95 (Nov. 1957).

7.60. S. L. MEYERS, Thermal coefficient of expansion of Portland cement—Long-time tests, *Industrial and Engineering Chemistry*, **32**, *No.* 8, pp. 1107–112 (Easton, Pa., 1940).

7.61. U.S. ARMY CORPS OF ENGINEERS, *Handbook for Concrete and Cement* (Vicksburg, Mississippi, 1949).

7.62. H. L. MALHOTRA, The effect of temperature on the compressive strength of concrete, *Mag. Concr. Res.*, **8**, No. 23, pp. 85–94 (Aug. 1956).

7.63. G. E. MONFORE and A. E. LENTZ, Physical properties of concrete at very low temperatures, *J. Portl. Cem. Assoc. Research and Development Laboratories*, **4**, No. 2, pp. 33–9 (May 1962).

7.64. U.S. ARMY ENGINEER WATERWAYS EXPERIMENT STATION, Permeability and triaxial tests of lean mass concrete, *Technical Memorandum No.* 6–380 (Vicksburg, Miss., March 1954).

7.65. R. W. CARLSON, Permeability, pore pressure, and uplift in gravity dams, *Trans. A.S.C.E.*, **122**, pp. 587–613 (1957).

7.66. E. C. HIGGINSON, Effect of steam curing on the important properties of concrete, *J. Amer. Concr. Inst.*, **58**, pp. 281–98 (Sept. 1961).

7.67. T. C. WATERS, Reinforced concrete as a material for containment, *Proc. of Symp. on Nuclear Reactor Containment Buildings and Pressure Vessels*, pp. 50–60 (Butterworths, London, 1960).

7.68. T. YOSHII, H. MORI and M. KANDA, Air-permeability of concrete (in Japanese), *Semento Gijutsu Nenpo*, **12**, pp. 339–43 (1958).

7.69.   O. GRAF, *Die Eigenschaften des Betons* (Berlin, Springer-Verlag, 1960).

7.70.   ANON, Air-permeability of concrete, *Concrete and Constructional Engineering*, p. 166 (London, May 1965).

7.71.   R. L. HENRY and G. K. KURTZ, Water vapor transmission of concrete and of aggregates, *U.S. Naval Civil Engineering Laboratory, Port Hueneme, California*, pp. 71 (June 1963).

7.72.   H.-J. WIERIG, Die Wasserdamfdurchlaessigkeit von Zementmoertel und Beton, *Zement-Kalk-Gips*, **18**, *No.* 9, pp. 471–82 (Sept. 1965).

7.73.   W. WOODSIDE, Water vapor permeability of porous media, *Canadian J. Physics*, **37**, No. 4, pp. 413–16 (April 1959).

7.74.   A. M. NEVILLE, Behaviour of concrete in saturated and weak solutions of magnesium sulphate and calcium chloride, *J. Mat.*, *A.S.T.M.*, **4**, No. 4, pp. 781–816 (Dec. 1969).

7.75.   M. G. DAVIDSON, *A new cold weather concrete technology (Potash as a frost-resistant admixture)* (Moscow, Lenizdat, 1966).

7.76.   A. G. LOUDON and E. F. STACY, The thermal and acoustic properties of lightweight concretes, *Structural Concrete*, **3**, No. 2, pp. 58–95 (London, March–April 1966).

7.77.   T. HARADA, J. TAKEDA, S. YAMANE and F. FURUMURA, Strength, elasticity and the thermal properties of concrete subjected to elevated temperatures, *Int. Seminar on Concrete for Nuclear Reactors*, Amer. Concr. Inst. Sp. Publicn. No. 34, **1**, pp. 377–406 (1972).

7.78.   H. W. BREWER, General relation of heat flow factors to the unit weight of concrete, *J. Portl. Cem. Assoc. Research and Development Laboratories*, **9**, No. 1, pp. 48–60 (Jan. 1967).

7.79.   R. A. HELMUTH, Dimensional changes of hardened Portland cement pastes caused by temperature changes, *Proc. Highw. Res. Board*, **40**, pp. 315–36 (1961).

7.80.   G. TOGNON, Behaviour of mortars and concretes in the temperature range from + 20°C to − 196°C, *Proc. 5th Int. Symp. on the Chemistry of Cement*, pp. 229–48 (Tokyo, The Cement Assoc. of Japan, 1969).

7.81.   D. J. HANNANT, Effects of heat on concrete strength, *Engineering*, **197**, p. 302 (London, Feb. 21, 1964).

7.82.   K. W. NASSER and A. M. NEVILLE, Creep of concrete at elevated temperatures, *J. Amer. Concr. Inst.*, **62**, pp. 1567–79 (Dec. 1965).

7.83.   M. S. ABRAMS and A. H. GUSTAFERRO, Fire endurance of concrete slabs as influenced by thickness, aggregate type, and moisture, *J. Portl. Cem. Assoc. Research and Development Laboratories*, **10**, No. 2, pp. 9–24 (May 1968).

7.84.   J. C. MARÉCHAL, Variations in the modulus of elasticity and Poisson's ratio with temperature, *Int. Seminar on Concrete for Nuclear Reactors*, Amer. Concr. Inst. Sp. Publicn. No. 34, **1**, pp. 495–503 (1972).

7.85.   M. E. PRIOR, Abrasion resistance, *Significance of Tests and Properties of Concrete and Concrete-making Materials*, *A.S.T.M. Sp. Tech. Publicn. No. 169–A*, pp. 246–60 (1966).

7.86.   A.C.I. COMMITTEE 210, Erosion resistance of concrete in hydraulic structures, *J. Amer. Concr. Inst.*, **52**, pp. 259–71 (Nov. 1955).

7.87.   U.S. ARMY CORPS OF ENGINEERS, Concrete abrasion study, Bonneville Spillway Dam, *Report* 15–1 (Bonneville, Or., Oct. 1943).

7.88.   J. M. HOBBS, Current ideas on cavitation erosion, *Pumping*, **5**, No. 51, pp. 142–9 (March 1963).

7.89.   M. J. KENN, Cavitating eddies and their incipient damage to concrete, *Civil Engineering*, **61**, No. 724, pp. 1404–5 (London, Nov. 1966).

7.90. S. P. KOZIREV, Cavitation and cavitation-abrasive wear caused by the flow of liquid carrying abrasive particles over rough surfaces, *Translation* (The British Hydromechanics Research Association, Feb. 1965).

7.91. M. J. KENN, Factors influencing the erosion of concrete by cavitation, *C.I.R.I.A.*, pp. 15 (London, July 1968).

7.92. K. K. SHAL'NEV, N. P. ROZANOV, P. A. PSHERITSYN, YU. P. INOZEMTSEV and V. I. SAKHAROV, Mechanism of cavitational erosion of cements and polymeric concretes, Reviewed in the *J. of Applied Chemistry*, **16**, No. 6 (London, June 1966).

7.93. F. L. SMITH, Effect of aggregate quality on resistance of concrete to abrasion, *A.S.T.M. Sp. Tech. Publicn. No.* 205, pp. 91–105 (1958).

7.94. W. H. PRICE, Erosion of concrete by cavitation and solids in flowing water, *J. Amer. Concr. Inst.*, **43**, pp. 1009–23 (1947).

7.95. D. A. HAUSMANN, Electrochemical behaviour of steel in concrete, *J. Amer. Concr. Inst.*, **61**, No. 2, pp. 171–88 (Feb. 1964).

7.96. G. E. MONFORE, The electrical resistivity of concrete, *J. Portl. Cem. Assoc. Research and Development Laboratories*, **10**, No. 2, pp. 35–48 (May 1968).

7.97. R. CIGNA, Measurement of the electrical conductivity of cement mortars, *Annali di Chimica*, **66**, pp. 483–94 (Jan. 1966).

7.98. R. L. HENRY, Water vapor transmission and electrical resistivity of concrete, *Technical Report R-244* (U.S. Naval Civil Engineering Laboratory, Port Hueneme, California, June 30, 1963).

7.99. V. P. GANIN, Electrical resistance of concrete as a function of its composition, *Beton i Zhelezobeton, No.* 10, pp. 462–5 (1964).

7.100. E. HAMMOND and T. D. ROBSON, Comparison of electrical properties of various cements and concretes, *The Engineer*, **199**, pp. 78–80 (Jan. 21, 1955); pp. 114–15, (Jan. 28, 1955).

7.101. ANON, Electrical properties of concrete, *Concrete and Constructional Engineering*, **58**, No. 5, p. 195 (London, May 1963).

7.102. G. L. KALOUSEK, L. C. PORTER and E. J. BENTON, Concrete for long-time service in sulphate environment, *Cement and Concrete Research*, **2**, No. 1, pp. 79–89 (1972).

7.103. J. VUORINEN, On the use of dilation factor and degree of saturation in testing concrete for frost resistance, *Nordisk Betong, No.* 1, pp. 37–64 (1970).

7.104. RILEM WINTER CONSTRUCTION COMMITTEE, Recommendations pour le bétonnage en hiver, *Supplément aux Annales de l'Institut Technique du Bâtiment et des Travaux Publics, No.* 190, *Béton, Béton Armé No.* 72, pp. 1012–1037 (Oct. 1963).

7.105. U. TRÜB, *Baustoff Beton* (Wildegg, Technische Forschungsund Beratungsstelle der Schweizerischen Zementindustrie, 1968).

7.106. M. A. WARD, A. M. NEVILLE and S. P. SINGH, Creep of air-entrained concrete, *Mag. Concr. Res.*, **21**, No. 69, pp. 205–210 (Dec. 1969).

8 Testing of Hardened Concrete

WE have seen that the properties of concrete are a function of time and ambient humidity, and this is why, in order to be of value, tests on concrete have to be performed under specified or known conditions. The most common of all tests on hardened concrete is the compressive strength test, partly because it is an easy test to make, and partly because many, though not all, of the desirable characteristics of concrete are qualitatively related to its strength; but mainly because of the intrinsic importance of the compressive strength of concrete in construction.

The strength tests can be broadly classified into mechanical tests to destruction and non-destructive tests which allow repeated testing of the same specimen and thus make possible a study of the variation in properties with time.

The tests to destruction have been in use for a great number of years, but no universally accepted standard test is available. Different methods and techniques are used in different countries and sometimes even in the same country. As many of these tests are encountered in laboratory work, and especially in research, a knowledge of the influence of the test methods on strength as measured is desirable.

Tests can be made for different purposes but the main two objectives of tests are control of quality and compliance with specifications. Additional tests can be made for specific purposes, e.g. compressive strength tests to determine the strength of concrete at transfer of prestress or at the time of striking the formwork. It should be remembered that tests are not an end in themselves. In the case of concrete they seldom lend themselves to a neat, concise interpretation, so that in order to be of real value tests should always be used against the background of experience.

Compression Tests

Three types of compression test specimens are used: cubes, cylinders, and prisms. Cubes are used in Great Britain, Germany, and many other countries in Europe. Cylinders are the standard specimens in the United States, France, Canada, Australia and New Zealand. In Scandinavia tests are made on both cubes and cylinders.

The tendency nowadays, especially in research, is to use cylinders in preference to cubes, but before comparing the two types of specimens the various tests should be considered in detail.

Cube Test

The specimens are cast in steel or cast-iron moulds, generally 150 mm (or 6 in.) cube, which should conform to the cubical shape, prescribed dimensions and planeness within narrow tolerances. It is preferable that the mould and its base be clamped together during casting as this reduces leakage of mortar. The use of a rigidly connected base is essential when compaction is effected by means of vibration.

Before assembling the mould its mating surfaces should be covered with mineral oil and a thin layer of similar oil must be applied to the inside surfaces of the mould in order to prevent the development of bond between the mould and the concrete.

The standard practice prescribed by B.S. 1881 : 1970 is to fill the mould in three layers. It is advisable, although rarely practised in England, to use a filling hopper mounted above the mould: the mould is then filled to overflowing, and after compaction excess concrete is removed by a sawing motion of a steel rule, and the surface is finished by means of a trowel. This method of filling results in a better homogeneity of the concrete compared with filling without a hopper.

Each layer of concrete is compacted by not less than 35 strokes of a 25 mm (1 in.) square steel punner. Ramming should continue until sufficient compaction has been achieved, for it is essential that the concrete in the cube be fully compacted if the compressive test is to be representative of the properties of fully-compacted concrete. If, on the other hand, a check on the properties of the concrete *as placed* is required, then the degree of compaction of the concrete in the cube should simulate that of the concrete in the structure. Thus in the case of precast members compacted on a vibrated table the test cube and the member may be vibrated simultaneously, but the disparity of the two masses makes the achievement of the same degree of compaction extremely difficult and this method is not recommended. Compaction of standard cubes by means of an electric or pneumatic hammer is permitted by B.S. 1881 : 1970.

After the top surface of the cube has been finished by means of a trowel, the cube is stored undisturbed for 24 hours at a temperature of 18 to 22°C (64 to 72°F) and a relative humidity of not less than 90 per cent. At the end of this period the mould is stripped and the cube is further cured in water at 19 to 21°C (66 to 70°F). This is the B.S. 1881 : 1970 standard method of curing, used in the laboratory and, with wider limits for temperature, on site when it is desired to determine the *potential* quality of the concrete. Of course, the concrete in the structure may actually be inferior, owing to inadequate compaction, segregation, or adverse curing conditions. The effects of the latter factor are of interest if we want to know when the falsework may be removed, or when further construction may continue, or the structure be put into service. For this purpose, cubes are cured under conditions as nearly similar as possible to those existing in the actual structure. Even then the effects of temperature and moisture would not be the same in a cube as in a relatively large mass of concrete: the use of large size test specimens is thus desirable.

The age at which *service* cubes are tested is governed by the information required. On the other hand, *standard* cubes are tested at prescribed ages, generally 28 days, with additional tests often made at 3 and 7 days. Test specimens for the various purposes are recognized by the appropriate codes of practice.

In the compression test, the cube is placed with the cast faces in contact with the platens of the testing machine, i.e. the position of the cube when tested is at right angles to that as-cast. According to B.S. 1881:1970 the load on the cube should be applied at the rate of 15 MN/m^2/min (2,200 lb/in^2/min). Because of the non-linearity of the stress/strain relationship for concrete at high stresses, the rate of increase in strain must be increased progressively as failure is approached, i.e. the speed of the movement of the head of the testing machine has to be increased. This can be done only with a hydraulically operated machine.

The crushing strength is generally reported to the nearest 0·5 MN/m^2 or 50 lb/in^2; a greater "accuracy" is usually only apparent.

Cylinder Test

The standard cylinder is 6 in. in diameter, 12 in. long (or 150 by 300 mm) and is cast in a mould generally made of steel or cast iron, preferably with a clamped base. Non-reusable cardboard moulds are sometimes used, but they result in an apparent lowering of strength of the order of a few per cent[8.28] possibly due to the expansion of the mould during setting. Paper and lightgauge metal moulds are covered by A.S.T.M. Standard 470–71T.

Cylindrical specimens are made in a similar way to the cubes but are compacted either in three layers using a 16 mm ($\frac{5}{8}$ in.) diameter bullet-nosed rod or in two layers by means of an immersion vibrator. Details of procedure are prescribed in A.S.T.M. Standard C 192–69. The top surface of a cylinder finished with a float is not smooth enough for testing and requires further preparation: this is the greatest disadvantage of this type of specimen as normally used.

To overcome this difficulty Thaulow[8.1] has suggested an alternative means of casting cylindrical specimens in a special mould, consisting of a cylinder, a collar with a handle, top and bottom plates, and a yoke (Fig. 8.1). The collar is clamped on and the assembly is filled and rodded in the usual manner. The cylinder is then placed on a timber base (cleats in the bottom plate being pressed into the wood) and the concrete is further compacted by allowing the handle to fall freely against the side of the mould; the number of blows is prescribed depending on the slump of the concrete. Once the compaction has been completed, the collar is removed, excess concrete is struck off and the top plate is worked into

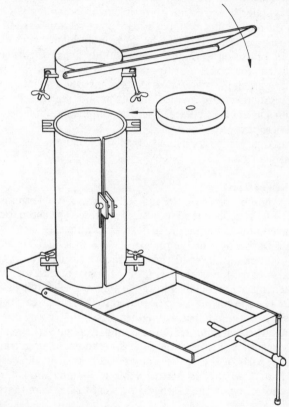

Fig. 8.1. *Thaulow's*[8.1] *cylinder mould*

position by a sliding and rotary motion. The plate is then secured by the yoke, and the specimen is placed in a horizontal position with the split uppermost and is lightly tapped to ensure good contact between concrete and the end plates. Setting in a horizontal position results in both end surfaces of the specimen being plane and square to the axis of the cylinder, i.e. satisfactory for testing. Bleeding would mainly affect the area beneath the split in the mould and is thus of little influence on the compressive strength. Thaulow[8.1] found that his method gave reliable results, but

Bloem's[8.2] tests on cylinders cast in a horizontal position showed a 15 per cent loss in strength compared with cylinders cast in the standard manner.

Prism Test

A test specimen which has approximately the height to cross-section ratio of a cylinder but presents no problems in obtaining good testing surfaces is the rectangular prism of square cross-section, used in France. Prisms are cast with their longer sides horizontal so that, like cubes, they are tested in a position normal to the position as cast. It was suggested in the past that the position as-tested relative to as-cast influences the observed strength of the material but tests[8.3] have shown that the strength of concrete is unaffected, provided the mix has not segregated and excessive bleeding has not taken place. These conditions should be satisfied by every well designed mix.

Prisms used in France are usually 70 by 70 by 350 mm and 100 by 100 by 500 mm.

Equivalent Cube Test

Sometimes the compressive strength of concrete is determined using parts of a beam tested for the modulus of rupture. The end parts of such a beam are left intact after failure in flexure, and since the beam is usually of square cross-section an "equivalent" or "modified" cube can be obtained by applying the load through square steel plates of the same size as the cross-section of the beam. It is important that the two plates be accurately placed vertically above one another; a suitable jig is shown in Fig. 8.2. The specimen should be placed so that the as-cast top surface of the beam is not in contact with either plate.

The strength of a modified cube is approximately the same as the strength of a standard cube of the same size: actually the restraint of the overhanging parts of the "cube" may result in a slight increase in ultimate strength[8.4] so that it is reasonable to assume the strength of a modified cube to be on the average 5 per cent higher than that of a cast cube of the same size.

Effect of End Condition of Specimen and Capping

When tested in compression the top surface of the test cylinder is brought into contact with the platen of the testing machine and, since (except in Thaulow's method) this surface is not obtained by casting against a machined plane but finished by means of a float, the top surface is somewhat rough and not truly plane. Under such circumstances stress concentrations are introduced and the apparent strength of the concrete is greatly reduced. Lack of planeness of 0·25 mm (0·01 in.) can lower the strength by one-third. Convex end surfaces cause a greater reduction than concave ones as they generally lead to higher stress concentrations. The loss in strength is particularly high in high-strength concrete.[8.5]

To avoid this loss of strength, plane end surfaces are essential: A.S.T.M. Standard C 617–71 requires the end surfaces of a cylinder to be plane within 0·05 mm (0·002 in.), as determined by a straight edge and

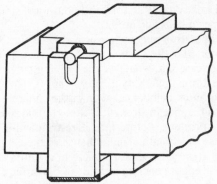

Fig. 8.2. *Jig for testing equivalent cubes*

a feeler gauge. In order to achieve this degree of planeness capping is usually necessary. A like limitation is placed on the platens of the testing machine.

In addition to the absence of "high spots," the contact surfaces should be free from grains of sand or other debris (from a previous test), which would lead to premature failure and in extreme cases possibly to violent splitting.

There are three means of overcoming the ill-effects of an uneven end surface of the specimen: capping, grinding, and packing with a bedding material.

Packing is nowadays rarely used as it results in an appreciable lowering of the apparent mean strength of concrete (compared with capped and often even with smooth-trowelled specimens). At the same time, the scatter of strength results is appreciably reduced since the influence of the defects in planeness (responsible for the great variation in strength) is eliminated.

The reduction in strength introduced by packing, usually of softboard, cardboard, or lead, arises from lateral strains induced in the cylinder by the Poisson's ratio effect in the packing material. Poisson's ratio of this material is generally higher than that of concrete so that splitting is induced. This effect is similar to, although usually greater than, that of lubricating the ends of the cylinder in order to eliminate the restraining influence of the friction between the specimen and the platen on lateral spread of the concrete. Such lubrication has been found to reduce the strength of the specimen.

Capping with a suitable material does not adversely affect strength and reduces its scatter compared with uncapped specimens. An ideal capping material should have strength and elastic properties similar to those of the concrete in the specimen; there is then no enhanced tendency to splitting, and a reasonably uniform distribution of stress over the cross-section of the specimen is achieved.

The capping operation may be performed either just before testing or alternatively soon after the specimen has been cast. Different materials are used in either case but, whatever the capping material, it is essential that the cap be thin, preferably 1·5 to 3 mm ($\frac{1}{16}$ to $\frac{1}{8}$ in.) thick. The capping material must be no weaker than the concrete in the specimen but too great a difference in strength is thought undesirable since a very strong cap may produce a large lateral restraint and thus lead to an apparent increase in strength. The influence of the capping material on strength is much greater in the case of high- or medium-strength concrete than when low-strength concrete is used; in the latter case, the capping material rarely causes a reduction in strength of more than 5 to 10 per cent[8.6] (Fig. 8.3).

When the capping operation is to be performed soon after casting, neat cement is generally used. It is preferable to allow two to four hours' delay after casting so that the plastic shrinkage of the concrete and the resulting subsidence of the top surface of the material in the mould can take place. It is convenient to finish the original concrete about 1·5 to 3 mm ($\frac{1}{16}$ to $\frac{1}{8}$ in.) short of the top of the mould. During capping this space is filled with a cement paste that preferably has been allowed partially to shrink, and by working down a glass or machined steel plate a plane surface is obtained. Experience is necessary to make this operation successful and particularly to obtain a clean break between the cement paste and the plate: greasing the plate with a mixture of lard oil and paraffin[8.7] or covering with a thin film of graphite grease[8.6] has been found helpful.

The alternative method is to cap the cylinder shortly before it is tested: the actual time depends on the hardening properties of the capping material. Suitable materials are aluminous cement, a mixture of Portland and aluminous cements, high strength dental plaster, and a molten sulphur mixture, but other materials have also been used.

Aluminous cement requires 8 to 18 hours to harden and should be applied to a moist specimen, as dry concrete would remove some of the mixing water by suction and thus cause dusting and cracking of the cap. Wet curing of the cap is also essential. L'Hermite[8.8] has used mixtures of three parts of Portland cement to two parts of aluminous cement with a water/cement ratio of 0·30 and obtained sufficient strength after 7 hours.

Dental plaster hardens sufficiently within one hour and, since it has the consistence of a viscous fluid when applied, a plane surface is easily obtained by allowing the capping material to harden against a plate of glass covered with absorbent paper (e.g. newspaper) to prevent bond. Plaster of Paris is not satisfactory as it has a rather low strength and

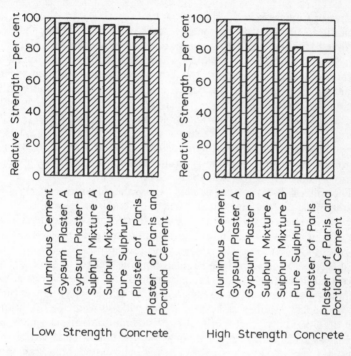

Fig. 8.3. *Influence of the type of capping material on the strength of low- and high-strength concretes*[8.6]

should be allowed to dry: the specimen must thus be kept out of the curing room, and while the plaster hardens the concrete dries out.

The sulphur mixture consists of sulphur and a granular material such as milled fired clay. The mixture is applied in a molten state and allowed to harden with the specimen in a jig which ensures a plane and square end surface. The sulphur mixture from tested cylinders can be re-used.

An alternative to capping is to grind the bearing surface of the specimen until plane and square. This method produces very satisfactory results but is rather expensive.

For research purposes, the application of a genuinely uniform compressive stress may be desirable. This has been achieved by loading through a mat of thin rubber strips with gaps in between,[8.12] or through a stiff wire brush.[8.56] A brush "platen" consists of filaments, about 5 by 3 mm (0·20 by 0·12 in.) in cross-section with gaps 0·2 mm (0·008 in.) wide. The combination allows the free lateral deformation of concrete to develop but the filaments do not buckle.

Testing of Compression Specimens

In addition to being plane, the end surfaces of the cylinder should be normal to its axis, and this guarantees also that the end planes are parallel to one another. A small tolerance is permitted, as an inclination of the axis of the specimen to the axis of the testing machine of 6 mm in 300 mm ($\frac{1}{4}$ in. in 12 in.) has been found to cause no loss of strength.[8.5] The axis of the specimen when placed in the testing machine should be as near the axis of the platen as possible, but errors up to 6 mm ($\frac{1}{4}$ in.) do not affect the strength.[8.5] Likewise, a small lack of parallelism between the end surfaces of the specimen does not adversely affect its strength, provided the testing machine is equipped with a spherically mounted seating, as prescribed by B.S. 1881: 1970.

The spherical seat can act not only when the platens are brought into contact with the specimen but also when the load is being applied. At this stage some parts of the specimen may deform more than others. This is the case in a cube in which, due to bleeding, the properties of different layers (as-cast) are not the same. In the testing position the cube is at right angles to the as-cast position so that the weaker and the stronger parts (parallel to one another) extend from platen to platen. Under load, the weaker concrete, having a lower modulus of elasticity, deforms more. With an effective spherical seat the platen will follow the deformation so that the stress on all parts of the cube is the same and failure occurs when this stress reaches the strength of the weaker part of the cube. On the other hand, if the platen does not change its inclination under load (i.e. moves parallel to itself) a greater load is carried by the stronger part of the cube. The weaker part still fails first, but the maximum load on the cube is reached only when the stronger part of the cube carries its maximum load too: thus the total load on the cube is greater than when the platen is free to rotate. This behaviour was confirmed experimentally by Tarrant.[8.9]

To make the spherical seat of a testing machine effective under load a highly polar lubricant has to be used to reduce the coefficient of friction to a value as low as 0·04 (compared with 0·15 when a graphite lubricant is used).[8.10] It is not clear, however, whether making this movement of the platen possible results in the observed strength being more characteristic of the concrete under test. There are indications that a machine with a platen that does not change inclination under load gives more reproducible results when nominally similar cubes are tested.[8.11] In any case, the observed strength is seriously affected by the friction at the surface of the ball seat, so that for tests to be comparable it is essential to maintain this surface in a standardized condition.

B.S. 1881: Part 4: 1970 specifies that the spherical seat must not move under load and forbids the use of high pressure lubricants.

The loading of a platen through a spherical seat induces bending and distortion of the platen, which depend on the thickness of the platen. A.S.T.M. Standard C 39–71 prescribes the size of the spherical seat in relation to the thickness of the platen.

Fig. 8.4 (a) indicates schematically the normal stress distribution at

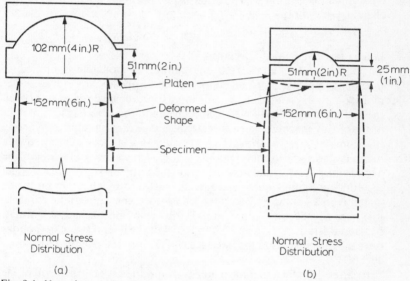

Fig. 8.4. *Normal stress distribution near ends of specimens when tested in a machine with: (a) hard platens, (b) soft platens*

the platen–concrete interface when a "hard" platen is used: the compressive stress is then higher near the perimeter than at the centre of the specimen. The same distribution exists when the specimen or the platen are slightly concave.

Conversely, when a "soft" platen is used (Fig. 8.4(b)), the compressive stress is higher near the centre of the specimen than around the perimeter. This condition is also produced by a slightly convex specimen face or platen.

In addition to these stress distributions, some local variations in stress exist owing to the heterogeneity of concrete, and specifically due to the presence of coarse aggregate particles near the end faces.

A description of the different types of testing machines is outside the scope of this book but it ought to be mentioned that the failure of the specimen is affected by the design of the machine, especially by the energy stored in it. With a very rigid machine the high deformation of the specimen under loads approaching the ultimate is not followed by the movement of the machine head, so that the rate at which the load is applied decreases and a higher strength is recorded. On the other hand, in a less rigid machine the load follows more nearly the load–deformation curve for the specimen, and when cracking commences the energy stored by the machine is released rapidly. This leads to failure under a lower load than would occur in a more rigid machine, often accompanied by a violent explosion.[8.8] The exact behaviour depends on the detailed characteristics of the machine (cf. p. 462).

Failure of Compression Specimens

On page 246 we considered the failure of concrete subjected to uniaxial compression. The compression test imposes, however, a rather more complex system of stress, tangential forces being developed between the end surfaces of the concrete specimen and the adjacent steel platens of the testing machine. In each material, the vertical compression acting (the nominal stress on the specimen) results in a lateral expansion owing to the Poisson's ratio effect. But the modulus of elasticity of steel is some 5 to 15 times greater, and Poisson's ratio no more than twice greater, than the corresponding values for concrete, so that the lateral strain in the platen is small compared with the transverse expansion of the concrete if it were *free* to move. For instance, Newman and Lachance[8.57] found the lateral strain in a steel platen to be 0·4 of the lateral strain in the concrete at a distance from the interface sufficient to remove the restraining effect.

It can be seen then that the platen restrains the lateral expansion of the concrete in the parts of the specimen near its ends: the degree of restraint exercised depends on the friction actually developed. When the friction is eliminated, e.g. by applying a layer of graphite or paraffin wax to the bearing surfaces, the specimen exhibits a large lateral expansion and eventually splits along its full length.

With friction acting, i.e. under normal conditions of test, an element within the specimen is subjected to a shearing stress as well as to compression. The magnitude of the shearing stress decreases and the lateral expansion increases with an increase in distance from the platen. As a result of the restraint, in a specimen tested to destruction there is a relatively undamaged cone or pyramid of height approximately equal to $\frac{\sqrt{3}}{2} d$ (where d is the lateral dimension of the specimen).[8.4] But if the specimen is longer than about $1·7 d$ a part of it will be free from the restraining effect of the platens. Note that specimens whose length is less than 1·5 *d* show a considerably higher strength than those with a greater length (*see* Fig. 8.6).

It seems then, that when a shearing stress acts in addition to the uniaxial compression, failure is delayed, and it can, therefore, be inferred that it is not the principal compressive stress that induces cracking and failure but probably the lateral tensile strain. The actual collapse may be due, at least in some cases, to the disintegration of the core of the specimen. The lateral strain is induced by the Poisson's ratio effect and, assuming this ratio to be approximately 0·2, the lateral strain is $\frac{1}{5}$ of the axial compressive strain. Now, we do not know the exact criteria of failure of concrete but there are strong indications that failure occurs at a limiting strain of 0·002 to 0·004 in compression or 0·0001 to 0·0002 in tension. Since the ratio of the latter of these strains to the former is less than Poisson's ratio of concrete it follows that conditions of failure in circumferential tension are achieved before the limiting compressive strain has been reached.

Vertical splitting has been observed in numerous tests on cylinders, particularly in high-strength specimens made of mortar or neat cement paste, since coarse aggregate provides lateral continuity.[8.4] The presence of vertical cracks has also been confirmed by measurements of ultrasonic pulse velocity along and across the specimen.[8.13]

These observations do not necessarily detract from the value of the compression test as a comparative test, but we should be wary of interpreting it as a true measure of the compressive strength of concrete.

Effect of Height/Diameter Ratio on Strength

Standard cylinders are of height *h* equal to twice the diameter *d*, but sometimes specimens of other proportions are encountered. This is particularly the case with cores cut from *in situ* concrete: the diameter depends on the size of the core-cutting tool while the height of the core varies with the thickness of the slab or member. If the core is too long it can be trimmed to the *h/d* ratio of 2 before testing but with too short a core it is necessary to estimate the strength of the same concrete as if it had been determined on a specimen with *h/d* = 2.

A.S.T.M. Standard C 42–68 and B.S. 1881: 1970 give correction factors (Table 8.1) but Murdock and Kesler[8.14] found that the correction depends also on the level of strength of the concrete (Fig. 8.5). High-strength concrete is less affected by variations in the proportions of the specimen, and such a concrete is also less influenced by the shape of the specimen (*see* Table 8.2); the two factors should be related as there is comparatively little difference between the strengths of a cube and of a cylinder with *h/d* ratio of unity.

The influence of strength on the conversion factor is of practical significance in the case of low-strength concrete, when we test cores with *h/d* smaller than 2. Using the A.S.T.M. C 42–68 or B.S. 1881: 1970 factors, we over-estimate the strength that would be obtained with an h/d ratio of 2 and it is in the case of concrete of low strength, or suspected of having too low a strength, that a correct estimate of strength is particularly important.

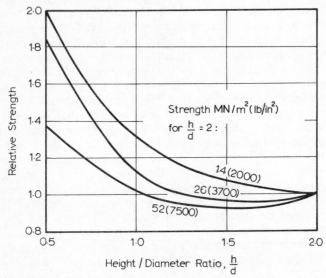

Fig. 8.5. *Influence of the height/diameter ratio on the apparent strength of a cylinder for different strength levels*[8.14]

Table 8.1: *Standard Correction Factors for Strength of Cylinders with different Ratios of Height to Diameter*

Height to diameter ratio $\left(\dfrac{h}{d}\right)$	Strength correction factor	
	A.S.T.M. C 42–68	B.S. 1881: 1970
2·00	1·00	1·00
1·75	0·99	0·98
1·50	0·97	0·96
1·25	0·94	0·94
1·00	0·91	0·92

The general pattern of influence of h/d on strength is shown in Fig. 8.6. For values of h/d smaller than 1·5 the measured strength increases rapidly owing to the restraining effect of the platens of the testing machine. When h/d varies between about 1·5 and 4, strength is affected only little, and for

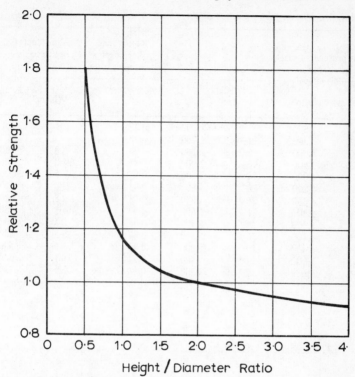

Fig. 8.6. *General pattern of influence of the height/diameter ratio on the apparent strength of a cylinder*[8.40]

h/d values between 1·5 and 2·5 strength is within 5 per cent of the strength of standard specimens ($h/d = 2$). For values of h/d above 5, strength falls off more rapidly, the effect of the slenderness ratio becoming apparent.

It seems thus that the choice of the standard height/diameter ratio of 2 is suitable, not only because the end effect is largely eliminated and a zone of uniaxial compression exists within the specimen, but also because a slight departure from this ratio does not seriously affect the measured value of strength.

The influence on strength of the ratio of height to the least lateral dimension applies also in the case of prisms.

Of course, if the end friction is eliminated, the effect of h/d on strength disappears but this is very difficult to achieve in a routine test. The general pattern of the influence of packing between the platen and the specimen is shown in Fig. 8.7.

The end effect decreases more rapidly the more homogeneous the

Table 8.2: *Strength of Cubes and Cylinders*[8.16]

Compressive strength				Ratio of strengths cylinder/cube	Difference of strengths (cube–cylinders)	
Cube MN/m²	lb/in²	Cylinder MN/m²	lb/in²		MN/m²	lb/in²
9·0	1,300	6·9	1,000	0·77	2·1	300
15·2	2,200	11·7	1,700	0·77	3·5	500
20·0	2,900	15·2	2,200	0·76	4·8	700
24·8	3,600	20·0	2,900	0·81	4·8	700
27·6	4,000	24·1	3,500	0·87	3·5	500
29·0	4,200	26·2	3,800	0·91	2·8	400
29·6	4,300	26·9	3,900	0·91	2·8	400
35·8	5,200	31·7	4,600	0·89	4·1	600
36·5	5,300	34·5	5,000	0·94	2·1	300
42·1	6,100	36·5	5,300	0·87	5·5	800
44·1	6,400	40·7	5,900	0·92	3·5	500
48·3	7,000	44·1	6,400	0·91	4·1	600
52·4	7,600	50·3	7,300	0.96	2·1	300

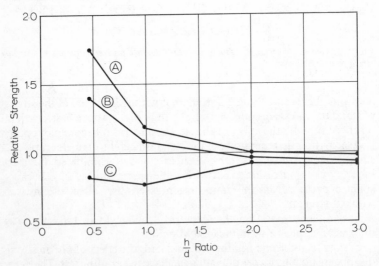

Fig. 8.7. *Relative strength of cylinders of different height–diameter ratios with various types of packing between the platens and the specimen.*[8.58] (*Strength of a cylinder with h/d = 2 and no packing taken as 1·0*)
A, no packing; B, 8 mm ($\frac{5}{16}$ in.) soft wallboard; C, 25 mm (1 in.) plastic board

material; it is thus less noticeable in mortars and probably also in light-weight aggregate concrete of low or moderate strength where a lower heterogeneity arises from the smaller difference between the elastic moduli of the cement paste and the aggregate than is the case with normal weight aggregate. It has been found that with lightweight aggregate concrete the value of the ratio of strengths of a standard cylinder to a cylinder with a height–diameter ratio of unity is 0·95 to 0·97[8.15,8.60] (cf. the data on p. 472). This has, however, not been confirmed in Russian tests on expanded clay concrete where a ratio of about 0·77 has been reported.[8.59]

Comparison of Strengths of Cubes and Cylinders

We have seen that the restraining effect of the platens of the testing machine extends over the entire height of a cube but leaves unaffected a part of a test cylinder. It is, therefore, to be expected that the strengths of cubes and cylinders made from the same concrete differ from one another.

According to B.S. 1881: 1970 the strength of a cylinder is equal to four-fifths of the strength of a cube, but experiments have shown that there is no unique relation between the strengths of the specimens of the two shapes. The ratio cylinder-strength/cube-strength depends primarily on the level of strength of the concrete, and is higher the higher the strength of concrete, as shown by Evans' data in Table 8.2. L'Hermite[8.8] suggested that the ratio of the strengths of a cylinder and a cube be taken as

$$0·76 + 0·2 \log_{10} \frac{\sigma_{cu}}{2,840}$$

where σ_{cu} is the strength of the cube in pounds per square inch. However, some secondary factors affecting strength may influence the strength of the two specimens to a different degree: for instance, the coarser the aggregate grading the lower the ratio cylinder-strength/cube-strength[8.17] (all mixes having a constant cement content and workability). The reasons for this are not clear, but the observation illustrates the fact that the relation between the cube- and cylinder-strengths is not a simple function of strength alone. The moisture condition of the specimen at the time of testing has also been found to affect the ratio of strengths of the two types of specimens.

It is difficult to say which type of specimen is better but there seems to be a tendency, at least for research purposes, to use cylinders rather than cubes, and this has been recommended by RILEM*—an international organization of testing laboratories. Cylinders are believed to give a

---

* Réunion Internationale des Laboratoires d'Essais et de Recherches sur les Matériaux et les Constructions.

greater uniformity of results for nominally similar specimens as their failure is less affected by the end restraint of the specimen; their strength is less influenced by the properties of the coarse aggregate used in the mix; and the stress distribution on horizontal planes in a cylinder is more uniform than on a specimen of square cross-section.

It may be recalled that cylinders are cast and tested in the same position, while in a cube the line of action of the load is at right angles to the axis of the cube as-cast. In structural compression members the situation is similar to that existing in a test cylinder, and it has been suggested that for this reason tests on cylinders are more realistic. The relation between the directions as-cast and as-tested has, however, been shown not to affect appreciably the strength of cubes made with unsegregated and homogeneous concrete[8.3] (Fig. 8.8). Moreover, as shown earlier, the

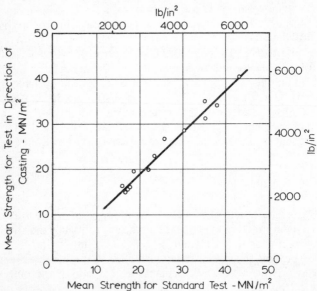

Fig. 8.8. *Relation between mean strength of concrete cubes loaded in the direction of casting and in the standard manner*[8.3]

stress distribution in any compression test is such that the test is only comparative and offers no quantitative data on the strength of a structural member.

Flexure Test

Although concrete is not normally designed to resist direct tension, the knowledge of tensile strength is of value in estimating the load under which cracking will develop. The absence of cracking is of considerable importance in maintaining the continuity of a concrete structure and in many cases in the prevention of corrosion of reinforcement. Cracking problems occur, for instance, when high-tensile steel reinforcement is used, or when diagonal tension arising from shearing stresses develops, but the most frequent case of cracking is due to restrained shrinkage (*see* p. 334) and temperature gradients. An appreciation of the tensile strength of concrete helps in understanding the behaviour of reinforced concrete even though the actual design calculations do not in many cases explicitly take the tensile strength into account.

A direct application of a pure tension force, free from eccentricity, is difficult, and is further complicated by secondary stresses induced by the grips or by embedded studs. There exists therefore no standard test using direct tension. However, some success has been achieved with the use of epoxy bonded end pieces and with lazy-tong grips.[8.19]

Because of these difficulties it is preferable to measure the tensile strength of concrete by subjecting a plain concrete beam to flexure. This is in fact the only standard tension test. The theoretical maximum tensile stress reached in the bottom fibre of the test beam is known as the modulus of rupture. The qualification "theoretical" refers to the assumption in the calculation of the modulus of rupture that stress is proportional to the distance from the neutral axis of the beam while the shape of the actual stress block under loads nearing failure is known to be not triangular but parabolic (cf. Fig. 6.6). The modulus of rupture thus overestimates the tensile strength of concrete and gives a higher value than would be obtained in a direct tension test on a briquette or a bobbin made of the same concrete. Nevertheless, the test is very useful, especially in relation to the design of road slabs and airfield runways because the flexure tension there is a critical factor.

The value of the modulus of rupture depends on the dimensions of the beam and, above all, on the arrangement of loading. Two systems are used: a central point load, which gives a triangular bending moment distribution so that the maximum stress occurs at one section of the beam only; and symmetrical two-point loading, which produces a constant bending moment between the load points. With the latter arrangement, a part of the bottom surface of the beam—usually one-third of the span—is subjected to the maximum stress, and the critical crack may start at any section not strong enough to resist this stress. On the other hand, with a central point load, failure will generally occur only when the strength of the fibre immediately under the load point is exhausted. This statement is not strictly correct, for a fibre subjected to a stress lower than the maximum acting on the beam may also be weak enough to fail. However, it can be seen that the probability of a weak element (of any specified strength) being subjected to the critical stress is considerably greater under two-point loading than when a central load acts. Since concrete consists of elements of varying strength (*see* p. 245), it is to be expected that two-point loading will yield a lower value of the modulus of rupture than when one point load is applied. The difference can be gauged from Wright's[8.20] data plotted in Fig. 8.9. The centre-point loading has been discontinued both in the U.K. and the U.S.

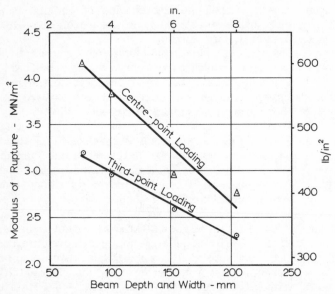

Fig. 8.9. *Modulus of rupture of beams of different sizes subjected to centre-point and third-point loading*[8.20]   *(Crown copyright)*

B.S. 1881 : 1970 prescribes third-point loading on 150 by 150 by 750 mm (6 by 6 by 30 in.) beams supported over a span of 600 mm (24 in.), but when the maximum size of aggregate is not more than 25 mm (1 in.), 100 by 100 by 500 mm (4 by 4 by 20 in.) beams with a span of 400 mm (16 in.) may be used.

There are four possible reasons why the modulus of rupture test yields a higher value of strength than a direct tensile test made on the same concrete. The first one is related to the assumption of the shape of the stress block, mentioned earlier. The second one is that accidental eccentricity in a direct tensile test results in a lower apparent strength of the concrete. The third is offered by an argument similar to that justifying the influence of the loading arrangement on the value of the modulus of rupture: under direct tension the entire volume of the specimen is subjected to the maximum stress, so that the probability of a weak element occurring is high. Fourthly, in the flexure test, the maximum fibre stress reached may be higher than in direct tension because the propagation of a crack is blocked by less stressed material nearer to the neutral axis. Thus the energy available is below that necessary for the formation of new crack surfaces. These four reasons for the difference between the modulus of rupture and the direct tensile strength are not all of equal importance.

Fig. 8.10 shows a typical relation between the direct tensile strength and the modulus of rupture, but actual values may vary depending on the properties of the mix.

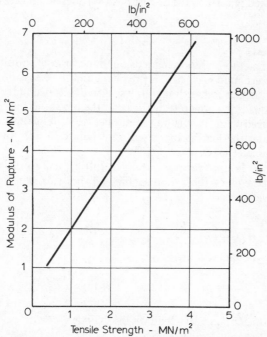

Fig. 8.10. *Relation between modulus of rupture and strength in direct tension*[8.21]

The requirements of A.S.T.M. Standard C 78–64 (re-approved 1972) are similar to those of B.S. 1881: 1970. If fracture occurs within the central one-third of the beam the modulus of rupture is calculated on the basis of ordinary elastic theory, and is therefore equal to $\dfrac{PL}{bd^2}$

where P = the maximum total load on the beam

L = span

b = width of the beam

d = depth of the beam.

If, however, fracture occurs outside the load points, say, at a distance *a* from the near support, *a* being measured along the centre line of the tension surface of the beam, then the modulus of rupture is given by $3Pa/bd^2$. This means that the maximum stress at the critical section, and not the maximum stress on the beam, is considered in the calculations. Both the British and American standards stipulate that tests in which failure occurs at a section such that $L/3 - a > 0.05L$ should be disregarded.

Beams are normally tested on their side in relation to the as-cast position but, provided the concrete is unsegregated, the position of the beam as tested relative to the as-cast position does not affect the modulus of rupture.[8.22, 8.23]

Splitting Test

An indirect method of applying tension in the form of splitting was suggested by Fernando Carneiro, a Brazilian, and the test is often referred to as the Brazilian test, although it was also developed independently in Japan. In this test a concrete cylinder, of the type used for compression tests, is placed with its axis horizontal between the platens of a testing machine, and the load is increased until failure by splitting along the vertical diameter takes place.

If the load is applied along the generatrix then an element on the vertical diameter of the cylinder (Fig. 8.11) is subjected to a vertical compressive stress of

$$\frac{2P}{\pi LD}\left[\frac{D^2}{r(D-r)} - 1\right]$$

and a horizontal tensile stress of

$$\frac{2P}{\pi LD}$$

where *P* is the compressive load on the cylinder,
   L is the length of the cylinder,
   D is its diameter,
and   *r* and $(D - r)$ are the distances of the element from the two loads respectively.

However, immediately under the load a high compressive stress would be induced, and in practice narrow strips of a packing material, such as plywood, are interposed between the cylinder and the platens. These strips are usually 3 mm ($\frac{1}{8}$ in.) thick, and it is convenient to make their width equal to $\frac{1}{12}$ of the diameter of the cylinder; A.S.T.M. Standard C 496–71 prescribes a width of 25 mm (1 in.). Under such circumstances

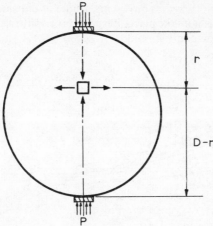

Fig. 8.11. *The splitting test*

the horizontal stress on a section containing the vertical diameter is as shown in Fig. 8.12. The stress is expressed in terms of $2P/\pi LD$ and it can be seen that a high horizontal compressive stress exists in the vicinity of the loads but, as this is accompanied by a vertical compressive stress of

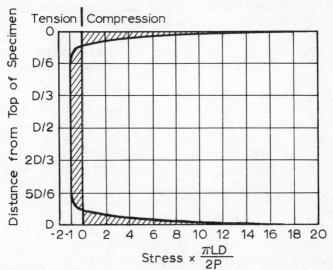

Fig. 8.12. *Distribution of horizontal stress in a cylinder loaded over a width equal to $\frac{1}{12}$ of the diameter*[8.24]

(*Crown copyright*)

comparable magnitude thus producing a state of biaxial stress, failure in compression does not take place. Results of the splitting test on different concretes are shown in Fig. 8.13.

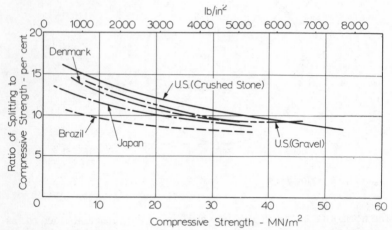

Fig. 8.13. *Tensile splitting strength of cylinders of varying compressive strength*[8.25]

During the splitting test the platens of the testing machine should not be allowed to rotate in a plane perpendicular to the axis of the cylinder, but a slight movement in the vertical plane containing the axis should be permitted in order to accommodate a possible non-parallelism of the generatrices of the cylinder. This can be achieved by means of a simple roller arrangement interposed between one platen and the cylinder. The rate of loading is prescribed by A.S.T.M. Standard C496–71. The test is also covered by B.S. 1881: 1970.

Cubes can also be subjected to the splitting test, the load being applied through semi-cylindrical pieces resting against the cube on centre lines of two opposing faces; alternatively, the load can be applied through two diagonally opposite edges of the cube. The former method yields the same result as the splitting test on a cylinder,[8.61] viz. the horizontal tensile stress is equal to

$$\frac{2P}{\pi a^2}$$

where a is the side of the cube. This means that only the concrete within a cylinder inscribed in the cube resists the applied load.

The diagonal splitting of a cube is less reliable, probably owing to uneven stress distribution. The splitting cube tests are of interest primarily only in countries where the cube and not the cylinder is used as a standard compression specimen; few data are available on the performance of this test.

The splitting test is simple to perform and gives more uniform results than other tension tests (Table 8.3). The strength determined in the

Table 8.3: *Variability of Results of Tests on Tensile Strength of Concrete*[8.24]

Type of test	Mean Strength		Standard deviation within batches		Coefficient of variation
	MN/m²	lb/in²	MN/m²	lb/in²	per cent
Splitting test	2·79	405	0·14	20	5
Direct tensile test	1·90	275	0·13	19	7
Modulus of rupture	4·17	605	0·25	36	6
Compression cube test	41·23	5,980	1·43	207	3½

splitting test is believed to be closer to the true tensile strength of concrete than the modulus of rupture; the splitting strength is 5 to 12 per cent higher than the direct tensile strength. Another advantage of the splitting test is that the same type of specimen can be used for both the compression and the tension tests.

As a variation of the splitting test, a new test has recently been devised by Malhotra *et al.*[8.62] who apply hydrostatic pressure against the inside of a concrete ring specimen. Failure takes place by radial cracking, with the computed tensile strength lying somewhat above the strength in direct tension.

Influence of Rate of Application of Load on Strength

In the range of speeds at which a load can be applied to a specimen, the rate of application has a considerable effect on the apparent strength of concrete: the lower the rate at which stress increases the lower the recorded strength. This is probably due to the increase in strain with time owing to creep, and when limiting strain is reached failure takes place largely independently of the value of the stress applied. Loading over a period of 30 to 240 minutes has been found to cause failure at 84 to 88 per cent of the ultimate strength obtained when the load is applied at the rate of approximately 12 MN/m²/min (30 lb/in²/s).[8.27] Concrete can withstand indefinitely only stresses up to about 70 per cent of the strength under a load applied at the rate of 12 MN/m²/min (30 lb/in²/s).[8.28]

In Fig. 8.14 are plotted the results of tests by a number of investigators and it can be seen that increasing the rate of application of load from 0·04 MN/m²/min to 4×10^6 MN/m²/min (0·1 lb/in²/s to 10^7 lb/in²/s) doubles the apparent strength of concrete. This has not, however, been confirmed by Evans,[8.29] who found no influence of the rate of loading for rates below 4×10^3 MN/m²/min (10^4 lb/in²/s); his data are included in Fig. 8.14.

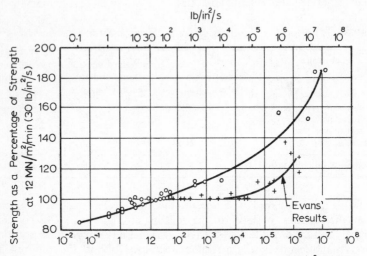

Fig. 8.14. *Influence of the rate of application of load on the compressive strength of concrete*[8.27]

The increase in strength with more rapid loading is proportionately greater the leaner the mix. Given below are the values, found by Evans,[8.29] when the time of application of load was reduced from 0·1 to 0·002 s.

Aggregate/cement ratio	Percentage increase in strength
3	17
10	28
14	31
18	35

These results were obtained using compressed-air testing machines. With ordinary laboratory machines the practical range of speed of loading is between 4 and 400 MN/m²/min (10 and 100 lb/in²/s), and within this range the recorded strength varies only between 97 and 103 per cent of the strength at 12 MN/m²/min (30 lb/in²/s). The latter is a common speed and is near the 15 MN/m²/min (36 lb/in²/s) prescribed by B.S. 1881: 1970. A.S.T.M. Standard C 39–71 stipulates a speed of 138 to 345 kN/m²/s (20 to 50 lb/in²/s), but permits the application of one-half of the anticipated load at a higher rate than standard, as the rate of loading during the first half of the load has no effect on the ultimate strength.

It is clear that for test results to be comparable, stress has to be applied at a standardized rate. In modern testing machines this is facilitated by pacing discs. When the failure of the specimen is approaching, the flow of hydraulic fluid to the cylinder of the machine must be greatly increased as the rate of deformation of the specimen becomes very high. This is usually done in England but A.S.T.M. Standard C 39–71 says that no adjustment shall be made in the controls of the testing machine while a specimen is yielding rapidly immediately before failure. Under such circumstances a slightly lower strength is recorded than would be obtained using the method of B.S. 1881: 1970.

The results of flexure tests are affected by the speed of loading in a way similar to compression tests. Wright's tests[8.20] show that increasing the rate of increase in stress in the extreme fibre from the $0.15 \text{ MN/m}^2/\text{min}$ to $7.8 \text{ MN/m}^2/\text{min}$ ($20 \text{ lb/in}^2/\text{min}$ to $1,140 \text{ lb/in}^2/\text{min}$) increases the modulus of rupture by about 15 per cent. Fig. 8.15 shows that there is a straight-line relation between the modulus of rupture and the logarithm of the

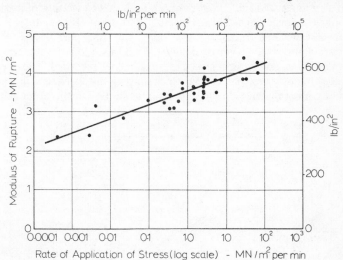

Fig. 8.15. *Influence of the rate of application of load on the modulus of rupture of concrete*[8.63]

rate of application of stress; this again is similar to the behaviour under compressive stress (Fig. 8.14). B.S. 1881: 1970 prescribes a speed of loading of $1.6 \text{ MN/m}^2/\text{min}$ ($230 \text{ lb/in}^2/\text{min}$) and A.S.T.M. Standard C78–64 (reapproved 1972) specifies a speed of no more than $1.03 \text{ MN/m}^2/\text{min}$ ($150 \text{ lb/in}^2/\text{min}$).

Influence of Moisture Condition during Test

The modulus of rupture of concrete which has been allowed to dry is lower than the modulus of a similar specimen in a saturated condition. This difference is due to the tensile stresses induced by restrained and non-uniform shrinkage prior to the application of the load. The magnitude of the apparent loss of strength depends on the rate at which moisture evaporates from the surface of the specimen.

If, however, the test specimen is small and drying takes place very slowly, so that internal stresses can be redistributed and alleviated by creep, an increase in strength is observed. This was found in tests on mortar briquettes[8.30] and concrete beams.[8.31] Conversely, wetting of dry specimens prior to testing reduces their strength.[8.31]

The strength of compression test specimens also increases on drying. This is of interest since both compression and tension specimens develop tensile cracks under load so that the influence of drying would be expected to be similar. In the past, however, many flexure test beams were dried non-uniformly, and contradictory data on the influence of drying on strength were reported.

Mills[8.32] suggested that the loss of strength due to wetting of a compression test specimen is caused by the dilation of the cement gel by adsorbed water: the forces of cohesion of the solid particles are then decreased. Conversely, when on drying the wedge-action of water ceases, an apparent increase in strength of the specimen is recorded. The effects of water are not merely superficial since dipping the specimens in water has much less influence on strength than soaking. On the other hand,

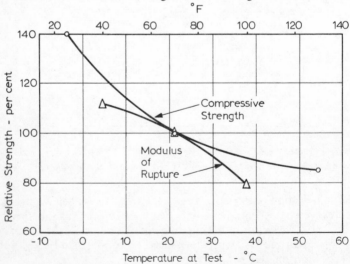

Fig. 8.16. *Influence of temperature at the time of testing on strength*

soaking concrete in benzene and paraffin, known not to be adsorbed by the cement gel, has no influence on strength. Re-soaking oven-dried specimens in water reduces their strength to the value of continuously wet-cured specimens, provided they have hydrated to the same degree.[8.32] The variation in strength due to drying appears thus to be a reversible phenomenon.

The quantitative influence of drying varies: with 34 MN/m² (5,000 lb/in²) concrete an increase in compressive strength up to 10 per cent has been reported[8.33] on thorough drying, but if the drying period is less than 6 hours, the increase is generally less than 5 per cent. Drying of cylinders used in the splitting test results in a proportionately greater variation in strength.[8.33]

According to B.S. 1881: 1970, cubes and flexure specimens should be tested immediately on removal from water. This condition has the advantage of being better reproducible than a "dry condition" which includes greatly varying degrees of dryness.

The temperature of the specimen at the time of testing (as distinct from the curing temperature) affects the strength, a higher temperature leading to a lower indicated strength, in the case of both compression (Fig. 8.16) and flexure specimens.

Influence of Size of Specimen on Strength

Since concrete is composed of elements of variable strength (*see* p. 245) it is reasonable to assume that the larger the volume of the concrete subjected to stress the more likely it is to contain an element of a given extreme (low) strength. As a result, the measured strength of a specimen decreases with increase in its size, and so does the variability in strength of nominally similar specimens. Since the influence of size on strength depends on the standard deviation of strength (Fig. 8.17) it follows that the size effects are smaller the greater the homogeneity of the concrete. Thus the size effect in lightweight concrete should be smaller but this has not been confirmed with any degree of certainty, although there is some support for this suggestion in the available data.[8.76] Fig. 8.17 can also explain why the size effect virtually disappears beyond a certain size of the specimen: for instance, for each successive tenfold increase in size of the specimen it loses progressively a smaller amount of strength.

On page 245 the concept of the weakest link was discussed; to use this concept we require the knowledge of the distribution of extreme values in samples of a given size, drawn at random from a parent population with a given distribution of strength. This distribution is generally not known, and certain assumptions regarding its form have to be made. Here it will suffice to give Tippett's[8.34] data on the variation in strength and standard deviation of samples of size *n* in terms of the strength and standard deviation of a sample of unit size, when the unit sample has a *normal* distribution of strength. Fig. 8.17 shows this variation in strength for samples when *n* equals 10, 10^2, 10^3 and 10^5.

In the case of tests on the strength of concrete we are interested in the averages of extremes as a function of the size of the specimen. Average values of samples chosen at random tend to have a normal distribution,

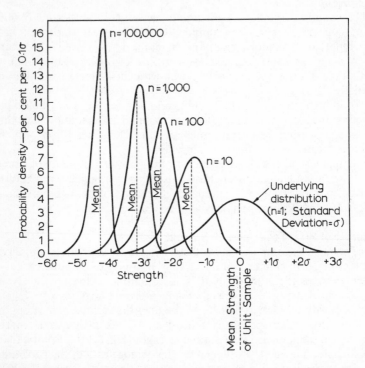

Fig. 8.17. *Strength distribution in samples of size n for an underlying normal distribution*[8.34]

so that the assumption of this type of distribution, when average values of samples are used, does not introduce serious error, and has the advantage of simplifying the computations. In some practical cases a skewness of distribution has been observed; this may not be due to any "natural" properties of concrete but to the rejection of poor quality concrete on the site so that such concrete never reaches the testing stage.[8.35] A full treatment of the statistical aspects of testing is outside the scope of this book.*

---

* See A. M. NEVILLE and J. B. KENNEDY, *Basic Statistical Methods for Engineers and Scientists* (New York and London, Intertext) 1964.

Fig. 8.17 shows that both the mean strength and dispersion decrease with an increase in the size of the specimen. Experimental results for the modulus of rupture are shown in Fig. 8.9, based on Wright's[8.20] tests, and Fig. 8.18 illustrates the influence of size on the scatter of results.[8.23] Similar behaviour was found in specimens tested in pure tension[8.19] and

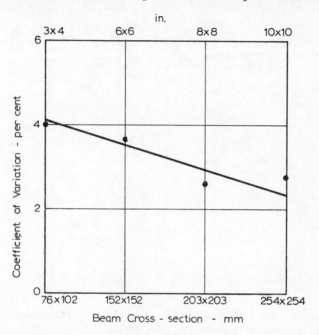

Fig. 8.18. *Coefficient of variation of the modulus of rupture for beams of different sizes*[8.23]

Table 8.4: *Standard Deviation of Cubes of Different Sizes*[8.18]

| Group | Standard deviation for cubes of size | | | | | |
| | 70·6 mm (2·78 in.) | | 127 mm (5 in.) | | 152 mm (6 in.) | |
	MN/m²	lb/in²	MN/m²	lb/in²	MN/m²	lb/in²
A	2·75	399	2·09	303	1·39	201
B	1·50	218	1·12	162	0·97	140
C	1·45	210	1·03	150	0·97	140
D	1·74	253	1·36	197	1·05	153

in indirect tension.[8.64] Fig. 8.19 shows the relation between mean strength and specimen size for cubes, and Table 8·4 gives the relevant values for

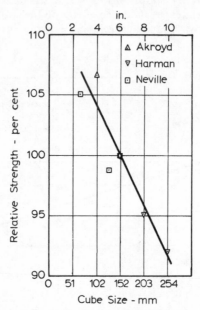

Fig. 8.19. *Compressive strength of cubes of different sizes*[8.35]

standard deviation. Prisms[8.36, 8.37] and cylinders exhibit a similar behaviour (Fig. 8.20). The size effects are, of course, not limited to concrete, and have been found also in anhydrite[8.39] and other materials.

It is interesting to note that the size effect disappears beyond a certain size so that a further increase in the size of a member does not lead to a decrease in strength. According to the Bureau of Reclamation,[8.77] the strength curve becomes parallel to the size axis at a diameter of 457 mm (18 in.), i.e. cylinders of 457 mm (18 in.), 610 mm (24 in.), and 914 mm (36 in.) diameter all have the same strength. The same investigation indicates that the decrease in strength with an increase in size of the specimen is less pronounced in lean mixes than in rich ones. For instance, the strength of 457 mm (18 in.) and 610 mm (24 in.) cylinders relative to 152 mm (6 in.) cylinders is 85 per cent for rich mixes but 93 per cent for lean (167 kg/m³ (282 lb/yd³)) mixes (*cf.* Fig. 8.20).

These experimental data are of interest as it could be speculated that, if the size effect is extrapolated to very large structures, a dangerously low strength might be expected. Evidently this is not so.

The various test results on the size effect are of interest because in the

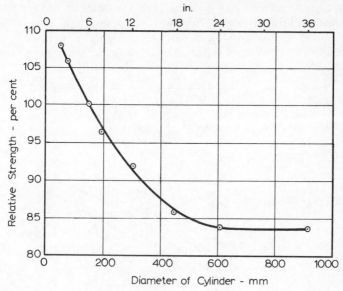

Fig. 8.20. *Compressive strength of cylinders of different sizes*[8.38]

past size effects have been ascribed to a variety of causes: the wall-effect; the ratio of the specimen size to the maximum aggregate size; the internal stresses caused by the difference in temperature and humidity between the surface and the interior of the specimen; the tangential stress at the contact surface between the platen of the testing machine and the specimen due to friction or bending of the platen; and the difference in the effectiveness of curing. The last suggestion, for instance, is disproved by Gonnerman's[8.40] results (Fig. 8.21) which show that specimens of different size and shape gain strength at the same rate.

Within the range of sizes of specimens normally used, the effect of size on strength is not large, but is significant and should not be ignored in work of high accuracy or in research. Recent work[8.65] has suggested a general relation between the strength of concrete and the shape and size of the specimen in terms of $V/hd + h/d$, where V = volume of specimen; h = its height; and d = its least lateral dimension. Fig. 8.22 indicates the fit of experimental data to the relation postulated.

The use of smaller specimens offers some advantages. They are easier to handle and are less likely to be accidentally damaged, a lower capacity testing machine is needed, and less concrete is used, which in the laboratory means less storage and curing space and also a smaller quantity of aggregate to be processed.[8.41] On the other hand, because of the higher scatter of results obtained with smaller specimens, they have to be used in a greater number to give the same precision of the mean: five to six 100 mm (4-in.) concrete cubes would be required instead of three 150 mm (6-in.) cubes;[8.42] or five 13 mm ($\frac{1}{2}$-in.) mortar cubes instead of two 100 mm (4-in.) cubes from the same batch or four 100 mm (4-in.) cubes from nominally similar batches.[8.43]

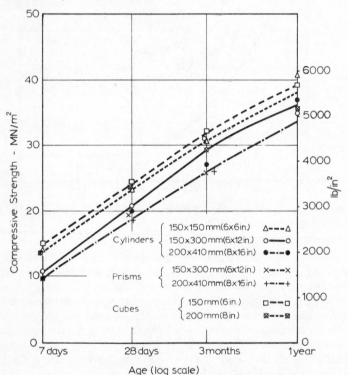

Fig. 8.21. *Effect of age on the compressive strength of specimens of different shape and size*[8.40] (*mix 1:5 by volume*)

Specimen Size and Aggregate Size

It is clear that a test specimen has to be appreciably larger than the largest size of the aggregate particles in the concrete. Various authorities recommend different values for the ratio of the minimum dimension of the test specimen to the maximum aggregate size. For instance, B.S. 1881: 1970 prescribes a test cube not smaller than 100 mm (4 in.) when 25 mm (1 in.) aggregate is used, i.e. a ratio of 4. A.S.T.M. Standard C 192–69 limits the ratio of the diameter of the cylinder to the maximum aggregate size to 3, and the U.S. Bureau of Reclamation to 4. A value of between 3 and 4 is generally accepted as satisfactory.

The limitation of size arises from the "wall effect" (*l'effet de paroi*):

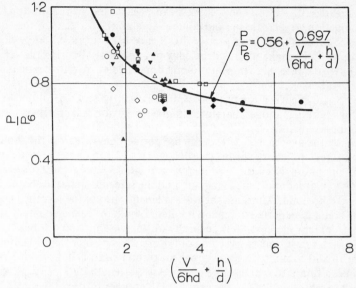

Fig. 8.22. *General relation between ratio of strength of concrete specimens P to strength of 6 in. cube P_6, and $V/6hd + h/d$, where V is volume of specimen, h its height, and d is its least lateral dimension (All dimensions in inches)*

the wall influences the packing of the concrete, since the quantity of mortar required to fill the space between the particles of the coarse aggregate and the wall is greater than that necessary in the interior of the mass, and therefore in excess of the mortar available in a well proportioned mix (Fig 8.23). In tests on concrete made with 19·05 mm ($\frac{3}{4}$ in.)

Fig. 8.23. *"Wall effect"*

aggregate, 101·6 mm (4 in.) cubes have been found to require for full compaction an increase in sand content equal to 10 per cent of the total weight of aggregate, compared with a mix used in an infinitely large section.[8.44] To make up this deficiency of fine material during the actual making of specimens, mortar is sometimes added from the remainder of the mix.

The wall effect is more pronounced the larger the surface/volume ratio of the specimen and is, therefore, smaller in flexure test specimens than in cubes or cylinders.

When the aggregate size exceeds the permissible value for the mould used, screening out of the large-size aggregate is sometimes resorted to. This operation is called *wet screening*. The screening must be done quickly in order to avoid drying out, and the screened material should be remixed by hand. Although the water/cement ratio of the screened concrete could be expected to remain unaltered, both the cement content and water content increase, and generally an increase in strength has been observed. For instance, screening out of particles greater than 19·05 mm ($\frac{3}{4}$ in.) from a mix with an original maximum size of 38·1 mm ($1\frac{1}{2}$ in.) has been found to increase the compressive strength by 7 per cent, and the flexural strength by 15 per cent.[8.45] On another project, the screening of the 38·1 to 152·4 mm ($1\frac{1}{2}$ to 6 in.) fraction has resulted in an increase in compressive strength of 17 to 29 per cent.[8.7] With air-entrained concrete, wet screening produces some loss of air, and this causes an increase in strength.

These data reflect not only the effect of the change in the composition of the mix but also the influence of the maximum size of aggregate *per se* (*see* p. 176).

Test Cores

The main purpose of measuring the strength of concrete test specimens is to estimate the strength of concrete in the actual structure. The emphasis is on the word "estimate", and indeed it is not possible to obtain more than an indication of the strength of concrete in a structure as this is dependent, *inter alia*, on adequacy of compaction and curing. As already mentioned, the strength of a test specimen depends on its shape, proportions and size so that a test result does not give the value of the intrinsic strength of the concrete. Nevertheless, if of two sets of similar specimens made from two concretes one set is stronger (at a statistically significant level), it is reasonable to conclude that the concrete represented by this specimen is stronger, too. There exist some methods of determining the strength of concrete *in situ*, but the limitations on the interpretation of test results must be remembered.

If the strength of the compression test specimens is found to be below the specified minimum then either the concrete in the actual structure is too weak too, or else the cubes are not truly representative of the concrete in the structure. This latter suggestion is often put forward in disputes on the acceptance or otherwise of a doubtful part of the structure: the cubes may have been disturbed while setting, they may have been exposed to frost before they hardened sufficiently or have otherwise been improperly cured, or simply the results of the crushing test are doubted.

The argument is often resolved by testing a sample of concrete taken from the suspect member. Usually, a core is cut by means of a rotary cutting tool with diamond bits. In this manner a cylindrical specimen is obtained, sometimes containing embedded fragments of reinforcement, and usually with end surfaces far from plane and square. The core should be soaked in water, capped, and tested in compression in a moist condition.

The influence of the height/diameter ratio of the core on the recorded strength was considered on page 471. If possible this ratio should be near 2; cores with height/diameter ratios lower than 1 yield unreliable results, and B.S. 1881: 1970 prescribes a minimum value of 0·95. The same standard specifies the use of 150 mm (6 in.) or 100 mm (4 in.) cores; however, cores as small as 50 mm (2 in.) have been successfully used in Switzerland[8.66] and are permitted in Swiss standards.

Cores are cut to determine the strength of concrete and can also be used to detect segregation or honeycombing or to check the bond at construction joints.

In some cases, beam specimens can be sawn from road or airfield slabs, using a diamond or carborundum saw. Such specimens are tested in flexure but, at least when siliceous aggregate is used, sawn specimens give appreciably lower strengths than comparable moulded beams.[8.23] Cutting of beams is not much used.

The strength of cores is generally lower than that of standard cylinders because site curing is invariably inferior to curing under standard moist conditions. Furthermore, the ratio of core strength to standard cylinder strength (at the same age) is not constant but decreases with an increase in the cylinder strength level.[8.67] The drop in the value of the ratio is about 0·2 over a strength range of 28 MN/m² (4,000 lb/in²). Approximate values of the ratio of core strength to standard cylinder strength are: 1·0 when the cylinder strength is 19 MN/m² (2,800 lb/in²), and 0·7 when the cylinder strength is 59 MN/m² (8,500 lb/in²). We should note that these values were obtained for structures cured in accordance with the recommended practice applicable in each case.

A further factor in the strength of cores is the position of the cut-out concrete in the structure. Cores usually have the lowest strength near the top surface of the structure, be it a column, a wall, a beam, or even a slab. With an increase in depth below the top surface, the strength of cores increases,[8.67] but at depths greater than about 300 mm (12 in.) there is no further increase.

Accelerated Curing Test

Concrete is usually placed in a structure in stages or lifts, one on top of another. Thus by the time the results of the 28-day test, or even the 7-day test, are available a considerable amount of concrete may overlay that represented by the test cubes in question. It is then rather late for remedial measures if the concrete is too weak; if it is too strong this indicates that the mix used was uneconomical.

It is clear that it would be a tremendous advantage to be able to predict the 28-day strength within a few hours of casting. The strength of concrete at 24 hours is an unreliable guide in this respect, not only because different cements gain strength at varying rates, but also because even small variations in temperature during the first few hours after casting have a considerable effect on the early strength. It is, therefore, necessary for the concrete to have achieved a greater proportion of its potential strength before testing, and a successful test based on accelerated curing was developed by King.[8.46]

In this test, standard concrete cubes are made but the moulds are immediately covered by top plates, sealed with grease on the metal surfaces of contact in order to prevent drying. Within 30 minutes of adding the mixing water, the cubes, in their covered moulds, are placed in an airtight oven, which is then switched on. The oven temperature should reach 93°C (200°F) in about an hour and the cubes are kept at this temperature for a further 5 hours, making a total of 6 hours in the oven. At the end of this period the cubes are removed from the oven, stripped, allowed to cool, and tested in compression in the standard manner, the time allowed for these operations being 30 minutes.

Thus the strength of the concrete is determined within 7 hours of casting, and this accelerated strength shows good correlation with the 7- and 28-day strengths of normally cured concrete. King's[8.46] curves are shown in Fig. 8.24. The reliability of the results so obtained is high but, because of the variation in the rate of gain of strength of different cements, it is not possible to use the 7-day strengths to predict the 28-day values.

If the 7-hour period is inconvenient, because, for instance, it may fall outside the normal working hours, alternative schemes can be used: the same accelerated strength is achieved if the cubes are kept in a closed tank over water at 16°C (60°F) for 18 to 24 hours and are then heated for about 4 hours instead of 6 (Fig. 8.25).

In a more recent variant of King's test, recommended by a committee of the Institution of Civil Engineers,[8.68] the specimens (in their moulds) are placed (at the age of 30 minutes) in a water bath at 55°C (131°F) and

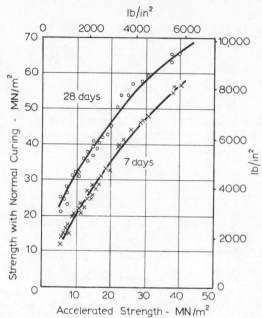

Fig. 8.24. *The relation between strength determined by King's accelerated curing test and the 7- and 28-day strengths of concrete moist-cured at 20°C (68°F)*[8.46]

are kept in it for 23 hours. They are then taken out, de-moulded and tested in the usual manner at the age of 24 hours.

In an alternative procedure,[8.69] developed in Canada, the cylinders are placed complete with moulds and cover plates in boiling water 20 minutes after the concrete has reached a needle penetration resistance of 24 MN/m² (3,500 lb/in²); this occurs 6 to 8 hours after mixing. The cylinders are maintained in the water bath for 16 hours, then are de-moulded, allowed to cool, and capped. They are tested one hour after removal from the boiling water.

It has been found that concrete must have set to a sufficient degree before boiling as otherwise a lower strength is obtained. To measure the set, A.S.T.M. Standard C 403–70 may be used, and the boiling commences at a fixed time after the given degree of set has been achieved. Since such a procedure takes into account the setting time of concrete, variations due to different cements and admixtures are allowed for, but the determination of setting time introduces a complication.

It should be noted that the acceleration by boiling leads to a lower accelerated strength (by, on the average, 10 MN/m² (1,400 lb/in²)) than King's accelerated strength for a given 28-day strength with normal curing.

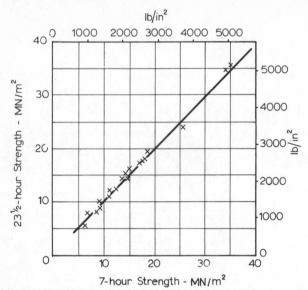

Fig. 8.25. *Correlation of accelerated strength at 23½ hours (19½ hours' delay and 3½ hours in the oven) with the 7-hour accelerated strength (½ hour's delay and 6 hours in the oven)*[8.46]

A rather different approach to accelerated curing utilizes the heat generated by the hydration of cement, i.e. uses autogenous curing. Under adiabatic conditions, this leads to a substantial rise in temperature; for instance, a 1:2:4 mix with a water/cement ratio of 0·6 would show a temperature rise of 40°C (72°F) in 48 hours (*see* p. 38). In the proposed procedure,[8.70] the concrete test specimen (in a covered mould) is sealed in a plastic bag one hour after mixing and then placed in a polyurethane container for 46 hours. The specimen is then de-moulded, allowed to cool for 30 minutes and tested at the age of 48 hours. It would be expected that the strength recorded is affected by the richness of the mix in a manner different from that under normal curing, but apparently a reliable relation between the accelerated strength and the 28-day normal curing strength has been obtained. This is of the form: 28-day strength = accelerated strength plus a constant.

In 1970, the Canadian Standard CSA A23.1 for Concrete Materials and Methods of Concrete Construction introduced three accelerated curing test methods as standard tests; one of these uses autogenous curing, one is related to the fixed set of concrete, and one is of the boiling type. In 1971, the A.S.T.M. prescribed a Tentative Method C 684–71T, which provides for the autogenous curing and the boiling tests, and also for a warm water (35 ± 3°C (95 ± 5°F)) test. In 1972, an addendum to the

British Standard B.S. 1881: Part 3 introduced an accelerated curing test, in which test cubes in covered moulds (after a waiting period of $1\frac{1}{2}$ to $3\frac{1}{2}$ hours from the time of mixing) are immersed in water at $55 \pm 2°C$ ($131°F$). Twenty hours later, the specimens are de-moulded and placed in water at $20 \pm 5°C$ ($68 \pm 9°F$) for 1 to 2 hours, and tested immediately afterwards.

From the preceding discussion it is evident that the different tests give different relations. However, this is not of fundamental importance, as it is generally preferable to obtain empirical curves relating the accelerated strength to the strength of normally cured specimens for the given construction materials and conditions. In this manner, a rapid and reliable means of strength control is achieved.

The argument can be extended: the accelerated strength test can be considered as a test in its own right and not merely as a means of predicting the 28-day strength. The basis for this is that, since a compression test specimen offers only a presumed measure of the strength of concrete in the structure, there is no inherent value in the standard 28-day test over some other tests. From this base, Smith and Chojnacki[8.69] suggested that "a suitable accelerated curing procedure can offer a more convenient and realistic way of ascertaining if the concrete will satisfy the purpose for which it was designed." Probably, the greatest usefulness of the accelerated test is as a quality control test but the acceptance of a design mix would be based on conventional tests; accelerated tests would be made at the same time and would form the basis for control.

It is of course possible to go even further and to evaluate concrete solely in terms of its accelerated strength; in such a case, the design "thinking" would have to be entirely in accelerated strength values. Since these are lower than the 28-day normal curing strength values, there is a certain amount of reluctance to accept the new "numbers". Nevertheless, there is nothing intrinsically fundamental about the 28-day strengths and it is possible that at some future date the strength of concrete will be based on the accelerated test values.

Rebound Hammer Test

The difficulties of core cutting, and indeed the entire procedure of making, curing and testing of standard test specimens would all be avoided if concrete could be tested *in situ* in a manner harmless to the part tested. Various attempts to devise non-destructive tests have been made but few have been successful. One method that has found practical application within a limited scope is the rebound hammer test, devised by Ernst Schmidt. It is also known as the impact hammer, or sclerometer, test.

The test is based on the principle that the rebound of an elastic mass depends on the hardness of the surface against which the mass impinges.

In the rebound hammer test (Fig. 8.26) a spring-loaded mass has a fixed amount of energy imparted to it by extending a spring to a fixed position; this is achieved by pressing the plunger against the surface of the concrete under test. Upon release the mass rebounds from the plunger, still in contact with the concrete surface, and the distance travelled by the mass, expressed as a percentage of the initial extension of the spring, is called the rebound number; it is indicated by a rider moving along a graduated scale. The rebound number is an arbitrary measure since it depends on the energy stored in the given spring and on the size of the mass.

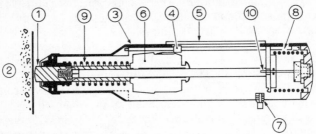

Fig. 8.26. *Rebound hammer*

1. plunger   2. concrete   3. tubular housing   4. rider   5. scale   6. mass   7. release button   8. spring   9. spring   10. catch

The hammer has to be used against a smooth surface, preferably a formed one. Open-textured concrete cannot therefore be tested. Trowelled surfaces should be rubbed smooth using a carborundum stone. If the concrete under test does not form part of a larger mass, it has to be supported in an unyielding manner, as jerking during the test would result in a lowering of the rebound number recorded.

The test is sensitive to local variations in the concrete; for instance, the presence of a large piece of aggregate immediately underneath the plunger would result in an abnormally high rebound number; conversely, the presence of a void in a similar position would lead to a very low result. For this reason it is desirable to take 10 to 12 readings spread over the area to be tested, and to assume their average value as representative of the concrete. The standard error of the mean is higher than if the strength were determined by the compression test, but the saving in effort, time and cost is considerable.

The plunger must always be normal to the surface of the concrete under test, but the position of the hammer relative to the vertical will affect the rebound number. This is due to the action of gravity on the travel of the mass in the hammer. Thus the rebound number of a floor would be smaller than that of a soffit, and inclined and vertical surfaces would yield intermediate values; the actual variation is best determined experimentally (Fig. 8.27).

The test determines in reality the hardness of the concrete surface and, although there is no unique relation between hardness and strength of concrete, empirical relationships can be determined for similar concretes cured in such a manner that both the surfaces tested by the hammer and the central regions, in whose strength we are interested, have the same

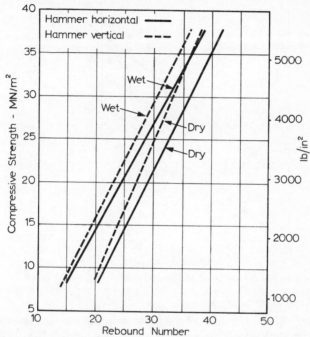

Fig. 8.27. *Relation between compressive strength of cylinders and rebound number for readings with the hammer horizontal and vertical on a dry and a wet surface of concrete*[8.47]

strength. Changes affecting only the surface of the concrete, such as the degree of saturation at the surface or carbonation, would be misleading as far as the properties of the concrete within the structure are concerned (*see* Fig. 8.27).

The type of aggregate used affects the rebound number (Fig. 8.27) so that the relation between the rebound number and strength should be determined experimentally for every concrete used on a site.

We can see then that the test is of a comparative nature only, and the claims of the manufacturers that the rebound number can be directly converted into a value of compressive strength are not justified. In particular, the hardness of concrete depends on the elastic properties of the aggregate used, and may also be affected by large differences in mix proportions and also by carbonation. Nevertheless, the test is useful as a measure of uniformity of concrete and is of great value in checking the quality of the material throughout a structure, or in the manufacture of a number of similar products such as precast sections. The actual strength of the concrete with which successive members are being compared has

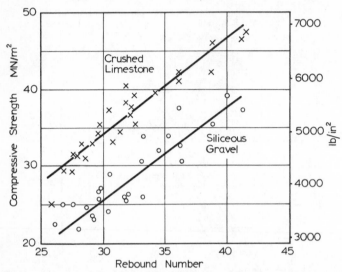

Fig. 8.28. *Relation between compressive strength and rebound number for concrete cylinders made with different aggregates.*[8,48] *Readings taken on the side of a cylinder with the hammer horizontal.*

to be determined by means of the usual crushing test. Methods of calibration and guidance for use of the hammer are given in B.S. 4408: Part 4: 1971.

A concrete structure can also be tested by means of the hammer to see whether the rebound number has reached a value known to correspond to the desired strength. This is of help in deciding when to remove falsework or to put the structure into service. Another use of the hammer is to check whether the strength development of a given concrete has been affected by frost at an early age.

Penetration Resistance Test

A novel test, known commercially as the Windsor probe test, estimates the strength of concrete from the depth of penetration by a metal rod driven by a standard charge of powder. The underlying principle is that, for standard test conditions, the penetration is inversely proportional to the compressive strength of concrete but the relation depends on the hardness of the aggregate. Thus, the hardness of aggregate on Moh's scale has to be determined but this presents no difficulty. Charts of strength versus penetration (or length of exposed probe) are available for aggregates with hardness between 3 and 7 on Moh's scale but in practice the penetration resistance should be correlated with the compressive strength of standard test specimens or cores of the actual concrete used. A typical relation is shown in Fig. 8.29. It should be

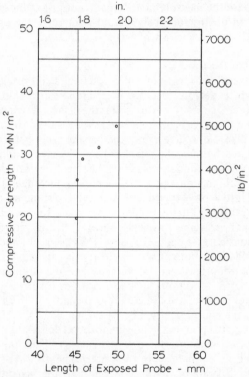

Fig. 8.29. *Relation between length of exposed probe and strength of 152 mm (6 in.) sawn cubes at the age of 35 days*[8.71]

remembered that the test basically measures hardness and cannot yield absolute values of strength but is very useful in determining relative strength, i.e. in comparisons.[8.78]

The probes are driven in sets of three in close vicinity, and the average penetration is used in estimating the strength. A set costs about £2·50; this may be thought expensive but, unlike the case of test cores, there is no additional cost of handling, storing, or testing in the machine, and of course core cutting and making good are costly.

The penetration resistance test can be considered almost non-destructive as the damage to concrete made by 6·3 mm ($\frac{1}{4}$ in.) probes is only local and re-testing in the vicinity is possible.

An evaluation of the test is not yet available as it has not been used extensively but an A.S.T.M. Method is currently being prepared. It is likely that the penetration resistance test will, at least in part, supplant the rebound hammer test to which it is under many circumstances superior because the measurement is made not just at the surface of the concrete but in depth: the probe actually fractures the aggregate and compresses the material into which it is driven.

Ultrasonic Pulse Test

The standard tests of strength of concrete are made on specially prepared specimens, which perforce are not true samples of the concrete in the actual structure. One result of this is that the degree of compaction of the concrete in the structure is not reflected in the results of the strength test, and it is not possible to determine whether the potential strength of the mix, as indicated by the cube test, has been actually developed. Admittedly, it is possible to cut a sample from the structure itself but this would necessarily result in damaging the member concerned. Furthermore, such a procedure is too expensive to be used as a standard method.

For these reasons, attempts have been made to measure in a non-destructive manner some physical property of concrete related to its strength. A considerable degree of success has been achieved in the determination of the longitudinal wave velocity in concrete.[8.49] There is no unique relationship between this velocity and the strength of concrete but under specified conditions the two quantities are directly related. The common factor is the density of concrete: a change in the density results in a change in the pulse velocity. Likewise, for a given mix, the ratio of the actual density to the potential (fully compacted) density and the resulting strength are also closely related (*see* Fig. 4.1). Thus the lowering of density caused by an increase in the water/cement ratio decreases both the compressive strength of the concrete and the velocity of a pulse transmitted through it.

The apparatus for high precision measurement of the velocity of ultrasonic pulse in concrete, represented diagrammatically in Fig. 8.30, can be obtained commercially and is being increasingly used. Apparatus with a digital output is also available. The technique is covered by B.S. 4408: Part 5: 1971.

The wave velocity is not determined direct but is calculated from the time taken by a pulse to travel a measured distance. This ultrasonic pulse —hence the name of the method of measurement—is obtained by applying a rapid change of potential from a transmitter driver (Fig. 8.30) to a piezo-electric crystal transducer emitting vibrations at its fundamental frequency. Barium titanate and lead zirconate titanate transducers have been found to be suitable. The transducer is in contact with the concrete, so that the vibrations travel through it and are picked up by another transducer in contact with the opposite face of the specimen under test. The transducers generate an electrical signal, which is fed through an

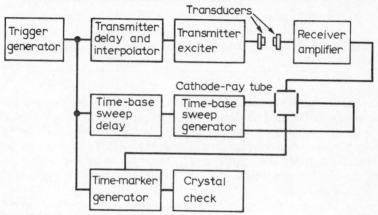

Fig. 8.30. *Layout of the ultrasonic pulse apparatus*[8.49]

(*Crown copyright*)

amplifier to a plate of a cathode-ray tube. A second plate supplies timing marks at fixed intervals. Thus from the measurement of the displacement of the pulse signal relative to the position when the transducers are in direct contact with one another, the time taken by the pulse to travel through the concrete can be measured with an accuracy of $\pm$ 0·1 microsecond. With a transmission time of 30 to 45 microseconds for a 150 mm (6-inch) thickness of concrete the velocity is determined within less than 0·5 per cent. However, with an increase in the length of the path travelled, the sharpness of the onset waveform decreases so that no gain in accuracy is achieved. Concrete from 0·1 to 2·5 m (4 in to 8 ft) thick can normally be tested, but tests have been made on concrete up to 15 m (50 ft) thick.[8.50]

The choice of frequency of the ultrasonic vibrations is governed, on the one hand, by the fact that the higher the frequency the smaller the spread of the direction along which the waves travel, and therefore the higher the energy received; on the other hand, the higher the frequency the higher the attenuation of energy. Transducers with natural frequencies of between 50 and 200 kHz are generally used; the lower end of the scale is more common.

When access to two opposite sides of a concrete member is not possible, the pulse velocity can be measured along a path parallel to the surface of the concrete. The transducers are then placed on the same face of the

Table 8.5 *Classification of the Quality of Concrete on the Basis of Pulse Velocity.*[8.50]

Longitudinal pulse velocity		Quality of concrete
km/s	10^3ft/s	
> 4·5	> 15	excellent
3·5–4·5	12–15	good
3·0–3·5	10–12	doubtful
2·0–3·0	7–10	poor
< 2·0	< 7	very poor

member, a known distance apart. The energy received is, however, considerably lower, and the accuracy of the reading is correspondingly decreased. Better results are obtained by placing the emitting transducer over the edge of the member on a face perpendicular to the main surface. Measurement of pulse velocity along the surface refers to the properties of the surface layer only and gives no indication of the strength of the concrete at depth.

The ultrasonic pulse velocity technique is used as a means of quality control of products which are supposed to be made of similar concrete: both lack of compaction and a change in the water/cement ratio would be easily detected. The technique cannot, however, be employed for the determination of strength of concretes made of different materials in unknown proportions. It is true that there is a broad tendency for concrete of higher density to have a higher strength (provided the specific gravity of the aggregate is constant) so that a general classification of the quality of concrete on the basis of the pulse velocity is possible.[8.49] Some figures suggested by Whitehurst[8.50] for concrete with a density of approximately 2,400 kg/m^3 (150 lb/ft^3) are given in Table 8.5. According to Jones,[8.49] however, the lower limit for good quality concrete is between 4·1 and 4·7 km/s (13,500 and 15,500 ft/s).

This discrepancy, and the generally wide variation in the pulse velocity of concretes of a given quality are due to the influence of the coarse aggregate. Both its quantity and type affect the pulse velocity while, for a constant water/cement ratio, the influence of the coarse aggregate on strength is comparatively small. Thus for different mix proportions a different relation between strength and pulse velocity would be obtained; this is illustrated in Fig. 8.31 based on Jones's[8.49] results. However, Kaplan[8.72] found that in concretes of the same age the effects of aggregate/cement ratio and water/cement ratio balance one another so that at a given age and at a *constant workability* there is a unique relation between pulse velocity and strength of concrete.

In practical cases it is convenient to establish the relation between strength and pulse velocity by means of test cubes. The cubes should be in the same state of dryness as the concrete in the structure because of the considerable influence of moisture in the concrete on the pulse velocity. If the calibration is made on wet cubes and the concrete in the structure is dry, the strength of the latter can be underestimated by 10 to 15 per

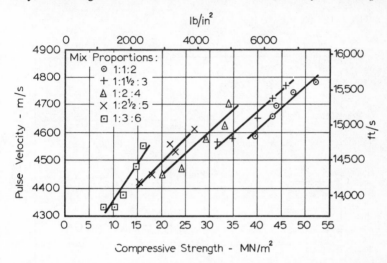

Fig. 8.31. *Relation between ultrasonic pulse velocity and compressive strength for concretes of different mix proportions*[8.49]

(*Crown copyright*)

cent,[8.51] possibly even more. It is only fair to add that the use of the ultrasonic pulse measurements as a means of quality control on a construction job is not practicable: for instance, no satisfactory correlation appears to exist between the variability of the compression test cubes and the variability of pulse measurements. Admittedly, the latter are affected by compaction and workmanship in general but even so unexplained discrepancies have been observed. Under laboratory conditions, a standard deviation of about 160 m/s (530 ft/s) in good quality concrete was measured.[8.73]

The relation between the pulse velocity and the dynamic modulus of elasticity of concrete is considered on page 319; this is of considerable importance. The relation between the static modulus and the pulse velocity can also be of value; Fig. 8.32 shows typical data for concretes made with one normal weight and one lightweight aggregate but a generalization cannot be made.

In addition to the control of the quality of concrete the ultrasonic pulse measurements can be used to detect the development of cracks in structures such as dams, and to check deterioration due to frost or chemical action. These are very important applications of the technique, which is suitable for the detection of any development of voids in concrete. Cracks with a component at right angles to the direction of propagation of the pulse cause the pulse to diffract round the crack. This increases the

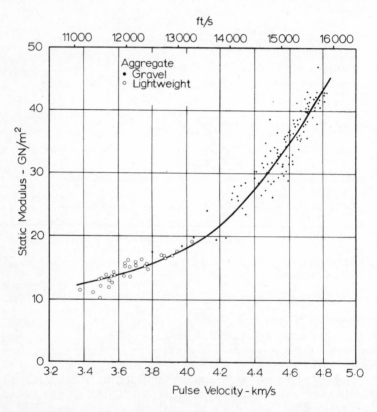

Fig. 8.32. *Relation between static modulus of elasticity and ultrasonic pulse velocity for concretes made with one normal weight and one lightweight aggregate*[8.74]

time of travel of the pulse and hence decreases the apparent velocity. When, however, the plane of the crack coincides with the direction of propagation of the pulse, it can pass on either side of the crack and the pulse velocity is not affected.[8.75] An echo type use of the ultrasonic pulse technique makes it possible to measure the thickness of concrete roads and similar slabs.[8.52, 8.79]

As a development of the ultrasonic pulse velocity method, Jones and

Mayhew[8.52] suggested that the wavelength and velocity of surface waves be measured at the surface of a concrete member (or at the surface of different layers of construction) at specific frequencies of 30 to 30,000 Hz. Using experimental data on the relation between the wave velocity and wavelength on samples of the same type of concrete, the strength and modulus of elasticity of the concrete in the actual member can be determined.

A number of specialized techniques for testing concrete are available. These range from the use of electromagnetic devices for the measurement of cover to reinforcement (B.S. 4408: Part 1: 1969) to gamma ray methods for the determination of the variation in density of concrete,[8.55] including the detection of faults in grout or honeycombing in concrete, and location of reinforcement; the technique is covered by B.S. 4408: Part 3: 1970. Because of the expense usually involved, the use of these techniques is not likely to become widespread in the near future.

Electrodynamic Determination of the Modulus of Elasticity*

In some cases it is desirable to determine the progressive changes in the strength of a concrete specimen, whether development or loss owing to the damaging action of frost or chemical attack. Such an investigation can be made using a large number of compression test specimens, a batch of them being tested at suitable time intervals. It is clear, however, that it would be preferable to observe the variation in the quality of the same specimen over the entire duration of the investigation.

To this end, advantage can be taken of the general relation between strength of concrete and its modulus of elasticity, since the modulus can be determined without damaging the specimen. This relation between strength and modulus of elasticity depends on the aggregate used, the mix proportions, and the curing conditions, but for a given specimen the variation in the value of the modulus would give a good indication of the variation in the strength of the concrete.

The modulus of elasticity is determined on laboratory specimens subjected to longitudinal vibration at their natural frequency, and is therefore known as the dynamic modulus. B.S. 1881: 1970 prescribes the use of specimens similar to those employed in the determination of the modulus of rupture, i.e. beams 150 by 150 by 750 mm (6 by 6 by 30 in.), or 100 by 100 by 500 mm (4 by 4 by 20 in.). The specimen is clamped at its centre (Fig. 8.33) with an electro-magnetic exciter unit placed against one end face of the specimen and a pick-up against the other. The exciter is driven by a variable frequency oscillator with a range of 100 to 10,000 Hz. The vibrations propagated within the specimen are received

---

* Also known as resonance method.

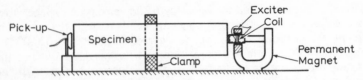

Fig. 8.33. *B.S. 1881: 1970 test arrangement for the determination of the dynamic modulus of elasticity* (*longitudinal vibration*)

by the pick-up, amplified, and their amplitude is measured by an appropriate indicator (Fig. 8.34). The frequency of excitation is varied until resonance is obtained at the fundamental (i.e. lowest) frequency of the specimen; this is indicated by the maximum deflexion of the indicator. If this frequency is n Hz, L is the length of the specimen, and ρ its density, then the dynamic modulus of elasticity is—

$$E = Kn^2L^2\rho$$

where K is a constant.

The length of the beam and its density have to be determined very accurately. If L is measured in inches, and ρ is in lb/ft^3, then

$$E = 6 \times 10^{-6}\, n^2L^2\rho \quad \text{lb/in}^2.$$

If L is measured in millimetres and ρ in kg/m^3, then

$$E = 4 \times 10^{-12}\, n^2L^2\rho \quad \text{MN/m}^2.$$

In addition to the test based on the longitudinal resonance frequency, tests for the transverse (flexural) frequency (Fig. 8.34, case 2) and the torsional frequency (case 3) are also used. These, as well as the longitudinal test, are described in A.S.T.M. Standard C 215–60 (re-approved 1970). The dynamic modulus test is valuable in many laboratory investigations, but in any general comparisons the limitations of the strength–modulus relationship must not be forgotten.

Tests on the Composition of Hardened Concrete

In some disputes about the quality of hardened concrete the question is raised whether the composition of concrete was as specified, and to answer this chemical and physical tests are made on a sample of hardened concrete.

Cement Content

A method of determining the cement content is prescribed by A.S.T.M. Standard C 85–66. This is based on the fact that the silicates in Portland cement are much more readily decomposed and made soluble in dilute hydrochloric acid than are the silica compounds normally contained in aggregate. The same applies to the relative solubilities of the lime compounds in the cement and in aggregate (excepting, however, limestone aggregates), so that there exists also a soluble calcium oxide method.

Briefly, the procedure is to crush a representative sample and to dehydrate it at 550°C (1022°F) for 3 hours. Small portions of the sample are treated with 1:3 hydrochloric acid, and this liberates the silica contained in the cement. The quantity of silica is determined by standard chemical methods.

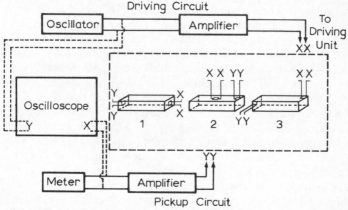

Fig. 8.34. *Layout of apparatus for the determination of the dynamic modulus of elasticity: (1) longitudinal vibration, (2) transverse vibration, (3) torsional vibration. (A.S.T.M. Standard C 215–60)*

The filtrate from the silica determination contains soluble calcium oxide from the aggregate and the cement, and further calculations depend on whether or not the aggregate is largely siliceous. If the original aggregate is available its solubility should be tested.

From the contents of soluble silica and calcium oxide the cement content in the original volume of the sample can be calculated, using the figures of A.S.T.M. Standard C 85–66. These results are reliable and can be used to check the cement content of different parts of a structure, e.g. when it is desired to establish whether or not segregation of cement has taken place. The accuracy of the test is, however, lowest for mixes with low cement contents, and it is often in this type of mix that the exact value of cement content is required. Furthermore, the test depends on the knowledge of the chemical composition of aggregate, which may not be available for testing. When large amounts of both soluble silica and calcium oxide are liberated from the aggregate, the method is not reliable.

More complex techniques are prescribed by B.S. 1881: Part 6: 1971 but it should be noted that chemical tests are rather expensive and are used only in resolving disputes, and not as a means of control of the quality of concrete.

Determination of the Original Water/Cement Ratio

A method of estimating the water/cement ratio that existed at the time of placing of a concrete mix, now hardened, has been developed by Brown.[8.53] In essence, this involves determining the volume of the capillary pores and the weight of cement and combined water.

A sample of concrete is oven-dried at 105°C (221°F) and the air is removed from the pores under vacuum. The pores are then refilled with carbon tetrachloride, whose weight is measured, and hence the weight of the water which originally occupied the pores can be calculated. Since the voids formed through air entrainment are discontinuous they remain filled with air when the vacuum is applied, and no water is absorbed in them. The result of the test is thus unaffected by entrained air.

The sample is now broken up, the carbon tetrachloride having been allowed to evaporate, and the aggregate is separated out and weighed. The loss on ignition and the CO_2 content of the remaining fine material are determined, and from these two quantities the weight of combined water can be calculated.

The sum of the combined water and the pore water gives the original mixing water. The quantity of anhydrous cement can also be determined, either in conjunction with this test or by the method described on page 511, and hence the water/cement ratio of the mix can be calculated within about 0·02 of the true value. The technique has been developed into a standard method of B.S. 1881: Part 6: 1971.

Physical Methods

Polivka[8.54] has successfully used a "point-count" method on a sawn and varnished surface of a dried concrete specimen to determine its cement content, total aggregate content, and fine/coarse aggregate ratio. The basis of the method is the fact that the relative volumes of the constituents of a heterogeneous solid are directly proportional to their relative areas in a plane section, and also to intercepts of these areas along a random line. Furthermore, the frequency with which a constituent occurs at a given number of equally spaced points along a random line is a direct measure of the relative volume of that constituent in the solid. Thus a point-count by means of a stereomicroscope can rapidly give the volumetric proportions of a hardened concrete specimen.

The aggregate and the voids (containing air or evaporable water) can be identified, the remainder being assumed to be hydrated cement. In order to convert the quantity of the latter to the volume of unhydrated cement, we have to know the specific gravity of dry cement and the non-evaporable water content of hydrated cement; these are known from Powers' studies (*see* p. 35).

The test determines the cement content of the concrete within 10 per cent, but the original water content or voids ratio cannot be estimated since no distinction is made in the test between air and water voids.

Variation in Test Results

The variation in strength of nominally similar test specimens has been mentioned, but this should be considered in terms of statistics. Some of the simpler statistical terms will first be introduced.

Distribution of Strength

Let us suppose that we have measured the compressive strength of 100 test cubes, all made from similar concrete. This concrete can be imagined to be a collection of units all of which could be tested; such a collection is referred to as the population, and the portion of concrete in the actual test cubes is called the sample. It is the purpose of the tests on the sample to supply information on the properties of the parent population.

From the nature of the strength of concrete (p. 245) it would be expected that the recorded strengths will be different for different cubes, i.e. the results will show a scatter. Let us assume that the strengths vary between 4,600 and 6,800 lb/in². A good picture of the distribution of these strengths can be obtained by grouping the actual strengths in intervals of 200 lb/in², so that we now have a certain number of cubes whose strength falls within each interval, for example—

Strength interval, lb/in²	Number of cubes in interval
4,600—4,800	1
4,800—5,000	2
5,000—5,200	6
5,200—5,400	13
5,400—5,600	21
5,600—5,800	23
5,800—6,000	18
6,000—6,200	8
6,200—6,400	5
6,400—6,600	1
6,600—6,800	2
	Total = 100

If we now plot the (constant) strength interval as abscissae and the number of cubes in each interval (known as frequency) as ordinates we

514 *Properties of Concrete*

obtain a histogram. The area under the curve represents, of course, the total number of cubes to an appropriate scale. Sometimes, it is more convenient to express the frequency as a percentage of the total number of specimens, i.e. to use relative frequency.

The histogram for the above-mentioned data is plotted in Fig. 8.35,

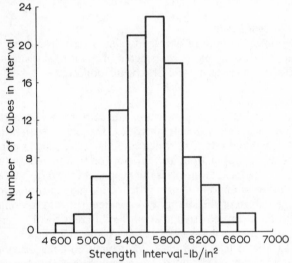

Fig. 8.35. *Histogram for the strength values of page* 513

and it can be seen that it gives a clear visual picture of the scatter of results, or, more accurately, of the distribution of strength within the sample tested.

Another simple measure of dispersion is given by the range of values, i.e. the difference between the highest and the lowest strengths—2,200 lb/in² in the above case. The range is, of course, calculated extremely rapidly, but it is a rather crude measure: it depends on two values only, and furthermore in a large sample these values are of low frequency; thus range increases with sample size for the same underlying distribution. The theoretical relation between range and standard deviation is shown in Fig. 8.36, together with data obtained in practice.

If the number of specimens is increased infinitely and at the same time the size of the interval is decreased to a limiting value of zero, the histogram would become a continuous curve, known as the distribution curve. For the strength of a certain type of material this curve would have a characteristic shape, and there are in fact several "type" curves whose properties have been calculated in detail and are listed in standard statistical tables.

One such type of distribution is the so-called normal or Gaussian distribution. The applicability of this type of distribution to the strength of concrete was mentioned on page 488; the assumption of normal distribution is sufficiently close to reality to be an extremely useful tool in computations.

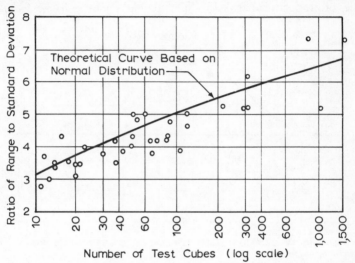

Fig. 8.36. *The ratio of range to standard deviation for samples of different size*[8.26]

(*Crown copyright*)

The equation to the normal curve, which depends only on the values of the mean, μ, and standard deviation, σ, is

$$y = \frac{1}{\sigma\sqrt{2\pi}}\ e^{\frac{-(x-\mu)}{\sigma^2}}$$

The standard deviation is defined in the subsequent section. This equation is represented graphically in Fig. 8.37 and it can be seen that the curve is symmetrical about the mean value and extends to plus and minus infinity. This is sometimes mentioned as a criticism of the use of a normal distribution for strength, but the extremely low probability of the occurrence of the very high or very low values is of little practical significance.

The area under the curve between certain values of strength (measured in terms of standard deviation) represents, in a manner similar to the histogram, the number of specimens between the given limits of strength. Since, however, the curve refers to an infinite population of specimens, and we deal with a limited number of them, the area under the curve between given ordinates, expressed as a fraction of the total area under the curve (and known therefore as proportional area), measures the chance that the strength of an individual drawn at random, x, will lie between the given limits. This chance multiplied by 100 gives the percentage of specimens that may be expected in the long run to have a strength between the two limits considered. Statistical tables give the values of proportional areas for different values of $(x - \mu)/\sigma$.

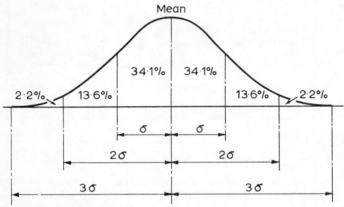

Fig. 8.37. *Normal distribution curve; percentage of specimens in intervals of one standard deviation shown*

Standard Deviation

It can be seen from the foregoing discussion of probability that the dispersion of strength about the mean is a fixed function of the standard deviation. This is defined as the root-mean-square deviation, i.e.

$$\sigma = \sqrt{\left(\frac{\Sigma(x - \mu)^2}{n}\right)}$$

where x represents the values of strength of all n specimens, and μ is the arithmetic mean of these strengths, i.e.

$$\mu = \frac{\Sigma x}{n}.$$

In practice, we deal with a limited number of specimens, and *their* mean, $\bar{x}$, is our estimate of the true (population) mean μ.* We calculate

---

* The accuracy with which $\bar{x}$ estimates the value of the true mean μ is governed by the standard deviation of the mean, known as the standard error σ_n where

$$\sigma_n = \frac{\sigma}{\sqrt{n}}.$$

Thus there is a probability of 0·68 that x is within the interval $\mu \pm \sigma_n$.

the deviations from $\bar{x}$ and not from μ, and therefore put $(n - 1)$ instead of n in the denominator of the expression for the estimate of σ. The reason for this correction of $n/(n - 1)$, known as Bessel's correction, is that the sum of squares of deviations has a minimum value when taken about the sample mean, $\bar{x}$, and is therefore smaller than it would be if taken about the population mean, μ. Thus, the estimate of σ is

$$\sigma = \sqrt{\left(\frac{\Sigma(x - \bar{x})^2}{n - 1} \right)}.$$

Bessel's correction is, of course, unimportant when n is large.

An important practical point is that one value (e.g. the result of one cube test) yields no information whatever about the standard deviation and therefore about the reliability or possible "error" of the value obtained.

The calculation of σ by the formula given above is laborious, and it is therefore more convenient to use another form of this expression, namely

$$\sigma = \sqrt{\left(\frac{(\Sigma x^2)}{n} - \bar{x}^2 \right)} = \frac{1}{n} \sqrt{\left(n\Sigma(x^2) - (\Sigma x)^2 \right)}.$$

Thus, the sum of x^2 is obtained without first finding the differences $(x - \bar{x})$; tables of squares or a calculating machine make this operation quite rapid. Other simplifications, such as subtracting a fixed quantity from all values, further aid computation. To find s (the standard deviation) Bessel's correction is applied—

$$s = \sigma\sqrt{\left(\frac{n}{n - 1} \right)}.$$

The standard deviation is expressed in the same units as the original variate, x, but for many purposes it is convenient to express the scatter of results on a percentage basis. We take then the ratio $\sigma/\bar{x} \times 100$, and this is called the coefficient of variation. It is a dimensionless quantity.

The graphical representation of the standard deviation (*see* Fig. 8.37) is the horizontal distance from the mean to the point of contraflexure of the normal distribution curve. Since the curve is symmetrical, the area under the curve contained between abscissae $\mu - \sigma$ and $\mu + \sigma$ is 68 per cent of the total area under the curve. In other words, the probability that the strength of a cube chosen at random lies within the range $\mu \pm \sigma$ is 0.68. The probabilities for other deviations from the mean are indicated in Fig. 8.37.

For a given mean strength, the standard deviation characterizes fully the distribution, assumed to be of the normal type; the variation in the value of the standard deviation determines the spread of strengths in pounds per square inch. Fig. 10.3 shows distribution curves for values of standard deviation of 2·5, 3·8, and 6·2 MN/m² (350, 560, and 900 lb/in²). The value of the standard deviation affects the (mean) strength that has to be aimed at in mix design for a given "minimum" or characteristic strength specified by the designer of the concrete structure. This problem is discussed fully in Chapter 10. Details of statistical methods applicable to testing, particularly data on the choice of sample size, have to be sought in specialized books.*

REFERENCES

8.1.   S. THAULOW, Field testing of concrete, *Norsk Cementforening* (Oslo, 1952).
8.2.   D. L. BLOEM, *Study of horizontal mould for concrete cylinders* (Series D–58) (National Ready Mixed Concrete Assoc., Washington D.C., 16th September 1958).
8.3.   A. M. NEVILLE, The influence of the direction of loading on the strength of concrete test cubes, *A.S.T.M. Bul. No.* 239, pp. 63–5 (July 1959).
8.4.   A. M. NEVILLE, The failure of concrete compression test specimens, *Civil Engineering*, **52**, No. 613, pp. 773–4 (London, July 1957).
8.5.   H. F. GONNERMAN, Effect of end condition of cylinder on compressive strength of concrete, *Proc. A.S.T.M.*, **24**, Part II, p. 1036 (1924).
8.6.   G. WERNER, The effect of type of capping material on the compressive strength of concrete cylinders, *Proc. A.S.T.M.*, **58**, pp. 1166–81 (1958).
8.7.   U.S. BUREAU OF RECLAMATION, *Concrete Manual*, 6th Ed. (Denver, 1956).
8.8.   R. L'HERMITE, Idées actuelles sur la technologie du béton. *Documentation Technique du Bâtiment et des Travaux Publics* (Paris, 1955).
8.9.   A. G. TARRANT, Frictional difficulty in concrete testing, *The Engineer*, **198**, No. 5159, pp. 801–2 (London, 1954).
8.10. A. G. TARRANT, Measurement of friction at very low speeds, *The Engineer*, **198**, No. 5143, pp. 262–3 (London, 1954).
8.11. P. J. F. WRIGHT, Compression testing machines for concrete, *The Engineer*, **201**, pp. 639–41 (London, 26th April 1957).
8.12. J. W. H. KING, Discussion on Properties of concrete under complex states of stress, in *The Proc. Int. Conf. on the Structure of Concrete*, p. 293 (London, Cement and Concrete Assoc., 1968).
8.13. R. JONES, A method of studying the formation of cracks in a material subjected to stress, *British Journal of Applied Physics*, **3**, pp. 229–32 (London, 1952).
8.14. J. W. MURDOCK and C. E. KESLER, Effect of length to diameter ratio of specimen on the apparent compressive strength of concrete, *A.S.T.M. Bul.*, pp. 68–73 (April 1957).
8.15. K. NEWMAN, Concrete control tests as measures of the properties of concrete, *Proc. of a Symposium on Concrete Quality*, pp. 120–38 (London, Cement and Concrete Assoc., 1964).
8.16. R. H. EVANS, The plastic theories for the ultimate strength of reinforced concrete beams, *J. Inst. C. E.*, **21**, pp. 98–121 (London, 1943–44).
       See also Discussion, **22**, pp. 383–98 (London, 1943–44).
8.17. T. GYENGO, Effect of type of test specimen and gradation of aggregate on compressive strength of concrete, *J. Amer. Concr. Inst.*, **34**, pp. 269–82 (Jan.- Feb. 1938).

---

* e.g. A. M. Neville and J. B. Kennedy, *Basic Statistical Methods for Engineers and Scientists* (New York and London, Intertext) 1964. Second edition 1976.

8.18. A. M. NEVILLE, The influence of size of concrete test cubes on mean strength and standard deviation, *Mag. Concr. Res.*, **8**, No. 23, pp. 101–10 (Aug. 1956).

8.19. D. P. O'CLEARY and J. G. BYRNE, Testing concrete and mortar in tension, *Engineering*, pp. 384–5 (London, 18th March 1960).

8.20. P. J. F. WRIGHT, The effect of the method of test on the flexural strength of concrete, *Mag. Concr. Res.*, **4**, No. 11, pp. 67–76.

8.21. H. GONNERMAN and E. C. SHUMAN, Compression, flexure, and tension tests of plain concrete, *Proc. A.S.T.M.*, **28**, Part II, pp. 527–73 (1928).

8.22. B. W. SHACKLOCK and P. W. KEENE, The comparison of compressive and flexural strengths of concrete with and without entrained air, *Cement Concr. Assoc. Tech. Report TRA/283* (London, Dec. 1957).

8.23. S. WALKER and D. L. BLOEM, Studies of flexural strength of concrete—Part 3: Effects of variations in testing procedures, *Proc. A.S.T.M.*, **57**, pp. 1122–39 (1957).

8.24. P. J. F. WRIGHT, Comments on an indirect tensile test on concrete cylinders, *Mag. Concr. Res.*, **7**, No. 20, pp. 87–96 (July 1955).

8.25. S. THAULOW, Tensile splitting test and high strength concrete test cylinder, *J. Amer. Concr. Inst.*, **53**, pp. 699–706 (Jan. 1957).

8.26. P. J. F. WRIGHT, Variations in the strength of Portland cement, *Mag. Concr. Res.*, **10**, No. 30, pp. 123–32 (Nov. 1958).

8.27. D. McHENRY and J. J. SHIDELER, Review of data on effect of speed in mechanical testing of concrete, *A.S.T.M. Sp. Tech. Publicn.*, No. 185, pp. 72–82 (1956).

8.28. W. H. PRICE, Factors influencing concrete strength, *J. Amer. Concr. Inst.*, **47**, pp. 417–32 (Feb. 1951).

8.29. R. H. EVANS, Effect of rate of loading on some mechanical properties of concrete, *Mechanical Properties of Non-metallic Brittle Materials*, pp. 175–90 (London, Butterworth, 1958).

8.30. T. WATERS, The effect of allowing concrete to dry before it has fully cured, *Mag. Concr. Res.*, **7**, No. 20, pp. 79–82 (July 1955).

8.31. S. WALKER and D. L. BLOEM, Effects of curing and moisture distribution on measured strength of concrete, *Proc. Highw. Res. Bd.*, **36**, pp. 334–46 (1957).

8.32. R. H. MILLS, Strength–maturity relationship for concrete which is allowed to dry, *R.I.L.E.M. Int. Symp. on Concrete and Reinforced Concrete in Hot Countries* (Haifa, 1960).

8.33. W. S. BUTCHER, The effect of air drying before test: 28-day strength of concrete, *Constructional Review*, pp. 31–32 (Sydney, Dec. 1958).

8.34. L. H. C. TIPPETT, On the extreme individuals and the range of samples taken from a normal population, *Biometrika*, **17**, pp. 364–87 (Cambridge and London, 1925).

8.35. A. M. NEVILLE, Some aspects of the strength of concrete, *Civil Engineering* (London) **54**, Part 1, pp. 1153–56 (Oct. 1959); Part 2, pp. 1308–11 (Nov. 1959); Part 3, pp. 1435–39 (Dec. 1959).

8.36. H. RÜSCH, Versuche zur Festigkeit der Biegedruckzone, *Deutscher Ausschuss für Stahlbeton, No.* 120 (1955).

8.37. M. PRÔT, Essais statistiques sur mortiers et betons, *Annales de l'Institut Technique de Bâtiment et des Travaux Publics, No.* 81, July–Aug. 1949, Béton, Béton Armé No. 8.

8.38. R. F. BLANKS and C. C. McNAMARA, Mass concrete tests in large cylinders, *J. Amer. Concr. Inst.*, **31**, pp. 280–303 (Jan.–Feb. 1935).

8.39. W. J. SKINNER, Experiments on the compressive strength of anhydrite, *The Engineer*, **207**, Part 1, pp. 255–59 (13th Feb. 1959); Part 2, pp. 288–92 (London, 20th Feb. 1959).

8.40. H. F. GONNERMAN, Effect of size and shape of test specimen on compressive strength of concrete, *Proc. A.S.T.M.*, **25**, Part II, pp. 237–50 (1925).

8.41. A. M. NEVILLE, The use of 4-inch concrete compression test cubes, *Civil Engineering*, **51**, No. 605, pp. 1251–52 (London, Nov. 1956).

8.42. A. M. NEVILLE, Concrete compression test cubes, *Civil Engineering*, **52**, No. 615, p. 1045 (London, Sept. 1957).

8.43. R. A. KEEN and J. DILLY, The precision of tests for compressive strength made on ½-inch cubes of vibrated mortar, *Cement Concr. Assoc. Tech. Report TRA/314* (London, Feb. 1959).

8.44. B. W. SHACKLOCK, Comparison of gap- and continuously-graded concrete mixes, *Cement Concr. Assoc. Tech. Report TRA/240* (London, Sept. 1959).

8.45. S. WALKER, D. L. BLOEM, and R. D. GAYNOR, Relationships of concrete strength to maximum size of aggregate, *Proc. Highw. Res. Bd.*, **38**, pp. 367–79 (Washington D.C., 1959).

8.46. J. W. H. KING, Further notes on the accelerated test for concrete, *Chartered Civil Engineer*, pp. 15–19 (London, May 1957).

8.47. C. H. WILLETTS, Investigation of the Schmidt concrete test hammer, *Miscellaneous Paper No. 6–267* (U.S. Army Engineer Waterways Experiment Station, Vicksburg, Miss., June 1958).

8.48. W. E. GRIEB, Use of the Swiss Hammer for estimating the compressive strength of hardened concrete, *Public Roads*, **30**, No. 2, pp. 45–50 (Washington D.C., June 1958).

8.49. R. JONES and E. N. GATFIELD, Testing concrete by an ultrasonic pulse technique, *D.S.I.R. Road Research Tech. Paper No.* 34 (London, H.M.S.O., 1955).

8.50. E. A. WHITEHURST, Soniscope tests concrete structures, *J. Amer. Concr. Inst.*, **47**, pp. 433–44 (Feb. 1951).

8.51. R. JONES and P. J. F. WRIGHT, Some problems involved in destructive and non-destructive testing of concrete, *Proc. of a Symposium on Mix Design and Quality Control of Concrete*, pp. 441–50 (London, Cement and Concrete Assoc., 1954).

8.52. R. JONES and H. C. MAYHEW, Thickness and quality of cemented surfacings and bases—measuring by a non-destructive surface wave method, *Civil Engineering*, **60**, No. 705, pp. 523–29 (London, April 1965).

8.53. A. W. BROWN, A tentative method for the determination of the original water/cement ratio of hardened concrete, *J. of Applied Chemistry*, **7**, pp. 565–72 (London, Oct. 1957).

8.54. M. POLIVKA, J. W. KELLY and C. H. BEST, A physical method for determining the composition of hardened concrete, *A.S.T.M. Sp. Tech. Publcn. No.* 205, pp. 135–52 (1958).

8.55. D. G. HARLAND, A radio-active method for measuring variations in density in concrete cores, cubes and beams, *Mag. Concr. Res.*, **18**, No. 55, pp. 95–101 (June 1966).

8.56. H. KUPFER, H. K. HILSDORF and H. RÜSCH, Behaviour of concrete under biaxial stresses, *J. Amer. Concr. Inst.*, **66**, pp. 656–66 (Aug. 1969).

8.57. K. NEWMAN and L. LACHANCE, The testing of brittle materials under uniform uniaxial compressive stress, *Proc. A.S.T.M.*, **64**, pp. 1044–67 (1964).

8.58. H. HANSEN, A. KIELLAND, K. E. C. NIELSEN and S. THAULOW, Compressive strength of concrete—cube or cylinder? *R.I.L.E.M. Bul. No.* 17, pp. 23–30 (Paris, Dec. 1962).

8.59. B. L. RADKEVICH, Shrinkage and creep of expanded clay–concrete units in compression, *C.S.I.R.O. Translation No.* 5910 *from Beton i Zhelezobeton, No.* 8, pp. 364–9 (1961).

8.60. Z. PIATEK, Własności wytrzymałościowe i reologiczne keramzytobetonu konstrukcyjnego, *Arch. Inz. Ladowej*, **16**, No. 4, pp. 711–29 (Warsaw, 1970).

8.61. S. NILSSON, The tensile strength of concrete determined by splitting tests on cubes, *R.I.L.E.M. Bul. No.* 11, pp. 63–7 (Paris, June 1961).

8.62. V. M. MALHOTRA, N. G. ZOLDNERS and H. M. WOODROOFFE, Ring test for determining the tensile strength of concrete, *A.S.T.M. Mat. Res. and Stand.*, **6**, No. 1, pp. 2–12 (Jan. 1966).
8.63. D. J. MCNEELY and S. D. LASH, Tensile strength of concrete, *J. Amer. Concr. Inst.*, **60**, pp. 751–61 (June 1963).
8.64. V. M. MALHOTRA, Effect of specimen size on tensile strength of concrete, *J. Amer. Concr. Inst.*, **67**, pp. 467–9 (June 1970).
8.65. A. M. NEVILLE, A general relation for strengths of concrete specimens of different shapes and sizes, *J. Amer. Concr. Inst.*, **63**, pp. 1095–109 (Oct. 1966).
8.66. BULLETIN DU CIMENT, Carottes de petit diamètre, **38**, No. 3 (Wildegg, Switzerland, March 1970).
8.67. N. PETERSONS, Should standard cube test specimens be replaced by test specimens taken from structures, *Materials and Structures*, **1**, No. 5, pp. 425–35 (Paris, Sept.–Oct. 1968).
8.68. I.C.E. COMMITTEE, An accelerated test for concrete, *Proc. Inst. C. E.*, **40**, pp. 125–33 (London, May 1968).
8.69. P. SMITH and B. CHOJNACKI, Accelerated strength testing of concrete cylinders, *Proc. A.S.T.M.*, **63**, pp. 1079–101 (1963).
8.70. P. SMITH and H. TIEDE, Earlier determination of concrete strength potential, *Report No. RR*124 (Department of Highways, Ontario, Jan. 1967).
8.71. V. M. MALHOTRA, Preliminary evaluation of Windsor probe equipment for estimating the compressive strength of concrete, *Mines Branch Investigation Report IR* 71-1 (Department of Energy, Mines and Resources, Ottawa, Dec. 1970).
8.72. M. F. KAPLAN, The relation between ultrasonic pulse velocity and the compressive strength of concretes having the same workability but different mix proportions, *Mag. Concr. Res.*, **12**, No. 34, pp. 3–8 (March 1960).
8.73. W. K. TSO and I. M. ZELMAN, Concrete strength variation in actual structures, *J. Amer. Concr. Inst.*, **67**, No. 12, pp. 981–8 (Dec. 1970).
8.74. R. H. ELVERY and J. A. FORRESTER, Non-destructive testing of concrete, *Progress in Construction, Science and Technology*, pp. 175–216 (Aylesbury, Medical and Technical Publishing, 1971).
8.75. R. JONES, The development of microcracks in concrete, *R.I.L.E.M. Bul. No. 9*, pp. 110–14 (Paris, Dec. 1960).
8.76. E. C. HIGGINSON, G. B. WALLACE and E. L. ORE, Effect of maximum size of aggregate on compressive strength of mass concrete, *Symp. on Mass Concrete*, *Amer. Concr. Inst. Sp. Publicn. SP*-6, pp. 219–56 (1963).
8.77. U.S. BUREAU OF RECLAMATION, Effect of maximum size of aggregate upon compressive strength of concrete, *Laboratory Report No. C*–1052 (Denver, Colorado, June 3, 1963).
8.78. V. M. MALHOTRA, and K. PAINTER, Evaluation of the Windsor probe test for estimating compressive strength of concrete, *Mines Branch Investigation Report*, *IR* 71-50 (Department of Energy, Mines and Resources, Ottawa, July 1971).
8.79. H. MAILER, Pavement thickness measurement using ultrasonic techniques, *Highway Research Record*, 378, pp. 20–28 (1972).

9. Lightweight and High-density Concretes

THIS chapter deals with insulating concrete and with structural concretes whose density (unit weight) is appreciably lower or higher than the usual range of 2,200 to 2,600 kg/m³ (140 to 160 lb/ft³). High-density concrete is used mainly in the construction of biological shields, while the use of lightweight concrete is governed primarily by economic considerations.

The relative brevity of the chapter does not imply that lightweight concrete is not an important material; it is merely that only those features of lightweight concrete which distinguish it from normal weight concrete are specifically considered here.

In concrete construction, self-weight represents a very large proportion of the total load on the structure, and there are clearly considerable advantages in reducing the density of concrete. The chief of these are the use of smaller sections and the corresponding reduction in the size of foundations. Furthermore, with lighter concrete the formwork need withstand a lower pressure than would be the case with ordinary concrete, and also the total weight of materials to be handled is reduced with a consequent increase in productivity. Lightweight concrete also gives better thermal insulation than ordinary concrete (*see* Fig. 9.2). The practical range of densities of lightweight concrete is between about 300 and 1,850 kg/m³ (20 and 115 lb/ft³).

Classification of Lightweight Concretes

There are three broad methods of producing lightweight concrete. In the first, porous lightweight aggregate of low apparent specific gravity is used instead of ordinary aggregate whose specific gravity is approximately 2·6. The resultant concrete is generally known by the name of the lightweight aggregate used.

The second method of producing lightweight concrete relies on introducing large voids within the concrete or mortar mass. These voids should be clearly distinguished from the extremely fine voids produced by air entraining. This type of concrete is variously known as aerated, cellular, foamed, or gas concrete.

The third means of obtaining lightweight concrete is by simply omitting the fine aggregate from the mix so that a large number of interstitial voids is present. Coarse aggregate of ordinary weight is generally used. This concrete is described succinctly by the term no-fines concrete.

In essence then, the decrease in density is obtained in each case by the presence of voids, either in the aggregate, or in the mortar, or in the interstices between the coarse particles. It is clear that the presence of these voids reduces the strength of lightweight concrete compared with ordinary concrete, but in many applications high strength is not essential. Lightweight concrete provides a very good thermal insulation and has a satisfactory durability but is not highly resistant to abrasion. In general, lightweight concrete is more expensive than ordinary concrete, and mixing, handling, and placing also require considerably more care and attention than ordinary concrete. However, for many purposes the advantages of lightweight concrete outweigh its disadvantages, and there is a continuing world-wide trend towards using more lightweight concrete and also towards using it in new applications, including prestressed concrete, high-rise buildings and even shell roofs.

Lightweight concrete can also be classified according to the purpose for which it is to be used: we distinguish between structural lightweight concrete and concrete used in non-load bearing walls, for insulation purposes, and the like. In the past, structural lightweight concrete tended to be a close-textured concrete made with lightweight aggregate but, since this is not always the case now, it is preferable to base the classification of structural lightweight concrete on a minimum compressive strength. In the United States, for instance, it is considered that structural lightweight concrete should have a compressive strength, measured on a standard cylinder at 28 days, of not less than 17 MN/m^2 (2,500 lb/in^2). The density of such concrete, determined in the dry state, should not exceed 1840 kg/m^3 (115 lb/ft^3) and is usually between 1,400 and 1,800 kg/m^3 (90 and 110 lb/ft^3). Insulating concrete generally has a density lower than 800 kg/m^3 (50 lb/ft^3) and a strength between 0·7 and 7 MN/m^2 (100 and 1,000 lb/in^2).

Lightweight Aggregates
The essential characteristic of lightweight aggregate is its high porosity, which results in a low apparent specific gravity. Some lightweight aggregates occur naturally; others are manufactured.

Natural Aggregates
The main aggregates in this category are diatomite, pumice, scoria, volcanic cinders and tuff; except for diatomite, all of these are of volcanic origin.

Because they are found only in some areas, natural lightweight aggregates are not extensively used. Pumice is more widely employed than any of the others but none is found in the United Kingdom, though some is imported from Germany and Italy.

Pumice is a light-coloured, froth-like volcanic glass with a bulk density in the region of 500 to 900 kg/m³ (30 to 55 lb/ft³). Those varieties of pumice which are not too weak structurally make a satisfactory concrete with a density of 700 to 1,400 kg/m³ (45 to 90 lb/ft³), and with good insulating characteristics, but having high absorption and shrinkage.

Scoria, which is a vesicular glassy rock, rather like industrial cinders, makes a concrete of similar properties.

Artificial Aggregates

Artificial aggregates are often known by a variety of trade names, but are best classified on the basis of the raw material used and the method of manufacture.

In the first group are included the aggregates produced by the application of heat in order to expand clay, shale, slate, diatomaceous shale, perlite, obsidian and vermiculite. The second type is obtained by special cooling processes through which an expansion of blast-furnace slag is obtained. Industrial cinders form the third and last type.

Expanded clay, *shale*, and *slate* are obtained by heating suitable raw materials in a rotary kiln to incipient fusion (temperature of 1,000 to 1,200°C) when expansion of the material takes place due to the generation of gases which become entrapped in a viscous pyroplastic mass. This porous structure is retained on cooling so that the apparent specific gravity of the expanded material is lower than before heating. Often the raw material is reduced to the desired size before calcining, but crushing after expansion may also be applied. Expansion can also be achieved by the use of a sinter strand. Here, the moistened material is carried by a travelling grate under burners so that burning gradually penetrates the full depth of the bed of the material. Its viscosity is such that gases are entrapped. As with the rotary kiln, the cooled mass is crushed or initially pelletized material can be used.*

The use of pelletized material produces particles with a smooth shell or "coating" over the cellular interior. These nearly spherical particles with a semi-impervious glaze have a lower water absorption than uncoated particles, whose absorption ranges from about 12 to 34 per cent. Coated particles are easier to handle and to mix, and produce concrete of higher workability, but are generally dearer than the uncoated angular aggregate.

Expanded shale and clay aggregates made by the sinter strand process

---

* Expanded clay produced by the sinter strand method is known as *Aglite*; the rotary kiln product is marketed as *Leca*. In Eastern Europe, expanded shale etc., aggregate prepared on a sinter strand is known as *agloporite* while the rotary kiln aggregate is called *keramzite*.

have a density of 650 to 900 kg/m³ (40 to 55 lb/ft³), and 300 to 650 kg/m³ (20 to 40 lb/ft³) when made in a rotary kiln. They produce concrete with a density usually within the range of 1,400 to 1,800 kg/m³ (85 to 110 lb/ft³), although values as low as 800 kg/m³ (50 lb/ft³) have been obtained. Concrete made with expanded shale and clay aggregates generally has a higher strength than when any other lightweight aggregate is used.

Perlite is a glassy volcanic rock found in Ulster, Italy, America and elsewhere. When heated rapidly to the point of incipient fusion (900 to 1,100°C) it expands owing to the evolution of steam and forms a cellular material with a bulk density as low as 30 to 240 kg/m³ (2 to 15 lb/ft³). Concrete made with perlite has a very low strength, a very high shrinkage and is used primarily for insulating purposes. An advantage of such concrete is that it is fast drying and can be finished rapidly.

Vermiculite is a material with a platy structure, somewhat similar to that of mica, and is found in America and Africa. When heated to a temperature of 650 to 1,000°C vermiculite expands to several, or even as many as 30 times, its original volume by exfoliation of its thin plates. As a result, the bulk density of exfoliated vermiculite is only 60 to 130 kg/m³ (4 to 8 lb/ft³) and the concrete made with it is of very low strength and exhibits high shrinkage but is an excellent heat insulator.

Expanded blast-furnace slag is produced in two ways. In one, a limited amount of water in the form of a spray comes into contact with the molten slag when being discharged from the furnace (in the production of pig-iron). Steam is generated and it bloats the still plastic slag, so that the slag hardens in a porous form, rather similar to pumice. This is the water-jet process. In the machine process, the molten slag is rapidly agitated with a controlled amount of water. Steam is entrapped and there is also some formation of gases due to chemical reactions of some slag constituents with water vapour. Expanded, or foamed, slag has been used in Great Britain for many years and is produced with a bulk density varying between 300 and 1,100 kg/m³ (20 and 70 lb/ft³), depending on the details of the cooling process and to a certain degree on the particle size and grading. Foamed slag for aggregate is covered by B.S. 877: 1967, while B.S. 1047: 1952 specifies the non-expanded air-cooled slag aggregate. Concrete made with foamed slag has a density of 950 to 1,750 kg/m³ (60 to 110 lb/ft³).

Clinker aggregate, known in the U.S. as cinders, is made from well-burnt residue of industrial high-temperature furnaces, fused or sintered into lumps. It is important that the clinker be free from harmful varieties of unburnt coal which may undergo expansion in the concrete, thus causing unsoundness. B.S. 1165: 1966 lays down the limits of loss on ignition and of soluble sulphate content for different concretes: plain concrete for general purposes, *in situ* interior concrete, and precast clinker concrete blocks. The standard does not recommend the use of clinker aggregate in reinforced concrete or in concrete required to have a specially high durability.

Iron or pyrites in the clinker may result in staining of surfaces, and should therefore be removed. Unsoundness due to hard-burnt lime can be avoided by allowing the clinker to stand wet for a period of several weeks: the lime will become slaked and will not expand in the concrete.

Breeze is the name given to a material similar to clinker but more lightly sintered and less well burnt. There is no clear-cut demarcation between breeze and clinker.

When cinders are used as both fine and coarse aggregate, concrete with a density of about 1,100 to 1,400 kg/m³ (70 to 85 lb/ft³) is obtained, but often natural sand is used in order to improve the workability of the mix: the density of the resulting concrete is then 1,750 to 1,850 kg/m³ (110 to 115 lb/ft³).

Pulverized-fuel ash, or fly ash, is a finely divided residue from the combustion of powdered coal in modern boiler plants such as power stations. The ash is moistened, made into pellets and then sintered in a suitable furnace: the small amount of unburnt fuel present in the ash will usually maintain this process without an addition of fuel. The sintered nodules provide a very good, rounded aggregate, known as *Lytag*, with a bulk density of about 1,000 kg/m³ (60 lb/ft³), the fine fraction may reach 1,200 kg/m³ (75 lb/ft³).

Typical ranges of densities of concretes made with various lightweight aggregates, largely based on A.C.I. classification,[9.13] are shown in Fig. 9.1. Some general requirements for lightweight aggregates are prescribed by A.S.T.M. Standards C 330–69, C 331–69, C 332–66 and by B.S. 3797: 1964.

Under special circumstances the range of raw materials from which lightweight aggregate can be made is wide. For instance, in Berlin refuse sintered into slag has been used in making concrete.[9.36] With a cement content of 230 to 360 kg/m³ (390 to 600 lb/yd³) a density of 1,900 kg/m³ (120 lb/ft³) and strengths of between 16 and 30 MN/m² (2,300 to 4,300 lb/in²) were obtained.

Lightweight Aggregate Concrete

We have seen that lightweight aggregate concrete covers an extremely wide field: using appropriate materials and methods the density of concrete can be varied between little over 300 and about 1,850 kg/m³ (20 to 115 lb/ft³) and the corresponding strength range is between 0·3 and sometimes above 40 MN/m² (50 to 6,000 lb/in²). Strengths up to 60 MN/m² (9,000 lb/in²) can be obtained with very high cement contents (560 kg/m³ (950 lb/yd³)). For any particular aggregate, strength increases with density but, depending on the type of aggregate, 20 MN/m² (3,000 lb/in²) concrete may require between 240 and 400 kg of cement per cubic metre (400 to 680 lb/yd³) of concrete; the corresponding range for 30 MN/m² (4,500 lb/in²) concrete is 330 to 500 kg/m³ (560 to 840 lb/yd³). Generally, with lightweight aggregate, the cement content varies from the same to two-thirds more than with normal aggregate for the same strength of concrete.

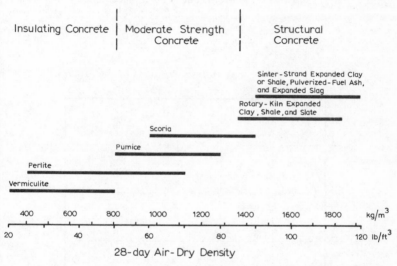

Fig. 9.1. *Typical ranges of densities of concretes made with various lightweight aggregates*

There is no simple correlation between the strength of the aggregate as such and the strength of the concrete made with the given aggregate but there is a strength ceiling above which an increase in cement content does little to increase the strength of the concrete.

Lightweight aggregate, even similar in appearance, may produce concretes varying widely in structural properties so that a careful check on the performance of each new aggregate is necessary. Classification of concrete according to the type of aggregate used is difficult as the properties of concrete are affected also by the grading of the aggregate, the cement content, the water/cement ratio, and the degree of compaction. Typical properties are listed in Table 9.3.

The main points to watch concern the workability of concrete, its drying shrinkage and moisture movement. Other factors, such as strength and thermal conductivity (*see* Fig. 9.2 and Table 9.3), both closely related to density, and cost, must also, of course be considered.

Many lightweight aggregates are angular and have a rough surface, and produce harsh mixes which are more suited to factory production of building blocks than to *in situ* work. Workability may, however, be improved by using fine aggregate of ordinary weight, but the density and the thermal conductivity of such concretes are higher than when light-weight fine aggregate is used. However, because of improved workability, there is a decrease in the water requirement of the mix with natural sand and often also in the cement content. The replacement of lightweight fines by sand is usually made on an equal volume basis; partial or total

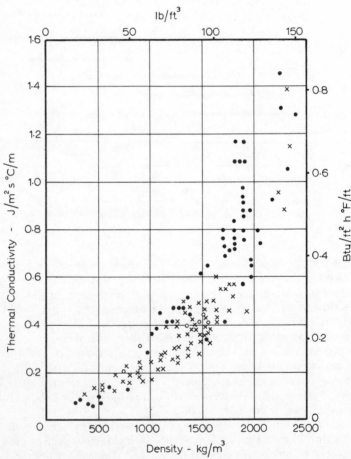

Fig. 9.2. *Thermal conductivity of lightweight aggregate concretes of various types*[9.12]

replacement is used. In the latter case, a reduction in water content of 12 to 24 per cent, compared with an all-lightweight aggregate mix, has been reported.[9.14] Concrete with total sand replacement has a higher modulus of elasticity than an all-lightweight aggregate concrete by 10 to 30 per cent, depending on the aggregate used and on the strength level of the mix.[9.14] Shrinkage is reduced by 15 to 35 per cent.[9.14] Partial sand replacement affects these properties in proportion to the fraction of fine aggregate replaced.

Workability of lightweight aggregate mixes which tend to be harsh can also be considerably improved by air entrainment: water requirement is reduced, and so is the tendency to bleeding and segregation. The usual *total* air contents by volume are: for 19 mm ($\frac{3}{4}$ in.) maximum size of aggregate, 4 to 8 per cent; for 9·5 mm ($\frac{3}{8}$ in.) maximum size of aggregate, 5 to 9 per cent.

Air content in excess of these values lowers the compressive strength by about 1 MN/m^2 (150 lb/in^2) for each additional percentage point of air.[9.13]

The majority of lightweight aggregates have a high and rapid absorption but it is possible to waterproof the aggregate by a coating of bitumen, using a special process. If this is not done, a considerable amount of water may be absorbed by the aggregate during mixing: this will cause an increase in the density of concrete and a fall in its thermal insulation value.

When lightweight aggregate is used in reinforced concrete, special care should be given to the protection of reinforcement from corrosion. Measurements of the depth of carbonation, i.e. the depth within which corrosion can occur under suitable conditions, have shown that with some lightweight aggregates this depth can be up to twice that with normal aggregate.[9.15] The behaviour of different aggregates varies considerably but generally with lightweight aggregate a larger cover to reinforcement is desirable. Alternatively, the use of a rendered finish or coating of reinforcement with rich mortar has been found useful. In the case of concrete made with clinker aggregate there is the additional danger of corrosion due to the presence of sulphur in the clinker, and coating of steel is necessary, but the use of clinker aggregate in reinforced concrete is not common.

All concretes made with lightweight aggregate exhibit a higher moisture movement than is the case with normal weight concrete. They have a high initial drying shrinkage, about 5 to 40 per cent higher than ordinary concrete, but the total shrinkage with some lightweight aggregates may be even higher; the concretes made with expanded clay and shale and foamed slag are in the lower shrinkage range.* In view of the compara-

---

* Limiting values of shrinkage prescribed by B.S. 2028: 1968 are given on page 333.

tively low tensile strength* of lightweight concrete there is a considerable danger of shrinkage cracking, although some compensation is afforded by a lower modulus of elasticity and a greater extensibility of lightweight concrete. Contraction joints should generally be provided, and with suitable precautions trouble due to moisture movement can usually be avoided.

As far as other general characteristics of lightweight concrete are concerned it seems that creep, taken on the basis of the stress/strength ratio, is of the same order as for ordinary concrete.† It has been suggested that long term creep of lightweight concrete is somewhat higher, but this has not been confirmed. Poisson's ratio appears to be similar to that for ordinary concrete. The modulus of elasticity is about $\frac{1}{2}$ to $\frac{3}{4}$ of that of ordinary concrete of the same strength:[9.1] some typical values are given in Fig. 6.5. In general, the modulus varies with density according to the expression given on page 312, but at strengths in excess of about 30 MN/m² (4,500 lb/in²) the modulus is lower than predicted by the expression.[9.16] The rate of gain of strength of lightweight concrete is similar to that of ordinary concrete under the same conditions of curing but is generally less sensitive to absence of moist curing. Owing to the nature of the aggregate, the abrasion resistance of lightweight aggregate concrete is not very good. On the other hand, the frost resistance, except when the aggregate was saturated before mixing, is excellent.

The sound *absorption* of lightweight concrete can be rated as good because the air-borne sound energy becomes converted into heat in the minute channels of the concrete, so that the absorption coefficient of sound is about twice that for ordinary concrete. A rendered surface, however, would offer a much higher reflection of sound. Lightweight concrete does not possess particularly good sound *insulation* properties as this insulation is better the higher the density of the material (*see* p. 453).

Some values of the coefficient of thermal expansion are given in Table 9.1. From a comparison with Fig. 7.26 it can be seen that lightweight aggregate concrete has generally a lower thermal expansion than ordinary concrete. This can produce some problems when lightweight and ordinary concretes are used side by side. It can be noted that the low thermal expansion of lightweight aggregate concrete reduces the tendency to warp or buckle when the two faces of a concrete element are exposed to different temperatures. Also, the low thermal conductivity of the concrete reduces the temperature rise of the embedded steel.

---

* But the ratio of tensile to compressive strengths of lightweight aggregate concrete, except for very high strength concretes, is at least the same as in ordinary concrete. In aerated concrete the ratio is higher: about 0·3. However, diagonal cracking in a beam can occur at a stress lower than the tensile strength.

† Allowance must be made for the generally lower modulus of elasticity of lightweight aggregate (see p. 314).

Table 9.1: *Coefficient of Thermal Expansion of Concretes made with Lightweight Aggregate*[9.2,9.5]

Type of aggregate used	Linear coefficient of thermal expansion (determined over a range of −22°C to 52°C (−7°F to 125°F))	
	10^{-6} per °C	10^{-6} per °F
Pumice	9·4 to 10·8	5·2 to 6·0
Perlite	7·6 to 11·0	4·2 to 6·1
Vermiculite	8·3 to 14·2	4·6 to 7·9
Cinders	about 3·8	about 2·1
Expanded shale	6·5 to 8·1	3·6 to 4·5
Expanded slag	7·0 to 11·2	3·9 to 6·2

The fire resistance of lightweight aggregate concrete is, as a rule, greater than when ordinary aggregate is used (Table 9.2) because lightweight aggregate concrete has a lesser tendency to spall; the concrete also

Table 9.2: *Estimated Fire Resistance of Hollow Masonry Walls*[9.2]

Type of aggregate used	Minimum equivalent thickness for ratings of—							
	4 h		3 h		2 h		1 h	
	mm	in.	mm	in.	mm	in.	mm	in.
Expanded slag or pumice	119	4·7	102	4·0	81	3·2	53	2·1
Expanded clay or shale	145	5·7	122	4·8	96	3·8	66	2·6
Limestone, cinders or unexpanded slag	150	5·9	127	5·0	102	4·0	69	2·7
Calcareous gravel	157	6·2	135	5·3	107	4·2	71	2·8
Siliceous gravel	170	6·7	145	5·7	114	4·5	76	3·0

loses a lesser proportion of its original strength with a rise in temperature. Table 9.3 gives a summary of properties of different lightweight concretes. It should be stressed that the values listed are typical, but not necessarily limiting ones. Furthermore, the density of concrete is quoted in an oven-dry condition, which is convenient for comparison purposes as it is reproducible, but is not the condition of the concrete in practice.

As distinct from the generally close-textured structural lightweight concrete, mentioned earlier, precast blocks and *in situ* walls for non-structural purposes are made of concrete honeycombed in texture. This is obtained by using a much larger quantity of coarse lightweight aggregate in the mix, with a corresponding reduction in the quantity of the fine aggregate. This type of concrete is somewhat similar to the no-fines concrete.

Table 9.3: *Typical Properties of Lightweight Concretes*

Type of concrete		Bulk density of aggregate		Mix proportions by volume cement: aggregate	Dry density of concrete		Compressive strength		Drying shrinkage	Thermal conductivity	
		kg/m³	lb/ft³		kg/m³	lb/ft³	MN/m²	lb/in²	10^{-6}	Jm/m²s°C	Btu/ft²h°F/ft
Aerated	p.f.a.*	950	60	1:3	750	47	3	500	700	0·19	0·11
	sand	1,600	100	1:3	900	55	6	800		0·22	0·13
Autoclaved aerated		—	—	—	800	55	4	600	800	0·25	0·14
Foamed slag	fine	900	50	1:8	1,700	105	7	1,000	400	0·45	0·26
				1:6	1,850	115	21	3,000	500	0·69	0·40
	coarse	650	40	1:3·5	2,100	130	41	6,000	600	0·76	0·44
Rotary-kiln expanded clay	fine	700	45	1:11	650–1,000	42–62	3–4	400–600	—	0·17	0·10
				1:6	1,100	70	14	2,000	550	0·31	0·18
	coarse	400	25	1:5	1,200	75	17	2,500	600	0·38	0·22
				1:4	1,300	80	19	2,800	700	0·40	0·23
Rotary-kiln expanded clay with natural sand	coarse	400	25	1:5	1,350–1,500	85–95	17	2,500	—	0·57	0·33

Table 9.3—contd.

Material											
Sinter-strand expanded clay	fine	1,050	65	1:5	1,500	95	24	3,500	600	0·55	0·32
	coarse	650	40	1:4	1,600	100	31	4,500	750	0·61	0·35
Rotary-kiln expanded slate	fine	950	60	1:6	1,700	105	28	4,000	400	0·61	0·35
	coarse	700	45	1:4·5	1,750	110	35	5,000	450	0·69	0·40
Sintered pulverized-fuel ash	fine	1,050	65	1:6	1,450	90	28	4,000	400	0·47	0·27
	coarse	800	50	1:4·5	1,500	95	36	5,200	500	0·49	0·28
				1:3·5	1,550	97	41	6,000	600	0·50	0·29
				1:2·6	1,600	100	52	7,500	700	0·52	0·30
Sintered pulverized-fuel ash with natural sand	coarse	800	50	1:6·4	1,650	104	30	4,400	400	0·55	0·32
				1:4·5	1,700	106	38	5,500	500	0·57	0·33
				1:3·8	1,750	109	45	6,600	600	0·57	0·33
				1:3·0	1,800	112	52	7,600	700	0·59	0·34
Pumice		500–800	30–50	1:6	1,200	74	14	2,000	1,200	—	—
				1:4	1,250	77	19	2,800	1,000	0·14	0·08
				1:2	1,450	90	29	4,200	—	—	—
Exfoliated vermiculite		65–130	4–8	1:6	300–500	20–20	2	300	3,000	0·10	0·06
Perlite		95–130	6–8	1:6	—	—	—	—	2,000	0·05	0·03

* p.f.a. = pulverized fuel ash (fly ash).

It is essential that all particles of aggregate are uniformly coated with a film of cement paste, but an assessment of the workability and the determination of the necessary water content are not easy. A simple though hardly scientific way of judging consistence of the mix is to grip a handful of concrete tightly in one's hand, then throw the concrete away and observe the pattern of grout on the palm: if the hand is well spotted the mix is correct. If there are only a few specks the mix is too dry and crumbly; conversely, if the palm is covered in grout the mix is too wet, and consequently has a higher density and lower thermal insulating properties. With experience, consistence can be judged by eye as well, but different aggregates will result in a different appearance of the mix.

If workability is taken to mean ease of compaction, then, for equal workability, lightweight aggregate concrete registers a lower slump and compacting factor than normal weight concrete.[9.17] The reason for this is that the work done by gravity is smaller in the case of the lighter material. Since the Kelly ball penetration is independent of gravity, the value recorded in the Kelly ball test is not affected by the aggregate.[9.18] If, however, gravity is to effect compaction by itself—a practice certainly not recommended—then slump or compacting factor give the correct indication.

From the foregoing it can be seen that a slump of 50 to 75 mm (2 to 3 in.) represents a high workability (cf. Table 4.2), and a compacting factor not smaller than 0·8 or a Vebe time of less than 12 sec correspond approximately to a medium workability.[9.17] A slump in excess of 75 to 100 mm (3 to 4 in.) may cause segregation with the light large aggregate particles floating to the top. Likewise, prolonged vibration may lead to segregation much more readily than with normal weight aggregate.

If at the time of placing in the mixer the aggregates are dry, they will rapidly absorb water and the workability of the mix will quickly decrease. Therefore, when concrete is made with lightweight aggregates which have a high rate of absorption but a low initial moisture content, it is desirable first to mix the aggregate with at least one-half of the mixing water and only then to add the cement into the mixer.[9.7] This procedure prevents the balling-up of cement and a loss of slump.

The strength of lightweight concrete used for non-structural purposes is not of primary importance; the main requirements are: thermal insulation, a good surface for rendering, and not too high a shrinkage. Often, nail-holding properties and, in the case of precast blocks, ease of cutting are important.

Design of Lightweight Aggregate Mixes
The water/cement ratio "law" applies to concrete made with lightweight aggregate in the same way as to normal aggregate concrete, and it is possible to follow the usual procedure of mix design when lightweight

aggregate is employed. It is very difficult, however, to determine how much of the total water in the mix is absorbed by the aggregate and how much actually occupies space within the concrete, i.e. forms part of the cement paste. This difficulty is caused not only by the very high value of the water absorption of lightweight aggregates—in some cases up to 20 per cent—but also by the fact that the absorption varies widely in rate, and with some aggregates may continue at an appreciable rate for several days. A reliable determination of the saturated, surface-dry bulk specific gravity seems therefore extremely difficult.

Thus the nett water/cement ratio would depend on the rate of absorption at the time of mixing, and not only on the moisture content of the aggregate. Hence the use of the water/cement ratio in mix design calculations is rather difficult. For this reason, proportioning on the basis of the cement content is preferable, although in the case of rounded lightweight aggregate with a coated or sealed surface and a relatively low absorption the standard method of mix design is directly applicable.

Lightweight aggregate produced artificially is usually bone-dry, and is rather prone to segregation. If the aggregate is saturated before mixing, the strength of the resulting concrete is about 5 to 10 per cent lower than when dry aggregate is used, for the same cement content and workability. This is due to the fact that in the latter case some of the mixing water is absorbed prior to setting, this water having contributed to the workability at the time of placing (cf. vacuum-processed concrete). Furthermore, the density of concrete made with a saturated aggregate is higher, and the durability of such concrete, especially its resistance to frost, is impaired. On the other hand, when aggregate with a high absorption is used, it is difficult to obtain a sufficiently workable and yet cohesive mix, and generally aggregates with absorption of over 10 per cent are pre-soaked.

It is interesting to note that initially damp lightweight aggregate usually contains more total absorbed water after a short immersion in water than initially dry aggregate immersed for the same length of time. The reason for this is probably that a small amount of water just moistening an aggregate particle does not remain in the surface pores but diffuses inward and fills the small pores inside. According to Hanson,[9.19] this clears the larger surface pores of water so that, upon immersion, these are open to an influx of water almost as great as when the aggregate contains no initially absorbed water.

The American Concrete Institute method of mix design[9.7] can be used for aggregate in any moisture condition. The method does not require the determination of absorption or specific gravity of the lightweight aggregate, as trial mixes form the basis of the design. The actual moisture content of the aggregate must be known, however, and should remain constant for all mixes.

For many lightweight aggregates the apparent specific gravity of the particles varies with their size, the finer particles being heavier than the large ones. Thus on a weight basis the percentage of the finer material has to be higher than would be the case with ordinary aggregate. As far as proportioning is concerned, it is the volume occupied by each size fraction, and not its weight, that determines the final volume of voids, the paste content, and the workability of the mix. A well graded aggregate with a minimum volume of voids will require only a moderate amount of cement, and is likely to produce concrete with a comparatively small drying shrinkage and thermal movement.

When the maximum size of aggregate is 19 mm ($\frac{3}{4}$ in.) the fine aggregate normally forms 40 to 60 per cent of the total volume of the aggregate, measured on a dry loose basis. It is often convenient to start with equal volumes of fine and coarse aggregates and to make adjustments as necessary. A trial mix of a required workability and with a given cement content is made. Too wet a consistence (say, a slump of 75 mm (3 in.)) is inadvisable since segregation can then occur either by the separation of mortar or by the large aggregate floating toward the surface. From the knowledge of the moisture content of the aggregate and from the density of the concrete, the mix proportions can be worked out. It is usual to make three trial mixes, each with a different cement content but all of the required workability. Hence a relationship between cement content and strength, for the given consistence, can be obtained. A general idea of this relation is given in Table 9.4 but the actual values for different aggregates vary widely.

Table 9.4: *Approximate Relationship Between Strength of Lightweight Aggregate Concrete and Cement Content*[9.7]

Compressive strength of standard cylinders		Cement content	
MN/m²	lb/in²	kg/m³	lb/yd³
14	2,000	230 to 390	380 to 660
21	3,000	280 to 450	470 to 750
28	4,000	330 to 510	560 to 850
34	5,000	390 to 560	660 to 940

The proportions of the various trial mixes can be related to the first mix of satisfactory workability, using a so-called specific gravity factor,[9.8] which is a ratio of the weight of the dry aggregate to the space occupied by it. This space is the volume of the concrete less the absolute volume of the cement and less the volume of the water, including that absorbed during mixing. It is in not allowing for the absorption of water that the difference between a specific gravity factor and the actual specific gravity lies; but for a given aggregate in a given moisture condition the volume of water absorbed during mixing is approximately constant. Thus the specific gravity factor can be used as if it were the actual specific gravity. The factor has, of course, a different value for fine and coarse aggregate.

Example

Assume that the dry loose densities of the fine and coarse aggregates are 56 and 44 lb/ft³ respectively.

The sum of the bulk volumes of the two aggregates in 1 yd³ of concrete is usually between 30 and 33 ft³. Assume that 32 ft³ is required in our case, 16 ft³ of each size being used.

Then the weights required for a concrete with a cement content of 550 lb/yd³ are—

cement	=	550 lb
fine aggregate = 16 × 56	=	895 lb
coarse aggregate = 16 × 44	=	705 lb
total water to obtain the required workability, say	=	485 lb
Total		2,635 lb

The density of the resulting concrete is then $\dfrac{2,635}{27} = 97 \cdot 6$ lb/ft³

The volumes of the ingredients are:

$$\text{absolute volume of cement} = \frac{550}{3 \cdot 15 \times 62 \cdot 4} = 2 \cdot 80 \text{ ft}^3$$

$$\text{volume of water} = \frac{485}{62 \cdot 4} = 7 \cdot 78 \text{ ft}^3$$

Then, ignoring any entrapped air, the volume of aggregate is $27 - (2 \cdot 80 + 7 \cdot 78) = 16 \cdot 42$ ft³, i.e. $8 \cdot 21$ ft³ of fine and coarse aggregate each.

The specific gravity factors are then—

$$\text{for the fine aggregate:} \frac{895}{8 \cdot 21 \times 62 \cdot 4} = 1 \cdot 75$$

$$\text{for the coarse aggregate:} \frac{705}{8 \cdot 21 \times 62 \cdot 4} = 1 \cdot 38$$

To proportion now a mix with a cement content of 750 lb per cubic yard of concrete we assume the total water requirement to be unaltered, and we keep the weight of the coarse aggregate the same. The fine to coarse aggregate ratio will therefore become lower, but the increased amount of cement provides some fine material. (This argument is of course applicable to normal aggregate as well.)

The quantities required are then:

$$\text{absolute volume of cement} = \frac{750}{3 \cdot 15 \times 62 \cdot 4} = 3 \cdot 81 \text{ ft}^3$$

volume of water	$= 7 \cdot 78 \text{ ft}^3$, as before
volume of coarse aggregate	$= 8 \cdot 21 \text{ ft}^3$, as before
Total	$= 19 \cdot 80 \text{ ft}^3$

Thus the volume of fine aggregate required is $27 - 19 \cdot 80 = 7 \cdot 20 \text{ ft}^3$, and its weight is $7 \cdot 20 \times 1 \cdot 75 \times 62 \cdot 4 = 785 \text{ lb}$.

A trial mix would show whether any adjustment in the water content is necessary, and an actual measurement of the density of the concrete would show whether it differs from the expected value of

$$\frac{750 + 485 + 705 + 785}{27} = 101 \text{ lb/ft}^3.$$

The preceding mix design method was altered in the 1969 revision of the A.C.I. Standard 211.2–69,[9.7] mainly with respect to the determination of the specific gravity factor. In the new method, the factor is determined for each aggregate (fine or coarse) by a direct pycnometer method. The determination is made at all the moisture contents expected to be encountered. The mix proportions, or adjustments to them, are then calculated on the basis of the aggregates in their actual moisture condition, using the appropriate specific gravity factor, and for the actual amount of added water. Since the numerical values of the "old" and "new" specific gravity factors are not the same, it is important to refer to the latter as the pycnometer specific gravity factor.

This factor is the ratio of the weight of the aggregate as introduced into the mixer to the effective volume displaced by the aggregate. The weight of the aggregate thus includes any moisture, absorbed or free, at the time of placing the aggregate in the mixer.

The pycnometer measurements give the pycnometer specific gravity factor as

$$S = \frac{C}{B - A + C}$$

where A = weight of the pycnometer with the sample and topped up with water (usually after a 10-minute immersion),

B = weight of the pycnometer full of water,

and C = weight of the aggregate tested in the given condition, moist or dry (cf. p. 122).

Table 9.5: *B.S. 3797: 1964 Grading Limits of Coarse Lightweight Aggregate*

B.S. sieve size		Percentage by weight passing B.S. sieves					
		Nominal size of graded aggregate			Nominal size of single-size aggregate		
		$\frac{3}{4}$ in. to $\frac{3}{16}$ in. (19 mm–5 mm)	$\frac{1}{2}$ in. to $\frac{3}{16}$ in. (13 mm–5 mm)	$\frac{3}{8}$ in. to $\frac{1}{8}$ in. (10 mm–3 mm)	$\frac{3}{4}$ in. (19 mm)	$\frac{1}{2}$ in. (13 mm)	$\frac{3}{8}$ in. (10 mm)
$1\frac{1}{2}$ in.	38·10 mm	100	—	—	100	—	—
$\frac{3}{4}$ in.	19·05 mm	95–100	100	—	85–100	100	—
$\frac{1}{2}$ in.	12·70 mm	—	90–100	100	—	85–100	100
$\frac{3}{8}$ in.	9·52 mm	25–55	40–85	80–100	0–20	0–30	85–100
$\frac{3}{16}$ in.	4·76 mm	0–10	0–10	5–40	0–5	0–10	0–20
7	2·40 mm	—	—	0–15	—	—	0–5

Using the pycnometer specific gravity factors, it is possible to allow for the fact that the total water requirement actually changes with the changes in the moisture condition of the aggregates at the time of mixing (*see* p. 592).

In the British practice, mix design procedures similar to those for normal weight concrete are generally used. They are helped somewhat by the grading requirements laid down by B.S. 3797: 1964. This standard prescribes the grading of nominal size and single size coarse aggregates (*see* Table 9.5) and allows two grading zones of fine aggregate (Table 9.6).

Table 9.6: *B.S. 3797: 1964 Grading Zones of Fine Light-weight Aggregate*

B.S. sieve size		Percentage by weight passing B.S. sieves	
		Grading Zone L1	Grading Zone L2
$\frac{3}{8}$ in.	9·52 mm	100	100
$\frac{3}{16}$ in.	4·76 mm	90–100	90–100
7	2·40 mm	55–95	60–100
14	1·20 mm	35–70	40–80
25	600 μm	20–50	30–60
52	300 μm	10–30	25–40
100	150 μm	5–19	20–35

Note: Certain departures from a zone are permitted.

The standard recognizes that different types of lightweight concrete call for different aggregate gradings and that a particular grading which is suitable for use with one type of lightweight aggregate may not be suitable with another type. The ratio of fine to coarse lightweight aggregate should nevertheless be generally lower for aggregates in Zone L2 than in Zone L1 of Table 9.6, and also for mixes with higher cement contents. The selection of the correct ratio is particularly important as the grading of the fine aggregate approaches the lower limit of Zone L1 or the upper limit of Zone L2.

It may be noted that the material passing the 150 μm (No. 100) sieve represents the boundary of the two zones. This material can be beneficial in reducing segregation, particularly in lean mixes.

For comparison, grading requirements of A.S.T.M. C 330–69 are given in Table 9.7.

It should be noted that the values of Tables 9.5 to 9.7 are based on weight; because of the increase in specific gravity of any lightweight aggregate with a decrease in particle size, the percentages of fine material are not as high as would at first appear.

Table 9.7: *A.S.T.M. C 330–69 Grading Requirements for Lightweight Aggregates for Structural Concrete*

A.S.T.M. sieve size		Percentage by weight passing A.S.T.M. sieves							
		Nominal size of graded coarse aggregate					Fine Aggregate	Nominal size of combined fine and coarse aggregate	
		1 in. to ½ in. (25 mm–13 mm)	1 in. to N°4 (25 mm–5 mm)	¾ in. to N°4 (19 mm–5 mm)	½ in. to N°4 (13 mm–5 mm)	⅜ in. to N°8 (10 mm–2 mm)		½ in. (13 mm)	⅜ in. (10 mm)
1 in.	25·40 mm	95–100	95–100	100	—	—	—	—	—
¾ in.	19·05 mm	—	—	90–100	100	—	—	100	—
½ in.	12·70 mm	0–10	25–60	—	90–100	100	—	95–100	100
⅜ in.	9·52 mm	—	—	20–60	40–80	80–100	100	—	90–100
4	4·76 mm	—	0–10	0–10	0–20	5–40	85–100	50–80	65–90
8	2·38 mm	—	—	—	0–10	0–20	—	—	35–65
16	1·19 mm	—	—	—	—	—	40–80	—	—
50	297 μm	—	—	—	—	—	10–35	5–20	10–25
100	149 μm	—	—	—	—	—	5–25	2–15	5–15

Aerated Concrete

As mentioned earlier, one means of obtaining lightweight concrete is by introducing gas bubbles into the plastic cement mix in order to produce a material with a cellular structure, somewhat similar to sponge rubber. For this reason the resulting concrete is known as cellular or aerated concrete. There are two basic methods of producing aeration, an appropriate name being given to each end product.

Gas concrete is obtained by a chemical reaction generating a gas in fresh mortar, so that when it sets it contains a large number of gas bubbles. The mortar must be of the right consistence so that the gas can expand the mortar but does not escape. Thus the speed of gas evolution, consistence of mortar and its setting time must be matched. Finely divided aluminium powder is most commonly used, its proportion being of the order of 0·2 per cent of the weight of cement.* The reaction of the active powder with a hydroxide of calcium or alkali liberates hydrogen, which forms the bubbles. Powdered zinc or aluminium alloy can also be used. Sometimes hydrogen peroxide is employed; this generates oxygen.

Foamed concrete is produced by adding to the mix a foaming agent (usually some form of hydrolyzed protein or resin soap) which introduces and stabilizes air bubbles during mixing at high speed. In some processes a stable pre-formed foam is added to the mortar during mixing in an ordinary mixer.

Aerated concrete may or may not contain aggregate, the latter generally being the case with non-structural concrete required for heat insulation when a density of 300 kg/m³ (20 lb/ft³) and exceptionally as low as 200 kg/m³ (12 lb/ft³) can be obtained. More usual mixes have densities between 500 and 1,100 kg/m³ (30 and 70 lb/ft³) when a mixture of cement and very fine or ground sand is used. As in other lightweight concretes, strength varies in proportion to density, and so does the thermal conductivity. For instance, a concrete with a density of 500 kg/m³ (30 lb/ft³) would have strength in the region of 3 to 4 MN/m² (450 to 600 lb/in²) and a thermal conductivity of about 0·1 Jm/m²s°C (0·06 Btu/ft²h°F/ft). For a concrete with a density of 1,400 kg/m³ (90 lb/ft³) the corresponding values would be approximately 12 to 14 MN/m² (1,800 to 2,000 lb/in²) and 0·4 Jm/m²s°C (0·23 Btu/ft²h°F/ft). By comparison, the conductivity of ordinary concrete is about 10 times larger. The modulus of elasticity of these concretes is between 1·7 and 3·5 GN/m² (0·25 and 0·5 × 10⁶ lb/in²). Creep, expressed on the basis of stress/strength ratio, is sensibly the same as for ordinary concrete.

Aerated concrete has a high thermal movement and a high shrinkage and moisture movement (rather higher than lightweight aggregate concrete of the same strength) but these may be reduced by high pressure

---

* Aluminium powder is also used in grout for post-tensioned concrete in order to ensure complete filling of the cavity by the expanding grout.

steam curing, which improves also the compressive strength (*see* Fig. 9.3). When concrete is to be high-pressure steam-cured, pozzolanic material is generally added to the mix, or pulverized fuel ash (fly ash) is used instead of fine aggregate; blast-furnace slag is also used. Lime may be used in the mix as for example in the Swedish *Ytong*; a similar product, *Siporex*, uses a cement base.

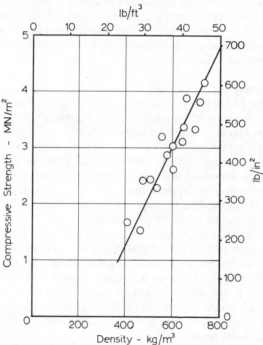

Fig. 9.3. *Relation between strength and density of high-pressure steam-cured aerated concrete*[9.4]

Aerated concrete is mostly used for partitions for heat insulation purposes because of its low thermal conductivity, and for fireproofing as it offers better fire resistance than ordinary concrete. Structurally, it is used mostly in the form of high-pressure steam-cured blocks or precast members but it can also be used for floor construction instead of a hollow tile floor. Aerated concrete can be sawn and it holds nails. It is reasonably durable for, although its water absorption is high, the rate of water penetration through aerated concrete is low as the larger pores will not fill by suction. For this reason, aerated concrete has a comparatively good resistance to frost, and if rendered can be used in wall construction. There exist some special foaming agents that impart a water-repellent character to the concrete but the behaviour of these over long periods is yet to be tested. Untreated aerated concrete should not be exposed to an aggressive atmosphere.

Unprotected reinforcement in aerated concrete would be vulnerable to corrosion even when the external attack is not very severe. The reinforcement should therefore be treated by dipping in a suitable anticorrosive liquid: bituminous solutions and epoxy resins have been found successful. Rubber latex cement is suitable when the concrete is to be autoclaved as the protective layer becomes vulcanized; the bond of the steel is then particularly good,[9.20] while some other anticorrosion coatings affect bond adversely.

No-fines Concrete

This is a form of lightweight concrete obtained when fine aggregate is omitted, i.e. consisting of cement, water, and coarse aggregate only.

No-fines concrete is thus an agglomeration of coarse aggregate particles, each surrounded by a coating of cement paste up to about 1·3 mm (0·05 in.) thick. There exist, therefore, large pores within the body of the concrete which are responsible for its low strength, but their large size means that no capillary movement of water can take place.

Although the strength of no-fines concrete is considerably lower than that of normal-weight concrete, this strength, coupled with the lower dead load of the structure, is sufficient in buildings up to about 20 storeys high and in many other applications. Since no-fines concrete does not segregate, it can be dropped from a considerable height and placed in very high lifts. The cost of no-fines concrete is comparatively low as the cement content is low: in lean mixes it can be as little as 70 to 130 kg of cement per cubic metre (120 to 220 lb/yd³) of concrete. This is due to the absence of a large surface area of sand particles that would have to be coated with cement paste.

The density of no-fines concrete depends primarily on the grading of the aggregate. Since well-graded aggregate packs to a higher bulk density than when the particles are all of one size, low density no-fines concrete is obtained with one-size aggregate. The usual size is 9·5 to 19 mm ($\frac{3}{8}$ to $\frac{3}{4}$ in.),* but particles up to 50 mm (2 in.) have been used. For a given specific gravity of aggregate, graded aggregate would result in a density some 10 per cent higher than when one-size aggregate is used. With normal aggregate the density of no-fines concrete varies between 1,600 and 2,000 kg/m³ (100 and 125 lb/ft³) (*see* Table 9.8), but using lightweight aggregate no-fines concrete weighing as little as 640 kg/m³ (40 lb/ft³) can be obtained. The use of sharp-edged crushed aggregate is not recommended as local crushing can take place under load.

---

* This allows 5 per cent oversize and 10 per cent undersize, but no material smaller than 4·76 mm ($\frac{3}{16}$ in.).

Table 9.8: *Typical Data for 9·5–19 mm ($\frac{3}{8}$–$\frac{3}{4}$ inch) No-fines Concrete*[9.6]

Aggregate/cement ratio by volume	Water/cement ratio by weight	Density		28-day compressive strength	
		kg/m³	lb/ft³	MN/m²	lb/in²
6	0·38	2,020	126	14	2,100
7	0·40	1,970	123	12	1,700
8	0·41	1,940	121	10	1,450
10	0·45	1,870	117	7	1,000

The density of no-fines concrete is calculated simply as the sum of the *bulk* density of the aggregate (in the appropriate state of compaction) plus the cement content in kg/m³ (lb/ft³) plus the water content in kg/m³ (lb/ft³). This is because no-fines concrete compacts very little and in fact vibration can be applied for very short periods only as otherwise the cement paste would run off. Rodding is not recommended as it can lead to high local density.

The strength of no-fines concrete varies generally between 1·4 and 14 MN/m² (200 and 2,000 lb/in²), depending mainly on its density which is governed by the cement content. (Fig. 9.4). The water/cement ratio as such is not the main controlling factor and in fact there is a narrow optimum water/cement ratio for any given aggregate. A water/cement ratio higher than the optimum would make the cement paste drain away from the aggregate particles, while with too low a water/cement ratio the paste would not be sufficiently adhesive and proper compaction could not be achieved.

It is rather difficult to predict the optimum water/cement ratio, particularly since it is affected by the absorption of the aggregate, but as a general guide the water content of the mix can be taken as 180 kg per cubic metre (300 lb/yd³) of concrete. The water/cement ratio will then depend on the cement content necessary for a sufficient coating of the aggregate. The resulting strength has to be determined by test. The increase in strength with time is of the same form as in normal concrete.

Practical mixes vary rather widely, with the lean limit between a 1:10 cement/aggregate ratio by volume (corresponding to a cement content of approximately 130 kg/m³ (220 lb/yd³)) and a 1:20 mix by volume (with a cement content of 70 kg/m³ (120 lb/yd³)).

Comparatively little is known about various physical properties of no-fines concrete. Typical values of the modulus of elasticity for concretes of different strengths are given in the table following—

| Compressive strength | | Modulus of elasticity | |
MN/m²	lb/in²	GN/m²	10⁶ lb/in²
4·8	700	10·3	1·5
3·4	500	9·0	1·3
2·4	350	6·9	1·0

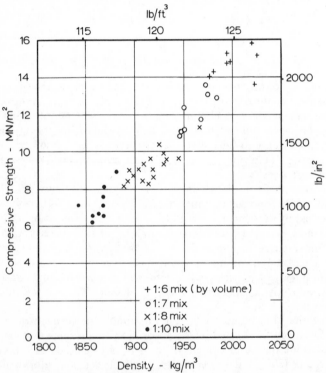

Fig. 9.4. *Compressive strength of no-fines concrete at the age of 28 days as a function of its density at the time of testing*[9.6]

Shrinkage of no-fines concrete is considerably lower than that of normal concrete: a typical value is 120×10^{-6}. This is because the cement paste is present as a thin coating only and contraction on drying is largely restrained by the aggregate. Because the paste has a large surface area exposed to air the rate of shrinkage is very high: the total movement may be completed in little over a month, and half the shrinkage may take place in 10 days.

The thermal movement of no-fines concrete is about 0·6 to 0·8 of that of normal concrete, but the actual value of the coefficient of thermal expansion depends of course on the type of aggregate used.

Because of the absence of capillaries, no-fines concrete is highly resistant to frost, provided of course that the pores are not saturated, in which case freezing would cause a rapid disintegration. High absorption of water, however, makes no-fines concrete unsuitable for use in foundations and in situations where it may be in contact with water. The maximum absorption may be as high as 25 per cent by volume, or half that amount by weight, but under normal conditions the absorbed water does not exceed one-fifth of the maximum. Nevertheless, external walls have to be rendered on both sides; this has also the effect of reducing the permeability to air. Rendering and painting reduce the sound-absorbing properties of no-fines concrete (through closing of the pores) so that where the acoustic properties are considered to be of paramount importance one side of a wall should not be rendered.

The coefficient of thermal conductivity of no-fines concrete is between 0·69 and 0·94 Jm/m²s°C (0·40 and 0·54 Btu/ft²h°F/ft) when ordinary aggregate is used but only about 0·22 Jm/m²s°C (0·13 Btu/ft²h°F/ft) with lightweight aggregate. However, a high moisture content in the concrete very appreciably increases the thermal conductivity.

No-fines concrete is not normally used in reinforced concrete, but if this is required the reinforcement has to be coated with a thin layer (about 3 mm ($\frac{1}{8}$ in.)) of cement paste in order to improve the bond characteristics and to prevent corrosion. The easiest way to coat the reinforcement is by guniting.

Sawdust Concrete

It is sometimes required to make *nailing* concrete and this may be achieved by using sawdust as aggregate. Nailing concrete is a material into which nails can be driven and in which they are firmly held. The last stipulation is made because, for instance, in some of the lighter lightweight concretes nails, although easily driven in, fail to hold. The nailing properties are required in some types of roof construction and in precast units for houses, etc. Because of its very large moisture movement, sawdust concrete should not be used in situations where it is exposed to moisture.

Sawdust concrete consists of roughly equal parts by volume of Portland cement, sand, and pine sawdust, with water to give a slump of 25 to 50 mm (1 to 2 in.). Such a concrete bonds well to ordinary concrete and is a good insulator. The sawdust should be clean and without a large amount of bark as this introduces a high organic content and upsets the reactions of hydration. Chemical treatment of sawdust is advisable to avoid adverse effect on setting and hydration, to prevent the sawdust rotting and to reduce its moisture movement. Best results are obtained with sawdust size between 6·3 mm ($\frac{1}{4}$ in.) and a 1·18 mm (No. 14) B.S. test sieve, but because of the variable behaviour of different kinds of sawdust the use of a trial mix is recommended. Sawdust concrete has a density of between 650 and 1,600 kg/m³ (40 and 100 lb/ft³).

Nailing concrete can also be made with some other aggregates, such as expanded slag, pumice, scoria, and perlite.

Other wood waste, such as splinters and shavings, suitably treated chemically, have also been used to make non-load-bearing concrete with a density of 800 to 1,200 kg/m³ (50 to 75 lb/ft³). In some countries, rice husks and even wheat have been used to make concrete but these are not materials of widespread interest.

Synthetic organic materials have also been used, e.g. expanded polystyrene. This has a bulk density of below 10 kg/m³ (well below 1 lb/ft³) and produces concrete with particularly good insulating properties. A mix with 410 kg of cement per cubic metre (700 lb/yd³) has a density of 550 kg/m³ (35 lb/ft³) and a strength of 2 MN/m² (300 lb/in²). However, because of a wide disparity in the density of the mix ingredients, mixing is difficult and the use of a large volume of entrained air, up to 15 per cent, may be required. Fully satisfactory placement techniques have not yet been developed.

Properties of Concrete as a Radiation Shield

Concrete is commonly used as a shielding material against high energy X-rays, gamma rays and neutrons because it combines radiation absorption properties with good mechanical characteristics, durability, and economic structural use. Much of such concrete has high density, and this is why the properties of concrete relevant to its behaviour as a radiation shield are considered in this chapter.

The X- and gamma rays, called photons, can be described by either wavelength or frequency (inversely proportional to wavelength) or energy (also inversely proportional to wavelength), the latter being normally used in shielding considerations. Although the X- and gamma rays are fundamentally the same in nature (electromagnetic high frequency waves), they differ in origin and in properties. X-rays result from an atomic process outside the atomic nucleus, while gamma rays usually stem from a nuclear process. Gamma rays given off when neutrons are absorbed by nuclei often complicate the design of shields which must attenuate neutrons as well as gamma rays.[9.21]

Concrete shields against gamma radiation are primarily effective through their Compton scattering effect (one of the mechanisms through which photons interact with matter), in which photons undergo an elastic collision with an atomic electron in such a way that energy and momentum are conserved.[9.21] Gamma ray absorption requires therefore a high density of the shield; the relation between these two quantities is shown in Fig. 9.5. In addition to absorbing gamma rays, reactor shields have to attenuate neutrons, which are the heavy particles of the atomic nuclei. These particles, because they have no electric charge, are slowed

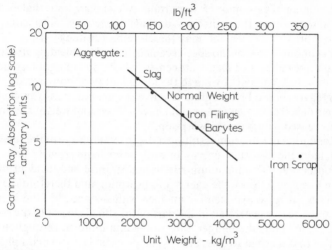

Fig. 9.5. *Relation between gamma ray absorption and density (unit weight) of concrete*[9.22]

down only on collision with atomic nuclei. Neutrons are classified into slow (thermal) (energy below 100 eV), intermediate (up to 100 keV), and fast (up to 10 MeV). For efficient neutron shielding, the material must contain some heavy elements (to slow down the fast neutrons), and a quantity of hydrogen (to slow down the intermediate neutrons and to absorb the slow neutrons). Because, on absorption of slow neutrons, hydrogen emits gamma rays, the presence of boron (which is more effective) is desirable. However, care is required as water-soluble boron compounds reduce the development of strength of concrete; on the other hand, the addition of water-insoluble boron compounds provides the same radiation protection without an adverse effect on the mechanical properties of concrete.[9.23]

It seems thus that the attenuation of neutrons and absorption of gamma rays require the presence of different elements in the shielding material, and concrete satisfies both these requirements as it has a high density and contains a large number of hydrogen atoms.

The hydrogen content is satisfactory at about 0·45 per cent by weight for normal density concrete, and this is reached by a water content representing approximately 4 per cent of the weight of the concrete.[9.24] Some of this water is chemically combined in the hydrated cement paste, some may be free water (the thick shield probably never drying out completely), and some may be in the form of the water of crystallization in aggregate. Fully dried out concrete has a hydrogen content of about 0·25 per cent by weight,[9.25] so that nearly one-half of the hydrogen required has to be found outside the chemically combined water. For this reason, the choice of suitable aggregate may be important. We should note that oxygen helps in moderating the neutrons: some aggregates, e.g. silica, contain a large quantity of oxygen and may therefore be advantageous.

For economic reasons, no special cement is used in shielding structures but it may be mentioned that white cement has the advantage of a reduced neutron-induced activity owing to the low quantity of iron present.[9.26]

Aggregate should be well graded to avoid segregation: any pockets of concrete deficient in aggregate would have a lower unit weight and therefore decreased gamma ray absorption.

Attenuation of radiation results in a rise in temperature of the shielding concrete as the absorbed energy of radiation is converted to heat. The relation between the maximum rise in temperature and incident flux is shown in Fig. 9.6. Since the energy of absorption and therefore the heat vary in an inverse exponential manner with distance,[9.27] the greatest amount of heat is generated in the part of the shield closest to the source of radiation. Thus the temperature rise is non-uniform throughout the shield so that thermal stresses arise. (The problem is further complicated by the forced cooling of the shield surfaces.) To avoid structural failure, or even local damage, it is necessary therefore to relate the maximum incident energy flux to the allowable compressive or tensile stress in concrete (*see* Fig. 9.7). For instance, Thomas[9.27] found that a 1·4 m (4·5 ft) thick shield with reinforcement to resist thermal stresses can absorb an incident energy flux of 268 J/m²s (85 Btu/h ft²) with a maximum temperature rise of 53°C (95°F) when cooled by air on both the internal and external surfaces. This situation is practically independent of the nature of the radiation, be it gamma rays or neutrons. Without reinforcement, a flux of 33 J/m²s (10·5 Btu/h ft²), producing a temperature rise of 9°C (16°F), leads to cracking on the outer surface. From Fig. 9.6 and 9.7 it can be seen that in the shield considered, a temperature rise of 64°C (116°F) produces a compressive stress of 7 MN/m² (1,000 lb/in²).

Typical relations between incident energy flux and shield thickness for different values of the maximum stress in compression and in tension are shown in Figs. 9.8 and 9.9; in the case of the limiting compressive stress, the total tangential tension was assumed to be resisted by the reinforcement at the outer surface of the shield. It is important to remember, however, that thermal conditions at the shield surfaces (such as insulation or convection) may affect the temperature and stress distributions more significantly than the shield thickness.

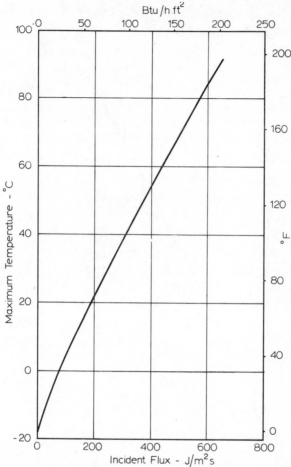

Fig. 9.6. *Relation between the maximum rise in temperature and incident flux in a shield with internal diameter of 4·8 m (16 ft), wall thickness of 1·4 m (4·5 ft), and cooling by convection in air of the same temperature on both shield surfaces*[9.27]

The temperature distribution within the shield is affected not only by the distribution of the absorbed energy but also by the thermal properties of concrete. The desirable thermal properties of shielding concrete are: high thermal conductivity (to minimize high local rise in temperature), low coefficient of thermal expansion (to minimize strains due to temperature gradients) and low drying shrinkage (to minimize differential strains). While high creep is not necessarily disadvantageous, it is important to be able to predict its magnitude. The influence of irradiation on creep is considered on page 555.

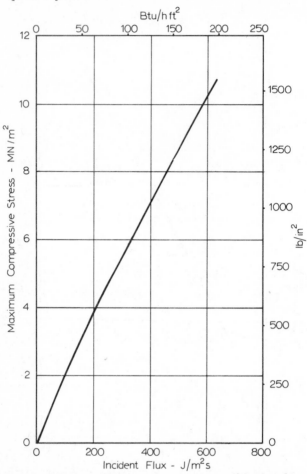

Fig. 9.7. *Relation between maximum compressive stress and incident flux in a shield with internal diameter of 4·8 m (16 ft), wall thickness of 1·4 m (4·5 ft), and cooling by convection in air of the same temperature on both shield surfaces*[9.27]

Thermal stresses are not the only reason for limiting the temperature rise in a shield: at elevated temperatures, the strength of concrete may be adversely affected, creep may be increased, and the attenuation of neutrons may also be reduced as water is driven out of the concrete: a reduction in attenuation as high as 30 per cent has been observed.[9.28] For these reasons, a limit on the maximum temperature should be imposed in design. However, the shielding effectiveness is only slightly impaired up to 300°C (570°F), and only in few reactors is the temperature of 320°C (610°F) reached in any part.

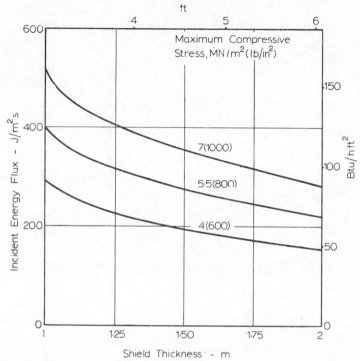

Fig. 9.8. *Relation between the shield thickness and incident energy flux for different values of maximum compressive stress*[9.27]

The permissible stresses in a concrete shield are invariably low as it is important to ensure that no local cracking or deterioration of concrete takes place.

Although shields are usually cast in place, precast concrete may also be used. To compensate for greater radiation transmission through mortar joints, it may be thought that a greater thickness of a precast shield is required than with *in situ* concrete, but this is made up for, at least in part, by the more uniform properties of the concrete block.

Influence of Irradiation on Strength and Creep

Tests[9.29] have shown that irradiation reduces the tensile strength of concrete but the effects of irradiation were not separated from those of high temperature. After six months' exposure to a flux of 10^{10} neutrons/mm²s at 50°C (122°F), the strength fell to about two-thirds of the original value but there was no further loss up to three years. Oven storage at 200°C (392°F) resulted in the same decrease in strength so that it is possible that

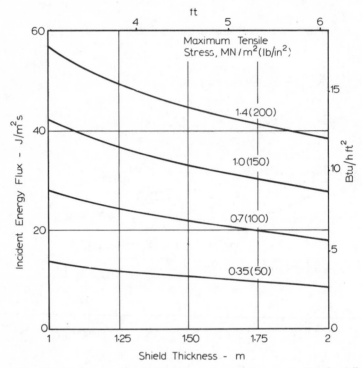

Fig. 9.9. *Relation between the shield thickness and incident energy flux for different values of maximum tensile stress*[9.27]

irradiation in itself is not an important factor. These values were obtained for a 1:3 mortar with a water/cement ratio of 0.45 made with ordinary Portland cement. When aluminous cement was used, the strength was reduced to about one-third of the original value; this high loss may be related to the chemical conversion of this type of cement.

Concrete exposed to either 3×10^{18} roentgens of gamma ray or 10^{16} neutrons per mm² at 77°C (171°F) has been found to undergo no significant change in compressive strength and density.[9.30] The same was found in Russian tests[9.31] under a total flux of 5×10^{17} neutrons per mm² (fast neutron flux of 5×10^{15} neutrons per mm²) except when crystalline quartz aggregate was used; this material becomes amorphous during irradiation with a resulting failure of bond. Some tests[9.22] have shown small damage at 2.4×10^{19} neutrons per mm² at a temperature between 205 and 538°C (400 to 1,000°F) but such exposure is beyond that expected in actual shielding or in a containment vessel.

Tests on the influence of irradiation on creep are not easy as, to perform the experiment in a reasonably short time, a high radiation flux has to be applied and this would raise the temperature in the core of the concrete specimen and induce differential thermal stresses. Indirect evidence has, therefore, to be obtained from the effects of radiation on strength and elasticity of concrete. From the data given earlier, it seems reasonable to expect that creep of concrete is unaffected by radiation except in so far as temperature is increased, and this enhances creep and accelerates drying. It is possible, however, that displacement of water molecules or hydrogen atoms by irradiation causes some damage.

The practical significance of any influence of irradiation on creep is not very large as the damaging action is reduced by about 90 per cent in a 300 mm (12 in.) thickness of concrete, so that any effects on the structure would be limited to its inner face.

High-density Concrete

While much concrete used in radiation protection is of normal weight, the use of high-density concrete is necessary where thickness of the shield is governed by the space available. In order to increase the density of concrete some or all of the ordinary aggregate can be replaced by a material of very much higher specific gravity, usually of over 4·0 (compared with the specific gravity of ordinary aggregate of about 2·6). Natural and artificial heavy aggregates are used. These are covered by A.S.T.M. Specifications C 637–72 and C 638–72.

One of the more common natural aggregates is barytes (barium sulphate). It has a specific gravity of 4·1, and occurs as a natural rock with a purity of about 95 per cent.

Barytes behaves rather like ordinary crushed aggregate and does not present any special problems as far as proportioning of mixes is concerned. The aggregate tends, however, to break up and dust so that care must be exercised in handling and processing, and over-mixing should be avoided. Trial mixes are advisable since some fine barytes aggregates delay setting and hardening.

Barytes concrete does not stand up well to weathering but for most applications of high-density concrete this is of little importance. The modulus of elasticity and Poisson's ratio of barytes concrete are approximately the same as for ordinary concrete,[9,10] but shrinkage is reduced by about $\frac{1}{4}$ or $\frac{1}{3}$. The coefficient of thermal expansion of barytes concrete, measured in the range of 4 to 38°C (40 to 100°F), is about twice that of ordinary concrete; the specific heat, thermal conductivity and diffusivity are all considerably lower than the corresponding values when ordinary aggregate is used.

The density of the concrete varies somewhat with mix proportions: a value of 3,700 kg/m³ (230 lb/ft³) has been obtained using a 1:4·6:6·4 mix with a water/cement ratio of 0·58. The strength of this concrete, measured on standard cylinders, has been found to be 42 MN/m² (6,100 lb/in²) at 28 days. For a water/cement ratio of 0·90 a strength of 24 MN/m² (3,500 lb/in²) was obtained.[9.10] Further typical results are presented in Fig. 9.10. Clearly, no entrained air should be present in the mix, and sometimes a detraining agent is therefore used. It may be noted that owing to the high specific gravity of the aggregate high-density concretes appear to be lean on a weight basis. In order to compare the actual volumetric proportions with ordinary concrete the weight of barytes should be multiplied by a factor of 2·6/4·1.

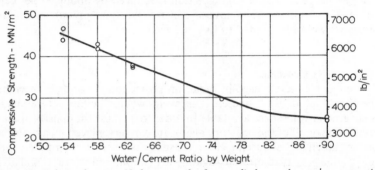

Fig. 9.10. *Relation between 28-day strength of test cylinders and water/cement ratio for concrete made with barytes aggregate*[9.10]

Another type of natural heavy aggregate is iron ore: magnetite, limonite, hematite, and goethite have been used, and ilmenite ($FeTiO_3$) has been employed as fine aggregate. In general, iron ore tends to dust and has a high water absorption but, provided the ore is not unsound, satisfactory concrete with densities of between 3,000 and 3,900 kg/m³ (190 and 245 lb/ft³) can be made. The water of crystallization present in limonite and goethite forms a reliable source of hydrogen for neutron attenuation, provided the temperature does not exceed 200°C (390°F).[9.11] Using magnetite or ilmenite coarse and fine aggregate in mixes containing 490 to 580 kg of cement per cubic metre (820 to 970 lb/yd³) of concrete and having a water/cement ratio of 0·30 to 0·35, 28-day strength of 75 MN/m² (11,000 lb/in²) has been obtained at a density of 3,700 kg/m³ (230 lb/ft³).[9.3]

Artificial heavy aggregates are also used, mostly steel, ferrophosphorus (iron phosphides obtained in the production of phosphorus), and sometimes lead. Steel shot makes concrete with a very high density (up to 5,500 kg/m³ (345 lb/ft³)) but this type of aggregate is about six times dearer than natural heavy aggregates. Concrete made with steel shot is rather prone to segregation because of the great disparity in the specific gravities of steel (used as coarse aggregate) and of ordinary sand, which is usually employed as fine aggregate.

Steel punchings and sheared bars are also used and these may be somewhat cheaper than the shot, but because of their shape additional difficulties in mixing the concrete are encountered. It must be remembered that all steel must be free from oil, which would interfere with bond.

High-density concrete can be placed using the preplaced aggregate method (*see* p. 223): this reduces segregation, particularly when fine aggregate of ordinary weight is used, and the heavy wear of the mixer is also avoided.

It should not be forgotten that, in addition to its functions in relation to radiation and neutron attenuation, concrete used in shielding must also have good structural properties under operating conditions at high temperatures: these properties include satisfactory performance in regard to thermal conductivity, shrinkage, thermal expansion, and creep. As sufficient data on all of these may not be available, in many practical cases where space is not critical the alternative of a thick shield made with ordinary aggregate should be considered.

Specialized Concretes

This book is concerned with concrete for use in structures of a wide variety but there exist also other specialized concretes of interest for particular applications. Several broad varieties can be distinguished, and this chapter is probably best suited to deal with them.

The first of these is *fibre-reinforced concrete*. This is a composite material consisting of cement paste, mortar or concrete with fibres of asbestos, glass, plastic or steel. Such a fibre-reinforced concrete may be useful when a large amount of energy has to be absorbed (e.g. in explosive loading), where a high tensile strength and reduced cracking are desirable, or even when conventional reinforcement cannot be placed because of the shape of the member. Various problems are associated with the different fibres and specialized knowledge is required for their use. A review of the material written in 1968 by Monfore[9.32] may be of interest.

The second variety can be broadly termed *resin concrete* and mortar. This is a material in which aggregate is bonded by a synthetic resin instead of hydraulic cement. Epoxy resins, known also as epoxide resins, are the most common binder but polyester resins or even phenolics are also used.

The binder consists, strictly speaking, of a resin and a hardener which interact and harden rapidly. Epoxy resins are thermosetting but are not suitable for exposure to fire. The chief advantages of epoxy concretes are a high strength rapidly developed, dimensional stability, toughness, excellent adhesion to a wide range of materials, high resistance to chemical attack by many acids, alkalis and solvents, and to weathering and abrasion, and also good thermal and electrical insulation properties.

Epoxy concrete is therefore used largely in repair of conventional concrete, in lengthening piles, in bonding structural concrete elements to one another, and in surface protection. There is also a possibility of using composite reinforced members consisting of conventional concrete with epoxy concrete in the tension zone to achieve a greater flexural strength and protection of reinforcement.[9.33]

The chief disadvantage of epoxy concrete is its very high cost so that the material is unlikely to become of general structural interest except under very severe corrosion conditions. However, in specialized applications it is extremely valuable so that a detailed knowledge of its potential is desirable (*see*, for instance, references [9.34] and [9.35]).

The third variety of specialized concrete is the *polymer concrete*. This is conventional concrete made with Portland cement, wet cured, and subsequently impregnated with a liquid or gaseous monomer and polymerized by gamma radiation or by chemically initiated means. The impregnation may be aided by drying the concrete at a higher temperature, by evacuation at, say, 75 mm (3 in.) of mercury, and by soaking in the monomer under pressure. The use of gamma radiation necessitates shielding but produces a better concrete than chemical initiators.

The polymerized product has much higher compressive, tensile and impact strengths than before treatment, a higher resistance to freezing and thawing, abrasion and to chemical attack.[9.9] All of these are due to the very low void content of the polymer concrete, so that the permeability is reduced almost a hundred-fold. The improvement in the product is generally directly related to the quantity of the polymer present.

It is likely that there may be difficulties in achieving a controlled polymerization by gamma radiation of concrete members of considerable thickness. Because of the high absorption of gamma rays by concrete (*see* p. 548) there is a danger of local over- or under-irradiation. The former would lead to depolymerization and the latter would leave an unreacted monomer which might eventually decompose and possibly react with the hydrated cement paste. However, for many applications, polymerization of the surface layer of concrete may be adequate, such a layer offering protection from chemical or mechanical attack.

The chief drawback of polymer concrete is cost. Thus, the material is of interest primarily in pipes carrying aggressive waters or in desalination plants.

REFERENCES

9.1. R. W. KLUGE, Structural lightweight aggregate concrete, *J. Amer. Concr. Inst.*, **53**, pp. 383–402 (Oct. 1956).

9.2. C. C. CARLSON, Lightweight aggregates for concrete masonry units, *ibid.*, pp. 491–508.

9.3. K. MATHER, High strength, high density concrete, *J. Amer. Concr. Inst.*, **62**, pp. 951–62 (Aug. 1965).

9.4. A. SHORT and W. KINNIBURGH, The structural use of aerated concrete, *The Struct. E.*, **39**, No. 1, pp. 1–16 (London, Jan. 1961).

9.5. R. C. VALORE, Insulating concretes, *J. Amer. Concr. Inst.*, **53**, pp. 509–32 (Nov. 1956).

9.6. R. H. McINTOSH, J. D. BOTTON and C. H. D. MUIR, No-fines concrete as a structural material, *Proc. Inst. C. E.*, Part I, **5**, No. 6, pp. 677–94 (London, Nov. 1956).

9.7. AMERICAN CONCRETE INSTITUTE COMMITTEE 613, Recommended practice for selecting proportions for structural lightweight concrete (A.C.I. 211.2—69), *J. Amer. Concr. Inst.*, **55**, pp. 305–14 (Sept. 1958) and **65**, pp. 1–19 (Jan. 1968).

9.8. G. H. NELSON and O. C. FREI, Lightweight structural concrete proportioning and control, *J. Amer. Concr. Inst.*, **54**, pp. 605–21 (Jan. 1958).

9.9. J. T. DIKEOU, L. E. KUKACKA, J. E. BACKSTROM and M. STEINBERG, Polymerization makes tougher concrete, *J. Amer. Concr. Inst.*, **66**, pp. 829–39 (Oct. 1969).

9.10. L. P. WITTE and J. E. BACKSTROM, Properties of heavy concrete made with barite aggregates, *J. Amer. Concr. Inst.*, **51**, pp. 65–88.

9.11. H. S. DAVIS, F. L. BROWNE and H. C. WITTER, Properties of high-density concrete made with iron aggregate, *J. Amer. Concr. Inst.*, **52**, pp. 705–26 (March 1956).

9.12. N. DAVEY, Concrete mixes for various building purposes, *Proc. of a Symposium on Mix Design and Quality Control of Concrete*, pp. 28–41 (London, Cement and Concrete Assoc., 1954).

9.13. A.C.I. COMMITTEE 213, Guide for structural lightweight aggregate concrete, *J. Amer. Concr. Inst.*, **64**, pp. 433–69 (Aug. 1967).

9.14. J. A. HANSON, Replacement of lightweight aggregate fines with natural sand in structural concrete, *J. Amer. Concr. Inst.*, **61**, pp. 779–93 (July 1964).

9.15. C. HOBBS, Physical properties of lightweight aggregates and concretes, *Chemistry and Industry*, **1**, No. 14, pp. 594–600 (London, April 1964).

9.16. M. POLIVKA, D. PIRTZ and C. CAPANOGLU, Influence of rate of loading on strength and elastic properties of structural lightweight concretes, *Proc. R.I.L.E.M. Int. Symp. on Testing and Design Methods of Lightweight Aggregate Concretes*, pp. 649–66 (Budapest, March 1967).

9.17. D. C. TEYCHENNÉ, Lightweight aggregates: their properties and use in the United Kingdom, *Proc. 1st Int. Congress on Lightweight Concrete, Volume 1: Papers*, pp. 23–37 (London, Cement and Concrete Assoc., May 1968).

9.18. J. MURATA, Design method of mix proportion of lightweight aggregate concrete, *Proc. R.I.L.E.M. Int. Symp. on Testing and Design Methods of Lightweight Aggregate Concretes*, pp. 131–46 (Budapest, March 1967).

9.19. J. A. HANSON, American practice in proportioning lightweight-aggregate concrete, *Proc. 1st Int. Congress on Lightweight Concrete, Volume 1: Papers*, pp. 39–54 (London, Cement and Concrete Assoc., May 1968).

9.20. H. SCHÄFFLER, Rostschutz der Bewehrung (Protection of the reinforcement against corrosion), *Proc. R.I.L.E.M. Int. Symp. on Lightweight Concrete*, pp. 269–72 (Gothenburg, 1960).

9.21. B. E. FOSTER, Attenuation of X-rays and gamma rays in concrete, *A.S.T.M. Mat. Res. and Stand.*, **8**, No. 3, pp. 19–24 (March 1968).

9.22. J. JANSS, Étude de bétons spéciaux de protection contre les radiations nucléaires, *Revue Universelle des Mines*, No. 8, pp. 7 (Liège, Aug. 1962).

9.23. V. B. DUBROVSKII, SH. SH. IBRAGIMOV, A. YA. LADYGIN and B. K. PERGAMEN-SHCHIK, Effect of neutron irradiation on some properties of heat-resistant concretes, *English Translation in Soviet Atomic Energy*, **21**, No. 2, pp. 740–4 (Aug. 1966).

9.24. S. GILL, *Structures for nuclear power* (London, C. R. Books Ltd., 1964).

9.25. R. B. HYDE, Fulfilling requirements for materials for nuclear power projects, *J. Brit. Nuclear Energy Soc.*, **3**, No. 1, pp. 61–4 (Jan. 1964).

9.26. W. E. BINGHAM, Low activation shielding materials for nuclear reactor environmental test chambers, *Nuclear Structural Engineering*, **2**, pp. 243–7 (Amsterdam, 1965).

9.27. D. R. THOMAS, Temperature and thermal stress distributions in concrete primary shields for nuclear reactors, *Nuclear Structural Engineering*, **1**, pp. 368–84 (Amsterdam, 1965).

9.28. F. SEBOK, Role of water content in concrete structures shielding against atomic radiations, *Hungarian Technical Abstracts*, **16**, No. 3, p. 94 (1964) (*Mélyépit-éstudományi Szemle*, **13**, No. 12, pp. 566–70, 1968).

9.29. A. W. C. BATTEN, *Effect of irradiation on the strength of concrete* (U.K. Atomic Energy Authority, Harwell, pp. 10, 1960).

9.30. C. R. TIPTON, (Ed.), *Reactor Handbook, Vol. I: Materials* (New York. Interscience, 1960).

9.31. V. B. DUBROVSKII, SH. SH. IBRAGIMOV, M. YA. KULAKOVSKII, A. YA. LADYGIN and B. K. PERGAMENSHCHIK, Radiation damage in ordinary concrete, *English Translation in Soviet Atomic Energy*, **23**, No. 4, pp. 1053–8 (Oct. 1967).

9.32. G. E. MONFORE, A review of fiber reinforcement of Portland cement paste, mortar, and concrete, *J. Portl. Cem. Assoc. Research and Development Laboratories*, **10**, No. 3, pp. 43–9 (Sept. 1968).

9.33. H. G. GEYMAYER, Use of epoxy or polyester resin concrete in tensile zone of composite concrete beams, *Technical Report C–69–4* (U.S. Army Engineer Waterways Experiment Station, Vicksburg, Miss., March 1969).

9.34. AMERICAN CONCRETE INSTITUTE, Epoxies with concrete, *Sp. Publicn. No. 21* (Detroit, 1968).

9.35. P. C. KREIJGER, Improvement of concretes and mortars by adding resins, *Materials and Structures*, **1**, No. 3, pp. 187–225 (Paris, May–June 1968).

9.36. F. PILNY and W. HIESE, Schwer-und Leichtbeton aus Muell-Schlackensinter, *Die Bautechnik*, **44**, No. 7, pp. 230–9 (Berlin, 1967).

10. Mix Design

It may be said that the properties of concrete are studied primarily for the purpose of mix design, and it is in this light that the various properties of concrete will be considered in this Chapter.

The required properties of hardened concrete are specified by the designer of the structure and the properties of fresh concrete are governed by the type of construction and by the techniques of placing and transporting. These two sets of requirements enable the engineer to determine the composition of the mix, bearing in mind the degree of control exercised on the site. Mix design can thus be defined as the process of selecting suitable ingredients of concrete and determining their relative quantities with the object of producing as economically as possible concrete of certain minimum properties, notably consistence, strength, and durability.

Basic Considerations

The definition in the preceding paragraph stresses two points: that the concrete is to have certain specified minimum properties, and that it is to be produced as economically as possible—a common enough requirement in engineering.

Cost

The cost of concreting, as of any other type of construction, is made up of the costs of the materials, plant, and labour. The variation in the cost of materials arises from the fact that cement is several times dearer than aggregate, so that it is natural in mix design to aim at as lean a mix as possible. The use of comparatively lean mixes confers also considerable technical advantages, not only in the case of mass concrete where the evolution of excessive heat of hydration may cause cracking, but also in structural concrete where a rich mix may lead to high shrinkage and cracking. It is therefore clear that to err on the side of rich mixes is not desirable, even if the cost aspect is ignored.

In estimating the cost of concrete it is essential to consider also the variability of its strength because it is the "minimum" strength that is specified by the designer of the structure and is indeed the criterion of acceptance of the concrete, while the actual cost of the concrete is related to the materials producing a certain mean strength. This touches very closely on the problem of quality control, but there is no doubt that quality control represents expenditure both on supervision and on batching equipment and there are occasions when careful mix design and quality control may not be justified. The decision on the extent of quality control, often an economic compromise, will thus depend on the size and the type of job. It is essential that the degree of control is estimated at the outset of mix design calculations so that the difference between the mean and the minimum or characteristic strengths is known.

The cost of labour is influenced greatly by the workability of the mix: workability inadequate for the available means of compaction results in a high cost of labour (or in insufficiently compacted concrete). Even with efficient mechanical equipment, e.g. in road construction, the cost of placing extremely dry mixes is high. The exact cost of labour depends on the details of organization of the job and the type of equipment used, but these factors are not considered in this discussion of mix design.

Specifications

This large topic cannot be dealt with in this book and will be considered only in so far as the type of specification affects the mix design.[10.1]

In the past, and sometimes even today, the specification for concrete prescribed the proportions of cement, and fine and coarse aggregate. Certain traditional mixes were thus produced but, owing to the variability of the mix ingredients, concretes having fixed cement–aggregate proportions and a given workability vary widely in strength. For this reason, the minimum compressive strength was later included in many specifications. This makes the specification unduly restrictive when good quality materials are available, but elsewhere it may not be possible to achieve an adequate strength using the prescribed mix proportions. This is why, sometimes, clauses prescribing the grading of aggregate and the shape of the particles have been added to the other requirements. However, the distribution of natural aggregates in many countries is such that these restrictions are often uneconomic. Furthermore, compliance with the requirements of strength, mix proportions, and aggregate shape and grading leaves no room for economies in the mix design, and makes progress in the production of cheap and satisfactory mixes on the basis of a study of the properties of concrete impossible.

It is not surprising, therefore, that the modern tendency is for specifications to be less restrictive. They lay down limiting values but often give also as a guide the traditional mix proportions for the benefit of the contractor who does not wish to use a high degree of control. The limiting values may cover a range of properties; the more usual ones are—

1. "minimum" compressive strength necessary from structural considerations;

2. maximum water/cement ratio and/or minimum cement content, and in certain climates a minimum content of entrained air to give adequate durability;

3. maximum cement content to avoid cracking due to the temperature cycle in mass concrete;

4. maximum cement content to avoid shrinkage cracking under conditions of exposure to a very low humidity; and

5. minimum density for gravity dams and similar structures.

These various requirements must then be satisfied in the mix design calculations and they form, in fact, the basis of selection and proportioning of mix ingredients.

The British Code of Practice for the Structural Use of Concrete CP 110: 1972 has moved further than its predecessors towards designed mixes. These are virtually the norm nowadays, and standard mixes (prescribed in the Code by quantities of the dry ingredients per cubic metre and by slump) may be used only on very small jobs, when the strength of concrete does not exceed 30 MN/m^2 (4,400 lb/in^2) and when no air entrainment is used. No control testing is necessary, reliance being made on the weights of the ingredients.

The same code lays down the following minimum strengths of concrete for various uses:

7 MN/m^2 (1,000 lb/in^2) for plain concrete,
15 MN/m^2 (2,200 lb/in^2) for reinforced concrete with
lightweight aggregate,
20 MN/m^2 (2,900 lb/in^2) for reinforced concrete with normal aggregate,
30 MN/m^2 (4,400 lb/in^2) for post-tensioned concrete, and
40 MN/m^2 (5,800 lb/in^2) for pre-tensioned concrete.

The Process of Mix Design

The basic factors that have to be considered in determining the mix proportions are represented schematically in Fig. 10.1. The sequence of decisions is also shown down to the quantity of each ingredient per batch. There are, of course, variations in the exact method of selecting the mix proportions. For instance, in the excellent method of the American Concrete Institute[10.2] (*see* p. 612), the water content in kilogrammes per cubic metre or pounds per cubic yard of concrete is determined direct from the workability of the mix (given the maximum size of aggregate) instead of being found indirectly from the water/cement and aggregate/cement ratios, as is done in the method of Road Note No. 4[10.3] used below.

It should be explained that a design in the strict sense of the word is not possible: the materials used are essentially variable and many of their properties cannot be assessed truly quantitatively, so that we are really making no more than an intelligent guess at the optimum combinations of the ingredients on the basis of the relationships established in the earlier chapters. It is not surprising, therefore, that in order to obtain a satisfactory mix we not only have to calculate or estimate the proportions

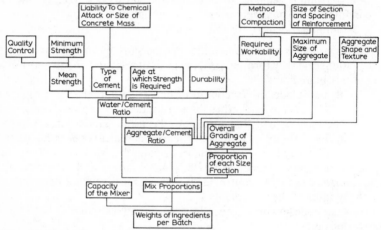

Fig. 10.1. *Basic factors in the process of mix design*

of the available materials but must also make trial mixes. The properties of these mixes are checked and adjustments in the mix proportions are made and are followed by further trial mixes until a fully satisfactory mix is obtained.

Furthermore, a laboratory trial mix does not provide the final answer even when the moisture condition of aggregate is taken into account. Only a mix made and used on the site can guarantee that all the properties of the concrete are satisfactory in every detail for the particular job in hand. To justify this statement two points may be mentioned. Firstly, the mixer used in the laboratory is generally different in type and performance from that employed on the site. Secondly, the wall-effect (arising from the surface to volume ratio) in laboratory test specimens is larger than in the full-size structure, so that the sand content of the mix as determined in the laboratory may be unnecessarily high (*see* p. 492).

Other factors, such as effects of handling, transporting, delay in placing, and weather conditions may also influence the properties of concrete on the site but these are generally secondary and necessitate no more than minor adjustments in the mix proportions during the progress of work.

It can be seen then that mix design requires both a knowledge of the properties of concrete and empirical data or experience.

Factors in the Choice of Mix Proportions

It may be convenient at this stage to re-state the basic problem: we are to determine the proportions of the cheapest concrete mix that will be satisfactory both in the fresh and in the hardened state.

We shall now consider the various factors of Fig. 10.1 and follow the sequence of decisions right down to the final choice of mix proportions.

Strength

This is one of the most important properties of concrete, both *per se* and in so far as it influences many other desirable properties of hardened concrete. Basically, the *mean* compressive strength required at a specified age, usually 28 days, determines the nominal water/cement ratio of the mix. Fig. 10.2 gives this relation for Portland cements used in concrete cured at normal temperatures, but modern cements tend to have higher strengths than indicated. If, however, one batch of cement is to be used throughout the job (e.g. the cement is obtained in bulk and stored in a silo) it is possible to take advantage of the actual strength of the given cement, i.e. to use an experimental relation between strength and the water/cement ratio.

If curves of the type shown in Fig. 10.2 are used the type of cement must be known since the rate of hardening of cements of different types varies (*see* Chapter 2); however, beyond the age of one or two years the strength of concretes made with different cements tends to be approximately the same.

In the case of high-strength mixes (say over 40 MN/m^2 (6,000 lb/in^2)) the aggregate/cement ratio also influences the strength (*see* p. 253) and the design of such mixes is considered separately.

Relation Between Mean and "Minimum" Strengths

Structural design is based on the assumption of a certain *minimum* strength of concrete, but the actual strength of concrete produced, whether on the site, or in the laboratory, is a variable quantity (*see* p. 513). In designing a concrete mix we must therefore aim at a *mean* strength higher than the minimum.

The modern approach of some structural codes (e.g. the British Code for the Structural Use of Concrete CP 110: 1972) is based on the *characteristic* strength of concrete, which is defined as the strength exceeded by 95 per cent of test specimens. In Swiss standards the concept of "nominal" strength is used. This is defined as a value exceeded by 84 per cent of test specimens; reference to Fig. 8.37 shows that this strength is one standard deviation below the mean strength.

The distribution of strength of test specimens can be described by the mean and the standard deviation. From the knowledge of the probability of a specimen having a strength differing from the mean by a given amount (Table 10.1) we can define the "minimum" strength of a given mix. No absolute minimum can be specified because from the statistical viewpoint there is always a certain probability of a test result falling below a minimum however low it is set; to make this probability extremely

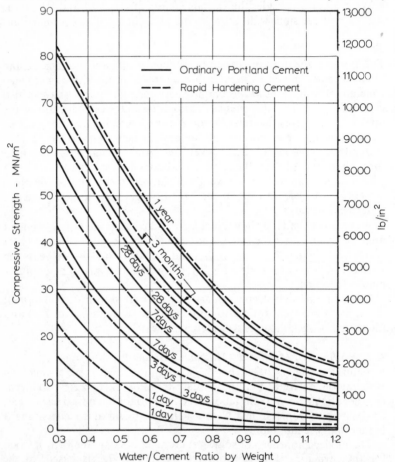

Fig. 10.2. *Relation between compressive strength and water/cement ratio for 100 mm (4-in.) cubes of fully compacted concrete for mixes of various proportions.*[10.3] *Modern cements tend to have higher strengths.*

low would be uneconomical. It is, therefore, usual to define the "minimum" as a value to be exceeded by a predetermined proportion of all test results, usually 95 or 99 per cent. As already mentioned, such a value is called the characteristic strength, and in British practice is based on the

95 per cent proportion. If the characteristic strength approach is not used in an explicit way, it is important to state clearly whether the percentage requirement is to be applied to single cubes or to averages of each group of cubes tested at one time: the probability of the group mean falling below the characteristic value is lower than when single cubes are considered.

Table 10.1: *Percentage of Specimens having a Strength lower than (Mean − k × Standard Deviation)*

k	Percentage of specimens having strength below $(\bar{x} - k\sigma)$
1·00	15·9
1·50	6·7
1·96	2·5
2·33	1·0
2·50	0·6
3·09	0·1

The approach of the 1971 Building Code of the American Concrete Institute is based, in essence, on two requirements for the "minimum" strength, f_{cr}, in relation to the design strength, f_c'. First, there is a probability of 0·01 that the average of three consecutive tests is smaller than the design strength. Second, there is a probability of 0·01 that an individual test result falls below the design strength by more than 500 lb/in² (3·5 MN/m²). In terms of standard deviation, σ, the first of these can be written as

$$f_{cr} = f_c' + \frac{2·326\,\sigma}{\sqrt{3}} = f_c' + 1·343\,\sigma$$

and the second as

$$f_{cr} = f_c' - 500 + 2·326\,\sigma.$$

The two conditions are equivalent when the standard deviation σ is approximately 500 lb/in² (3·5 MN/m²). When it is greater, the first condition is the more severe of the two.

We should note that no absolute bar is laid down: the approach is probabilistic so that failure to meet these requirements once in a 100 times is inherent in the system. Such failure should not be a reason for rejection of the concrete. We may add that all schemes imply a risk of wrong rejection or wrong acceptance: it is the two risks that have to be judiciously balanced.[10.23]

A specific proportion of failures is not practicable when the number of test results is small. Thus, the British Code of Practice CP 114 (1957) stipulated that the mean of 3 cubes tested at the same age must not be less than the minimum, but individual cubes may fall below minimum provided the difference between the highest and the lowest result is not more than 20 per cent of the test mean.

The exact requirements of specifications vary; for instance, a specification may require[10.4]—

(*a*) the average of any five consecutive tests to be not less than the stipulated "minimum";

(*b*) 90 per cent of all tests to exceed the "minimum"; and

(*c*) no test to fall below 80 per cent of the "minimum."

The last clause clearly shows that the engineer's experience and sense of proportion must overrule a rigid application of statistical theory. In the British Code of Practice for the Structural Use of Concrete CP 110: 1972, concrete is assumed to comply with the specified strength when single cubes yield results such that—

(*a*) the average of any four consecutive tests exceeds the specified characteristic strength by 7·5 MN/m² (1,100 lb/in²) for concretes with a characteristic strength of at least 20 MN/m² (2,900 lb/in²) or by one-third of the specified strength for weaker concretes;

(*b*) no test falls below 85 per cent of the specified characteristic strength.

The characteristic strength is defined as a value expected to be exceeded by 95 per cent of all test results.

We remember (p. 517) that the abscissa of any point on the normal distribution curve can be expressed in terms of the standard deviation σ, and the number of specimens whose strength differs from the mean by more than $k\sigma$ is represented by the appropriate proportional area under the normal curve and is given in statistical tables (Table 10.1).

Thus, if the mean strength of a sample of test cubes is $\bar{x}$, and the percentage of cubes whose strength may fall below a certain value $(\bar{x} - k\sigma)$ is specified, then the value of k can be found from statistical tables, and the actual difference between the mean and the minimum, $k\sigma$, will depend only on the value of the standard deviation σ. This is illustrated in Fig. 10.3. Since the cement content of the mix is based on the mean strength, it can be seen that the greater the standard deviation the higher the cement content required for a given minimum strength.

The difference $(\bar{x} - k\sigma)$ can also be expressed in terms of the coefficient of variation, $C = \sigma/\bar{x}$, as $\bar{x}(1 - kC)$. The two methods of estimating the minimum strength are identical when applied to concrete of the same mean strength, but when the data obtained for one mix are used to predict the variability of a mix of different strength the result will depend on whether the standard deviation or the coefficient of variation is unaffected by the change in strength.

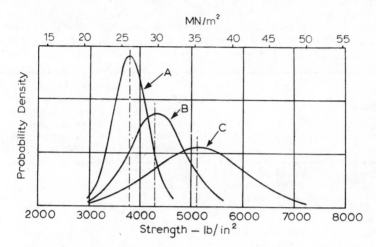

Fig. 10.3. *Normal distribution curves for concretes with a minimum strength (exceeded by 99 per cent of results) of 3,000 lb/in² (20·6 MN/m²)*
(*Crown copyright*)

	$\bar{x}$		σ	
	lb/in²	MN/m²	lb/in²	MN/m²
A	3,800	26·2	350	2·4
B	4,300	29·6	560	3·9
C	5,100	35·2	900	6·2

If a constant standard deviation is assumed then, knowing the estimated value of the standard deviation σ for one mix, we can calculate the mean strength of any other mix by adding a constant value $k\sigma$ to the minimum. This difference between the mean and the minimum would be constant for the same process of manufacture of concrete. On the other hand, if the coefficient of variation is assumed to be constant, the minimum strength would form a fixed proportion of the mean. This is illustrated by the following numerical example.

Let us assume that concrete produced and tested under a given set of conditions has a mean strength of 3,000 lb/in² with a standard deviation of 500 lb/in². This gives a coefficient of variation of 16·7 per cent, and would be classified as a somewhat better than "fair" control (*see* Table

10.2). The minimum strength, defined as a strength exceeded by 99 per cent of all results, would be

$$3,000 - 2\cdot33 \times 500 = 1,840 \text{ lb/in}^2$$

k being 2·33 (from Table 10.1).

Table 10.2: *Estimated Relation between the Minimum and Mean Compressive Strengths of Site Cubes (after Road Note No. 4[10.3]) with Additional Data on the Coefficient of Variation*

Degree of control	Conditions	Minimum strength as a percentage of mean strength	Coefficient of variation for the probability of a cube strength below the minimum occurring:	
			once in 100	once in 200
Very good	Weigh-batching; use of graded aggregates, moisture determinations on aggregates, etc. Constant supervision	75	10·7	9·7
Fair	Weigh-batching; use of two sizes of aggregate only; water content left to mixer-driver's judgment. Occasional supervision	60	17·2	15·5
Poor	Inaccurate volume batching of all-in aggregates. No supervision	40	25·8	23·3

(*Crown copyright*)

Imagine now that it is desired to produce under the *same* conditions a concrete of a minimum strength of, say, 4,500 lb/in². The mean strength aimed at, according to the "coefficient of variation method," would be—

$$\frac{4,500}{1 - 2\cdot33 \times 0\cdot167} = 7,360 \text{ lb/in}^2$$

while the figure given by the "standard deviation method" is—

$$4,500 + 2\cdot33 \times 500 = 5,660 \text{ lb/in}^2.$$

The practical significance of the difference between the two methods is clearly reflected in the cost of producing a 7,360 lb/in² concrete as compared with a 5,660 lb/in² concrete under the same control.

An estimate of the difference between the mean strength and the specified minimum must be made at the outset of every concrete mix design and, unless he has first-hand experience, the designer is guided by the available recommendations, such as those of Road Note No. 4.[10.3] This Note estimates the minimum crushing strength of works cubes as a percentage of the mean strength of such cubes for "different works conditions": the data are reproduced in Table 10.2. Since the method of Road Note No. 4 implies a constant coefficient of variation, its values have been added to Table 10.2, assuming that 1 and 0·5 per cent of all cubes tested may have a strength below the minimum. The Note represents a good approach to mix design, but it has been revised[10.24] to allow a greater choice in the use of fine aggregates. Some guidance is now given on the selection of an appropriate content of fines, taking into account such factors as type of aggregate, its maximum size and grading, cement content, and workability. Thus, instead of grading curves, the grading zones for sand of B.S. 882: 1965 are used. While the basic format on Road Note No. 4 has been retained, some substantial changes are introduced. Specifically, the mix proportions appear not as ratios, but as content in kilogrammes per cubic metre. As before, the mix is designed to have a certain strength and a specified workability. The latter is related to the total free water content of the mix, which is similar to the American practice. The strengths are given by a revision of the curves in Fig. 10.2. The difference between the "minimum" strength and the mean strength will be based on an expected standard deviation.

The problem of the constant standard deviation or constant coefficient of variation is still controversial, but for a constant degree of control, laboratory test data, as well as some results of actual site tests, have been shown[10.4] to support the suggestion of a constant coefficient of variation for well-compacted concretes of different mix proportions with strengths higher than about 10 MN/m² (1,500 lb/in²) (Fig. 10.4). However, surveys of test data on a large number of construction sites suggest that neither the assumption of a constant standard deviation or of a constant coefficient of variation at all ages is generally valid for site-made test specimens. From Newlon's review of the problem,[10.17] it appears that the coefficient of variation is constant up to some limiting value of strength but for higher strengths, the standard deviation remains constant (*see* Fig. 10.5). Different investigators have found different values of this limiting strength, which may well depend on site conditions and general construction practice.

It is possible, however, to suggest some generalizations. Fig. 10.5 indicates that for single cubes, the limiting strength is about 34 MN/m² (5,000 lb/in²); for averages of two cylinders, it is about 17 MN/m² (2,500 lb/in²); and an international survey involving both cubes and cylinders, tested singly and in pairs, gives an intermediate value of about 31 MN/m² (4,500 lb/in²). The factors accounting for these differences

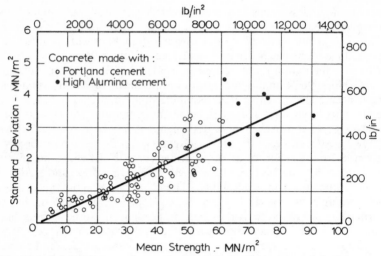

Fig. 10.4. *Relation between the standard deviation and mean strength for laboratory test cubes; regression line shown*[10.4]

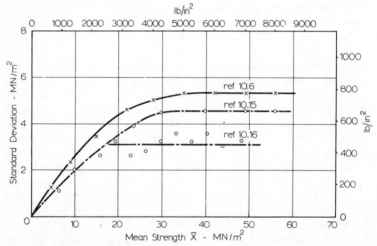

Fig. 10.5. *Relation between the standard deviation and the mean strength of test specimens obtained from surveys of site data*[10.17]

are not clear but the probably lower variability of cylinders compared with cubes (cf. p. 476) may be relevant. It may also be noted that for the same intrinsic strength, the cylinder strength is lower than the cube strength. All these data apply to tests at a given fixed age: for the same source of concrete, an increase in age leads to a reduction in the coefficient of variation, but the standard deviation increases; thus it is the strength level and not merely the concrete-making that is relevant.

In view of these observations, it seems that the practice of mix design on the basis of a constant coefficient of variation for all concretes made with a given degree of control, as recommended by Road Note No. 4 and by Walker[10.5] in the United States (Table 10.3), is not entirely satisfactory

Table 10.3: *Stanton Walker's*[10.5] *Suggested Values for the Coefficient of Variation for Different Degrees of Control*

Degree of control	Coefficient of variation, per cent
Probably attainable only in well-controlled laboratory tests	5
Excellent, approaches laboratory precision	10
Excellent	12
Good	15
Fair	18
Fair minus	20
Bad	25

and may require a modification for high strength concretes. The following is a reasonable approach. Where the plant performance has been established by at least 40 separate batches of concrete, the margin between the mean strength and the (95 per cent) characteristic strength should be at least 1·64 times the standard deviation (which corresponds to a probability of failure of 0·05), and preferably twice the standard deviation but not less than one-third of the specified strength for concrete below 20 MN/m² (2,900 lb/in²) and not less than 7·5 MN/m² (1,100 lb/in²) for stronger mixes. When reliable data on previous performance are not available the numerical values given above are doubled, i.e. the minimum margin is two-thirds the specified strength for concrete below 20 MN/m² (2,900 lb/in²) and 15 MN/m² (2,200 lb/in²) above 20 MN/m² (2,900 lb/in²). A test strength lower than 80 per cent of the specified strength may be grounds for rejection of concrete.

According to the 1971 Building Code of the American Concrete Institute, the standard deviation used in assessing the margin between the mean strength and the "minimum" strength for a specified probability of failure must be based on experience under similar conditions of manufacture to those expected. This margin is:

when the standard deviation is below 300 lb/in²: 400 lb/in²
when the standard deviation is 300 to 400 lb/in²: 550 lb/in²
when the standard deviation is 400 to 500 lb/in²: 700 lb/in²
when the standard deviation is 500 to 600 lb/in²: 900 lb/in²
when the standard deviation is greater than 600 lb/in²: 1,200 lb/in²

The latter value applies also when no reliable data on the previous performance are available.

The explanation of the Code[10.18] suggests that "there may be an increase in standard deviation as average strength level is raised, although the increment should be less than proportional to the strength increase." It may be interesting to note also the finding of the Road Research Laboratory[10.19] that the higher the specified strength of concrete the greater the proportion of defective concrete. This indirectly supports the suggestion that the variability (standard deviation) of stronger concrete is greater than the variability of low-strength concrete.

It is possible that for certain methods of production of concrete the relation between the variability of strength and its mean value may be such that the difference between the mean and the minimum strengths is a simple function of the water/cement ratio. Erntroy[10.6] suggested that the mean strength be taken as that corresponding to a fixed percentage of the water/cement ratio required for the minimum strength.

From the extensive discussion above it is apparent that the problem of variability of concretes of different strengths has not been fully resolved. Unless specifications require otherwise, it is best to establish empirical relations between the mean and "minimum" strengths under actual site conditions.

Quality Control

It is apparent from Fig. 10.3 that the lower the difference between the minimum strength and the mean strength of the mix the lower the cement content that need be used. The factor controlling this difference for concrete of a given level of strength is the quality control. By this is meant the control of variation in the properties of the mix ingredients and also control of accuracy of all those operations which affect the strength or consistence of concrete: batching, mixing, placing, curing, and testing.

The variation in the strength of cement was discussed in Chapter 5. On a big job it is possible to eliminate a large part of this variation by obtaining cement from one source only, when advantage can be taken of the actual strength of the cement to be used.

The influence of the variation in the grading of aggregate was stressed in Chapter 3, and this factor is particularly important when the mix is controlled by workability requirements: for the workability to be kept constant, a change in grading may require an increase in water content with a consequent drop in strength.

Variations in strength of concrete arise also from inadequate mixing, insufficient compaction, irregular curing, and variations in testing procedures—all discussed in the appropriate chapters. The need for control of these factors on the site is obvious.

Changes in the moisture content of aggregate, unless carefully compensated for by the amount of added water, also seriously affect the

Table 10.4: *Graham and Martin's*[10.7] *suggested Values for the Coefficient of Variation for Various Types of Control*

Coefficient of variation per cent	Specified minimum strength MN/m²	lb/in²	Controls* Materials	Water	Batching	Supervision
9 to 10	28 to 35	4,000 to 5,000	Three sizes of coarse aggregate plus fine aggregate, all with strict grading tolerances	Controlled water/cement ratio	Weigh-batching	Rigid
10 to 12	28 to 35	4,000 to 5,000	Ditto	Ditto	Accurate volumetric batching with cement weighed	Ditto
	28 to 35	4,000 to 5,000	Two sizes of coarse plus fine aggregate, all with strict grading tolerances	Ditto	Weigh-batching	Ditto
12 to 13	21 to 28	3,000 to 4,000	Ditto, with less strict grading tolerances	Ditto	Ditto	Very good
13 to 14	21 to 28	3,000 to 4,000	Ditto	Ditto	Accurate volumetric batching with cement weighed	Ditto
14 to 16	21 to 28	3,000 to 4,000	Coarse aggregate with fine aggregate	Ditto	Weigh-batching	Good
16 to 18	14 to 21	2,000 to 3,000	Ditto	Ditto	Accurate volumetric batching with cement weighed	Fair

* All concrete to be spread and compacted by mechanical means.

strength of concrete. To minimize these changes stockpiles should be arranged so that the aggregate is allowed to drain before use; also the mixer operator should be well trained in maintaining a constant workability of the mix.

A standard deviation can be ascribed to each factor separately, although in some cases the magnitude of the individual effects cannot be

Table 10.5: *Maximum Permissible Water/Cement Ratios for Different Types of Structures and Degrees of Exposure, Prescribed by A.C.I. Standard 613–54*[10.2]

Type of structure	Exposure conditions*					
	Severe wide range in temperature or frequent alternations of freezing and thawing (air-entrained concrete only)			Mild temperature rarely below freezing, or rainy, or arid		
	In air	At the water line or within the range of fluctuating water level or spray		In air	At the water line or within the range of fluctuating water level or spray	
		In fresh water	In sea water or in contact with sulphates †		In fresh water	In sea water or in contact with sulphates †
Thin sections, such as railings, kerbs, sills, ledges, ornamental or architectural concrete, reinforced piles, pipe, and all sections with less than 1 in. concrete cover to reinforcement	0·49	0·44	0·40‡	0·53	0·49	0·40‡
Moderate sections, such as retaining walls, abutments, piers, girders, beams	0·53	0·49	0·44‡	§	0·53	0·44‡
Exterior portions of heavy (mass) sections	0·58	0·49	0·44‡	§	0·53	0·44‡

Table 10.5: *(cont.)*

	Exposure conditions*					
	Severe wide range in temperature or frequent alternations of freezing and thawing (air-entrained concrete only)			Mild temperature rarely below freezing, or rainy, or arid		
Type of structure	In air	At the water line or within the range of fluctuating water level or spray		In air	At the water line or within the range of fluctuating water level or spray	
		In fresh water	In sea water or in contact with sulphates †		In fresh water	In sea water or in contact with sulphates †
Concrete deposited by tremie under water	—	0·44	0·44	—	0·44	0·44
Concrete slabs laid on the ground	0·53	—	—	§	—	—
Concrete protected from the weather, interiors of buildings, concrete below ground	§	—	—	§	—	—
Concrete which will later be protected by enclosure or backfill but which may be exposed to freezing and thawing for several years before such protection is offered	0·53	—	—	§	—	—

* Air-entrained concrete should be used under all conditions involving severe exposure and may be used under mild exposure conditions to improve workability of the mix.
† Soil or ground water containing sulphate concentrations of more than 0·2 per cent.
‡ When sulphate-resisting cement is used, maximum water/cement ratio may be increased by 0·04.
§ Water/cement ratio should be selected on basis of strength and workability requirements.

determined. The various standard deviations are additive in the root-square-form, so that if σ_1 and σ_2 are ascribed to two causes the resultant standard deviation is $\sigma = \sqrt{(\sigma_1{}^2 + \sigma_2{}^2)}$. This is important to remember as the assumption of arithmetic addition would lead to a gross over-estimate of the total standard deviation. The knowledge of individual contributions of various factors to the overall variation, obtained by statistical methods, is of value in deciding whether taking some measures to reduce variation is economic, or whether the reduction in variability is disproportionately small for the cost of improved control.

On the majority of large sites weigh-batching is used nowadays but, provided the operators are well trained and take all necessary precautions, volume-batching can also produce concrete of low variability.

Quality control is sometimes taken to be synonymous with production of high-strength concrete. This is certainly not so, as low strength concrete can be manufactured under good control, and this is indeed practised in the case of mass concrete construction where obtaining large quantities of lean concrete of low variability results in large economies.

The degree of control is ultimately evaluated by the variation in test results, usually in terms of the coefficient of variation; estimates of various degrees of control were given in Tables 10.2 and 10.3, but a more detailed set of values for concrete spread and compacted mechanically is given in Table 10.4.

Finally, it must not be forgotten that control includes supervision, and its absence, coupled with inferior workmanship, can nullify all the efforts of mix design and specification. As Glanville[10.8] said, "The difference between good and bad workmanship and supervision may well be represented by the difference between an almost indefinite life (of concrete) and a life of only a few years."

Durability

Concrete of reasonable strength, properly placed, is durable under ordinary conditions but when high strength is not necessary and the conditions of exposure are such that high durability is vital, it is the durability requirement that will determine the water/cement ratio to be used. Some recommended values of maximum water/cement ratio for different conditions of exposure were given in Table 7.7, and Table 10.5 gives the 1954 specification of the American Concrete Institute. The need for air entrainment under conditions of severe exposure should not be forgotten. The British Code of Practice for the Structural Use of Concrete CP 110: 1972 gives the minimum cement content for various conditions of exposure (*see* Table 10.6) and also the minimum cement content and maximum water/cement ratio for different concentrations of aggressive sulphate.

When concrete is to be subjected to chemical attack a suitable type of

Table 10.6: *Minimum Cement Content of Concretes with 20 mm ($\frac{3}{4}$ in.) Maximum Aggregate Size under Different Conditions of Exposure, Prescribed by the British Code of Practice for the Structural Use of Concrete CP 110: 1972*

| Conditions of exposure | Minimum cement content for | | | | | |
| | plain concrete | | reinforced concrete | | prestressed concrete | |
	kg/m³	lb/yd³	kg/m³	lb/yd³	kg/m³	lb/yd³
Non-corrosive	220	370	250	420	300	510
Buried or sheltered from rain and freezing	250	420	290	490	300	510
Exposed to alternating wetting and drying, or to freezing while wet, or to sea water	310	530	360	610	360	610
Subject to de-icing salt (use air-entrained concrete)	280	470	290	490	300	510

cement must be used but if resistance to freezing and thawing is the only durability requirement the choice of the type of cement is governed by other considerations, e.g. the development of early strength or a high heat of hydration for concreting in cold weather.

Since the type of cement affects the early development of strength, it may be necessary with some cements to use a low water/cement ratio to ensure a satisfactory strength at early ages. Thus strength, type of cement, and durability determine between them the water/cement ratio required —one of the essential quantities in the calculation of mix proportions.

Workability

So far we have considered the requirements for the concrete to be satisfactory in the hardened state but, as said before, properties when being handled and placed are equally important. One essential at this stage is a satisfactory workability.

The workability that is considered desirable depends on two factors. The first of these is the size of the section to be concreted and the amount and spacing of reinforcement; the second is the method of compaction to be used.

It is clear that when the section is narrow and complicated, or when there are numerous corners or inaccessible parts, the concrete must have a high workability so that full compaction can be achieved with a reasonable amount of effort. The same applies when embedded steel sections or fixtures are present, or when the amount and spacing of reinforcement make placing and compaction difficult. Since these features of the structure are determined during its design, the engineer designing the mix is presented with fixed requirements and has little choice. On the other hand, when no such limitations are present. workability may be chosen within fairly wide limits, but the means of compaction must be decided upon accordingly; it is important that the prescribed method of compaction is used during the entire progress of construction. A guide to workability for different types of construction is given in Tables 4.2 and 10.7.

Table 10.7: *Selected Values of Compacting Factor for Different Conditions of Placing of Concretes with 9·5 mm ($\frac{3}{8}$ in.) Maximum Size of Aggregate*[10.9, 10.10]

Conditions of placing	Degree of workability	Compacting factor	Slump mm	in.
Sections subjected to extremely intensive vibration: pressure may also be required	Extremely low	0·65	0	0
Sections subjected to intensive vibration	Very low	0·75	0–3	0–$\frac{1}{8}$
Simply reinforced sections with vibration	Low	0·83	3–6	$\frac{1}{8}$–$\frac{1}{4}$
Simply reinforced sections without vibration and heavily reinforced sections with vibration	Medium	0·90	6–25	$\frac{1}{4}$–1
Heavily reinforced sections (not normally suitable for vibration)	High	0·95	25–100	1–4

A property closely related to workability is cohesiveness. This depends largely on the proportion of fine particles in the mix, and especially in lean mixes attention must be paid to the grading of the aggregate at the fine end of the scale. It is generally necessary to make several trial mixes with different proportions of fine to coarse aggregate in order to find the mix with an adequate cohesiveness.

While every mix should be cohesive so that uniform and well-compacted concrete can be obtained, the exact importance of cohesiveness varies. For instance, where concrete has to be hauled a long distance and is handled down a chute, or has to pass through reinforcement, possibly to some inaccessible corner, it is essential that the mix be truly cohesive. In cases when the conditions leading to segregation are less likely to be encountered, cohesion is of smaller importance, but a mix which segregates easily must never be used.

Maximum Size of Aggregate

In reinforced concrete the maximum size of aggregate that can be used is governed by the width of the section and the spacing of the reinforcement. With this proviso, it has generally been considered desirable to use as large a maximum size of aggregate as possible. However, it now seems that the improvement in the properties of concrete with an increase in the size of aggregate does not extend beyond about 40 mm ($1\frac{1}{2}$ in.) so that the use of larger sizes may not be advantageous (*see* p. 176).

Furthermore, the use of a larger maximum size means that a greater number of stockpiles has to be maintained and the batching operations become correspondingly more complicated. This may be uneconomical on small sites, but where large quantities of concrete are to be placed the extra handling cost may be offset by a reduction in the cement content of the mix.

The choice of the maximum size may also be governed by the availability of material and by its cost. For instance, when various sizes are screened from a pit it is generally preferable not to reject the largest size, provided this is acceptable on technical grounds.

Grading and Type of Aggregate

Most of the remarks in the preceding paragraph apply equally to the considerations of aggregate grading, as it is often more economical to use the material available locally, even though it requires a richer mix (but provided it will produce concrete free from segregation) rather than to bring in a better graded aggregate from farther afield.

It has been stressed repeatedly that, although there are certain desiderata for a good grading curve, no ideal gradings exist, and excellent concrete can be made with a wide range of aggregate gradings.

The grading influences the mix proportions for a desired workability and water/cement ratio: the coarser the grading the leaner the mix which can be used, but this is true within limits only as a very lean mix will not be cohesive without a sufficient amount of fine material.

It is possible, however, to reverse the direction of choice: if the cement/aggregate ratio is fixed (e.g. a lean mix may be essential for mass concrete construction) then a grading must be chosen such that concrete of given water/cement/aggregate proportions and satisfactory workability can be made. Clearly, there are limits outside which it is not possible to make good concrete.

The influence of the type of aggregate should also be considered, as its surface texture, shape and allied properties influence strongly the aggregate/cement ratio for a desired workability and a given water/cement ratio. In designing a mix it is essential, therefore, to know at the outset what type of aggregate is available.

An important feature of satisfactory aggregate is the uniformity of its grading. In the case of coarse aggregate this is achieved comparatively easily by the use of separate stockpiles for each size fraction. However, considerable care is required in maintaining the uniformity of grading of fine aggregate, and this is especially important when the water content of the mix is controlled by the mixer operator on the basis of a constant workability: a sudden change toward finer grading requires additional water for the workability to be preserved, and this means a lower strength of the batch concerned. Also, an excess of fine aggregate may make full compaction impossible and thus lead to a drop in strength.

Thus, while narrow specification limits for aggregate grading may be unduly restrictive, it is essential that the grading of aggregate varies from batch to batch within prescribed limits only.

Aggregate/Cement Ratio

All the factors considered up to now, including water/cement ratio, will determine between them the aggregate/cement ratio of the mix. To obtain a clear picture of the various influences, Fig. 10.1 should once again be consulted.

The choice of the aggregate/cement ratio is made either on the basis of personal experience of the mix designer or alternatively from charts and tables prepared from comprehensive laboratory tests. The latter course is frequently followed, use, as a rule, being made of tables of Road Note No. 4 although in its revised form[10.24] aggregate and cement contents in kg/m³ are given, and the rounded and irregular aggregates are not distinguished. The ratios of Road Note No. 4 are reproduced in Table 10.8 for aggregate of 38·1 mm (1½ in.) maximum size and in Table 10.9 for 19·05 mm (¾ in.) aggregate. Data for 9·52 mm (⅜ in.) maximum size aggregate, obtained by McIntosh and Erntroy,[10.9] are given in Table 10.10. Aggregate of this size is used mainly in precast concrete.

It should be emphasized that these tables are no more than a guide to the mix proportions required, as they apply fully only to the actual aggregates used in their derivation. The Tables recognize four type gradings (*see* Figs. 3.14, 3.15, and 3.16) and three shapes of aggregate, but in practice many aggregates may have intermediate properties which, moreover, are not always easily evaluated. Interpolation is of course possible, remembering that the influence of fine aggregate is greater than that of coarse aggregate. This applies to shape (e.g. when natural sand is

Table 10.8: *Aggregate/Cement Ratio (by weight) required to give Four Degrees of Workability with Different Gradings of 38·1 mm (1½ in.) Irregular Aggregate*[10.3] *(See Fig. 3.15)*

Degree of workability	Very low				Low				Medium				High			
Grading curve No. on Fig. 3.15	1	2	3	4	1	2	3	4	1	2	3	4	1	2	3	4
0·35	4·0	3·9	3·5	3·2	3·4	3·3	3·2	2·9	2·9	2·8	2·6	2·5	2·7	2·5	2·3	2·3
0·40	5·3	5·3	4·7	4·3	4·5	4·5	4·2	3·8	3·8	3·8	3·7	3·4	3·5	3·5	3·3	3·1
0·45	6·5	6·5	5·9	5·3	5·6	5·6	5·3	4·8	4·6	4·7	4·6	4·3	4·1	4·4	4·3	4·0
0·50	7·7	7·7	7·1	6·3	6·7	6·6	6·3	5·7	5·4	5·7	5·5	5·1	4·8	5·2	5·1	4·8
0·55	—	—	8·1	7·3	7·6	7·6	7·2	6·6	6·2	6·5	6·3	5·8	×	5·9	6·0	5·5
0·60					—	—	—	7·4	7·0	7·3	7·1	6·6	×	×	6·7	6·2
0·65								8·1	7·8	8·1	7·8	7·2	×	×	7·3	6·9
0·70									—	—	—	7·9	×	×	—	7·4
0·75													×	×	—	8·0
0·80																—

Water/cement ratio by weight

(*Crown copyright*)

— Indicates that the mix was outside the range tested.

× Indicates that the mix would segregate.

These proportions are based on specific gravities of approximately 2·5 for the coarse aggregate and 2·6 for the fine aggregate.

Table 10.9: *Aggregate/Cement Ratio (by weight) required to give Four Degrees of Workability with Different Gradings and Shapes of 19·05 mm (¾ in.) Aggregate*[10.3] *(See Fig. 3.14)*

(a) Rounded Aggregate

Degree of workability	Very low				Low				Medium				High			
Grading curve No. on Fig. 3.14	1	2	3	4	1	2	3	4	1	2	3	4	1	2	3	4
0·35	4·5	4·5	3·5	3·2	3·8	3·6	3·2	3·1	3·1	3·0	2·8	2·7	2·8	2·8	2·6	2·5
0·40	6·6	6·3	5·3	4·5	5·3	5·1	4·5	4·1	4·2	4·2	3·9	3·7	3·6	3·7	3·5	3·3
0·45	8·0	7·7	6·7	5·8	6·9	6·6	5·9	5·1	5·3	5·3	5·0	4·5	4·6	4·8	4·5	4·1
0·50	—	—	8·0	7·0	8·2	8·0	7·0	6·0	6·3	6·3	5·9	5·4	5·5	5·7	5·3	4·8
0·55	—	—	—	8·1	—	—	8·2	6·9	7·3	7·3	7·4	6·4	6·3	6·5	6·1	5·5
0·60	—	—	—	—	—	—	—	7·7	—	—	8·0	7·2	×	7·2	6·8	6·1
0·65	—	—	—	—	—	—	—	8·5	—	—	—	7·8	×	7·7	7·4	6·6
0·70												—	×	—	7·9	7·2
0·75													×	—	—	7·6
0·80													×	—	—	—
0·85													×	—	—	—
0·90																

Water/cement ratio by weight

(Crown copyright)

— Indicates that the mix was outside the range tested.
× Indicates that the mix would segregate.
These proportions are based on specific gravities of approximately 2·5 for the coarse aggregate and 2·6 for the fine aggregate.

Table 10.9 (cont.)

(b) Irregular Aggregate

Degree of workability	Very low				Low				Medium				High			
Grading curve No. on Fig. 3.14	1	2	3	4	1	2	3	4	1	2	3	4	1	2	3	4
0·35	3·7	3·7	3·5	3·0	3·0	3·0	3·0	2·7	2·6	2·6	2·7	2·4	2·4	2·5	2·5	2·2
0·40	4·8	4·7	4·7	4·0	3·9	3·9	3·8	3·5	3·3	3·4	3·5	3·2	3·1	3·2	3·2	2·9
0·45	6·0	5·8	5·7	5·0	4·8	4·8	4·6	4·3	4·0	4·1	4·2	3·9	×	3·9	3·9	3·5
0·50	7·2	6·8	6·5	5·9	5·5	5·5	5·4	5·0	4·6	4·8	4·8	4·5	×	4·4	4·4	4·1
0·55	8·3	7·8	7·3	6·7	6·2	6·2	6·0	5·7	×	5·4	5·4	5·1	×	4·8	4·9	4·7
0·60	9·4	8·6	8·0	7·4	6·8	6·9	6·7	6·2	×	6·0	6·0	5·6	×	×	5·4	5·2
0·65	—	—	—	8·0	7·4	7·5	7·3	6·8	×	×	6·4	6·1	×	×	5·8	5·6
0·70	—	—	—	—	8·0	8·0	7·7	7·4	×	×	6·8	6·6	×	×	6·2	6·1
0·75					—	—	—	7·9	×	×	7·2	7·0	×	×	6·6	6·5
0·80					—	—	—	—	×	×	7·5	7·4	×	×	×	7·0
0·85					—	—			×	×	7·8	7·8	×	×	×	7·4
0·90									×	×	×	8·1	×	×	×	7·7
0·95									×	×	×	—	×	×	×	8·0
1·00													×	×	×	×

Water/cement ratio by weight

— Indicates that the mix was outside the range tested.

× Indicates that the mix would segregate.

These proportions are based on specific gravities of approximately 2·5 for the coarse aggregate and 2·6 for the fine aggregate.

Table 10.9 (cont.)

(c) Angular Aggregate

Degree of workability	Very low				Low				Medium				High			
Grading curve No. on Fig. 3.14	1	2	3	4	1	2	3	4	1	2	3	4	1	2	3	4
0·35	3·2	3·0	2·9	2·7	2·7	2·7	2·5	2·4	2·4	2·4	2·3	2·2	2·2	2·3	2·1	2·1
0·40	4·5	4·2	3·7	3·5	3·5	3·5	3·2	3·0	3·1	3·1	2·9	2·7	2·9	2·9	2·8	2·6
0·45	5·5	5·0	4·6	4·3	4·3	4·2	3·9	3·7	3·7	3·7	3·4	3·3	3·5	3·5	3·2	3·1
0·50	6·5	5·8	5·4	5·0	5·0	4·9	4·5	4·3	4·2	4·2	3·9	3·8	×	3·9	3·8	3·5
0·55	7·2	6·6	6·0	5·6	5·7	5·4	5·0	4·8	4·7	4·7	4·5	4·3	×	×	4·3	4·0
0·60	7·8	7·2	6·6	6·3	6·3	6·0	5·6	5·3	×	5·2	4·9	4·8	×	×	4·7	4·4
0·65	8·3	7·8	7·2	6·9	6·9	6·5	6·1	5·8	×	5·7	5·4	5·2	×	×	5·1	4·9
0·70	8·7	8·3	7·7	7·5	7·4	7·0	6·5	6·3	×	6·2	5·8	5·7	×	×	5·5	5·3
0·75	—	—	8·2	8·0	7·9	7·5	7·0	6·8	×	×	6·2	6·1	×	×	5·8	5·7
0·80	—	—	—	—	—	—	7·4	7·2	×	×	6·6	6·5	×	×	6·1	6·0
0·85					—	—	7·8	7·6	×	×	7·1	6·9	×	×	6·4	6·3
0·90					—	—	—	—	×	×	7·5	7·3	×	×	×	6·7
0·95									×	×	8·0	7·6	×	×	×	7·0
1·00									×	×	—	—	×	×	×	7·3

Water/cement ratio by weight (rows 0·35 to 1·00)

— Indicates that the mix was outside the range tested.

× Indicates that the mix would segregate.

These proportions are based on specific gravity of approximately 2·7 for both coarse and fine aggregate.

Table 10.10: *Aggregate/Cement Ratio (by weight) required to give Four Degrees of Workability with Different Gradings and Shapes of 9·52 mm (⅜ in.) Aggregate*[10.9] *(See Fig. 3.16)*

(a) Rounded Gravel Aggregate

Degree of workability	Very low				Low				Medium				High			
Grading curve No. on Fig. 3.16	1	2	3	4	1	2	3	4	1	2	3	4	1	2	3	4
Water/cement ratio by weight																
0·40	5·6	5·0	4·2	3·2	4·5	3·9	3·3	2·6	3·9	3·5	3·0	2·4	3·5	3·2	2·8	2·3
0·45	7·2	6·4	5·3	4·1	5·5	4·9	4·1	3·2	4·7	4·3	3·7	3·0	4·2	3·9	3·4	2·9
0·50	—	7·8	6·4	4·9	6·5	5·8	4·9	3·8	5·4	5·0	4·3	3·5	4·8	4·5	4·0	3·4
0·55	—	—	7·5	5·7	7·4	6·7	5·7	4·4	6·1	5·7	4·9	4·0	5·3	5·1	4·5	3·9
0·60	—	—	—	6·5	—	7·5	6·4	5·0	6·7	6·3	5·5	4·5	5·8	5·6	5·0	4·3
0·65	—	—	—	7·2	—	—	7·1	5·6	7·3	6·9	6·1	5·0	×	6·1	5·5	4·7
0·70					—	—	7·7	6·2	7·9	7·5	6·7	5·5	×	6·6	6·0	5·1
0·75					—	—	—	6·7	—	—	7·2	5·9	×	7·1	6·5	5·5
0·80					—	—	—	7·2	—	—	7·7	6·3	×	7·6	6·9	5·9
0·85									—	—	—	6·8	×	—	7·3	6·3
0·90									—	—	—	7·2	×	—	7·7	6·7
0·95													×	—	—	7·0
1·00													×	—	—	7·3

— Indicates that the mix was outside the range tested.
× Indicates that the mix would segregate.
These proportions are based on specific gravity of approximately 2·59 for both coarse and fine aggregate.

Table 10.10 (cont.)

(b) Irregular Gravel Aggregate

Degree of workability	Very low				Low				Medium				High			
Grading curve No. on Fig. 3.16	1	2	3	4	1	2	3	4	1	2	3	4	1	2	3	4
0·40	4·1	3·8	3·3	2·8	3·3	3·1	2·8	2·3	—	—	—	—	3·2	—	—	—
0·45	5·1	4·8	4·3	3·6	4·1	3·9	3·5	3·0	3·5	3·4	3·2	2·8	×	3·1	3·0	2·7
0·50	6·1	5·8	5·2	4·4	4·8	4·6	4·2	3·7	4·2	4·1	3·8	3·4	×	3·8	3·6	3·2
0·55	7·0	6·7	6·1	5·2	5·5	5·3	4·9	4·3	×	4·7	4·4	4·0	×	4·4	4·2	3·7
0·60	7·9	7·6	7·0	6·0	×	6·0	5·6	4·9	×	5·3	5·0	4·5	×	4·9	4·7	4·2
0·65	—	—	7·8	6·8	×	6·6	6·2	5·5	×	5·9	5·6	5·0	×	5·4	5·2	4·6
0·70					×	7·2	6·8	6·1	×	6·4	6·1	5·5	×	5·9	5·7	5·0
0·75					×	7·8	7·4	6·7	×	6·9	6·6	6·0	×	6·4	6·1	5·4
0·80					×	—	8·0	7·3	×	7·4	7·1	6·4	×	6·8	6·5	5·8
0·85									×	7·9	7·5	6·8	×	7·2	6·9	6·2
0·90									×	—	8·0	7·2	×	7·6	7·3	6·6
0·95													×	×	7·7	6·9
1·00													×	×	8·0	7·2

Water/cement ratio by weight

— Indicates that the mix was outside the range tested.

× Indicates that the mix would segregate.

These proportions are based on specific gravity of approximately 2·59 for both coarse and fine aggregate.

Table 10.10 (cont.)

(c) Crushed Granite Aggregate*

Water/cement ratio by weight	Very low				Low				Medium				High			
Degree of workability / Grading curve No. on Fig. 3.16	1	2	3	4	1	2	3	4	1	2	3	4	1	2	3	4
0·40	3·7	3·3	2·8	2·0	—	3·6	3·0	2·2	3·3	—	—	—	—	—	—	—
0·45	4·5	4·1	3·5	2·6	3·8	4·2	3·6	2·7	3·8	3·1	2·7	2·1	—	—	—	—
0·50	5·2	4·9	4·2	3·2	4·4	4·8	4·2	3·2	×	3·7	3·2	2·6	×	3·2	2·9	2·4
0·55	5·9	5·6	4·9	3·8	4·9	5·3	4·7	3·7	×	4·2	3·7	3·0	×	3·7	3·4	2·8
0·60	6·6	6·3	5·5	4·3	×	5·8	5·2	4·2	×	4·7	4·2	3·4	×	4·2	3·8	3·2
0·65	7·3	7·0	6·1	4·8	×	6·3	5·7	4·6	×	5·1	4·6	3·8	×	4·6	4·2	3·6
0·70	7·9	7·6	6·7	5·3	×	6·8	6·2	5·0	×	5·6	5·1	4·2	×	5·0	4·6	4·0
0·75	—	—	7·3	5·8	×	7·2	6·6	5·5	×	6·0	5·5	4·6	×	5·4	5·0	4·4
0·80	—	—	7·8	6·3	×	7·6	7·1	6·0	×	6·4	5·9	5·0	×	5·8	5·4	4·7
0·85	—	—	—	6·8	×	—	7·5	6·4	×	6·7	6·3	5·4	×	6·1	5·8	5·1
0·90	—	—	—	7·3	×	—	7·9	6·8	×	7·1	6·7	5·8	×	6·4	6·1	5·4
0·95					×	—	—	7·2	×	7·5	7·1	6·1	×	6·7	6·4	5·7
1·00					×	—	—	—	×	7·8	7·5	6·5	×	7·0	6·7	6·1
1·05									×	—	7·8	6·9	×	7·3	7·0	6·4
1·10										—	—	7·2	×	7·6	7·3	6·7
1·15													×	×	7·6	7·0
1·20													×	×	7·9	7·3

* With crushed aggregate of poorer shape than that tested segregation may occur at a lower aggregate/cement ratio.

— Indicates that the mix was outside the range tested.

× Indicates that the mix would segregate.

These proportions are based on specific gravity of approximately 2·66 for both coarse and fine aggregate.

used with crushed coarse aggregate) and also to grading: a deviation from the type grading curve over the finer range of sizes affects the workability more than the deviation at the coarse end of the scale. Road Note No. 4[10.3] recommends that when there is an excess of particles smaller than 600 μm (No. 25) B.S. sieve the quantity of material passing the 4·76 mm ($\frac{3}{16}$ in.) sieve should be reduced by an amount up to 10 per cent of the total aggregate. On the other hand, when there is an excess of particles in the 1·20 mm–4·76 mm (No. 14—$\frac{3}{16}$ in.) sieve range, the quantity of fine aggregate should be increased. However, fine aggregate with a large excess of particles between 1·20 mm (No. 14) and 4·76 mm ($\frac{3}{16}$ in.) sieves produces a harsh mix and may require a higher cement content for a satisfactory workability.

One difference between the data of Table 10.8 and 10.9 on the one hand, and those of Table 10.10 on the other, should be mentioned: the latter are based on the *total* water added to air-dry aggregate, while the former are based on the effective water, i.e. water in excess of that absorbed by the aggregate. For this reason, when applying the tabulated data to laboratory mixes an appropriate correction for absorption should be made.

It should also be noted that the relation between aggregate/cement ratio (by weight) and workability, given in Tables 10.8, 10.9, and 10.10 applies to aggregates of the given specific gravity, ρ, only (indicated at the bottom of each Table). The relation depends in fact on the gross apparent volume of the solid particles. Thus when the aggregate used has a specific gravity, ρ_1, the aggregate/cement ratio from the Tables should by multiplied by ρ_1/ρ; any variation in the specific gravity of cement can be ignored.

Mix Proportions and Weights per Batch

Knowing the water/cement ratio and the aggregate/cement ratio we now have no difficulty in determining the proportions of cement, water, and aggregate. In practice, the aggregate is supplied from at least two stockpiles, and the quantities of aggregate of each size have to be given separately. This presents no difficulty as in finding a suitable grading we already had to calculate the proportions of the different size fractions of aggregate. The details of calculation are given in the example on page 594.

For practical purposes the mix quantities are given in kilogrammes or pounds per batch. If cement is supplied in bulk we choose the batch quantities so that their sum is equal to the capacity of the mixer. When cement is supplied in bags, and there is no provision for weighing it, it is preferable to choose the batch quantities so that the weight of cement per batch is one bag or its multiple. The weight of cement is then known accurately. In exceptional cases a half-bag can be used, but other fractions are impossible to determine with any accuracy and should never be used.

A standard bag in Britain and many other countries contains 50 kg (112 lb) of cement, a U.S. sack is 94 lb, a Canadian sack is 80 lb, and some other weights are used elsewhere but 40 and 50 kg are becoming most common.

In some cases a mixer may thus be used at only a fraction of its capacity and if possible this should be avoided either by using a different mixer size or different mix proportions. The majority of mixers, however, are of such size that they can operate near their capacity, using whole bags, when the more common ("traditional") aggregate/cement ratios are used.

Method of Calculation by Absolute Volume

The procedure so far described leads to the determination of values of the water/cement ratio, aggregate/cement ratio, and also the relative proportions of the aggregates of various sizes, but does not give the volume of fully compacted concrete produced by these materials. This volume is obtained by a simple calculation, using the so-called absolute volume method, which assumes that the volume of compacted concrete is equal to the sum of the absolute volumes of all ingredients.

It is usual to calculate the quantities of ingredients to produce 1 cubic metre or cubic yard of concrete. Then, if W, C, A_1, and A_2 are the required weights of water, cement, fine aggregate, and coarse aggregate respectively, we have for the cubic yard

$$\frac{W}{62 \cdot 4} + \frac{C}{62 \cdot 4\rho_c} + \frac{A_1}{62 \cdot 4\rho_1} + \frac{A_2}{62 \cdot 4\rho_2} = 27$$

where ρ with the appropriate suffix represents the specific gravity of each material. Since the density of water (62·4) is expressed in pounds per cubic foot, the total volume (1 cubic yard) has also to be expressed in cubic feet (27). For the cubic metre, the corresponding equation is

$$\frac{W}{1,000} + \frac{C}{1,000\rho_c} + \frac{A_1}{1,000\rho_1} + \frac{A_2}{1,000\rho_2} = 1$$

The mix design calculations give the values of W/C, $C/(A_1 + A_2)$, and A_1/A_2, whence the values of W, C, A_1, and A_2 can be found.

When an additional ingredient, such as pozzolana, is present, or when the coarse or fine aggregate is in more than one stockpile, additional terms of similar form are added to the equation. When entrained air is present, and its percentage is, say, a per cent of the volume of concrete, the right-hand side of the "cubic yard" equation would read:

$$27 \left(1 - \frac{a}{100}\right).$$

For the "cubic metre" equation, 27 is replaced by 1.

C represents the cement content in kilogrammes per cubic metre or pounds per cubic yard of the concrete,* and W the water content in the same units; the latter must not be confused with the water/cement ratio.

In comparing various mixes, it is sometimes convenient to convert rapidly the aggregate/cement ratio into the cement content or vice versa: Fig. 10.6 makes such a conversion very easy.

If the aggregate contains free moisture whose weight is, say, m per cent of the weight of the dry aggregate then the weights of the *added* water W and of (wet) aggregate must be adjusted. The weight of free water in A lb of aggregate is x such that

$$\frac{m}{100} = \frac{x}{A - x}$$

whence,

$$x = A \frac{m}{100 + m}$$

This weight is added to A to give the weight of wet aggregate per batch, $A [1 + m/(100 + m)]$, and is subtracted from W to give the weight of added water, $W - Am/(100 + m)$.

Generally, each size fraction of aggregate has a different moisture content, and the correction should be applied to A_1, A_2, etc., with an appropriate value of m.

In the manufacture of concrete of low or medium strength the determination of the moisture content of aggregate can be dispensed with if the grading of aggregate is reasonably constant and weigh-batching is used. Under those circumstances a change in workability caused by a variation in the moisture content of aggregate can be prevented by an experienced mixer operator who can adjust the amount of added water so that the workability, as judged by eye, remains constant. The water/cement ratio remains then also sensibly constant.

In the case of volume batching no correction for moisture content need be made in the case of coarse aggregate, but the bulking of fine aggregate must be allowed for (*see* p. 130). The quantity of added water must be adjusted by the mixer operator as in the case of weigh-batching.

Combining Aggregates to Obtain a Type Grading

While there are no ideal gradings—a point repeatedly stressed—it may be desirable to proportion the available materials in such a way that the grading of the combined aggregate is similar to one of the type curves of Figs. 3.14 to 3.16. This can be done by calculation, or graphically using the method of Road Note No. 4.[10.3] Both procedures are best illustrated by means of examples.

---

* In the U.S. the cement content is sometimes expressed in sacks of cement per cubic yard of concrete, and is referred to as *cement factor*.

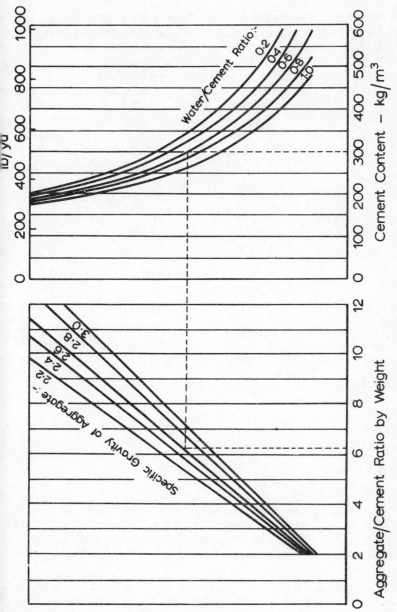

Fig. 10.6. *Conversion chart for aggregate/cement ratio and cement content (courtesy of Cement and Concrete Association)*

Suppose the gradings of the fine aggregate and the two coarse aggregate size fractions are as listed in Table 10.11, and we are to combine the materials so as to approximate to the coarsest grading of Fig. 3.15 (curve 1).* On this curve 24 per cent of the total aggregate passes the $\frac{3}{16}$ in. B.S. sieve, and 50 per cent passes the $\frac{3}{4}$ in. sieve.

Table 10.11: *Example of Combining Aggregates to Obtain a Type Grading*

B.S. sieve size		Cumulative percentage passing for							Grading of combined aggregate (7): 4.53
		Fine aggregate	$\frac{3}{4}-\frac{3}{16}$ in.	$1\frac{1}{2}-\frac{3}{4}$ in.	(1) × 1	(2) × 0.94	(3) × 2.59	(4) + (5) + (6)	
mm or μm	in. or No.	(1)	(2)	(3)	(4)	(5)	(6)	(7)	(8)
38.1	$1\frac{1}{2}$	100	100	100	100	94	259	453	100
19.05	$\frac{3}{4}$	100	99	13	100	93	34	227	50
9.52	$\frac{3}{8}$	100	33	8	100	31	21	152	34
4.76	$\frac{3}{16}$	99	5	2	99	5	5	109	24
2.40	7	76	0	0	76	0	0	76	17
1.20	14	58			58			58	13
600	25	40			40			40	9
300	52	12			12			12	3
150	100	2			2			2	$\frac{1}{2}$

Let x, y, z, be the proportions of fine, $\frac{3}{4}-\frac{3}{16}$ in., and $1\frac{1}{2}-\frac{3}{4}$ in. aggregates. Then to satisfy the condition that 50 per cent of the combined aggregate passes the $\frac{3}{4}$ B.S. sieve, we have

$$1\cdot0x + 0\cdot99y + 0\cdot13z = 0\cdot5\,(x + y + z).$$

The condition that 24 per cent of the combined aggregate passes the $\frac{3}{16}$ in. B.S. sieve can be written—

$$0\cdot99x + 0\cdot05y + 0\cdot02z = 0\cdot24\,(x + y + z).$$

From these two equations we find

$$x:y:z = 1:0\cdot94:2\cdot59,$$

i.e. the three aggregates are combined in the proportions $1:0\cdot94:2\cdot59$.

To find the grading of the combined aggregate we multiply columns (1), (2) and (3) of Table 10.11 by 1, 0·94, and 2·59 respectively, the products being shown in columns (4), (5) and (6). We now add these three

---

* The gradings are based on volumetric proportions and thus apply to aggregates all of the same specific gravity. If the specific gravities of the different size fractions differ appreciably from one another, the proportions required should be adjusted accordingly. Cf. design of lightweight aggregate mixes where lightweight coarse aggregate and natural sand are used.

columns (column 7) and divide the sum by $1 + 0.94 + 2.59 = 4.53$. The result, given in column (8), is the grading of the combined aggregate. The grading is given to the nearest per cent, as owing to the variability of the materials any higher apparent accuracy has no meaning.

Fig. 10.7 shows the grading of the combined aggregate, together with the type curve of Road Note No 4.[10.3] Deviations are apparent, and indeed unavoidable, as agreement with the type curve is generally possible only at specified points. The grading of our aggregate is slightly finer than that of the type aggregate, so that for any desired workability a slightly richer mix than indicated in Table 10.8 might be used.

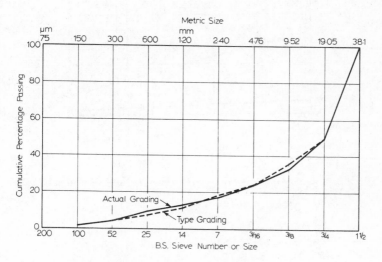

Fig. 10.7. *Grading of the aggregate for the example of Table 10.11*

The graphical method is shown in Fig. 10.8. The two coarse aggregates are combined first, using the percentage passing the $\frac{3}{4}$ in. sieve as a criterion. Percentage passing is marked along three sides of a square. The values for the two coarse aggregates are entered on two opposite sides, and the points corresponding to the same sieve size are joined by straight lines. A vertical line is now drawn through the point where the line joining the $\frac{3}{4}$ in. values intersects the horizontal line representing the correct percentage of aggregate smaller than $\frac{3}{4}$ in. In our case $(50 - 24)$ = 26 parts of aggregate coarser than $\frac{3}{16}$ in. are to pass the $\frac{3}{4}$ in. sieve while 50 parts are to be retained. The ratio is thus $26:(50 + 26)$, or 34 per cent of *all* coarse aggregate. A horizontal line is therefore drawn

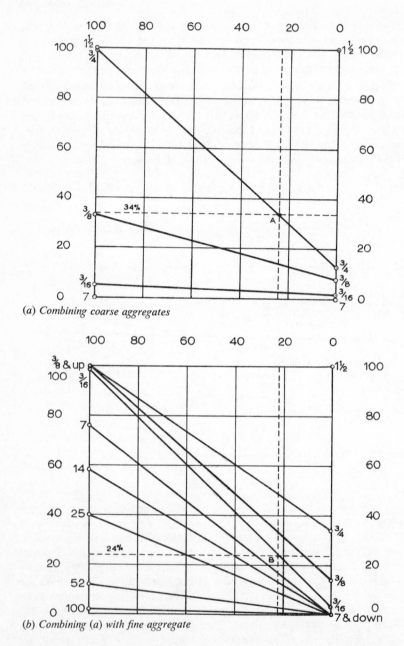

(a) Combining coarse aggregates

(b) Combining (a) with fine aggregate

Fig. 10.8. *Graphical method of combining aggregates* (*example of Table 10.11*)

through the 34 per cent point to intersect the $\frac{3}{4}$ in. line at A. A vertical through A gives the quantity of material $\frac{3}{4}-\frac{3}{16}$ in. as a percentage of the total coarse aggregate. In Fig. 10.8(a) this value is 24 per cent. The vertical line gives also the grading of the combined coarse aggregate, and this is combined with the fine aggregate in a similar manner to that already described (Fig. 10.8(b)). We find that 22 parts of fine aggregate are to be combined with 78 parts of aggregate coarser than $\frac{3}{16}$ in. sieve. The aggregate is thus to be proportioned as 22 : (24/100) × 78 : (76/100) × 78, or 1 : 0·85 : 1·69. The vertical line through B (Fig. 10.8(b)) gives the combined grading of aggregate obtained by proportioning the three aggregates in the ratio 1 : 0·85 : 2·69. This agrees with the grading obtained earlier by calculation but both methods are approximations based on quantities passing two specific sieve sizes. A method of combining four aggregates has also been developed.

Further examples of the graphical method are given in Road Note No. 4.[10.3] It is possible to draw (in a figure of the type of Fig. 10.8(b)) envelopes of standard gradings: since any vertical line represents a possible grading it is immediately apparent whether or not a grading within the envelope can be obtained; the range of proportions is then given by a point similar to B, corresponding to any chosen vertical line.

Simple Example of Mix Design

A mix with a minimum compressive strength of 4,200 lb/in² at 28 days is required for use in a road slab. Compaction will be effected by vibration, and good control is anticipated. Ordinary Portland cement and irregular aggregate with a grading of that in the example on page 594 will be used.

As no experimental data on the difference between the minimum and the mean strengths are available, the ratio of the two is taken from Table 10.2 as 0·75. Thus the mean strength is 4,200/0·75 = 5,600 lb/in². Since it is not normally possible to determine the strength properties of the cement to be used, recourse must be made to Fig. 10.2, from which the water/cement ratio is found to be 0·48. This water/cement ratio is suitable from the standpoint of durability.

Table 4.2 tells us that a mix of very low workability is likely to be required. For this workability and a water/cement ratio of 0·48 Table 10.8 shows an aggregate/cement ratio of about 7·2.

On page 594 we found that the fine, $\frac{3}{4}-\frac{3}{16}$ in., and $1\frac{1}{2}-\frac{3}{4}$ in. aggregates are proportioned in the ratio 1 : 0·94 : 2·59. Since the aggregate/cement ratio is 7·2, the mix proportions become: 1 part of cement to (1 : 0·94 : 2·59) × 7·2/4.53 for the three aggregates respectively, i.e. a 1 : 1·59 : 1·50 : 4·11 mix with a water/cement ratio of 0·48.

Assuming the specific gravity of cement to be 3·15, that of coarse aggregate 2·50, and of fine aggregate 2·60, we can find the cement content, C, in pounds per cubic yard from the expression—

$$\frac{0 \cdot 48C}{62 \cdot 4} + \frac{C}{3 \cdot 15 \times 62 \cdot 4} + \frac{1 \cdot 59C}{2 \cdot 60 \times 62 \cdot 4} + \frac{(1 \cdot 50 + 4 \cdot 11)C}{2 \cdot 50 \times 62 \cdot 4} = 27$$

Hence, $C = 462 \text{ lb/yd}^3$,

and the weights of the ingredients per cubic yard of concrete are—

Cement	=	462 lb
Water = 0·48 × 462	=	222 lb
Fine aggregate = 1·59 × 462	=	735 lb
$\frac{3}{4}-\frac{3}{16}$ in. aggregate = 1·50 × 462 =		693 lb
$1\frac{1}{2}-\frac{3}{4}$ in. aggregate = 4·11 × 462 =		1,900 lb

Total = 4,012 lb

Hence, the density of fresh concrete is $4{,}012/27 = 149 \text{ lb/ft}^3$.

In the United States the yield of concrete in cubic feet per sack of cement (94 lb) is sometimes quoted. For the mix just designed the yield is $\dfrac{27}{462/94} = 5 \cdot 5 \text{ ft}^3/\text{sack}$.

A trial mix should now be made, and the proportions adjusted as necessary. It is important to remember that if workability is to be changed, but the strength is to remain unaffected, the water/cement ratio must remain unaltered. Changes can be made in the aggregate/cement ratio or, if suitable aggregates are available, in the grading of the aggregate; the influence of grading on workability was discussed in Chapter 3. A study of Tables 10.8 to 10.10 will show the variation in workability caused by a change in grading or in aggregate/cement ratio.

Conversely, changes in strength but not in workability are made by varying the water/cement ratio with the water *content* of the mix unaltered. This means that a change in the water/cement ratio must be accompanied by a change in the aggregate/cement ratio so that the weight ratio

$$\frac{\text{water}}{\text{water} + \text{cement} + \text{aggregate}}$$

is approximately constant.

It will be remembered that the values of Tables 10.8 to 10.10 refer to mixes in which the coarse and the fine aggregate are of the same shape. In many cases natural sand is used with crushed coarse aggregate so that interpolation between the values of the relevant Tables is necessary. Let us suppose that in our example a well-rounded sand is to be used with irregular coarse aggregate. We would have to interpolate between aggregate/cement ratio of 7·2 for irregular coarse and fine aggregate and aggregate/cement ratio of x for rounded coarse and fine aggregate.

There is no table giving the value of x, but it can be obtained by a further interpolation; from Tables 10.8 and 10.9 we have the following data—

Maximum aggregate size	Aggregate shape	Aggregate/cement ratio for water/cement ratio of 0·48
¾ in.	rounded	8·7 approx.
¾ in.	irregular	6·7
1½ in.	rounded	x
1½ in.	irregular	7·2

Hence, $6·7/7·2 = 8·7/x$ or $x = 9·2$.

Interpolation between x and 7·2 would yield a value of about 8·4, since the influence of the shape of sand is greater than that of the shape of the coarse aggregate. It is clear that the value of 8·4 is no more than a guide for a trial mix.

Other examples of interpolation are given in Road Note No. 4.[10.3]

Design of High-strength Mixes
The properties of high-strength concrete (say, with a 28-day compressive strength above 40 or 50 MN/m² (6,000 or 7,000 lb/in²)) depend on factors additional to those considered in Fig. 10.1, so that the procedure of mix design outlined on page 597 must be modified. We are considering, of course, concrete whose high-strength is achieved by suitable proportioning and not by steam curing or application of pressure (*see* p. 27); the highest strengths recorded are in the vicinity of 110 MN/m² (16,000 lb/in²).[10.11]

The main difference lies in the fact that the workability of the mix and the type and maximum size of aggregate (assumed to have a sufficiently high ceiling of strength), as well as the strength requirement, influence the selection of the water/cement ratio.[10.10] The influence of richness of the mix on the strength of high-strength concrete was mentioned on page 253.

It follows that, in addition to the type of coarse aggregate, either the aggregate/cement ratio or the workability has to be known in order to choose the water/cement ratio for a required strength. Because of this, Erntroy and Shacklock[10.10] have prepared empirical graphs relating compressive strength to an arbitrary "reference number" for concretes made with irregular gravel and crushed granite coarse aggregates. The graphs for the two aggregates respectively are reproduced in Figs. 10.9

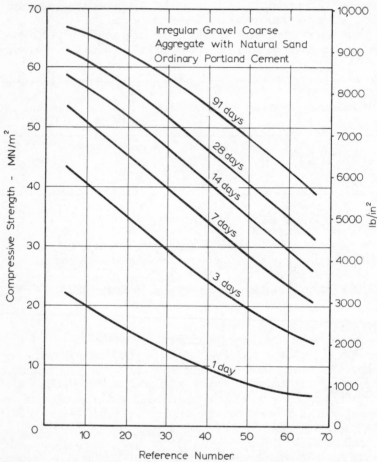

Fig. 10.9. *Relation between compressive strength and "reference number" for mixes containing irregular gravel coarse aggregate, natural sand, and ordinary Portland cement*[10.10]

and 10.10 for mixes with ordinary Portland cement, and in Figs. 10.11 and 10.12 for mixes with rapid hardening Portland cement.

Having obtained the reference number for the desired strength, the water/cement ratio to give the required workability is found with the aid of Figs. 10.13 and 10.14 for aggregates with a maximum size of 19·05 mm (¾ in.) and 9·52 mm (⅜ in.) respectively. The various degrees of workability are defined in Table 10.7; this table is similar to Table 4.2, but an "extremely low" workability, often used in the manufacture of precast

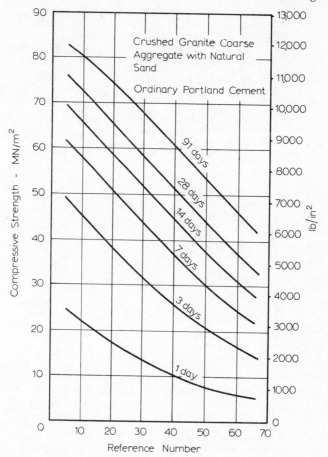

Fig. 10.10. *Relation between compressive strength and "reference number" for mixes containing crushed granite coarse aggregate, natural sand, and ordinary Portland cement*[10.10]

prestressed concrete, has been added,* and values of the compacting factor for 9·5 mm ($\frac{3}{8}$ in.) maximum size aggregate have also been included.

The aggregate/cement ratio can now be found from Tables 10.12 and 10.13. The values given in these tables were obtained with aggregates containing 30 per cent of material passing the 4·76 mm ($\frac{3}{16}$ in.) B.S. sieve, and for other gradings suitable adjustments must be made.

---

* Concrete with 19·05 mm ($\frac{3}{4}$ in.) aggregate and extremely low workability has a compacting factor of 0·68.

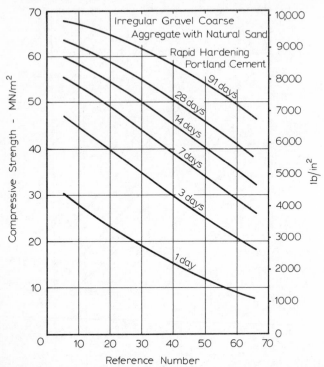

Fig. 10.11. *Relation between compressive strength and "reference number" for mixes containing irregular gravel coarse aggregate, natural sand, and rapid hardening Portland cement*[10.10]

Owing to a considerable variation in the properties of coarse aggregates, the data given in this section are even less precise than those of Road Note No. 4,[10.3] and trial mixes must be made in the preliminary stages of mix design.

Example
We require a mix of very low workability with a mean 28-day compressive strength of 59 MN/m² (8,500 lb/in²), using ordinary Portland cement and 19·05 mm (¾ in.) aggregate, either irregular gravel or crushed granite.

The table following gives the reference numbers from Figs. 10.9 and 10.10, and the corresponding values of water/cement ratio from Fig. 10.13. Table 10.12 gives then the aggregate/cement ratio, for each mix. Given the water/cement and aggregate/cement ratios and the specific gravities of the materials (say, 2·6 for all aggregates), the cement content of the mix can be calculated, using the formula of page 591.

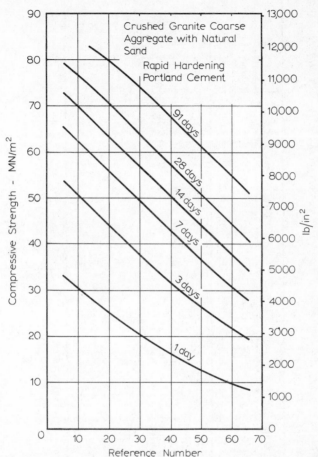

Fig. 10.12. *Relation between compressive strength and "reference number" for mixes containing crushed granite coarse aggregate, natural sand, and rapid hardening Portland cement*[10.10]

Type of coarse aggregate	Reference number	Water/cement ratio	Aggregate/cement ratio	Cement content	
				kg/m³	lb/yd³
Irregular gravel	21	0·35	3·2	526	886
Crushed granite	37	0·43	5·1	370	623

The use of crushed granite leads to a leaner mix, but the choice of aggregate would depend of course on its availability and cost. Trial

Table 10.12: *Aggregate/Cement Ratio (by weight) required to give Four Degrees of Workability with Different Water/Cement Ratios using Ordinary Portland Cement*[10.10]

Type of coarse aggregate*	Irregular gravel								Crushed granite							
Maximum size of aggregate	19·05 mm (¾ in.)				9·52 mm (⅜ in.)				19·05 mm (¾ in.)				9·52 mm (⅜ in.)			
Degree of workability†	EL	VL	L	M	EL	VL	L	M	EL	VL	L	M	EL	VL	L	M
0·30	3·0	—	—	—	2·4	—	—	—	3·3	—	—	—	2·9	—	—	—
0·32	3·8	2·5	—	—	3·2	—	—	—	4·0	2·6	—	—	3·6	2·3	—	—
0·34	4·5	3·0	2·5	—	3·9	2·6	—	—	4·6	3·2	2·6	—	4·2	2·8	2·3	2·3
0·36	5·2	3·5	3·0	2·5	4·6	3·1	2·6	—	5·2	3·6	3·1	2·6	4·7	3·2	2·7	2·6
0·38	—	4·0	3·4	2·9	5·2	3·5	3·0	2·5	—	4·1	3·5	2·9	5·2	3·6	3·0	2·9
0·40	—	4·4	3·8	3·2	—	3·9	3·3	2·7	—	4·5	3·8	3·2	—	4·0	3·3	2·9
0·42	—	4·9	4·1	3·5	—	4·3	3·6	3·0	—	4·9	4·2	3·5	—	4·4	3·6	3·1
0·44	—	5·3	4·5	3·8	—	4·7	3·9	3·3	—	5·3	4·5	3·7	—	4·8	3·9	3·3
0·46	—	—	4·8	4·1	—	5·1	4·2	3·6	—	—	4·8	4·0	—	5·1	4·2	3·6
0·48	—	—	5·2	4·4	—	5·4	4·5	3·8	—	—	5·1	4·2	—	5·5	4·5	3·8
0·50	—	—	5·5	4·7	—	—	4·8	4·1	—	—	5·4	4·5	—	—	4·7	4·0

Water/cement ratio by weight

* Natural sand used in combination with both types of coarse aggregate.

†EL = "Extremely Low" ⎱ as defined in
 VL = "Very Low" ⎰ Table 10.7
 L = "Low"
 M = "Medium"

Table 10.13: Aggregate/Cement Ratio (by weight) required to give Four Degrees of Workability with Different Water/Cement Ratios using Rapid-hardening Portland Cement[10.10]

Type of coarse aggregate*	Irregular gravel								Crushed granite							
Maximum size of aggregate	19·05 mm (¾ in.)				9·52 mm (⅜ in.)				19·05 mm (¾ in.)				9·52 mm (⅜ in.)			
Degree of workability†	EL	VL	L	M	EL	VL	L	M	EL	VL	L	M	EL	VL	L	M
Water/cement ratio by weight 0·32	2·6	—	—	—	—	—	—	—	2·9	—	—	—	2·5	—	—	—
0·34	3·4	2·2	—	—	2·8	—	—	—	3·6	2·4	—	—	3·2	—	—	—
0·36	4·1	2·7	2·3	—	3·5	2·4	—	—	4·3	2·9	2·4	—	3·9	2·5	—	—
0·38	4·8	3·2	2·8	2·3	4·2	2·9	2·4	—	4·9	3·4	2·9	2·4	4·5	3·0	2·5	—
0·40	5·5	3·7	3·2	2·7	4·9	3·3	2·8	2·3	5·5	3·9	3·3	2·7	5·0	3·4	2·9	2·4
0·42	—	4·2	3·6	3·0	—	3·7	3·1	2·6	—	4·2	3·6	3·0	5·5	3·8	3·2	2·7
0·44	—	4·6	4·0	3·4	—	4·1	3·5	2·9	—	4·7	4·0	3·3	—	4·2	3·5	3·0
0·46	—	5·0	4·3	3·7	—	4·5	3·8	3·2	—	5·1	4·3	3·6	—	4·6	3·8	3·2
0·48	—	5·5	4·7	4·0	—	4·9	4·1	3·5	—	5·5	4·6	3·9	—	5·0	4·1	3·4
0·50	—	—	5·0	4·3	—	5·2	4·4	3·7	—	—	4·9	4·1	—	5·3	4·4	3·7

* Natural sand used in combination with both types of coarse aggregate.

† EL = "Extremely Low"
VL = "Very Low" } as defined in
L = "Low"   Table 10·7
M = "Medium"

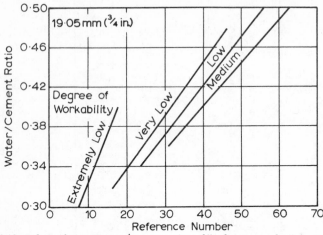

Fig. 10.13. *Relation between water/cement ratio and "reference number" for 19·05 mm (¾ in.) maximum size aggregate*[10.10]

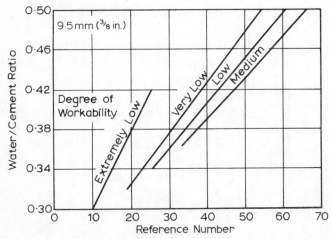

Fig. 10.14. *Relation between water/cement ratio and "reference number" and for 9·5 mm (⅜ in.) maximum size aggregate*[10.10]

mixes should now be made and followed by adjustments in grading and mix proportions as required.

Very High Strength Concrete
Very high strength concrete, well above the range considered so far, can be obtained by using specialized techniques but no mix design procedures

as such exist. Compacts of cement powder with strengths up to 375 MN/m² (54,500 lb/in²) were mentioned on page 27. Strengths up to 140 MN/m² (20,000 lb/in²) are practicable using commercial techniques of combined pressure and vibration. Use of cementitious aggregate also raises the strength of concrete by providing a better bond strength between the cement paste and the aggregate, but the cost is of course high.

Strengths up to 97 MN/m² (14,000 lb/in²) can be obtained more easily. High speed slurry mixing (*see* p. 204) is one means available; the extra cost involved is modest. Electromagnetic vibrators operated at frequencies up to 5,000 Hz may compact the cement and not only the aggregate to a greater density. But even conventional techniques and materials can produce such strengths: Parrott[10.20] used a 1:2 mix with 90 per cent of 9·52 to 4·76 mm ($\frac{3}{8}$ to $\frac{3}{16}$ in.) aggregate, 10 per cent fine aggregate, a rapid hardening Portland cement, and a water/cement ratio of 0·28 (*see* Fig. 10.15). It may be noted that the large difference between the strength of such a mix and a more conventional one is apparent already at one day; the later rate of gain of strength is not much higher. The tensile strength of such mixes is also greater but there appears to be a ceiling of about 6 MN/m² (900 lb/in²).

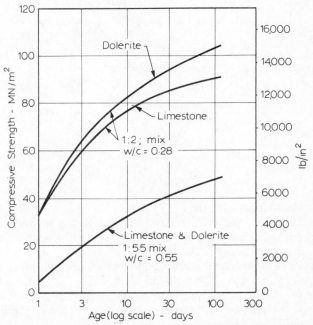

Fig. 10.15. *Relation between strength and age for very high strength and conventional concretes made with two aggregates*[10.20]

The use of very high strength concrete requires care in structural design because the strains involved are large. Also, the relation between the modulus of elasticity and strength is not the same in that the increase in strength is not accompanied by an increase in the modulus (*see* Fig. 10.16). Thus, a given stress/strength ratio would produce a higher deformation in very high strength concrete members than when conventional mixes are used.

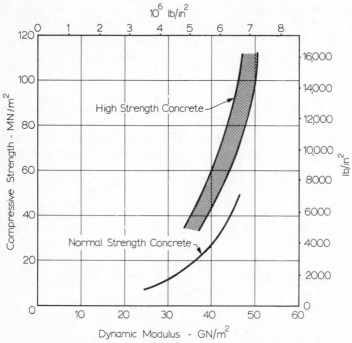

Fig. 10.16. *Relation between strength and the dynamic modulus of elasticity for very high strength concrete*[10.20]

Design for Strength in Flexure

In the discussion and calculations of the preceding sections we have considered the compressive strength of concrete, and this is indeed the usual criterion of strength. However, in some types of construction, such as roads or airport runways, mix design on the basis of flexural strength is frequently preferred. The procedure is basically the same, except that the shape of the coarse aggregate particles affects the relation between the water/cement ratio and the flexural strength in low-strength as well as in high-strength mixes.

Since with a given water/cement ratio angular aggregate produces a higher flexural strength than rounded or irregular aggregate, it follows that when designing a mix for a specified flexural strength a lower water/cement ratio has to be used with rounded or irregular than with angular aggregate. The use of a lower water/cement ratio means that a lower aggregate/cement ratio must be used, i.e. a richer mix is required. In other words, for the same flexural strength rounded aggregate requires a higher cement content. This is so despite the fact that for a given water/cement ratio, in order to produce the same workability, angular aggregate requires a richer mix than when rounded aggregate is used. The latter effect is dominant when compressive strength is the criterion, as the shape of aggregate *per se* does not influence the compressive strength (except at high strengths), and only the effects of water requirement are apparent.

These effects are clearly shown in a numerical example, based on Wright's[10.12] data. Suppose a concrete with a minimum modulus of rupture of 500 lb/in^2 at 28 days is required. The workability is to be "very low" and the aggregate is of $\frac{3}{4}$ in. maximum size. Irregular gravel or crushed rock angular aggregate may be used; the relation between the modulus of rupture and the water/cement ratio for concretes made with these two aggregates is shown in Fig. 10.17, based on Wright's[10.12] results whose approach was elaborated in a later paper.[10.21]

If the quality control of operations is "very good," the required minimum strength is given by Table 10.2 as 75 per cent of the mean strength. Thus the mean strength required is $500/0 \cdot 75 = 667$ lb/in^2. Fig. 10.17 shows that this strength can be achieved with values of water/cement ratio of 0·41 for the irregular aggregate, and 0·58 for the angular aggregate.

To find the aggregate/cement ratio that would produce a very low workability for the given values of water/cement ratio and the relevant types of aggregate we consult Table 10.9. Let us assume that the aggregate grading corresponds to curve No. 1 of Fig. 10.12. The aggregate/cement ratios required are 5·0 for the irregular aggregate at a water/cement ratio of 0·41, and 7·6 for the angular aggregate at a water/cement ratio of 0·58. It is clear that the use of angular aggregate leads to a leaner mix.

By way of contrast let us briefly calculate the proportions on the basis of a minimum compressive strength of 4,100 lb/in^2, other factors remaining unaltered. The mean strength is $4,100/0 \cdot 75 = 5,500$ lb/in^2 at 28 days, and using rapid hardening Portland cement, this corresponds to a water/cement ratio of 0·55 for any type of aggregate, as the compressive strength is sensibly independent of the shape of the aggregate particles. Table 10.9 gives the aggregate/cement ratio of 6·5 for irregular and 7·2 for angular aggregate, so that from the standpoint of compressive strength the latter requires a richer mix—a fact already mentioned.

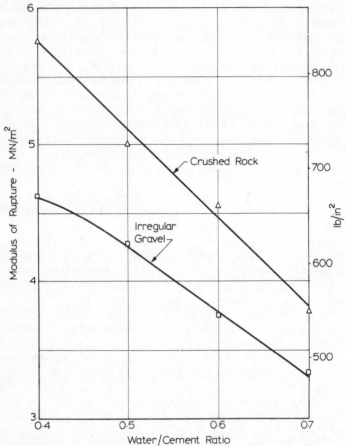

Fig. 10.17. *Relation between modulus of rupture at 28 days and water/cement ratio*[10.12]

Design of Mixes Containing Entrained Air

When entrained air is to be included in concrete the procedure outlined earlier needs modification to allow for the effect of entrained air on strength and workability.

If the strength of air-entrained concrete is to be the same as that of a non-air-entrained mix, it is necessary to base the initial mix design on a strength higher than actually desired. The loss of strength due to voids is about 5·5 per cent for each per cent of entrained air but, since the addition of air permits a lowering of the water content, it is not necessary to allow for the full loss of strength corresponding to the volume of entrained air. The actual value of the excess strength to be added varies with the richness of the mix but is usually about 10 per cent.[10.13] The water/cement ratio corresponding to the increased strength is now found in the usual manner and so are the proportions of the other ingredients.

These mix proportions are converted to absolute volumes and the volume occupied by each ingredient is expressed as a percentage of the total volume of the concrete, air being ignored at this stage. Since, however, entrained air occupies a part of the total volume of concrete, the volume of aggregate can be decreased by the volume of the entrained air. Wright,[10.14] who suggested this method of mix design, recommends that about 1 per cent of the total percentage volume of air be subtracted from the volume of the coarse aggregate, and the remainder from the volume of the fine aggregate, as air behaves like fine particles of aggregate.

Table 10.14: *Values of Wright's "W" for 19·05 mm ($\frac{3}{4}$ in.) Maximum Size Aggregate*[10.14]

Aggregate/ cement ratio	"W"—per cent		
	Rounded gravel aggregate	Irregular gravel aggregate	Crushed rock aggregate
6	0·325	0·375	0·425
7·5	0·40	0·45	0·50
9	0·45	0·50	0·55

But the addition of entrained air also affects the water requirement of the mix; the necessary correction can be made by reference to Fig. 4.2, or for each per cent of air the percentage volume of water can be reduced by a quantity *W*. Experimental values of *W*, determined by Wright,[10.14] are given in Table 10.14. Since, however, the total volume of concrete is to be kept constant, the reduction in the volume of water must be compensated for by a corresponding increase in the volume of aggregate. This is best done in such a manner that the ratio of fine to coarse aggregate remains constant.

The adjusted volumes of the mix ingredients are now converted back to weights and expressed as ratios in the usual manner.

There are other methods of proportioning air-entrained concrete with a more direct approach, but they require data on the workability of different mixes with varying amounts of entrained air (rather similar to Tables 10.8 and 10.9 for non-air-entrained concrete). No comprehensive data of this type are available, so that mix design of air-entrained concrete is largely based on experience.

American Method of Mix Design

The ACI 211.1–70 Standard of the American Concrete Institute[10.2] recommends a method of mix design which, because of its widespread use, deserves detailed consideration.

Table 10.15: *Approximate Mixing-water Requirements for Different Slumps and Maximum Sizes of Aggregates*[10.2]

Slump in.	Water requirement (lb per yd³ of concrete) for maximum size of aggregate of—							
	$\frac{3}{8}$ in.	$\frac{1}{2}$ in.	$\frac{3}{4}$ in.	1 in.	$1\frac{1}{2}$ in.	2 in.	3 in.	6 in.
	NON-AIR-ENTRAINED CONCRETE							
1 to 2	350	335	315	300	275	260	240	210
3 to 4	385	365	340	325	300	285	265	230
6 to 7	410	385	360	340	315	300	288	—
Approximate entrapped air content, per cent	3	2·5	2	1·5	1	0·5	0·3	0·2
	AIR-ENTRAINED CONCRETE							
1 to 2	305	295	280	270	250	240	225	200
3 to 4	340	325	305	295	275	265	250	220
6 to 7	365	345	325	310	290	280	270	—
Recommended total air content, per cent	8	7	6	5	4·5	4	3·5	3

The quantities given are maxima for reasonably well-shaped angular coarse aggregate.

Note: The slump values for concrete containing aggregate larger than $1\frac{1}{2}$ in. are based on slump tests made after wet screening of particles larger than $1\frac{1}{2}$ in.

Table 10.15: *Metric*

Slump mm	Water (kg per m³ of concrete) for maximum size of aggregate of—							
	10 mm	12·5 mm	20 mm	25 mm	40 mm	50 mm	70 mm	150 mm
	NON-AIR-ENTRAINED CONCRETE							
30 to 50	205	200	185	180	160	155	145	125
80 to 100	225	215	200	195	175	170	160	140
150 to 180	240	230	210	205	185	180	170	—
Approximate entrapped air content, per cent	3	2·5	2	1·5	1	0·5	0·3	0·2

Table 10.15: *Metric (cont.)*

Slump mm	Water (kg per m³ of concrete) for maximum size of aggregate of—							
	10 mm	12·5 mm	20 mm	25 mm	40 mm	50 mm	70 mm	150 mm
	AIR-ENTRAINED CONCRETE							
30 to 50	180	175	165	160	145	140	135	120
80 to 100	200	190	180	175	160	155	150	135
150 to 180	215	205	190	185	170	165	160	—
Recommended total air content, per cent	8	7	6	5	4·5	4	3·5	3

The quantities given are maxima for reasonably well-shaped angular coarse aggregate.

Note: The slump values for concrete containing aggregate larger than 40 mm are based on slump tests made after wet screening of particles larger than 40 mm.

The ACI method utilizes the fact that, for a given maximum size of aggregate, the water content in pounds per cubic yard determines the workability of the mix, largely independently of the mix proportions. It is thus possible to start the mix design by selecting the water content from Table 10.15, although there would be some variations due to differences in aggregate shape and texture.

A further assumption is made that the optimum ratio of the bulk volume of coarse aggregate to the total volume of concrete depends only on the maximum size of aggregate and on the grading of fine aggregate. The shape of the coarse aggregate particles does not directly enter the relation since, for instance, a crushed aggregate has a greater bulk volume for the same weight (i.e. a lower bulk density) than a well-rounded aggregate. Thus the shape factor is automatically taken into account in the determination of the bulk density. Table 10.16 gives values of the optimum volume of coarse aggregate when used with fine aggregates of different fineness moduli (*see* p. 151).

Thus, having chosen the maximum size and type of aggregate, in order to obtain concrete of a certain workability we use the water content from Table 10.15 and the bulk volume of coarse aggregate from Table 10.16. Given the specific gravity of coarse aggregate, its absolute volume is determined. The water/cement ratio is now chosen in the usual manner to satisfy both strength and durability requirements, and the cement content is computed by dividing the water content by the water/cement ratio. We have thus the absolute volumes of water, coarse aggregate, and cement, and by subtracting the sum of these from the total volume of concrete we find the absolute volume of fine aggregate that has to be

Table 10.16: *Bulk Volume of Coarse Aggregate per Unit Volume of Concrete*[10.2]

Maximum size of aggregate		Bulk volume of rodded coarse aggregate per unit volume of concrete for fineness modulus of sand of—			
in.	mm	2·40	2·60	2·80	3·00
$\frac{3}{8}$	10	0·50	0·48	0·46	0·44
$\frac{1}{2}$	12·5	0·59	0·57	0·55	0·53
$\frac{3}{4}$	20	0·66	0·64	0·62	0·60
1	25	0·71	0·69	0·67	0·65
$1\frac{1}{2}$	40	0·75	0·73	0·71	0·69
2	50	0·78	0·76	0·74	0·72
3	70	0·82	0·80	0·78	0·76
6	150	0·87	0·85	0·83	0·81

The values given will produce a mix with a workability suitable for reinforced concrete construction. For less workable concrete, e.g. that used in road construction, the values may be increased by about 10 per cent. For more workable concrete, such as may be required for placing by pumping, the values may be reduced by up to 10 per cent.

added to the mix. Multiplying this volume by the specific gravity of fine aggregate and by 62·4, the weight of sand in pounds is obtained. Alternatively, the weight of fine aggregate can be obtained direct by subtracting the total weight of other ingredients from the weight of a unit volume of concrete, if this can be estimated from experience. This approach is slightly less accurate than the absolute volume method.

If entrained air is used, allowance for its volume is made prior to calculating the volume of fine aggregate.

The method of the ACI Standard 211.1–70, just described, is applicable only to mixes whose slump is at least 1 in. (or 30 mm) (*see* Table 10.15). For mixes with a stiffer consistence, the approach of the American Concrete Institute Committee Standard 211–65[10.22] can be used. This is essentially an extension of the ACI Standard 211.1–70 with two differences: the workability is measured by means other than the slump test, and the coarse aggregate content used is much higher than with more workable mixes. Thus, instead of Table 10.15 we use Table 10.17; also, the volume of coarse aggregate calculated on the basis of Table 10.16 is multiplied by a factor given in Table 10.18, which is, however, no more than a guide for the first trial mix. The remainder of the mix design process is essentially unaltered.

The mix design method of the American Concrete Institute can be readily programmed for computer use.

Example

We require a mix with a *mean* compressive strength of 5,600 lb/in² and a slump of 2 inches, ordinary Portland cement being used. The maximum size of aggregate is $1\frac{1}{2}$ in., its bulk density is 100 lb/ft³, and its specific gravity is 2·64. The available fine aggregate has a fineness modulus of 2·60 and a specific gravity of 2·58.

From Table 10.15 the water requirement is 275 lb per cubic yard of concrete, and entrapped air is estimated to occupy 1 per cent of the volume of the concrete. The water/cement ratio is estimated from Fig. 10.2 to be 0·48. Hence, the cement content is 275/0·48 = 573 lb/yd³. Table 10.15 gives the bulk volume of coarse aggregate per unit volume of concrete (using the given fine aggregate) as 0·73. Hence, the weight of coarse aggregate per cubic yard of concrete is 0·73 × 100 × 27 = 1,970 lb.

We can now write the absolute volumes of the mix ingredients per cubic yard of concrete—

$$\text{Cement} = \frac{573}{3\cdot15 \times 62\cdot4} = 2\cdot92 \text{ ft}^3$$

$$\text{Water} = \frac{275}{62\cdot4} = 4\cdot41 \text{ ft}^3$$

$$\text{Coarse aggregate} = \frac{1{,}970}{2\cdot64 \times 62\cdot4} = 11\cdot96 \text{ ft}^3$$

$$\text{Entrapped air} = 0\cdot01 \times 27 = 0\cdot27 \text{ ft}^3$$

$$\text{Total} = 19\cdot56 \text{ ft}^3$$

Hence, the volume of fine aggregate required is 27 − 19·56 = 7·44 ft³. This corresponds to 7·44 × 2·58 × 62·4 = 1,200 lb.
Thus the weights of the materials per cubic yard of concrete are—

Cement	=	573 lb
Water	=	275 lb
Fine aggregate	=	1,200 lb
Coarse aggregate	=	1,970 lb
Total	=	4,018 lb

Hence, the density of concrete is 4,018/27 = 149 lb/ft³.

The corresponding calculations in the S.I. system of measurements are as follows. The mean compressive strength required is 39 MN/m², the slump is to be 50 mm, the maximum size of aggregate is 40 mm, and the bulk density of coarse aggregate is 1,600 kg/m³; its specific gravity is

Table 10.17: *Approximate Mixing-water Requirements for Mixes of Stiff Consistence with Different Maximum Sizes of Aggregates*[10.22]

Consistence	Slump in.	Slump mm	Vebe time sec	Compacting factor	Water requirement for maximum size of aggregate of—									
					3/8 in. lb/yd³	10 mm kg/m³	1/2 in. lb/yd³	12·5 mm kg/m³	3/4 in. lb/yd³	20 mm kg/m³	1 in. lb/yd³	25 mm kg/m³	1½ in. lb/yd³	40 mm kg/m³
NON-AIR-ENTRAINED CONCRETE														
Extremely dry	—	—	32–18	—	300	180	285	170	265	160	250	150	235	140
Very stiff	—	—	18–10	0·70	315	190	310	185	285	170	265	160	250	150
Stiff	0–1	0–30	10–5	0·75	335	200	325	195	300	180	285	170	265	160
Stiff plastic*	1–2	30–50	5–3	0·85	350	205	335	200	315	185	300	180	275	160
AIR-ENTRAINED CONCRETE														
Extremely dry	—	—	32–18	—	265	160	250	150	235	140	225	140	210	125
Very stiff	—	—	18–10	0·70	285	170	265	160	250	150	235	140	225	140
Stiff	0–1	0–30	10–5	0·75	300	180	285	170	265	160	250	150	235	140
Stiff plastic*	1–2	30–50	5–3	0·85	305	180	295	175	280	165	270	160	250	145

* These values are the same as in the appropriate line of Table 10.15. The quantities given are for reasonably well-shaped angular coarse aggregate.

Table 10.18: *Factors to be Applied to the Volume of Coarse Aggregate Calculated on the Basis of Table 10.16 for Mixes of Stiff Consistence*

Consistence	Slump		Vebe time sec	Compacting factor	Factor for maximum size of aggregate of—					
	in.	mm			$\frac{3}{8}$ in. 10 mm	$\frac{1}{2}$ in. 12·5 mm	$\frac{3}{4}$ in. 20 mm	1 in. 25 mm	$1\frac{1}{2}$ in. 40 mm	
Extremely dry	—	—	32–18	—	1·90	1·70	1·45	1·40	1·35	
Very stiff	—	—	18–10	0·70	1·60	1·45	1·30	1·25	1·25	
Stiff	0–1	0–30	10–5	0·75	1·35	1·30	1·15	1·15	1·20	
Stiff plastic	1–2	30–50	5–3	0·85	1·08	1·06	1·04	1·06	1·09	
Plastic	3–4	80–100	3–0	0·91	1·00	1·00	1·00	1·00	1·00	

2·64. As before, the fineness modulus of fine aggregate is 2·60 and its specific gravity, 2·58. The absolute density in kg/m^3 is numerically 1,000 times larger than the specific gravity.

From Table 10.15 Metric, the water requirement is 160 kg/m^3 and the entrapped air content is 1 per cent. From Fig. 10.2, the water/cement ratio is 0·48. Hence the cement content is $160/0·48 = 334$ kg/m^3. Table 10.16 gives the bulk volume of coarse aggregate per unit volume of concrete as 0·74. Hence, the weight of coarse aggregate per cubic metre of concrete is $0·74 \times 1,600 = 1,180$ kg.

The absolute volumes of mix ingredients per cubic metre of concrete are therefore:

$$\text{Cement} = \frac{334}{3·15 \times 1,000} = 0·106 \text{ m}^3$$

$$\text{Water} = \frac{160}{1,000} = 0·160 \text{ m}^3$$

$$\text{Coarse aggregate} = \frac{1,180}{2·64 \times 1,000} = 0·447 \text{ m}^3$$

$$\text{Entrapped air} = 0·01 \times 1 = 0·010 \text{ m}^3$$

$$\text{Total} = 0·723 \text{ m}^3$$

Hence, the volume of fine aggregate required is $1 - 0·723 = 0·277$ m^3. This corresponds to $0·277 \times 2·58 \times 1,000 = 715$ kg.

Thus the weights of materials per cubic metre of concrete are—

Cement	=	334 kg
Water	=	160 kg
Fine aggregate	=	715 kg
Coarse aggregate	=	1,180 kg
		2,389 kg

Hence, the density of concrete is 2,390 kg/m^3.

Concluding Remarks

The procedure of trial mixes and consecutive adjustments in all methods of mix design must seem empirical and gives the impression of being non-scientific, but the variability of the properties of both cement and aggregate is such that our calculations are really only guesses. However, the better our knowledge of the various properties of the ingredients of concrete the more accurate our guess can be. With this knowledge and experience in the use of the materials involved satisfactory mixes can be designed, although the process can never become automatic but is an art as much as a science.

(Note: If the reader is unable to design a satisfactory mix he should seriously consider the alternative of construction in steel.)

REFERENCES

10.1. J. D. McIntosh, Basic principles of concrete mix design, *Proc. of a Symposium on Mix Design and Quality Control of Concrete*, pp. 3–18 (London, Cement and Concrete Assoc., 1954).

10.2. A.C.I. Committee 211, Recommended practice for selecting proportions for normal weight concrete (ACI 211.1–70), *J. Amer. Concr. Inst.*, **66**, pp. 612–629 (Aug. 1969).

10.3. Road Research, Design of Concrete Mixes, *D.S.I.R. Road Note No. 4* (London, H.M.S.O., 1950).

10.4. A. M. Neville, The relation between standard deviation and mean strength of concrete test cubes, *Mag. Concr. Res.*, **11**, No. 32, pp. 75–84 (July 1959).

10.5. S. Walker, *Application of theory of probability to design of concrete for strength specifications* (Paper presented at 14th Annual Meeting of National Ready Mixed Concrete Association at New York, 27th January 1944) (Washington, N.R.M.C.A., 1955).

10.6. H. C. Erntroy, The variation of works test cubes, *Cement Concr. Assoc. Research Report No. 10* (London, Nov. 1960).

10.7. G. Graham and F. R. Martin, The construction of high-grade quality concrete paving for modern transport aircraft, *J. Inst. C. E.*, **26**, No. 6, pp. 117–90 (London, April 1946).

10.8. W. H. Glanville, Introductory Address, *Proc. of a Symposium on Mix Design and Quality Control of Concrete*, pp. xiii–xvi (London, Cement and Concrete Assoc., May 1954).

10.9. J. D. McIntosh and H. C. Erntroy, The workability of concrete mixes with ⅜ in. aggregates, *Cement Concr. Assoc. Res. Rep. No. 2* (London, June 1955).

10.10. H. C. Erntroy and B. W. Shacklock, Design of high-strength concrete mixes, *Proc. of a Symposium on Mix Design and Quality Control of Concrete*, pp. 55–65 (London, Cement and Concrete Assoc., May 1954).

10.11. A. R. Collins, *The principles of making high-strength concrete*, Report of eleven lectures on prestressed concrete given at the Building Exhibition, London, 17th to 30th Nov. 1949 (Cement and Concrete Assoc.).

10.12. P. J. F. Wright, The design of concrete mixes on the basis of flexural strength, *Proc. of a Symposium on Mix Design and Quality Control of Concrete*, pp. 74–76 (London, Cement and Concrete Assoc., May 1954).

10.13. A. R. Collins, Mix design for frost resistance, *ibid.*, pp. 92–96.

10.14. P. J. F. Wright, Entrained air in concrete, *Proc. Inst. C. E.*, Part I, **2**, No. 3, pp. 337–58 (London, May 1953).

10.15. H. Rüsch, Zur statistischen Qualitätskontrolle des Betons (On the statistical quality control of concrete), *Materialprüfung*, **6**, No. 11, pp. 387–394 (Nov. 1964).

10.16. A.C.I. Committee 214, Recommended practice for evaluation of compression test results of field concrete, *J. Amer. Concr. Inst.*, **61**, pp. 1057–72 (Sept. 1964).

10.17. H. H. Newlon, Variability of Portland cement concrete, *Proceedings*,

National Conf. on Statistical Quality Control Methodology in Highway and Airfield Construction, pp. 259–84 (University of Virginia School of General Studies, Charlottesville, 1966).

10.18. A.C.I. COMMITTEE 318, Explanation of the proposed revision of A.C.I. 318–63: Building code requirements for reinforced concrete, *J. Amer. Concr. Inst.*, **67**, pp. 147–86 (Feb. 1970).

10.19. J. B. METCALF, The specification of concrete strength, Part II: the distribution of strength of concrete for structures in current practice, *Report LR* 300 (Road Research Laboratory, Crowthorne, Berks., 1970).

10.20. L. J. PARROTT, The production and properties of high-strength concrete, *Concrete*, **3**, No. 11, pp. 443–8 (London, Nov. 1969).

10.21. P. J. F. WRIGHT, The flexural strength of plain concrete—its measurement and use in designing concrete mixes, *Road Research Technical Paper No. 67* (London, H.M.S.O., 1964).

10.22. A.C.I. COMMITTEE 211, Recommended practice for selecting proportions for no-slump concrete (A.C.I. 211–65), *Amer. Concr. Inst. Book of Standards*, 1966.

10.23. A. M. NEVILLE and J. B. KENNEDY, *Basic Statistical Methods for Engineers and Scientists*, (New York and London, Intertext 1964).

10.24. DEPARTMENT OF THE ENVIRONMENT, *Design of concrete mixes*, 31 pp (London H.M.S.O., 1975).

Appendix I Relevant British Standards

A. Cement

B.S.   12: 1958   Portland cement (ordinary and rapid-hardening) (metric version, 1971)
        146: 1958   Portland-blastfurnace cement (metric version, 1973)
       1370: 1958   Low heat Portland cement (metric version, 1974)
       4246: 1968   Low heat Portland-blastfurnace cement (metric version, 1974)
       4248: 1974   Supersulphated cement
        915: 1947   High alumina cement (metric version, 1972)
       1014: 1975   Pigments for cement, magnesium oxychloride and concrete

B. Aggregate

B.S. 882, 1201: 1965   Aggregates from natural sources for concrete (including granolithic) (metric version, 1973)
        812: 1975   Methods for the sampling and testing of mineral aggregates, sands and fillers
        877: 1967   Foamed or expanded blastfurnace slag lightweight aggregate for concrete (metric version, 1973)
       1047: 1952   Air-cooled blastfurnace slag coarse aggregate for concrete (metric version, 1974)
       1165: 1966   Clinker aggregate for concrete
        410: 1969   Test sieves
       3797: 1964   Lightweight aggregates for concrete
       3681: 1963   Methods for sampling and testing of lightweight aggregates for concrete (metric version, 1973)

C. Concrete

B.S. 1881: Part 1: 1970   Methods of sampling fresh concrete
     1881: Part 2: 1970   Methods of testing fresh concrete
     1881: Part 3: 1970   Methods of making and curing test specimens
     1881: Part 4: 1970   Methods of testing concrete for strength
     1881: Part 5: 1970   Methods of testing hardened concrete for other than strength
     1881: Part 6: 1971   Analysis of hardened concrete
     4408: Part 1: 1969   Electromagnetic cover measuring devices
     4408: Part 2: 1969   Strain gauges for concrete investigations
     4408: Part 3: 1970   Gamma radiography of concrete
     4408: Part 4: 1971   Surface hardness methods
     4408: Part 5: 1974   Measurement of the velocity of ultrasonic pulses in concrete
             1926: 1962   Ready-mixed concrete
             1305: 1974   Batch type concrete mixers
             3963: 1964   Method for testing the performance of batch type concrete mixers
              368: 1971   Precast concrete flags
       2028, 1364: 1968   Precast concrete blocks
             3148: 1959   Tests for water for making concrete

Appendix II Selected List of Relevant A.S.T.M. Standards*

A. Cement

C 150–74   Spec. for Portland Cement
C 595–74   Spec. for Blended Hydraulic Cements
C 115–74   Test for Fineness of Portland Cement by the Turbidimeter
C 186–73   Test for Heat of Hydration of Hydraulic Cement
C 151–74a   Test for Autoclave Expansion of Portland Cement

B. Admixtures

C 618–73   Spec. for Fly Ash and Raw or Calcined Natural Pozzolans for Use in Portland Cement Concrete
C 494–71   Spec. for Chemical Admixtures for Concrete
C 441–69   Test for Effectiveness of Mineral Admixtures in Preventing Excessive Expansion of Concrete Due to the Alkali-Aggregate Reaction
C 260–73   Spec. for Air-Entraining Admixtures for Concrete

C. Aggregate

C 294–69   Descriptive Nomenclature of Constituents of Natural Mineral Aggregates
C 33–74   Spec. for Concrete Aggregates
C 330–69   Spec. for Lightweight Aggregates for Structural Concrete
C 331–69   Spec. for Lightweight Aggregates for Concrete Masonry Units
C 332–66   (1971) Spec. for Lightweight Aggregates for Insulating Concrete
C 117–69   Test for Materials Finer than No. 200 (75-μm) Sieve in Mineral Aggregates by Washing
C 70–73   Test for Surface Moisture in Fine Aggregate
C 40–73   Test for Organic Impurities in Sands for Concrete
C 123–69   Test for Lightweight Pieces in Aggregate
C 88–73   Test for Soundness of Aggregates by Use of Sodium Sulfate or Magnesium Sulfate
C 131–69   Test for Resistance to Abrasion of Small Size Coarse Aggregate by Use of the Los Angeles Machine
C 289–71   Test for Potential Reactivity of Aggregates (Chemical Method)
C 227–71   Test for Potential Alkali Reactivity of Cement–Aggregate Combinations (Mortar-Bar Method)
C 586–69   Test for Potential Alkali Reactivity of Carbonate Rocks for Concrete Aggregates (Rock Cylinder Method)
C 638–73   Descriptive Nomenclature of Constituents of Aggregates for Radiation-Shielding Concrete
C 637–73   Spec. for Aggregates for Radiation-Shielding Concrete
E 11–70   Spec. for Wire-Cloth Sieves for Testing Purposes

---

* T denotes Tentative Standard. The two digits after the dash denote the year of publication.

D. Concrete

C 124–71   Test for Flow of Portland Cement Concrete by Use of the Flow Table (discontinued 1974)

C 143–71   Test for Slump of Portland Cement Concrete

C 360–63   (1968) Test for Ball Penetration in Fresh Portland Cement Concrete

C 403–70   Test for Time of Setting of Concrete Mixtures by Penetration Resistance

C 232–71   Test for Bleeding of Concrete

C 138–74   Test for Unit Weight, Yield, and Air Content (Gravimetric) of Concrete

C 173–73a Test of Air Content of Freshly Mixed Concrete by the Volumetric Method

C 231–73   Test for Air Content of Freshly Mixed Concrete by the Pressure Method

C 470–73T Spec. for Molds for Forming Concrete Test Cylinders Vertically Concrete Test Cylinders

C 192–69   Making and Curing Concrete Test Specimens in the Laboratory

C 39–72   Test for Compressive Strength of Cylindrical Concrete Specimens

C 617–73   Capping Cylindrical Concrete Specimens

C 78–64   (1972) Test for Flexural Strength of Concrete (Using Simple Beam with Third-Point Loading)

C 496–71   Test for Splitting Tensile Strength of Cylindrical Concrete Specimens

C 42–68   (1974) Obtaining and Testing Drilled Cores and Sawed Beams of Concrete

C 215–60   (1970) Test for Fundamental Transverse, Longitudinal, and Torsional Frequencies of Concrete Specimens

C 418–68   (1974) Test for Abrasion Resistance of Concrete

C 85–66   (1973) Test for Cement Content of Hardened Portland Cement Concrete

C 457–71   Rec. Practice for Microscopical Determination of Air-Void Content and Parameters of the Air-Void System in Hardened Concrete

C 666–73   Test for Resistance of Concrete to Rapid Freezing and Thawing

C 94–74   Spec. for Ready-mixed Concrete

C 156–74   Test for Water Retention by Concrete Curing Materials

Name Index

The first entry (Abrams, D. A.) can be used to illustrate the working of this index: the 27, 233 and 239 are the pages on which reference is made to Abrams; 4.23 is the reference number of a publication (as given in the REFERENCES at the end of each chapter) and the 205, 207 in parentheses are the pages on which it appears in the Text. *

Subject Index

3/104